T0233716

Developments in Mathematics

VOLUME 39

Series Editors:
Krishnaswami Alladi, *University of Florida, Gainesville, FL, USA*
Hershel M. Farkas, *Hebrew University of Jerusalem, Jerusalem, Israel*

More information about this series at http://www.springer.com/series/5834

Saïd Abbas • Mouffak Benchohra

Advanced Functional Evolution Equations and Inclusions

 Springer

Saïd Abbas
Laboratoire de Mathématiques
Université de Saïda
Saïda, Algeria

Mouffak Benchohra
Department of Mathematics
University of Sidi Bel Abbes
Sidi Bel Abbes, Algeria

ISSN 1389-2177
Developments in Mathematics
ISBN 978-3-319-36725-5
DOI 10.1007/978-3-319-17768-7

ISSN 2197-795X   (electronic)

ISBN 978-3-319-17768-7   (eBook)

Springer Cham Heidelberg New York Dordrecht London

Printed on acid-free paper

Springer International Publishing AG Switzerland is part of Springer Science+Business Media (www.
springer.com)

We dedicate this book to our family members. In particular, Saïd Abbas dedicates to the memory of his father Abdelkader Abbas; and Mouffak Benchohra makes his dedication to the memory of his father Yahia Benchohra

Preface

Functional differential equations and inclusions occur in a variety of areas of biological, physical, and engineering applications, and such equations have received much attention in recent years. This book is devoted to the existence of local and global mild solutions for some classes of functional differential evolution equations and inclusions, and other densely and non-densely defined functional differential equations and inclusions in separable Banach spaces or in Fréchet spaces. Some of these equations and inclusions present delay which may be finite, infinite, or state-dependent. Other equations are subject to impulses effect. The tools used include classical fixed point theorems and the measure of non-compactness (MNC). Each chapter concludes with a section devoted to notes and bibliographical remarks. All the presented abstract results are illustrated by examples.

The content of the book is new and complements the existing literature devoted to functional differential equations and inclusions. It is useful for researchers and graduate students for research, seminars, and advanced graduate courses, in pure and applied mathematics, engineering, biology, and all other applied sciences.

We are grateful to our colleagues and friends N. Abada, E. Alaidarous, S. Baghli, A. Baliki, M. Belmekki, K. Ezzinbi, H. Hammouche, J. Henderson, I. Medjedj, J.J. Nieto, and M. Ziane for their collaboration in research related to the problems considered in this book. Last but not least, we are grateful to Elizabeth Loew and Dahlia Fisch for their support and to Jeffin Thomas Varghese for his help during the production of the book.

Saïda, Algeria
Sidi Bel Abbes, Algeria

Saïd Abbas
Mouffak Benchohra

Introduction

Nonlinear evolution equations, i.e., partial differential equations with time t as one of the independent variables, arise not only from many fields of mathematics, but also from other branches of science such as physics, mechanics, and material science. For example, Navier–Stokes and Euler equations from fluid mechanics, nonlinear reaction-diffusion equations from heat transfers and biological sciences, nonlinear Klein–Gordon equations and nonlinear Schrödinger equations from quantum mechanics, and Cahn–Hilliard equations from material science, to name just a few, are special examples of nonlinear evolution equations. See the books [174, 176–178].

Functional differential equations and inclusions arise in a variety of areas of biological, physical, and engineering applications, and such equations have received much attention in recent years. A good guide to the literature for functional differential equations is the books by Hale [131], Hale and Verduyn Lunel [133], Kolmanovskii and Myshkis [148], and the references therein. During the last decades, existence and uniqueness of mild, strong, classical, almost periodic, almost automorphic solutions of semi-linear functional differential equations and inclusions has been studied extensively by many authors using the semigroup theory, fixed point argument, degree theory, and measures of non-compactness. We mention, for instance, the books by Ahmed [16], Diagana [103], Engel and Nagel [106], Kamenskii et al. [144], Pazy [168], Wu [184], Zheng [187], and the references therein. In recent years, there has been a significant development in evolution equations and inclusions; see the monograph of Perestyuk et al. [169], the papers of Baliki and Benchohra [33, 37], Benchohra and Medjedj [55, 56], Benchohra et al. [82], and the references therein.

Neutral functional differential equations arise in many areas of applied mathematics and such equations have received much attention in recent years. A good guide to the literature for neutral functional differential equations is the books by Hale [131], Hale and Verduyn Lunel [133], Kolmanovskii and Myshkis [148], and the references therein. Hernandez in [137] proved the existence of mild, strong, and periodic solutions for neutral equations. Fu in [117, 118] studies the controllability on a bounded interval of a class of neutral functional differential equations. Fu and

Ezzinbi [119] considered the existence of mild and classical solutions for a class of neutral partial functional differential equations with nonlocal conditions. Various classes of partial functional and neutral functional differential equations with infinite delay are studied by Adimy et al. [10–12], Belmekki et al. [52], and Ezzinbi [108]. Henriquez [136] and Hernandez [137, 138] studied the existence and regularity of solutions to functional and neutral functional differential equations with unbounded delay. Balachandran and Dauer have considered various classes of first and second order semi-linear ordinary, functional and neutral functional differential equations on Banach spaces in [43]. By means of fixed point arguments, Benchohra et al. have studied various classes of functional differential equations and inclusions and proposed some controllability results in [28, 33, 33, 37, 58, 72, 73, 75, 76, 80]. See also the works by Gatsori [120], Li et al. [155], Li and Xue [156], and Li and Yong [157].

Impulsive differential equations and inclusions appear frequently in applications such as physics, aeronautic, economics, engineering, and population dynamics; see the monographs of Bainov and Simeonov [39, 40], Benchohra et al. [81], Erbe and Krawcewicz [107], Graef et al. [127], Samoilenko and Perestyuk [172], and Perestyuk et al. [169], and the paper of Coldbeter et al. [95] where numerous properties of their solutions are studied. In this way, they make changes of states at certain moments of time between intervals of continuous evolution such changes can be reasonably well approximated as being instantaneous changes of this state which we will represent by impulses and then these processes are modeled by impulsive differential equations and for this reason the study of this type of equations has received great attention in the last years. There has been a significant development in impulsive theory especially in the area of impulsive differential equations with fixed moments. See, for instance, the monographs by Benchohra et al. [81], Lakshmikantham et al. [150], and Samoilenko and Perestyuk [172]. There exists an extensive literature devoted to the case where the impulses are absent (i.e., $I_k = 0, k = 1, \ldots, m$), see, for instance, the monograph by Liang and Xiao [158] and the paper by Schumacher [158]. We mention here also the use of impulsive differential equations in the study of oscillation and non-oscillation of impulsive dynamic equations, see, for instance, the papers of Graef et al. [124, 125], oscillation of dynamic equations with delay was considered in [13, 14]. During the last 10 years impulsive ordinary differential inclusions and functional differential equations and inclusions have attracted the attention of many mathematicians and are intensively studied. At present the foundations of the general theory and such kind of problems are already laid and many of them are investigated in detail in [58, 59, 63, 79, 81, 107] and the references therein.

It is well known that the issue of controllability plays an important role in control theory and engineering because they have close connections to pole assignment, structural decomposition, quadratic optimal control, observer design, etc. In recent years, the problem of controllability for various kinds of differential and impulsive differential systems has been extensively studied by many authors [71, 155–157, 186] using different approaches. Several authors have extended the controllability concept to infinite dimensional systems in Banach space with unbounded operators, see the monographs [85, 98, 157, 186] and the references therein. Sufficient conditions for controllability are established by Lasiecka and Triggiani [153]. Fu in [117, 118] studied the controllability on a bounded interval of a class of neutral functional differential equations. Fu and Ezzinbi [119] considered the existence of mild and classical solutions for a class of neutral partial functional differential equations with nonlocal conditions. Adimy et al. [10–12] studied some classes of partial functional and neutral functional differential equations with infinite delay. When the delay is infinite, the notion of the phase space $\mathcal{B}$ plays an important role in the study of both qualitative and quantitative theory. A usual choice is a semi-normed space satisfying suitable axioms, which was introduced by Hale and Kato in [132], see also Corduneanu and Lakshmikantham [97] and Kappel and Schappacher [145].

The literature related to ordinary and partial functional differential equations with delay is very extensive. On the other hand, functional differential equations with state-dependent delay appear frequently in applications as model of equations, and for this reason the study of this type of equations has received great attention in the last year, see, for instance [31, 183] and the references therein. The literature related to partial functional differential equations with state-dependent delay is limited; see [139, 171].

Several authors have considered extensively the problem

$$x'(t) = A(t)x(t) + f(t, x_t)$$

when $A(t) = A$. Existence of mild solutions is developed by Heikkila and Lakshmikantham [134], Kamenski et al. [144], and the pioneer Hino and Murakami paper [141] for some semi-linear functional differential equations with finite delay. By means of fixed point arguments, Benchohra and his collaborators have studied many classes of first and second order functional differential inclusions on a bounded interval with local and nonlocal conditions in [59, 60, 62, 64, 65, 77, 78, 121]. Extension to the semi-infinite interval is given by Benchohra and Ntouyas in [58, 61]. When A depends on time, Arara et al. [26, 28] considered a control multi-valued problem on a bounded interval. Uniqueness results of mild solutions for some classes of partial functional and neutral functional differential evolution equations on the semi-infinite interval $J = \mathbb{R}_+$ for a finite delay with local and nonlocal conditions were given in [33, 37]. When the delay is infinite, existence and uniqueness results for evolution problems are proposed in [33], and controllability result of mild solutions for the evolution equations are given in [15, 36]. The case when A is non-densely defined and generates an integrated semigroup was done by Benchohra et al. [80]. Some global existence results for impulsive differential equations and inclusions were obtained by Guo [129], Graef and Ouahab [126], and the references therein.

Partial functional evolution equations and inclusions with infinite and state-dependent delay, controllability on finite interval are our concerns. Our approach is based upon the fixed point theory for multi-valued condensing maps under assumptions expressed in terms of the MNC [144].

In the last three decades, the theory of C_0-semigroup has been developed extensively, and the achieved results have found many applications in the theory of partial differential equations, for instance see [106, 122, 168] and the papers of Arara et al. [26, 27] and Benchohra et al. [74]. Recently, increasing interest has been observed in applications to impulsive differential equations and inclusions, see Liu [69, 159]. The case where the generator of the semigroup is non-densely defined, the existence of integral solutions on compact intervals for differential equations and inclusions were studied by Adimy et al. [8–10], Arendt [29, 30], Ezzinbi and Liu [109, 110], and Henderson and Ouahab [135]. The model with multi-valued jump sizes arises in a control problem where we want to control the jump sizes in order to achieve given objectives. There are very few results for impulsive evolution inclusions with multi-valued jump operator, see [161]. We present the existence of solutions for both densely or non-densely defined impulsive functional differential inclusions.

The multi-valued jumps (i.e., the difference operator $\Delta x|_{t=t_k} \in \mathcal{I}_k(x(t_k^-))$) is a natural model of an impulsive system where the jump sizes are not deterministic as in [17–19, 161] but rather they are uncertain. However given the state x and time t_i, the set of possible jump sizes at this state is determined by the set $\mathcal{I}_k(x)$. The set-valued maps $\mathcal{I}_k$ may be given by the sub-differential of a lower semi-continuous convex functional ϕ_i. In this case, the system is governed by evolution inequations at the points of time t_k. Another situation that may give rise to such a dynamic model originates from the parametric uncertainty such as $\mathcal{I}_k(x) = \{I_k(t, x); t \in I\}$, where $\{I_k\}$ is a suitable family of functions $I \times E \to E$. To our knowledge, there are very few results for impulsive evolution inclusions with multi-valued jump operators; see [5, 19, 36]. The results of this book extend and complement those obtained in the absence of the impulse functions I_k, and for those with single-valued impulse functions I_k.

This book is arranged and organized as follows:

In Chap. 1, we introduce notations, definitions, and some preliminary notions. In *Sect. 1.1*, we give some notations from the theory of Banach spaces. *Section 1.2* is concerned to recall some basic definitions and some properties in Fréchet spaces. In *Sect. 1.3*, we recall some basic definitions and give some examples of Phase spaces. *Section 1.4* contains some properties of set-valued maps. In *Sect. 1.5*, we give some preliminaries about evolution systems. Some definitions and properties of the theory of semigroups are presented in *Sect. 1.6*. In *Sect. 1.6.3*, we give some properties of the extrapolation method. The last section (*Sect. 1.7*) contains some fixed point theorems.

In Chap. 2, we study some first order classes of partial functional, neutral functional, integro-differential, and neutral integro-differential evolution equations with finite delay on the positive real line. *Section 2.2* deals with the existence and uniqueness of mild solutions for some classes of partial evolution equations with local and nonlocal conditions. We give some results based on the fixed point theorem of Frigon in Fréchet spaces. An example will be presented at the last illustrating the abstract theory. In *Sect. 2.3*, we study some neutral differential evolution equations in Fréchet spaces. In *Sect. 2.4*, we give existence results for other classes of partial functional integro-differential evolution equations. *Section 2.5* deals with uniqueness results of neutral functional integro-differential evolution equations.

In Chap. 3, we provide sufficient conditions for the existence of the unique mild solution on the positive half-line $\mathbb{R}_+$ for some classes of first order partial functional and neutral functional differential evolution equations with infinite delay. In *Sect. 3.2*, we study the existence and uniqueness of mild solutions for partial functional evolution equations in Fréchet spaces. *Section 3.3* deals with the controllability of mild solutions on finite interval for partial evolution equations. In *Sect. 3.4*, we study the controllability of mild solutions on semi-infinite interval for partial evolution equations. *Section 3.5* deals with the existence of the unique mild solution of neutral functional evolution equations. In *Sect. 3.6*, we study the controllability of mild solutions on finite interval for neutral evolution equations. *Section 3.7* deals with the controllability of mild solutions on semi-infinite interval for neutral evolution equations.

In Chap. 4, we shall be concerned by perturbed partial functional and neutral functional evolution equations with finite and infinite delay on the semi-infinite interval $\mathbb{R}_+$. Our main tool is the nonlinear alternative proved by Avramescu (1.30) for the sum of contractions and completely continuous maps in Fréchet spaces [32], combined with semigroup theory. In *Sect. 4.2*, we study the existence of mild solutions for perturbed partial functional evolution equations with finite delay. *Section 4.3* deals with perturbed neutral functional evolution equations with finite delay. In *Sect. 4.4*, we study the existence of mild solutions for perturbed partial evolution equations with infinite delay.

In Chap. 5, we provide sufficient conditions for the existence of mild solutions on the semi-infinite interval $\mathbb{R}_+$ for some classes of first order partial functional and neutral functional differential evolution inclusions with finite delay. In *Sect. 5.2*, we study the existence of mild solutions for a class of functional partial evolution equations. *Section 5.3* deals with neutral partial evolution equations.

In Chap. 6, we study the existence of mild solutions on the semi-infinite interval $\mathbb{R}_+$ for some classes of first order partial functional and neutral functional differential evolution inclusions with infinite delay. In *Sect. 6.2*, we study functional partial evolution equations. *Section 6.3* deals with neutral partial evolution equations.

In Chap. 7, we are concerned by the existence of mild and extremal solutions of some first order classes of impulsive semi-linear functional differential inclusions with local and nonlocal conditions when the delay is finite in separable Banach spaces. Using a recent theorem due to Dhage combined with the semigroup theory, the existence of the mild and extremal mild solution are assured. The nonlocal case is studied too. In *Sect. 7.2*, we study the existence of mild solutions with local conditions. *Section 7.3* deals with the existence of mild solutions with nonlocal conditions. In *Sect. 7.4*, we give an application to the control theory.

In Chap. 8, we shall establish sufficient conditions for the existence of integral solutions and extremal integral solutions for some non-densely defined impulsive semi-linear functional differential inclusions in separable Banach spaces with local and nonlocal conditions. In *Sect. 8.2*, we give some results for integral solutions of non-densely defined functional differential inclusions with local conditions. *Section 8.3* deals with extremal integral solutions with local conditions, and *Sect. 8.4* deals with extremal integral solutions with nonlocal conditions. In *Sect. 8.5*, we give an application to the control theory.

In Chap. 9, we study the existence of mild solutions impulsive semi-linear functional differential equations. In *Sect. 9.2*, we study some existence results for semi-linear differential evolution equations with impulses and delay. *Section 9.3* is devoted to some classes of impulsive semi-linear functional differential equations with non-densely defined operators. In *Sect. 9.4*, we study impulsive semi-linear neutral functional differential equations with infinite delay. *Section 9.5* deals with integral solutions of non-densely defined impulsive semi-linear functional differential equations with state-dependent delay.

In Chap. 10, we shall establish sufficient conditions for the existence of mild, extremal mild, integral, and extremal integral solutions for some impulsive semi-linear neutral functional differential inclusions in separable Banach spaces. In *Sect. 10.2*, we study some densely defined impulsive functional differential inclusions. *Section 10.3* deals with the existence of mild solutions for non-densely defined impulsive neutral functional differential inclusions. In *Sect. 10.4*, we study the controllability of impulsive semi-linear differential inclusions in Fréchet spaces.

In Chap. 11, we study functional differential inclusions with multi-valued jumps. In *Sect. 11.2*, we study some existence of integral solutions for semi-linear functional differential inclusions with state-dependent delay and multi-valued jump. *Section 11.3* deals with impulsive evolution inclusions with infinite delay and multi-valued jumps. *Section 11.4* deals with impulsive semi-linear differential evolution inclusions with non-convex right-hand side. In *Sect. 11.5*, we study some impulsive evolution inclusions with state-dependent delay and multi-valued jumps. *Section 11.6* deals with the controllability of impulsive differential evolution inclusions with infinite delay.

In Chap. 12, we study functional differential equations and inclusions with delay. In *Sect. 12.2*, we prove some global existence for functional differential equations with state-dependent delay. *Section 12.3* deals with global existence results for neutral functional differential equations with state-dependent delay. In *Sect. 12.4*, we give some global existence results for functional differential inclusions with delay. *Section 12.4.1* deals with global existence results for functional differential inclusions with state-dependent delay.

In Chap. 13, we shall establish sufficient conditions for global existence results of second order functional differential equations with delay. In *Sect. 13.2*, we give some global existence results of second order functional differential equations with delay.

Keywords and Phrases: Evolution differential equations and inclusions, integro-differential equations, densely and non-densely defined differential equations, convex and non-convex valued multi-valued, mild solution, weak solution, initial value problem, nonlocal conditions, contraction, existence, uniqueness, measure of noncompactness, Banach space, Fréchet space, phase space, impulses, time delay, state-dependent delay, fixed point.

Contents

Chapter 1
Preliminary Background

In this chapter, we introduce notations, definitions, and preliminary facts which are used throughout this book.

1.1 Notations and Definitions

Let $\mathbb{R}_+ = [0, +\infty)$ be the positive real line, $H = [-r, 0]$ be an interval with $r > 0$, and $(E, |\cdot|)$ be a real Banach space.

By $\mathcal{C}(H, E)$ we denote the Banach space of continuous functions from H into E, with the norm

$$\|y\| = \sup_{t \in H} |y(t)|.$$

Let $B(E)$ be the space of all bounded linear operators from E into E, with the norm

$$\|N\|_{B(E)} = \sup_{|y|=1} |N(y)|.$$

A measurable function $y : \mathbb{R}_+ \to E$ is Bochner integrable if and only if $|y|$ is Lebesgue integrable. For properties of the Bochner integral, see for instance, Yosida [185].

© Springer International Publishing Switzerland 2015
S. Abbas, M. Benchohra, *Advanced Functional Evolution Equations
and Inclusions*, Developments in Mathematics 39,
DOI 10.1007/978-3-319-17768-7_1

As usual, by $L^1(\mathbb{R}_+, E)$ we denote the Banach space of measurable functions $y : \mathbb{R}_+ \to E$ which are Bochner integrable normed by

$$\|y\|_{L^1} = \int_0^{+\infty} |y(t)| dt.$$

Let $L^1_{loc}(\mathbb{R}_+, E)$ be the space of measurable functions which are locally Bochner integrable.

For any continuous function y defined on $[-r, +\infty)$ and any $t \in \mathbb{R}_+$, we denote by y_t the element of $C(H, E)$ defined by

$$y_t(\theta) = y(t + \theta) \quad \text{for } \theta \in H.$$

Here $y_t(\cdot)$ represents the history of the state from time $t - r$ up to the present time t.

Definition 1.1. A function $f : \mathbb{R}_+ \times E \to E$ is said to be an L^1-Carathéodory function if it satisfy:

  (i) for each $t \in \mathbb{R}_+$ the function $f(t, .) : E \to E$ is continuous;
 (ii) for each $y \in E$ the function $f(., y) : \mathbb{R}_+ \to E$ is measurable;
(iii) for every positive integer k there exists $h_k \in L^1(\mathbb{R}_+, \mathbb{R}_+)$ such that

$$|f(t, y)| \leq h_k(t) \quad \text{for all } |y| \leq k \quad \text{and almost each } t \in \mathbb{R}_+.$$

1.2   Some Properties in Fréchet Spaces

Let X be a Fréchet space with a family of semi-norms $\{\|\cdot\|_n\}_{n \in \mathbb{N}}$. Let $Y \subset X$, we say that F is bounded if for every $n \in \mathbb{N}$, there exists $M_n > 0$ such that

$$\|y\|_n \leq M_n \quad \text{for all } y \in Y.$$

To X we associate a sequence of Banach spaces $\{(X^n, \|\cdot\|_n)\}$ as follows: For every $n \in \mathbb{N}$, we consider the equivalence relation $\sim_n$ defined by: $x \sim_n y$ if and only if $\|x - y\|_n = 0$ for all $x, y \in X$. We denote $X^n = (X|_{\sim_n}, \|\cdot\|_n)$ the quotient space, the completion of X^n with respect to $\|\cdot\|_n$. To every $Y \subset X$, we associate a sequence $\{Y^n\}$ of subsets $Y^n \subset X^n$ as follows: For every $x \in X$, we denote $[x]_n$ the equivalence class of x of subset X^n and we define $Y^n = \{[x]_n : x \in Y\}$. We denote $\overline{Y^n}$, $int_n(Y^n)$ and $\partial_n Y^n$, respectively, the closure, the interior, and the boundary of Y^n with respect to $\|\cdot\|$ in X^n. We assume that the family of semi-norms $\{\|\cdot\|_n\}$ verifies:

$$\|x\|_1 \leq \|x\|_2 \leq \|x\|_3 \leq \ldots \quad \text{for every } x \in X.$$

Definition 1.2 ([116]). A function $f : X \to X$ is said to be *a contraction* if for each $n \in \mathbb{N}$ there exists $k_n \in [0, 1)$ such that:

$$\|f(x) - f(y)\|_n \leq k_n \|x - y\|_n \quad \text{for all } x, y \in X.$$

1.3 Phase Spaces

In this section, we will define the phase space $\mathcal{B}$ axiomatically, using ideas and notations developed by Hale and Kato [132] and follow the terminology used in [142], (see also Kapper and Schappacher [145] and Schumacher [173]). More precisely, $(\mathcal{B}, \|\cdot\|_\mathcal{B})$ will denote the vector space of functions defined from $(-\infty, 0]$ into E endowed with a semi norm denoted $\|.\|_\mathcal{B}$ and satisfying the following axioms:

(A_1) If $y : (-\infty, b) \to E, b > 0$, is continuous on $[0, b]$ and $y_0 \in \mathcal{B}$, then for every $t \in [0, b)$ the following conditions hold:

  (i) $y_t \in \mathcal{B}$;
  (ii) There exists a positive constant H such that $|y(t)| \leq H\|y_t\|_\mathcal{B}$;
  (iii) There exist two functions $K(\cdot), M(\cdot) : \mathbb{R}_+ \to \mathbb{R}_+$ independent of $y(t)$ with K continuous and M locally bounded such that:

$$\|y_t\|_\mathcal{B} \leq K(t) \sup\{ |y(s)| : 0 \leq s \leq t\} + M(t)\|y_0\|_\mathcal{B}.$$

Denote

$$K_b = \sup\{K(t) : t \in [0, b]\} \text{ and } M_b = \sup\{M(t) : t \in [0, b]\}.$$

(A_2) For the function $y(.)$ in (A_1), y_t is a $\mathcal{B}$-valued continuous function on $[0, b]$.
(A_3) The space $\mathcal{B}$ is complete.

Remark 1.3.   1. (ii) is equivalent to $|\phi(0)| \leq H\|\phi\|_\mathcal{B}$ for every $\phi \in \mathcal{B}$.
  2. Since $\|\cdot\|_\mathcal{B}$ is a seminorm, two elements $\phi, \psi \in \mathcal{B}$ can verify $\|\phi - \psi\|_\mathcal{B} = 0$ without necessarily $\phi(\theta) = \psi(\theta)$ for all $\theta \leq 0$.
  3. From the equivalence of (ii), we can see that for all $\phi, \psi \in \mathcal{B}$ such that $\|\phi - \psi\|_\mathcal{B} = 0$: This implies necessarily that $\phi(0) = \psi(0)$.

For any continuous function y and any $t \geq 0$, we denote by y_t the element of $\mathcal{B}$ defined by

$$y_t(\theta) = y(t + \theta) \qquad \text{for } \theta \in (-\infty, 0].$$

We assume that the histories y_t belong to some abstract *phase space* $\mathcal{B}$.

Consider the following space

$$B_{+\infty} = \left\{ y : \mathbb{R} \to E : y|_{\mathbb{R}_+} \in \mathcal{C}(\mathbb{R}_+, E), \ y_0 \in \mathcal{B} \right\},$$

where $y|_{\mathbb{R}_+)}$ is the restriction of y to $[0, +\infty)$.

Hereafter are some examples of phase spaces. For other details we refer, for instance, to the book by Hino et al. [142].

Example 1.4. The spaces BC, BUC, C^∞, and C^0. Let:

BC the space of bounded continuous functions defined from $(-\infty, 0]$ to E;

BUC the space of bounded uniformly continuous functions defined from $(-\infty, 0]$ to E;

$$C^\infty \quad := \left\{ \phi \in BC : \lim_{\theta \to -\infty} \phi(\theta) \text{ exists in } E \right\};$$

$$C^0 \quad := \left\{ \phi \in BC : \lim_{\theta \to -\infty} \phi(\theta) = 0 \right\}, \text{ endowed with the uniform norm}$$

$$\|\phi\| = \sup\{|\phi(\theta)| : \theta \le 0\}.$$

We have that the spaces BUC, C^∞, and C^0 satisfy conditions (A_1)–(A_3). BC satisfies (A_2), (A_3) but (A_1) is not satisfied.

Example 1.5. The spaces C_g, UC_g, C_g^∞, and C_g^0. Let g be a positive continuous function on $(-\infty, 0]$. We define:

$$C_g \quad := \left\{ \phi \in \mathcal{C}((-\infty, 0], E) : \frac{\phi(\theta)}{g(\theta)} \text{ is bounded on } (-\infty, 0] \right\};$$

$$C_g^0 \quad := \left\{ \phi \in C_g : \lim_{\theta \to -\infty} \frac{\phi(\theta)}{g(\theta)} = 0 \right\}, \text{ endowed with the uniform norm}$$

$$\|\phi\| = \sup \left\{ \frac{|\phi(\theta)|}{g(\theta)} : \theta \le 0 \right\}.$$

We consider the following condition on the function g.

(g_1) For all $a > 0$, $\displaystyle \sup_{0 \le t \le a} \sup \left\{ \frac{g(t + \theta)}{g(\theta)} : -\infty < \theta \le -t \right\} < \infty.$

Then we have that the spaces C_g and C_g^0 satisfy conditions (A_3). They satisfy conditions (A_1) and (A_2) if g_1 holds.

Example 1.6. The space C_γ. For any real constant γ, we define the functional space C_γ by

$$C_\gamma := \left\{ \phi \in \mathcal{C}((-\infty, 0], E) : \lim_{\theta \to -\infty} e^{\gamma \theta} \phi(\theta) \text{ exist in } E \right\}$$

endowed with the following norm

$$\|\phi\| = \sup\{e^{\gamma\theta}|\phi(\theta)| : \theta \leq 0\}.$$

Then in the space C_γ the axioms (A_1)–(A_3) are satisfied.

1.4 Set-Valued Maps

Let (X, d) be a metric space. We use the following notations:

$$\mathcal{P}_{cl}(X) = \{Y \in \mathcal{P}(X) : Y \text{ closed}\}, \quad \mathcal{P}_b(X) = \{Y \in \mathcal{P}(X) : Y \text{ bounded}\},$$

$$\mathcal{P}_{cv}(X) = \{Y \in \mathcal{P}(X) : Y \text{ convex}\}, \quad \mathcal{P}_{cp}(X) = \{Y \in \mathcal{P}(X) : Y \text{ compact}\}.$$

Consider $H_d : \mathcal{P}(X) \times \mathcal{P}(X) \longrightarrow \mathbb{R}_+ \cup \{\infty\}$, given by

$$H_d(\mathcal{A}, \mathcal{B}) = \max \left\{ \sup_{a \in \mathcal{A}} d(a, \mathcal{B}), \sup_{b \in \mathcal{B}} d(\mathcal{A}, b) \right\},$$

where $d(\mathcal{A}, b) = \inf_{a \in \mathcal{A}} d(a, b), d(a, \mathcal{B}) = \inf_{b \in \mathcal{B}} d(a, b)$. Then $(\mathcal{P}_{b,cl}(X), H_d)$ is a metric space and $(\mathcal{P}_{cl}(X), H_d)$ is a generalized (complete) metric space (see [147]).

Lemma 1.7. *If A and B are compact, then there exists either an $a \in A$ with $d(a, B) = H_d(A, B)$ or a $b \in B$ with $d(A, b) = H_d(A, B)$*

Definition 1.8. A multi-valued map $G : \mathbb{R}_+ \to \mathcal{P}_{cl}(X)$ is said to be *measurable* if for each $x \in E$, the function $Y : \mathbb{R}_+ \to X$ defined by

$$Y(t) = d(x, G(t)) = \inf\{|x - z| : z \in G(t)\}$$

is measurable where d is the metric induced by the normed Banach space X.

Definition 1.9. A function $F : \mathbb{R}_+ \times X \longrightarrow \mathcal{P}(X)$ is said to be an L^1_{loc}-Carathéodory multi-valued map if it satisfies:

(i) $x \mapsto F(t, y)$ is continuous for almost all $t \in \mathbb{R}_+$;
(ii) $t \mapsto F(t, y)$ is measurable for each $y \in X$;
(iii) for every positive constant k there exists $h_k \in L^1_{loc}(\mathbb{R}_+, \mathbb{R}_+)$ such that

$$\|F(t, y)\| \leq h_k(t) \quad \text{for all } \|y\|_{\mathcal{B}} \leq k \text{ and for almost all } t \in \mathbb{R}_+.$$

Let $(X, \|\cdot\|)$ be a Banach space. A multi-valued map $G : X \to \mathcal{P}(X)$ has *convex (closed) values* if $G(x)$ is convex (closed) for all $x \in X$. We say that G is *bounded on bounded sets* if $G(B)$ is bounded in X for each bounded set B of X, i.e.,

$$\sup_{x \in B} \{\sup\{\|y\| : y \in G(x)\}\} < \infty.$$

Finally, we say that G has a *fixed point* if there exists $x \in X$ such that $x \in G(x)$.

For each $y \in C(\mathbb{R}_+, E)$ let the set $S_{F,y}$ known as *the set of selectors* from F defined by

$$S_{F,y} = \{v \in L^1(\mathbb{R}_+, E) : v(t) \in F(t, y(t)), \ a.e. \ t \in \mathbb{R}_+\}.$$

For more details on multi-valued maps we refer to the books of Deimling [101], Djebali et al. [104], Górniewicz [123], Hu and Papageorgiou [143], and Tolstonogov [181].

Definition 1.10. A multi-valued map $F : X \to \mathcal{P}(X)$ is called an *admissible contraction* with constant $\{k_n\}_{n \in \mathbb{N}}$ if for each $n \in \mathbb{N}$ there exists $k_n \in [0, 1)$ such that

i) $H_d(F(x), F(y)) \le k_n \|x - y\|_n$ for all $x, y \in X$.
ii) for every $x \in X$ and every $\epsilon \in (0, \infty)^n$, there exists $y \in F(x)$ such that

$$\|x - y\|_n \le \|x - F(x)\|_n + \epsilon_n \text{ for every } n \in \mathbb{N}$$

Lemma 1.11 ([154]). *Let X be a Banach space. Let $F : [a, b] \times X \longrightarrow P_{cp,c}(X)$ be an L^1-Carathéodory multi-valued map with $S_{F,y} \ne \emptyset$ and let Γ be a linear continuous mapping from $L^1([a, b], X)$ into $C([a, b], X)$, then the operator*

$$\Gamma \circ S_F : C([a, b], X) \longrightarrow \mathcal{P}_{cp,c}(C([a, b], X)),$$
$$y \longmapsto (\Gamma \circ S_F)(y) := \Gamma(S_{F,y})$$

is a closed graph operator in $C([a, b], X) \times C([a, b], X)$.

Proposition 1.12 ([167]). *Let the space E be separable and the multi-function $\Phi : [0, b] \to \mathcal{P}(E)$ be integrable bounded and $\chi(\Phi(t)) \le q(t)$ for a.a $t \in [0, b]$ where $q(.) \in L^1([0, b], \mathbb{R}^+)$. Then*

$$\chi \left(\int_0^\tau \Phi(s)ds \right) \le \int_0^\tau q(s)ds, \quad \text{for all } \tau \in [0, b].$$

In particular, if the multi-function $\Phi : [0, b] \to \mathcal{P}_{cl}(E)$ is measurable and integrable bounded, then the function $\chi(\Phi(.))$ is integrable and

$$\chi \left(\int_0^\tau \Phi(s)ds \right) \le \int_0^\tau \chi(\Phi(s))ds, \quad \text{for all } \tau \in [0, b].$$

1.5 Evolution System

In what follows, for the family $\{A(t), t \geq 0\}$ of closed densely defined linear unbounded operators on the Banach space E we assume that it satisfies the following assumptions (see [16], p. 158).

(P1) The domain $D(A(t))$ is independent of t and is dense in E.

(P2) For $t \geq 0$, the resolvent $R(\lambda, A(t)) = (\lambda I - A(t))^{-1}$ exists for all λ with $Re\lambda \leq 0$, and there is a constant M independent of λ and t such that

$$\|R(t, A(t))\| \leq M(1 + |\lambda|)^{-1}, \text{ for } Re\lambda \leq 0.$$

(P3) There exist constants $L > 0$ and $0 < \alpha \leq 1$ such that

$$\|(A(t) - A(\theta))A^{-1}(\tau)\| \leq L|t - \tau|^{\alpha}, \text{ for } t, \theta, \tau \in J.$$

Lemma 1.13 ([16], p. 159). *Under assumptions (P1)–(P3), the Cauchy problem*

$$y'(t) - A(t)y(t) = 0, \ t \in J, \ y(0) = y_0,$$

has a unique evolution system $U(t, s), (t, s) \in \Delta := \{(t, s) \in J \times J : 0 \leq s \leq t < +\infty\}$ satisfying the following properties:

1. $U(t, t) = I$ *where I is the identity operator in E,*
2. $U(t, s) \, U(s, \tau) = U(t, \tau)$ *for $0 \leq \tau \leq s \leq t < +\infty$,*
3. $U(t, s) \in B(E)$ *the space of bounded linear operators on E, where for every $(t, s) \in \Delta$ and for each $y \in E$, the mapping $(t, s) \rightarrow U(t, s) \, y$ is continuous.*

More details on evolution systems and their properties can be found in the books of Ahmed [16], Engel and Nagel [106], and Pazy [168].

1.6 Semigroups

1.6.1 C_0-Semigroups

Let E be a Banach space and $B(E)$ be the Banach space of linear bounded operators on E.

Definition 1.14. A semigroup of class (C^0) is a one parameter family $\{T(t) \mid t \geq 0\} \subset B(E)$ satisfying the conditions:

(i) $T(0) = I$,
(ii) $T(t)T(s) = T(t + s)$, for $t, s \geq 0$,
(iii) the map $t \rightarrow T(t)(x)$ is strongly continuous, for each $x \in E$, i.e;

$$\lim_{t \to 0} T(t)x = x, \ \forall x \in E.$$

A semigroup of bounded linear operators $T(t)$, is uniformly continuous if

$$\lim_{t \to 0} \|T(t) - I\| = 0.$$

Here I denotes the identity operator in E.

We note that if a semigroup $T(t)$ is class (C_0), then satisfies the growth condition

$\|T(t)\|_{B(E)} \leq Me^{\beta t}$, for $0 \leq t < \infty$ with some constants $M > 0$ and β.

If, in particular $M = 1$ and $\beta = 0$, i.e; $\|T(t)\|_{B(E)} \leq 1$, for $t \geq 0$, then the semigroup $T(t)$ is called a *contraction semigroup* (C_0).

Definition 1.15. Let $T(t)$ be a semigroup of class (C_0) defined on E. The infinitesimal generator A of $T(t)$ is the linear operator defined by

$$A(x) = \lim_{h \to 0} \frac{T(h)x - x}{h}, \quad \text{for } x \in D(A),$$

where $D(A) = \{x \in E \mid \lim_{h \to 0} \frac{T(h)(x) - x}{h} \text{ exists in } E\}$.

Let us recall the following property:

Proposition 1.16. *The infinitesimal generator A is a closed, linear, and densely defined operator in E. If $x \in D(A)$, then $T(t)(x)$ is a C^1-map and*

$$\frac{d}{dt}T(t)(x) = A(T(t)(x)) = T(t)(A(x)) \quad \text{on } [0, \infty).$$

Theorem 1.17 (Hille and Yosida [168]). *Let A be a densely defined linear operator with domain and range in a Banach space E. Then A is the infinitesimal generator of uniquely determined semigroup $T(t)$ of class (C_0) satisfying*

$$\|T(t)\|_{B(E)} \leq Me^{\omega t}, \quad t \geq 0,$$

where $M > 0$ and $\omega \in \mathbb{R}$ if and only if $(\lambda I - A)^{-1} \in B(E)$ and

$$\|(\lambda I - A)^{-n}\| \leq M/(\lambda - \omega)^n, \ n = 1, 2, \ldots, \text{ for all } \lambda \in \mathbb{R}.$$

For more details on strongly continuous operators, we refer the reader to the books of Ahmed [16], Goldstein [122], Fattorini [111], Pazy [168], and the papers of Travis and Webb [179, 180].

1.6.2 Integrated Semigroups

Definition 1.18 ([29]). Let E be a Banach space. An integrated semigroup is a family of bounded linear operators $(S(t))_{t \geq 0}$ on E with the following properties:

(i) $S(0) = 0$;

(ii) $t \to S(t)$ is strongly continuous;

(iii) $S(s)S(t) = \int_0^s (S(t+r) - S(r))dr$, for all $t, s \geq 0$.

Definition 1.19 ([146]). An operator A is called a generator of an integrated semigroup if there exists $\omega \in \mathbb{R}$ such that $(\omega, \infty) \subset \rho(A)$ ($\rho(A)$, is the resolvent set of A) and there exists a strongly continuous exponentially bounded family $(S(t))_{t \geq 0}$ of bounded operators such that $S(0) = 0$ and $R(\lambda, A) := (\lambda I - A)^{-1} = \lambda \int_0^\infty e^{-\lambda t} S(t) dt$ exists for all λ with $\lambda > \omega$.

Proposition 1.20 ([29]). *Let A be the generator of an integrated semigroup $(S(t))_{t \geq 0}$. Then for all $x \in E$ and $t \geq 0$,*

$$\int_0^t S(s)xds \in D(A) \quad and \quad S(t)x = A \int_0^t S(s)xds + tx.$$

Definition 1.21 ([146]).

(i) An integrated semigroup $(S(t))_{t \geq 0}$ is called locally Lipschitz continuous if, for all $\tau > 0$ there exists a constant L such that

$$|S(t) - S(s)| \leq L|t - s|, \quad t, s \in [0, \tau].$$

(ii) An integrated semigroup $(S(t))_{t \geq 0}$ is called nondegenerate if $S(t)x = 0$, for all $t \geq 0$ implies that $x = 0$.

Definition 1.22. We say that the linear operator A satisfies the Hille–Yosida condition if there exists $M \geq 0$ and $\omega \in \mathbb{R}$ such that $(\omega, \infty) \subset \rho(A)$ and

$$\sup\{(\lambda - \omega)^n |(\lambda I - A)^{-n}| : n \in \mathbb{N}, \lambda > \omega\} \leq M.$$

Theorem 1.23 ([146]). *The following assertions are equivalent:*

(i) A is the generator of a nondegenerate, locally Lipschitz continuous integrated semigroup;

(ii) A satisfies the Hille–Yosida condition.

If A is the generator of an integrated semigroup $(S(t))_{t \geq 0}$ which is locally Lipschitz, then from [29], $S(\cdot)x$ is continuously differentiable if and only if $x \in \overline{D(A)}$ and $(S'(t))_{t \geq 0}$ is a C_0 semigroup on $\overline{D(A)}$.

1.6.3   Extrapolated Semigroups

Let A_0 be the part of A in $X_0 = \overline{D(A)}$ which is defined by

$$D(A_0) = \{x \in D(A) : Ax \in \overline{D(A)}\}, \text{ and } A_0x = Ax, \text{ for } x \in D(A_0).$$

Lemma 1.24 ([106]). *A_0 generates a strongly continuous semigroup $(T_0(t))_{t \geq 0}$ on X_0 and $|T_0(t)| \leq N_0 e^{\omega t}$, for $t \geq 0$. Moreover $\rho(A) \subset \rho(A_0)$ and $R(\lambda, A_0) = R(\lambda, A)/X_0$, for $\lambda \in \rho(A)$.*

For a fixed $\lambda_0 \in \rho(A)$, we introduce on X_0 a new norm defined by

$$\|x\|_1 = |R(\lambda_0, A_0)x| \text{ for } x \in \overline{D(A_0)}.$$

The completion X_1 of $(X_0, \|\cdot\|_1)$ is called the *extrapolation space of X associated with A*. Note that $\|\cdot\|_1$ and the norm on X_0 given by $|R(\lambda, A_0)x|$, for $\lambda \in \rho(A)$, are extensions $T_1(t)$ to the Banach space X_1, and $(T_1(t))_{t \geq 0}$ is a strongly continuous semigroup on X_1. $(T_1(t))_{t \geq 0}$ is called *the extrapolated semigroup of $(T_0(t))_{t \geq 0}$*, and we denote its generator by $(A_1, D(A_1))$.

Lemma 1.25 ([130]). *The following properties hold:*

(i) $|T(t)|_{B(X_1)} = |T_0(t)|_{L(X_0)}$.
(ii) $D(A_1) = X_0$.
(iii) $A_1 : X_0 \to X_1$ is the unique continuous extension of $A_0 : D(A_0) \subset (X_0, |.|) \to (X_0, \|.\|_1)$, and $(\lambda - A_1)^{-1}$ is an isometry from $(X_0, |.|)$ into $(X_0, \|.\|_1)$.
(iv) If $\lambda \in \rho(A_0)$, then $(\lambda - A_1)$ is invertible and $(\lambda - A_1)^{-1} \in B(X_1)$. In particular $\lambda \in \rho(A_1)$ and $R(\lambda, A_1)/X_0 = R(\lambda, A_0)$
(v) The space $X_0 = \overline{D(A)}$ is dense in $(X_1, \|\cdot\|_1)$. Hence the extrapolation space X_1 is also the completion of $(X, \|.\|_1)$ and $X \hookrightarrow X_1$.
(vi) The operator A_1 is an extension of A. In particular if $\lambda \in \rho(A)$, then $R(\lambda, A_1)/X = R(\lambda, A)$ and $(\lambda - A_1)X = D(A)$.

Abstract extrapolated spaces have been introduced by Da Prato and Grisvard [99] and Engel and Nagel [106] and used for various purposes [23–25, 160, 163, 164].

1.7   Some Fixed Point Theorems

First we will introduce the following compactness criteria in the space of continuous and bounded functions defined on the positive half line.

Lemma 1.26 (Corduneanu [96]). *Let $D \subset BC([0, +\infty), E)$. Then D is relatively compact if the following conditions hold:*

(a) *D is bounded in BC.*
(b) *The function belonging to D is almost equi-continuous on $[0, +\infty)$, i.e., equi-continuous on every compact of $[0, +\infty)$.*
(c) *The set $D(t) := \{y(t) : y \in D\}$ is relatively compact on every compact of $[0, +\infty)$.*
(d) *The function from D is equiconvergent, that is, given $\epsilon > 0$, responds $T(\epsilon) > 0$ such that $|u(t) - \lim_{t \to +\infty} u(t)| < \epsilon$, for any $t \geq T(\epsilon)$ and $u \in D$.*

Lemma 1.27 (Nonlinear Alternative [105]). *Let X be a Banach space with $C \subset X$ closed and convex. Assume U is a relatively open subset of C with $0 \in U$ and $G : \overline{U} \longrightarrow C$ is a compact map. Then either,*

(i) *G has a fixed point in $\overline{U}$; or*
(ii) *there is a point $u \in \partial U$ and $\lambda \in (0, 1)$ with $u = \lambda G(u)$.*

The multi-valued version of Nonlinear Alternative

Lemma 1.28 ([105]). *Let X be a Banach space with $C \subset X$ a convex. Assume U is a relatively open subset of C with $0 \in U$ and $G : X \to \mathcal{P}_{cp,c}(X)$ be an upper semi-continuous and compact map. Then either,*

(a) *there is a point $u \in \partial U$ and $\lambda \in (0, 1)$ with $u \in \lambda G(u)$, or*
(b) *G has a fixed point in $\overline{U}$.*

Theorem 1.29 (Nonlinear Alternative of Frigon and Granas [116]). *Let X be a Fréchet space and $Y \subset X$ a closed subset in Y and let $N : Y \to X$ be a contraction such that $N(Y)$ is bounded.*
Then one of the following statements holds:

(S1)   *N has a unique fixed point;*
(S2)   *There exists $\lambda \in [0, 1)$, $n \in \mathbb{N}$ and $x \in \partial_n Y^n$ such that $\|x - \lambda N(x)\|_n = 0$.*

The following nonlinear alternative is given by Avramescu in Fréchet spaces which is an extension of the same version given by Burton [87] and Burton and Kirk [88] in Banach spaces.

Theorem 1.30 (Nonlinear Alternative of Avramescu [32]). *Let X be a Fréchet space and let $A, B : X \to X$ be two operators satisfying:*

(1) *A is a compact operator,*
(2) *B is a contraction.*

Then either one of the following statements holds:

(S1)   *The operator $A + B$ has a fixed point;*

(S2)   *The set*

$$\left\{ x \in X, x = \lambda A(x) + \lambda B \left(\frac{x}{\lambda} \right) \right\}$$

is unbounded for $\lambda \in (0, 1)$.

Theorem 1.31 (Nonlinear Alternative of Frigon [114, 115]). *Let X be a* Fréchet
space and U an open neighborhood of the origin in X and let $N : \overline{U} \to \mathcal{P}(X)$ be an
admissible multi-valued contraction. *Assume that N is bounded.*
    Then one of the following statements holds:

(S1) N has a fixed point;
(S2)   There exists $\lambda \in [0, 1)$ and $x \in \partial U$ such that $x \in \lambda N(x)$.

The following fixed point theorem is due to Burton and Kirk.

Theorem 1.32 ([88]). *Let X be a Banach space, and A, B two operators satisfy-*
ing:

(i) A is a contraction, and
(ii) B is completely.

Then either

(a) the operator equation $y = A(y) + B(y)$ has a solution, or
(b) the set $\varepsilon = \{ u \in X : \lambda A(\frac{u}{\lambda}) + \lambda B(u) \}$ is unbounded for $\lambda \in (0, 1)$.

We need the following definitions in the sequel.

Definition 1.33. A nonempty closed subset C of a Banach space X is said to be a
cone if

  (i) $C + C \subset C$,
 (ii) $\lambda C \subset C$ for $\lambda > 0$, and,
(iii) $-C \cap C = \{0\}$.

A cone C is called normal if the norm $\| \cdot \|$ is semi-monotone on C, i.e., there
exists a constant $N > 0$ such that $\|x\| \leq N\|y\|$, whenever $x \leq y$. We equip the space
$X = C(J, E)$ with the order relation $\leq$ induced by a regular cone $\mathcal{C}$ in E, that is for
all $y, \overline{y} \in X : y \leq \overline{y}$ if and only if $\overline{y}(t) - y(t) \in \mathcal{C}, \quad \forall t \in J$. In what follows will
assume that the cone C is normal. Cones and their properties are detailed in [134].
Let $a, b \in X$ be such that $a \leq b$. Then, by an order interval $[a, b]$ we mean a set of
points in X given by

$$[a, b] = \{x \in X \mid a \leq x \leq b\}.$$

Definition 1.34. Let X be an ordered Banach space. A mapping $T : X \to X$ is
called isotone increasing if $T(x) \leq T(y)$ for any $x, y \in X$ with $x < y$. Similarly, T is
called isotone decreasing if $T(x) \geq T(y)$ whenever $x < y$.

Definition 1.35. We say that $x \in X$ is the least fixed point of G in X if $x = Gx$ and $x \leq y$ whenever $y \in X$ and $y = Gy$. The greatest fixed point of G in X is defined similarly by reversing the inequality. If both least and greatest fixed point of G in X exist, we call them extremal fixed point of G in X.

The following fixed point theorem is due to Heikkila and Lakshmikantham.

Theorem 1.36. *Let $[a, b]$ be an order interval in an order Banach space X and let $Q : [a, b] \to [a, b]$ be a nondecreasing mapping. If each sequence $(Qx_n) \subset Q([a, b])$ converges, whenever (x_n) is a monotone sequence in $[a, b]$, then the sequence of Q-iteration of a converges to the least fixed point x_* of Q and the sequence of Q-iteration of a converges to the greatest fixed point x_* of Q. Moreover*

$$x_* = \min\{y \in [a, b], y \geq Qy\} \text{ and } x^* = \max\{y \in [a, b], y \geq Qy\}.$$

As a consequence, Dhage and Henderson have proved the following fixed point theorem, which will be used to prove the existence of extremal solutions.

Theorem 1.37 ([102]). *Let $[a, b]$ be an order interval in a Banach space X and let $B_1, B_2 : [a, b] \to X$ be two functions satisfying:*

(a) B_1 is a contraction,
(b) B_2 is completely continuous,
(c) B_1 and B_2 are strictly monotone increasing, and
(d) $B_1(x) + B_2(x) \in [a, b], \ \forall x \in [a, b]$.
   Further if the cone C in X is normal, then the equation $x = B_1(x) + B_2(x)$ has a least fixed point x_* and a greatest fixed point $x^* \in [a, b]$. Moreover $x_* = \lim\limits_{n \to \infty} x_n$ and $x^* = \lim\limits_{n \to \infty} y_n$, where $\{x_n\}$ and $\{y_n\}$ are the sequences in $[a, b]$ defined by

$$x_{n+1} = B_1(x_n) + B_2(x_n), \ x_0 = a \text{ and } y_{n+1} = B_1(y_n) + B_2(y_n), \ y_0 = b.$$

Given a space X and metrics $d_\alpha, \alpha \in \bigwedge$, denote $\mathcal{P}(X) = \{Y \subset X : Y \neq \emptyset\}$, $\mathcal{P}_{cl}(X) = \{Y \in \mathcal{P}(X) : Y \text{ closed}\}$, $\mathcal{P}_b(X) = \{Y \in \mathcal{P}(X) : Y \text{ bounded}\}$. We denote by $D_\alpha, \ \alpha \in \bigwedge$, the Hausdorff pseudo-metric induced by d_α; that is, for $V, W \in \mathcal{P}(X)$,

$$D_\alpha(V, W) = \inf\Big\{\varepsilon > 0 : \forall x \in V, \ \forall y \in W, \ \exists \bar{x} \in V, \ \bar{y} \in W \text{ such that}$$

$$d_\alpha(x, \bar{y}) < \varepsilon, d_\alpha(\bar{x}, y) < \varepsilon\Big\},$$

with $\inf \emptyset = \infty$. In the particular case where X is a complete locally convex space, we say that a subset $V \subset X$ is bounded if $D_\alpha(\{0\}, V) < \infty$ for every $\alpha \in \bigwedge$.

Definition 1.38. A multi-valued map $F : X \to \mathcal{P}(E)$ is called an admissible contraction with constant $\{k_\alpha\}_{\alpha \in \bigwedge}$ if for each $\alpha \in \bigwedge$ there exists $k_\alpha \in (0, 1)$ such that

i) $D_\alpha(F(x), F(y)) \le k_\alpha d_\alpha(x, y)$ for all $x, y \in X$.

ii) for every $x \in X$ and every $\varepsilon \in (0, \infty)^\wedge$, there exists $y \in F(x)$ such that

$$d_\alpha(x, y) \le d_\alpha(x, F(x)) + \varepsilon_\alpha \text{ for every } \alpha \in \bigwedge.$$

Lemma 1.39 (Nonlinear Alternative, [113]). *Let E be a Fréchet space and U an open neighborhood of the origin in E, and let $N : \overline{U} \to \mathcal{P}(E)$ be an admissible multi-valued contraction. Assume that N is bounded. Then one of the following statements holds:*

(C1)   N has at least one fixed point;

(C2)   there exists $\lambda \in [0, 1)$ and $x \in \partial U$ such that $x \in \lambda N(x)$.

Lemma 1.40 ([144]). *If U is a closed convex subset of a Banach space E and $R : U \to \mathcal{P}_{cv,k}(E)$ is a closed β-condensing multi-function, where β is a nonsingular MNC defined on the subsets of U. Then R has a fixed point.*

The next results are concerned with the structure of solution sets for β-condensing u.s.c. multi-valued maps.

Lemma 1.41 ([144]). *Let W be a closed subset of a Banach space E and $R : W \to \mathcal{P}_{cv,k}(E)$ be a closed multi-function which is β-condensing on every bounded subset of W, where β is a monotone MNC. If the fixed points set $\mathcal{F}ixR$ is bounded, then it is compact.*

The following theorem is due to Mönch.

Theorem 1.42 ([162]). *Let E be a Banach space, U an open subset of E and $0 \in U$. Suppose that $N : U \to E$ is a continuous map which satisfies Mönch's condition (i.e., if $D \subseteq \overline{U}$ is countable and $D \subseteq \overline{co}(\{0\} \cup N(D))$, then $\overline{D}$ is compact) and assume that*

$$x \ne \lambda N(x), \quad \text{for } x \in \partial U \text{ and } \lambda \in (0, 1)$$

holds. Then N has a fixed point in $\overline{U}$.

Lemma 1.43 ([144, Theorem 2]). *The generalized Cauchy operator G satisfies the properties*

(G1)   there exists $\zeta \ge 0$ such that

$$\|Gf(t) - Gg(t)\| \le \zeta \int_0^t \|f(s) - g(s)\| ds, \text{ for every } f, g \in L^1(J, E), \, t \in J.$$

(G2)   for any compact $K \subseteq E$ and any sequence $(f_n)_{n \ge 1} \subset L^1(J, E)$ such that for all $n \ge 1$, $f_n(t) \in K$, a. e. $t \in J$, the weak convergence $f_n \rightharpoonup f_0$ in $L^1(J, E)$ implies the convergence $Gf_n \to Gf_0$ in $C(J, E)$.

Lemma 1.44 ([144]). *Let* $S : L^1(J,E) \to C(J,E)$ *be an operator satisfying condition (G2) and the following Lipschitz condition (weaker than (G1)).*

(G1')

$$\|Sf - Sg\|_{C(J,E)} \leq \zeta \|f - g\|_{L^1(J,E)}.$$

Then for every semi-compact set $\{f_n\}_{n=1}^{+\infty} \subset L^1(J,E)$ the set $\{Sf_n\}_{n=1}^{+\infty}$ is relatively compact in $C(J,E)$. Moreover, if $(f_n)_{n\geq 1}$ converges weakly to f_0 in $L^1(J,E)$ then $Sf_n \to Sf_0$ in $C(J,E)$.

Lemma 1.45 ([144]). *Let* $S : L^1(J,E) \to C(J,E)$ *be an operator satisfying conditions (G1), (G2) and let the set* $\{f_n\}_{n=1}^{\infty}$ *be integrable bounded with the property* $\chi(\{f_n(t) : n \geq 1\}) \leq \eta(t)$, *for a.e.* $t \in J$, *where* $\eta(.) \in L^1(J, \mathbb{R}^+)$ *and* χ *is the Hausdorff MNC. Then*

$$\chi(\{Sf_n(t) : n \geq 1\}) \leq 2\zeta \int_0^t \eta(s)ds, \text{ for all } t \in J,$$

where $\zeta \geq 0$ *is the constant in condition (G1).*

Let us recall the following result that will be used in the sequel.

Lemma 1.46 ([86]). *Let* E *be a separable metric space and let* $G : E \to \mathcal{P}$ $(L^1([0,b],E))$ *be a multi-valued operator which is lower semi-continuous and has nonempty closed and decomposable values. Then* G *has a continuous selection, i.e., there exists a continuous function* $f : E \to L^1([0,b],E)$ *such that* $f(y) \in G(y)$ *for every* $y \in E$.

Chapter 2
Partial Functional Evolution Equations with Finite Delay

2.1 Introduction

In this chapter, we study some first order classes of partial functional, neutral functional, integro-differential, and neutral integro-differential evolution equations on a positive line $\mathbb{R}_+$ with local and nonlocal conditions when the historical interval H is bounded, i.e., when the delay is finite. In the literature devoted to equations with finite delay, the phase space is much of time the space of all continuous functions on H for $r > 0$, endowed with the uniform norm topology. Using a recent nonlinear alternative of Leray–Schauder type for contractions in Fréchet spaces due to Frigon and Granas combined with the semigroup theory, the existence and uniqueness of the mild solution will be obtained. The method we are going to use is to reduce the existence of the unique mild solution to the search for the existence of the unique fixed point of an appropriate contraction operator in a Fréchet space.

The nonlocal Cauchy problem has been studied first by Byszewski in 1991 [90] (see also [89, 91, 92]). Then, Balachandran and his collaborators have considered various classes of nonlinear integro-differential systems [44].

2.2 Partial Functional Evolution Equations

2.2.1 Introduction

In this section, we consider partial functional evolution equations with local and nonlocal conditions where the existence of the unique mild solution is assured. Firstly, in Sect. 2.2.2 we consider the following partial functional evolution system

© Springer International Publishing Switzerland 2015
S. Abbas, M. Benchohra, *Advanced Functional Evolution Equations and Inclusions*, Developments in Mathematics 39,
DOI 10.1007/978-3-319-17768-7_2

$$y'(t) = A(t)y(t) + f(t, y_t), \quad \text{a.e.} \quad t \in J = \mathbb{R}_+ \tag{2.1}$$

$$y(t) = \varphi(t), \quad t \in H, \tag{2.2}$$

where $r > 0$, $f : J \times C(H, E) \to E$ and $\varphi \in C(H, E)$ are given functions and $\{A(t)\}_{t \geq 0}$ is a family of linear closed (not necessarily bounded) operators from E into E that generate an evolution system of bounded linear operators $\{U(t, s)\}_{(t,s) \in J \times J}$ for $0 \leq s \leq t < +\infty$ from E into E.

Later, we consider the functional evolution problem with a nonlocal condition of the form

$$y'(t) = A(t)y(t) + f(t, y_t), \quad \text{a.e.} \quad t \in J = \mathbb{R}_+ \tag{2.3}$$

$$y(t) + h_t(y) = \varphi(t), \quad t \in H, \tag{2.4}$$

where $A(\cdot), f$ and φ are as in evolution problem (2.1)–(2.2) and $h_t : C(H, E) \to E$ is a given function.

Using the fixed point argument, Frigon applied its own alternative to some differential and integral equations in [113]. In the literature devoted to equations with $A(\cdot) = A$ on a bounded interval, we can found the recent works by Benchohra and Ntouyas for semi-linear equations and inclusions [58, 59, 65], controllability results are established by Benchohra et al. in [26, 75, 76] and Li et al. in [156].

2.2.2 Main Result

Let us introduce the definition of the mild solution of the partial functional evolution system (2.1)–(2.2).

Definition 2.1. We say that the continuous function $y(\cdot) : [-r, +\infty) \to E$ is a mild solution of (2.1)–(2.2) if $y(t) = \varphi(t)$ for all $t \in H$ and y satisfies the following integral equation

$$y(t) = U(t, 0)\,\varphi(0) + \int_0^t U(t, s)\,f(s, y_s)\,ds, \quad \text{for each } t \in [0, +\infty).$$

We will need the following hypotheses which are assumed hereafter:

(2.1.1) There exists a constant $\widehat{M} \geq 1$ such that

$$\|U(t, s)\|_{B(E)} \leq \widehat{M}$$

for every $(t, s) \in \Delta := \{(t, s) \in J \times J : 0 \leq s \leq t < +\infty\};$

(2.1.2) There exist a continuous nondecreasing function $\psi : \mathbb{R}_+ \to (0, +\infty)$ and $p \in L^1_{\text{loc}}([0, +\infty), \mathbb{R}_+)$ such that

$$|f(t, u)| \leq p(t)\, \psi(\|u\|),$$

for a.e. $t \in [0, +\infty)$ and each $u \in C(H, E)$;

(2.1.3) For all $R > 0$, there exists $l_R \in L^1_{\text{loc}}([-r, +\infty), \mathbb{R}_+)$ such that

$$|f(t, u) - f(t, v)| \leq l_R(t)\, \|u - v\|$$

for all $u, v \in C(H, E)$ with $\|u\| \leq R$ and $\|v\| \leq R$.

For every $n \in \mathbb{N}$, we define in $C([-r, +\infty), E)$ the semi-norms by:

$$\|y\|_n := \sup \{ e^{-\tau\, L_n^*(t)}\, |y(t)| : t \in [0, n] \}$$

where $L_n^*(t) = \int_0^t \bar{l}_n(s)\, ds$, $\bar{l}_n(t) = \widehat{M}\, l_n(t)$ and l_n is the function from (2.1.3).

Then $C([-r, +\infty), E)$ is a Fréchet space with the family of semi-norms $\{\|\cdot\|_n\}_{n \in \mathbb{N}}$. In what follows we will choose $\tau > 1$.

Theorem 2.2 ([33]). *Suppose that hypotheses* (2.1.1)–(2.1.3) *are satisfied and moreover for $n > 0$*

$$\int_{c_1}^{+\infty} \frac{ds}{\psi(s)} > \widehat{M} \int_0^n p(s)\, ds, \tag{2.5}$$

where $c_1 = \widehat{M}\, \|\varphi\|$. Then the problem (2.1)–(2.2) *has a unique mild solution.*

Proof. Transform the problem (2.1)–(2.2) into a fixed point problem. Consider the operator $N : C([-r, +\infty), E) \to C([-r, +\infty), E)$ defined by:

$$(N_1 y)(t) = \begin{cases} \varphi(t), & \text{if } t \in H; \\ U(t, 0)\, \varphi(0) + \displaystyle\int_0^t U(t, s)\, f(s, y_s)\, ds, & \text{if } t \in \mathbb{R}_+. \end{cases}$$

Clearly, the fixed points of the operator N_1 are mild solutions of the problem (2.1)–(2.2).

Let y be a possible solution of the problem (2.1)–(2.2). Given $n \in \mathbb{N}$ and $t \leq n$, then from (2.1.1) and (2.1.2) we have:

$$|y(t)| \leq |U(t, 0)|\, |\varphi(0)| + \int_0^t \|U(t, s)\|_{B(E)}\, |f(s, y_s)| ds$$

$$\leq \widehat{M} \, \|\varphi\| + \widehat{M} \int_0^t p(s) \, \psi(\|y_s\|) ds.$$

We consider the function μ defined by

$$\mu(t) := \sup\{ |y(s)| \; : \; 0 \leq s \leq t \}, \quad 0 \leq t < +\infty.$$

Let $t^* \in [-r, t]$ be such that $\mu(t) = |y(t^*)|$. If $t^* \in [0, n]$, by the previous inequality we get

$$\mu(t) \leq \widehat{M} \, \|\varphi\| + \widehat{M} \int_0^t p(s) \, \psi(\mu(s)) \, ds, \quad t \in [0, n].$$

If $t^* \in H$, then $\mu(t) = \|\varphi\|$ and the previous inequality holds.

Let us take the right-hand side of the above inequality as $v(t)$. Then we have

$$\mu(t) \leq v(t) \; \text{ for all } t \in [0, n].$$

From the definition of v, we get

$$c_1 := v(0) = \widehat{M}\|\varphi\| \quad \text{and} \quad v'(t) = \widehat{M} \, p(t) \, \psi(\mu(t)) \quad \text{a.e. } t \in [0, n].$$

Using the nondecreasing character of ψ, we have

$$v'(t) \leq \widehat{M} \, p(t) \, \psi(v(t)) \quad \text{a.e. } t \in [0, n].$$

This implies that for each $t \in [0, n]$ and using (2.5) we get

$$\int_{c_1}^{v(t)} \frac{ds}{\psi(s)} \leq \widehat{M} \int_0^t p(s) \, ds$$

$$\leq \widehat{M} \int_0^n p(s) \, ds$$

$$< \int_{c_1}^{+\infty} \frac{ds}{\psi(s)}.$$

Thus there exists a constant Λ_n such that $v(t) \leq \Lambda_n$, $t \in [0, n]$ and hence $\mu(t) \leq \Lambda_n$, $t \in [0, n]$. Since for every $t \in [0, n]$, $\|y_t\| \leq \mu(t)$, we have

$$\|y\|_n \leq \max\{\|\varphi\|, \Lambda_n\} := \Delta_n.$$

Set

$$Y = \{ y \in \mathcal{C}([-r, +\infty), E) : \sup\{|y(t)| \; : 0 \leq t \leq n\} \leq \Delta_n + 1 \; \text{ for all } n \in \mathbb{N} \}.$$

Clearly, Y is a closed subset of $\mathcal{C}([-r, +\infty), E)$.

We shall show that $N_1 : Y \to \mathcal{C}([-r, +\infty), E)$ is a contraction operator. Indeed, consider $y, \bar{y} \in \mathcal{C}([-r, +\infty), E)$, thus using $(2, 1, 1)$ and $(2.1.3)$ for each $t \in [0, n]$ and $n \in \mathbb{N}$ we get

$$
\begin{aligned}
|(N_1 y)(t) - (N_1 \bar{y})(t)| &\leq \int_0^t \|U(t, s)\|_{B(E)} \, |f(s, y_s) - f(s, \bar{y}_s)| \, ds \\
&\leq \int_0^t \hat{M} \, l_n(s) \, \|y_s - \bar{y}_s\| \, ds \\
&\leq \int_0^t [\bar{l}_n(s) \, e^{\tau \, L_n^*(s)}] \, [e^{-\tau \, L_n^*(s)} \, \|y_s - \bar{y}_s\|] \, ds \\
&\leq \int_0^t \left[\frac{e^{\tau \, L_n^*(s)}}{\tau}\right]' \, ds \, \|y - \bar{y}\|_n \\
&\leq \frac{1}{\tau} \, e^{\tau \, L_n^*(t)} \, \|y - \bar{y}\|_n.
\end{aligned}
$$

Therefore,

$$
\|(N_1 y) - (N_1 \bar{y})\|_n \leq \frac{1}{\tau} \, \|y - \bar{y}\|_n.
$$

So, for $\tau > 1$, the operator N_1 is a contraction for all $n \in \mathbb{N}$. From the choice of Y there is no $y \in \partial Y^n$ such that $y = \lambda N_1(y)$ for some $\lambda \in (0, 1)$. Then the statement $(S2)$ in Theorem 1.29 does not hold. A consequence of the nonlinear alternative of Frigon and Granas that $(S1)$ holds, we deduce that the operator N_1 has a unique fixed point y^* which is the unique mild solution of the problem (2.1)–(2.2). □

2.2.3 An Example

As an application of Theorem 2.2, we consider the following partial functional differential equation

$$
\begin{cases}
\dfrac{\partial z}{\partial t}(t, x) = a(t, x)\dfrac{\partial^2 z}{\partial x^2}(t, x) + Q(t, z(t - r, x)) & t \in [0, +\infty), \quad x \in [0, \pi] \\[2ex]
z(t, 0) = z(t, \pi) = 0 & t \in [0, +\infty) \\[2ex]
z(t, x) = \Phi(t, x) & t \in H, \qquad x \in [0, \pi],
\end{cases}
$$

$$(2.6)$$

where $r > 0$, $a(t, x) : [0, \infty) \times [0, \pi] \to \mathbb{R}$ is a continuous function and is uniformly Hölder continuous in t, $Q : [0, +\infty) \times \mathbb{R} \to \mathbb{R}$ and $\Phi : H \times [0, \pi] \to \mathbb{R}$ are continuous functions.

Consider $E = L^2([0, \pi], \mathbb{R})$ and define $A(t)$ by $A(t)w = a(t, x)w''$ with domain

$$D(A) = \{w \in E : w, w' \text{ are absolutely continuous}, w'' \in E, w(0) = w(\pi) = 0\}.$$

Then $A(t)$ generates an evolution system $U(t, s)$ satisfying assumption (2.1.1) (see [112, 149]).

For $x \in [0, \pi]$, we set

$$y(t)(x) = z(t, x) \quad t \in \mathbb{R}_+,$$

$$f(t, y_t)(x) = Q(t, z(t - r, x)) \quad t \in \mathbb{R}_+$$

and

$$\varphi(t)(x) = \Phi(t, x) \quad -r \le t \le 0.$$

Thus, under the above definitions of f, φ and $A(\cdot)$, the system (2.6) can be represented by the abstract partial functional evolution problem (2.1)–(2.2). Furthermore, more appropriate conditions on Q ensure the existence of unique mild solution for (2.6) by Theorems 2.2 and 1.29.

2.2.4 Nonlocal Case

In this section, we extend the above results about the existence and uniqueness of mild solution to the partial functional evolution equations with nonlocal conditions (2.3)–(2.4). The nonlocal condition can be applied in physics with better effect than the classical initial condition $y(0) = y_0$. For example, $h_t(y)$ may be given by

$$h_t(y) = \sum_{i=1}^{p} c_i \, y(t_i + t), \quad t \in H$$

where c_i, $i = 1, \ldots, p$ are given constants and $0 < t_1 < \cdots < t_p < +\infty$.

At time $t = 0$, we have

$$h_0(y) = \sum_{i=1}^{p} c_i \, y(t_i).$$

Nonlocal conditions were initiated by Byszewski [90] to which we refer for motivation and other references.

Before giving the main result, we give first the definition of mild solution of the nonlocal partial functional evolution problem (2.3)–(2.4).

Definition 2.3. A function $y \in C([-r, +\infty), E)$ is said to be a mild solution of (2.3)–(2.4) if $y(t) = \varphi(t) - h_t(y)$ for all $t \in H$ and y satisfies the following integral equation

$$y(t) = U(t, 0) [\varphi(0) - h_0(y)] + \int_0^t U(t, s) f(s, y_s) \, ds, \quad \text{for each } t \in [0, +\infty).$$

We will need the following hypotheses on $h_t(\cdot)$ in the proof of the main result of this section.

(2.3.1) For all $n \geq 0$, there exists a constant $\sigma_n > 0$ such that

$$|h_t(u) - h_t(v)| \leq \sigma_n \|u - v\|$$

for all $t \in H$, $u, v \in C([-r, \infty), E)$ with $\|u\| \leq n$ and $\|v\| \leq n$;
(2.3.2) there exists $\sigma > 0$ such that

$$|h_t(u)| \leq \sigma \text{ for each } u \in C(H, E), \text{ and } t \in J.$$

Theorem 2.4 ([33]). *Assume that the hypotheses* (2.1.1)–(2.1.3), (2.3.1), *and* (2.3.2) *hold and moreover for $n > 0$*

$$\int_{c_2}^{+\infty} \frac{ds}{\psi(s)} > \widehat{M} \int_0^n p(s) \, ds, \tag{2.7}$$

where $c_2 = \widehat{M}(\|\varphi\| + \sigma)$. Then the nonlocal evolution problem (2.3)–(2.4) *has a unique mild solution.*

Proof. Transform the problem (2.3)–(2.4) into a fixed point problem. Consider the operator $N_2 : C([-r, +\infty), E) \to C([-r, +\infty), E)$ defined by:

$$(N_2 y)(t) = \begin{cases} \varphi(t) - h_t(y), & \text{if } t \in H; \\[2mm] U(t, 0) [\varphi(0) - h_0(y)] + \int_0^t U(t, s) f(s, y_s) \, ds, & \text{if } t \in \mathbb{R}_+. \end{cases}$$

Clearly, the fixed points of the operator N_2 are mild solutions of the problem (2.3)–(2.4).

Then, by parallel steps of Theorem 2.2's proof, we can easily show that the operator N_2 is a contraction which have a unique fixed point by statement (S1) in Theorem 1.29. The details are left to the reader. □

2.3   Neutral Functional Evolution Equations

2.3.1   Introduction

In this section, we investigate neutral functional evolution equations with local and nonlocal conditions. First, we study in Sect. 2.3.2 the following neutral functional evolution equations

$$\frac{d}{dt}[y(t) - g(t, y_t)] = A(t)y(t) + f(t, y_t); \text{ a.e. } t \in \mathbb{R}_+, \tag{2.8}$$

$$y(t) = \varphi(t); \ t \in H, \tag{2.9}$$

where $r > 0$; $f, g : J \times C(H, E) \to E$ and $\varphi \in C(H, E)$ are given functions and $\{A(t)\}_{t \geq 0}$ is a family of linear closed (not necessarily bounded) operators from E into E that generate an evolution system of operators $\{U(t, s)\}_{(t,s) \in J \times J}$ for $0 \leq s \leq t < +\infty$.

An extension of these existence results will be given in Sect. 2.3.4 for the following neutral functional evolution equation with nonlocal conditions

$$\frac{d}{dt}[y(t) - g(t, y_t)] = A(t)y(t) + f(t, y_t), \text{ a.e. } t \in J = \mathbb{R}_+ \tag{2.10}$$

$$y(t) + h_t(y) = \varphi(t), \ t \in H, \tag{2.11}$$

where $A(\cdot), f, g$, and φ are as in problem (2.8)–(2.9) and $h_t : C([-r, \infty), E) \to E$ is a given function.

Neutral equations have received much attention in recent years: existence and uniqueness of mild, strong, and classical solutions for semi-linear functional differential equations and inclusions has been studied extensively by many authors. Hernandez in [138] proved the existence of mild, strong, and periodic solutions for neutral equations. Fu in [117] studied the controllability on a bounded interval of a class of neutral functional differential equations. Fu and Ezzinbi [119] considered the existence of mild and classical solutions for a class of neutral partial functional differential equations with nonlocal conditions.

Here we are interesting to give existence and uniqueness of the mild solution for the neutral functional evolution equations (2.8)–(2.9) and the corresponding nonlocal problem (2.10)–(2.11).

2.3.2   Main Result

We give first the definition of the mild solution of (2.8)–(2.9).

Definition 2.5. We say that the continuous function $y(\cdot) : [-r, +\infty) \to E$ is a mild solution of (2.8)–(2.9) if $y(t) = \varphi(t)$ for all $t \in H$ and y satisfies the following integral equation

$$y(t) = U(t, 0) \left[\varphi(0) - g(0, \varphi) \right] + g(t, y_t) + \int_0^t U(t, s) A(s) g(s, y_s) \, ds$$

$$+ \int_0^t U(t, s) f(s, y_s) \, ds, \qquad \text{for each } t \in [0, +\infty).$$

We will need to introduce the following assumptions:

$(G1)$ There exists a constant $\overline{M}_0 > 0$ such that:

$$\|A^{-1}(t)\| \le \overline{M}_0 \quad \text{for all } t \ge 0.$$

$(G2)$ There exists a constant $0 < L < \dfrac{1}{\overline{M}_0}$, such that:

$$|A(t) g(t, \varphi)| \le L \left(\|\varphi\| + 1 \right) \text{ for all } t > 0 \text{ and } \varphi \in C(H, E).$$

$(G3)$ There exists a constant $L_* > 0$ such that:

$$|A(s) g(s, \varphi) - A(\overline{s}) g(\overline{s}, \overline{\varphi})| \le L_* \left(|s - \overline{s}| + \|\varphi - \overline{\varphi}\| \right)$$

for all $s, \overline{s} > 0$ and $\varphi, \overline{\varphi} \in C(H, E)$.

For every $n \in \mathbb{N}$, we define in $C([-r, +\infty), E)$ the semi-norms by:

$$\|y\|_n := \sup \left\{ e^{-\tau \, L_n^*(t)} \, |y(t)| : t \in [0, n] \right\}$$

where $L_n^*(t) = \int_0^t \overline{l}_n(s) \, ds$, $\overline{l}_n(t) = \widehat{M} [L_* + l_n(t)]$ and l_n is the function from (2.1.3).

Then $C([-r, +\infty), E)$ is a Fréchet space with the family of semi-norms $\{\| \cdot \|_n\}_{n \in \mathbb{N}}$. Let us fix $\tau > 0$ and assume $\left[\overline{M}_0 L_* + \dfrac{1}{\tau} \right] < 1$.

Theorem 2.6 ([37]). *Suppose that hypotheses (2.1.1)–(2.1.3) and the assumptions (G1)–(G3) are satisfied. If*

$$\int_{c_{3,n}}^{+\infty} \frac{ds}{s + \psi(s)} > \frac{\widehat{M}}{1 - \overline{M}_0 L} \int_0^n \max(L, p(s)) ds, \text{ for each } n > 0 \qquad (2.12)$$

with

$$c_{3,n} = \frac{\widehat{M}(1 + \overline{M}_0 L)\|\varphi\| + \overline{M}_0 L(\widehat{M} + 1) + \widehat{M} L n}{1 - \overline{M}_0 L}.$$

Then the problem (2.8)–(2.9) has a unique mild solution.

Proof. Transform the problem (2.8)–(2.9) into a fixed point problem. Consider the operator $N_3 : C([-r, +\infty), E) \to C([-r, +\infty), E)$ defined by:

$$(N_3 y)(t) = \begin{cases} \varphi(t), & \text{if } t \in H \\ U(t, 0)\,[\varphi(0) - g(0, \varphi)] + g(t, y_t) \\ \quad + \displaystyle\int_0^t U(t, s) A(s) g(s, y_s)\,ds + \int_0^t U(t, s) f(s, y_s)\,ds, & \text{if } t \in \mathbb{R}_+. \end{cases}$$

Clearly, the fixed points of the operator N_3 are mild solutions of the problem (2.8)–(2.9).

We are going to use Theorem 1.29 in the following way. (i) Define a set Y such that (2.8)–(2.9) does'nt have any solution out of Y. (ii) Show that (S2) of Theorem 1.29 doesn't hold under the above choice of Y, hence (S1) takes place.

Let y be such that $y = \lambda N_3(y)$ for $\lambda \in [0, 1]$. Given $n \in \mathbb{N}$ and $t \leq n$, then from (2.1.1), (2.1.2), (G1) and (G2) we have

$$|y(t)| \leq |U(t, 0)|\,|\varphi(0) - g(0, \varphi)| + |g(t, y_t)|$$

$$+ \int_0^t |U(t, s)|\,|A(s)\,g(s, y_s)|\,ds + \int_0^t |U(t, s)|\,|f(s, y_s)|\,ds$$

$$\leq \widehat{M}\,\|\varphi\| + \widehat{M}\,\|A^{-1}(0)\|\,|A(0)\,g(0, \varphi)| + \|A^{-1}(t)\|\|A(t)\,g(t, y_t)|$$

$$+ \widehat{M} \int_0^t |A(s)\,g(s, y_s)|\,ds + \widehat{M} \int_0^t p(s)\,\psi(\|y_s\|)\,ds$$

$$\leq \widehat{M}\,\|\varphi\| + \widehat{M}\,\overline{M}_0\,L(\|\varphi\| + 1) + \overline{M}_0\,L(\|y_t\| + 1)$$

$$+ \widehat{M}\,L \int_0^t (\|y_s\| + 1)\,ds + \widehat{M} \int_0^t p(s)\,\psi(\|y_s\|)\,ds.$$

We consider the function μ defined by

$$\mu(t) := \sup\{ |y(s)| \ : \ 0 \leq s \leq t \}, \quad 0 \leq t < +\infty.$$

Let $t^* \in [-r, t]$ be such that $\mu(t) = |y(t^*)|$. If $t^* \in [0, n]$, by the previous inequality we get for $t \in [0, n]$

$$\mu(t) \leq \widehat{M}\,\|\varphi\| + \widehat{M}\,\overline{M}_0\,L(\|\varphi\| + 1) + \overline{M}_0\,L(\mu(t) + 1)$$

$$+ \widehat{M} \int_0^t L\,(\mu(s) + 1)\,ds + \widehat{M} \int_0^t p(s)\,\psi(\mu(s))\,ds.$$

So

$$(1 - \overline{M}_0 L)\, \mu(t) \le \widehat{M}\, \|\varphi\| + \widehat{M}\, \overline{M}_0\, L\, \|\varphi\| + \widehat{M}\, \overline{M}_0\, L + \overline{M}_0\, L$$

$$+ \widehat{M}\, L\, n + \widehat{M} \int_0^t L\mu(s)ds + \widehat{M} \int_0^t p(s)\, \psi(\mu(s))\, ds.$$

Set

$$c_{3,n} := \frac{\widehat{M}(1 + \overline{M}_0 L)\|\varphi\| + \overline{M}_0 L(\widehat{M} + 1) + \widehat{M}Ln}{1 - \overline{M}_0 L},$$

then

$$\mu(t) \le \frac{1}{1 - \overline{M}_0 L}\left[\widehat{M}(1 + \overline{M}_0 L)\|\varphi\| + \overline{M}_0 L(\widehat{M} + 1) + \widehat{M}Ln\right]$$

$$+ \frac{\widehat{M}}{1 - \overline{M}_0 L}\left[\int_0^t L\mu(s)ds + \int_0^t p(s)\psi(\mu(s))ds\right]$$

$$= c_{3,n} + \frac{\widehat{M}}{1 - \overline{M}_0 L}\left[\int_0^t L\mu(s)ds + \int_0^t p(s)\psi(\mu(s))ds\right].$$

If $t^* \in H$, then $\mu(t) = \|\varphi\|$ and the previous inequality holds.

Let us take the right-hand side of the above inequality as $v(t)$. Then we have

$$\mu(t) \le v(t) \quad \text{for all } t \in [0, n].$$

From the definition of v, we get

$$v(0) = c_{3,n} \quad \text{and} \quad v'(t) = \frac{\widehat{M}}{1 - \overline{M}_0 L}[L\mu(t) + p(t)\psi(\mu(t))] \quad \text{a.e. } t \in [0, n].$$

Using the nondecreasing character of ψ, we have for a.e. $t \in [0, n]$

$$v'(t) \le \frac{\widehat{M}}{1 - \overline{M}_0 L}[Lv(t) + p(t)\psi(v(t))].$$

This implies that for each $t \in [0, n]$ and using (2.12) we get

$$\int_{c_{3,n}}^{v(t)} \frac{ds}{s + \psi(s)} \le \frac{\widehat{M}}{1 - \overline{M}_0 L} \int_0^t \max(L, p(s))ds$$

$$\le \frac{\widehat{M}}{1 - \overline{M}_0 L} \int_0^n \max(L, p(s))ds$$

$$< \int_{c_{3,n}}^{+\infty} \frac{ds}{s + \psi(s)}.$$

Thus, there exists a constant Λ_n such that $v(t) \leq \Lambda_n$, $t \in [0,n]$ and hence $\mu(t) \leq \Lambda_n$, $t \in [0,n]$. Since for every $t \in [0,n]$, $\|y_t\| \leq \mu(t)$, we have $\|y\|_n \leq \max\{\|\varphi\|, \Lambda_n\} := \Delta_n$. Set

$$Y = \{\, y \in \mathcal{C}([-r, +\infty), E) : \|y\|_n \leq \Delta_n + 1 \text{ for each } n \in \mathbb{N} \}.$$

Clearly, Y is an open subset of $\mathcal{C}([-r, +\infty), E)$.

We shall show that $N_3 : \overline{Y} \to \mathcal{C}([-r, +\infty), E)$ is a contraction operator. Indeed, consider $y, \overline{y} \in \mathcal{C}([-r, +\infty), E)$, thus using (2.1.1), (2.1.3), (G1) and (G3) for each $t \in [0,n]$ and $n \in \mathbb{N}$ we get

$$
\begin{aligned}
|(N_3 y)(t) - (N_3 \overline{y})(t)| \leq\; & |g(t, y_t) - g(t, \overline{y}_t)| \\
& + \int_0^t |U(t,s)|\,|A(s)\,(g(s, y_s) - g(s, \overline{y}_s))|\,ds \\
& + \int_0^t |U(t,s)|\,|f(s, y_s) - f(s, \overline{y}_s)|\,ds \\
\leq\; & \|A^{-1}(t)\|\,|A(t)\,g(t, y_t) - A(t)\,g(t, \overline{y}_t)| \\
& + \int_0^t \widehat{M}\,|A(s)\,g(s, y_s) - A(s)\,g(s, \overline{y}_s)|\,ds \\
& + \int_0^t \widehat{M}\,|f(s, y_s) - f(s, \overline{y}_s)|\,ds \\
\leq\; & \overline{M}_0\,L_*\,\|y_t - \overline{y}_t\| + \int_0^t \widehat{M}\,L_*\,\|y_s - \overline{y}_s\|\,ds \\
& + \int_0^t \widehat{M}\,l_n(s)\,\|y_s - \overline{y}_s\|\,ds \\
\leq\; & \overline{M}_0\,L_*\,\|y_t - \overline{y}_t\| + \int_0^t \left[\widehat{M}\,L_* + \widehat{M}\,l_n(s)\right]\|y_s - \overline{y}_s\|\,ds \\
\leq\; & \left[\overline{M}_0\,L_*\,e^{\tau\,L_n^*(t)}\right]\left[e^{-\tau\,L_n^*(t)}\,\|y_t - \overline{y}_t\|\right] \\
& + \int_0^t \left[\overline{l}_n(s)\,e^{\tau\,L_n^*(s)}\right]\left[e^{-\tau\,L_n^*(s)}\,\|y_s - \overline{y}_s\|\right]\,ds \\
\leq\; & \left[\overline{M}_0\,L_*\,e^{\tau\,L_n^*(t)}\right]\|y - \overline{y}\|_n + \int_0^t \left[\frac{e^{\tau\,L_n^*(s)}}{\tau}\right]'\,ds\,\|y - \overline{y}\|_n.
\end{aligned}
$$

Then

$$|(N_3 y)(t) - (N_3 \overline{y})(t)| \le \left[\overline{M}_0 L_* e^{\tau L_n^*(t)} \right] \|y - \overline{y}\|_n + \frac{1}{\tau} e^{\tau L_n^*(t)} \|y - \overline{y}\|_n$$

$$\le \left[\overline{M}_0 L_* + \frac{1}{\tau} \right] e^{\tau L_n^*(t)} \|y - \overline{y}\|_n.$$

Therefore,

$$\|N_3(y) - N_3(\overline{y})\|_n \le \left[\overline{M}_0 L_* + \frac{1}{\tau} \right] \|y - \overline{y}\|_n.$$

So, for $\left[\overline{M}_0 L_* + \frac{1}{\tau} \right] < 1$, the operator N_3 is a contraction for all $n \in \mathbb{N}$. From the choice of Y there is no $y \in \partial Y^n$ such that $y = \lambda N_3(y)$ for some $\lambda \in (0, 1)$. Then the statement $(S2)$ in Theorem 1.29 does'nt hold. Thus statement $(S1)$ holds, and hence the operator N_3 has a unique fixed point y^* in $\overline{Y}$, which is the unique mild solution of the neutral functional evolution problem (2.8)–(2.9). $\qquad \square$

2.3.3 An Example

As an application of our results we consider the following model

$$\begin{cases} \dfrac{\partial}{\partial t} \left[z(t, x) - \displaystyle\int_{-r}^{t} \int_0^{\pi} b(s - t, u, x) \, z(s, u) \, du \, ds \right] \\ \qquad = a(t, x) \dfrac{\partial^2 z}{\partial x^2}(t, x) + Q(t, z(t - r, x), \dfrac{\partial z}{\partial x}(t - r, x)), \ t \in [0, +\infty), \ x \in [0, \pi] \\ z(t, 0) = z(t, \pi) = 0, \qquad\qquad\qquad\qquad\qquad t \in [0, +\infty) \\ z(t, x) = \Phi(t, x), \qquad\qquad\qquad\qquad\qquad t \in H, \quad x \in [0, \pi] \end{cases}$$
$$\tag{2.13}$$

where $r > 0$; $a(t, x)$ is a continuous function and is uniformly Hölder continuous in t, $Q : [0, +\infty) \times \mathbb{R} \times \mathbb{R} \to \mathbb{R}$ and $\Phi : H \times [0, \pi] \to \mathbb{R}$ are continuous functions.

Let

$$y(t)(x) = z(t, x), \ t \in [0, \infty), \ x \in [0, \pi],$$

$$\varphi(\theta)(x) = \Phi(\theta, x), \ \theta \in H, \ x \in [0, \pi],$$

$$g(t, y_t)(x) = \int_{-r}^{t} \int_0^{\pi} b(s - t, u, x) z(s, u) \, du \, ds, \ x \in [0, \pi]$$

and

$$f(t, y_t)(x) = Q(t, z(\theta, x), \frac{\partial z}{\partial x}(\theta, x)), \ \theta \in H, \ x \in [0, \pi].$$

Consider $E = L^2([0, \pi], \mathbb{R})$ and define $A(t)$ by $A(t)w = a(t, x)w''$ with domain

$$D(A) = \{w \in E : w, w' \text{ are absolutely continuous, } w'' \in E, \ w(0) = w(\pi) = 0\}.$$

Then $A(t)$ generates an evolution system $U(t, s)$ satisfying assumptions (2.1.1) and (G1) (see [112, 149]).

Here we consider that $\varphi : H \rightarrow E$ such that φ is Lebesgue measurable and $h(s)|\varphi(s)|^2$ is Lebesgue integrable on H where $h : H \rightarrow \mathbb{R}$ is a positive integrable function. The norm is defined here by:

$$\|\varphi\| = |\Phi(0)| + \left(\int_{-r}^{0} h(s)|\varphi(s)|^2 \, ds \right)^{\frac{1}{2}}.$$

The function b is measurable on $[0, \infty) \times [0, \pi] \times [0, \pi]$,

$$b(s, u, 0) = b(s, u, \pi) = 0, \ (s, u) \in [0, \infty) \times [0, \pi],$$

$$\int_0^\pi \int_{-r}^t \int_0^\pi \frac{b^2(s, u, x)}{h(s)} ds du dx < \infty$$

and $\sup\limits_{t \in [0, \infty)} \mathcal{N}(t) < \infty$, where

$$\mathcal{N}(t) = \int_0^\pi \int_{-r}^t \int_0^\pi \frac{1}{h(s)} \left(a(s, x) \frac{\partial^2}{\partial x^2} b(s, u, x) \right)^2 ds du dx.$$

Thus, under the above definitions of f, g, and $A(\cdot)$, the system (2.13) can be represented by the abstract neutral functional evolution problem (2.8)–(2.9). Furthermore, more appropriate conditions on Q ensure the existence of at least one mild solution for (2.13) by Theorems 2.6 and 1.29.

2.3.4 Nonlocal Case

In this section we give existence and uniqueness results for the neutral functional evolution equation with nonlocal conditions (2.10)–(2.11). Nonlocal conditions were initiated by Byszewski [89]. Before giving the main result, we give first the definition of the mild solution.

Definition 2.7. A function $y \in C([-r, +\infty), E)$ is said to be a mild solution of (2.10)–(2.11) if $y(t) = \varphi(t) - h_t(y)$ for all $t \in H$ and y satisfies the following integral equation

$$
y(t) = U(t, 0) \left[\varphi(0) - h_0(y) - g(0, \varphi)\right] + g(t, y_t) +
$$
$$
+ \int_0^t U(t, s) A(s) g(s, y_s) \, ds + \int_0^t U(t, s) f(s, y_s) \, ds, \quad \text{for } t \in \mathbb{R}_+.
$$

We take here the same assumptions in Sect. 2.2.4 for the function $h_t(\cdot)$.

Theorem 2.8 ([37]). *Assume that the hypotheses (2.1.1)–(2.1.3), (G1)–(G3), (D1), and (D2) hold. If*

$$
\int_{c_{4,n}}^{+\infty} \frac{ds}{s + \psi(s)} > \frac{\widehat{M}}{1 - \overline{M}_0 L} \int_0^n \max(L, p(s)) \, ds, \text{ for each } n > 0 \qquad (2.14)
$$

with

$$
c_{4,n} = \frac{\widehat{M}\left[(1 + \overline{M}_0 L) \|\varphi\| + \sigma\right] + \overline{M}_0 L(\widehat{M} + 1) + \widehat{M} L n}{1 - \overline{M}_0 L};
$$

Then the nonlocal neutral functional evolution problem (2.10)–(2.11) has a unique mild solution.

Proof. Consider the operator $N_4 : C([-r, +\infty), E) \to C([-r, +\infty), E)$ defined by:

$$
(N_4 y)(t) = \begin{cases} \varphi(t) - h_t(y), & \text{if } t \in H; \\[2mm] U(t, 0) \left[\varphi(0) - h_0(y) - g(0, \varphi)\right] + g(t, y_t) \\[1mm] \quad + \int_0^t U(t, s) A(s) g(s, y_s) ds + \int_0^t U(t, s) f(s, y_s) \, ds, & \text{if } t \in \mathbb{R}_+. \end{cases}
$$

Clearly, the fixed points of the operator N_4 are mild solutions of the problem (2.10)–(2.11).

Then, by parallel steps of the Theorem 2.6's proof, we can prove that the operator N_4 is a contraction which have a fixed point by statement (S1) in Theorem 1.29. The details are left to the reader. □

2.4 Partial Functional Integro-Differential Evolution Equations

2.4.1 Introduction

In this section, we are interested by partial functional integro-differential evolution equations with local and nonlocal conditions. First, we look for the class of partial functional integro-differential evolution equations of the form

$$y'(t) = A(t)y(t) + \int_0^t \mathcal{K}(t, s) f(s, y_s) \, ds, \quad \text{a.e.} \quad t \in J = \mathbb{R}_+ \qquad (2.15)$$

$$y(t) = \varphi(t), \quad t \in H, \qquad (2.16)$$

where $\mathcal{K} : J \times J \to E, f : J \times C(H, E) \to E$ and $\varphi \in C(H, E)$ are given functions and $\{A(t)\}_{t \geq 0}$ is a family of linear closed (not necessarily bounded) operators from E into E that generate an evolution system of operators $\{U(t, s)\}_{(t,s) \in J \times J}$ for $0 \leq s \leq t < +\infty$.

Also, an extension of these existence results is given for the following partial functional integro-differential evolution problem with nonlocal conditions

$$y'(t) = A(t)y(t) + \int_0^t \mathcal{K}(t, s) f(s, y_s) \, ds, \quad \text{a.e.} \quad t \in J = \mathbb{R}_+ \qquad (2.17)$$

$$y(t) + h_t(y) = \varphi(t), \quad t \in H, \qquad (2.18)$$

where $A(\cdot), f, \mathcal{K}$, and φ are as in evolution problem (2.15)–(2.16) and $h_t : C(H, E) \to E$ is a given function.

The problem of proving the existence of mild solutions for integro-differential equations and inclusions in abstract spaces has been studied by several authors; see Balachandran and Anandhi [41, 42], Balachandran and Leelamani [45] and Benchohra et al. [72, 77], Benchohra and Ntouyas [60, 61, 66], Ntouyas [165].

Here we are interested to study the existence and uniqueness of the mild solution for the partial functional integro-differential evolution equations (2.15)–(2.16) and the corresponding nonlocal problem (2.17)–(2.18). The motivation of these problems is to look for the integro-differential equations considered in [33].

2.4.2 Main Result

Before stating and proving the main result, we give first the definition of mild solution of the partial functional integro-differential evolution problem (2.15)–(2.16).

Definition 2.9. We say that the function $y(\cdot) : [-r, +\infty) \to E$ is a mild solution of (2.15)–(2.16) if $y(t) = \varphi(t)$ for all $t \in H$ and y satisfies the following integral equation

$$y(t) = U(t, 0) \, \varphi(0) + \int_0^t U(t, s) \int_0^s \mathcal{K}(s, \tau) f(\tau, y_\tau) \, d\tau \, ds, \qquad \text{for each } t \in \mathbb{R}_+.$$

We will need to add the following assumption:

(2.9.1) For each $t \in J$, $\mathcal{K}(t, s)$ is measurable on $[0, t]$ and

$$\mathcal{K}(t) = ess \sup\{|\mathcal{K}(t, s)|; 0 \le s \le t\}$$

is bounded on $[0, n]$; let $\mathcal{S}_n := \sup\limits_{t \in [0,n]} \mathcal{K}(t)$.

For every $n \in \mathbb{N}$, we define in $\mathcal{C}([-r, +\infty), E)$ the semi-norms by:

$$\|y\|_n := \sup \{ e^{-\tau \, L_n^*(t)} \, |y(t)| : t \in [0, n] \}$$

where $L_n^*(t) = \int_0^t \bar{l}_n(s) \, ds$, $\bar{l}_n(t) = \widehat{M} \, n \, \mathcal{S}_n \, l_n(t)$ and l_n is the function from (2.1.3).

Then $\mathcal{C}([-r, +\infty), E)$ is a Fréchet space with the family of semi-norms $\{\| \cdot \|_n\}_{n \in \mathbb{N}}$. In what follows we will choose $\tau > 1$.

Theorem 2.10. *Suppose that hypotheses* (2.1.1)–(2.1.3) *are satisfied and the assumption* (2.9.1) *holds. If*

$$\int_{c_5}^{+\infty} \frac{ds}{\psi(s)} > \widehat{M} \, n \, \mathcal{S}_n \int_0^n p(s) \, ds, \qquad \text{for each } n > 0 \tag{2.19}$$

with $c_5 = \widehat{M} \, \|\varphi\|$. Then the problem (2.15)–(2.16) *has a unique mild solution.*

Proof. Transform the problem (2.15)–(2.16) into a fixed point problem. Consider the operator $N_5 : \mathcal{C}([-r, +\infty), E) \to \mathcal{C}([-r, +\infty), E)$ defined by:

$$(N_5 y)(t) = \begin{cases} \varphi(t), & \text{if } t \in H; \\ U(t, 0) \, \varphi(0) + \int_0^t U(t, s) \int_0^s \mathcal{K}(s, \tau) f(\tau, y_\tau) \, d\tau \, ds, & \text{if } t \in \mathbb{R}_+. \end{cases}$$

Clearly, the fixed points of the operator N_5 are mild solutions of the problem (2.15)–(2.16).

Let y be a possible solution of the problem (2.15)–(2.16). Given $n \in \mathbb{N}$ and $t \leq n$, then from (2.1.1), (2.1.2) and (2.9.1) we have:

$$|y(t)| \leq |U(t,0)|\,|\varphi(0)| + \int_0^t \|U(t,s)\|_{B(E)} \left| \int_0^s K(s,\tau)f(\tau, y_\tau)\, d\tau \right| ds$$

$$\leq \widehat{M}\,|\varphi(0)| + \widehat{M} \int_0^t \int_0^s |K(s,\tau)|\,|f(\tau, y_\tau)|\, d\tau\, ds$$

$$\leq \widehat{M}\,\|\varphi\| + \widehat{M} \int_0^t \int_0^s |K(s,\tau)|\, p(\tau)\, \psi(\|y_\tau\|)\, d\tau\, ds$$

$$\leq \widehat{M}\,\|\varphi\| + \widehat{M}\, n\, \mathcal{S}_n \int_0^t p(s)\, \psi(\|y_s\|)\, ds.$$

We consider the function μ defined by

$$\mu(t) := \sup\{\, |y(s)| \,:\, 0 \leq s \leq t \,\}, \quad 0 \leq t < +\infty.$$

Let $t^* \in [-r, t]$ be such that $\mu(t) = |y(t^*)|$. If $t^* \in [0, n]$, by the previous inequality we get

$$\mu(t) \leq \widehat{M}\,\|\varphi\| + \widehat{M}\, n\, \mathcal{S}_n \int_0^t p(s)\, \psi(\mu(s))\, ds, \quad t \in [0, n].$$

If $t^* \in H$, then $\mu(t) = \|\varphi\|$ and the previous inequality holds.

Let us take the right-hand side of the above inequality as $v(t)$. Then we have

$$\mu(t) \leq v(t) \quad \text{for all } t \in [0, n].$$

From the definition of v, we get

$$c_5 := v(0) = \widehat{M}\|\varphi\| \quad \text{and} \quad v'(t) = \widehat{M}\, n\, \mathcal{S}_n\, p(t)\, \psi(\mu(t)) \quad \text{a.e. } t \in [0, n].$$

Using the nondecreasing character of ψ, we have

$$v'(t) \leq \widehat{M}\, n\, \mathcal{S}_n\, p(t)\, \psi(v(t)) \quad \text{a.e. } t \in [0, n].$$

This implies that for each $t \in [0, n]$ and using (2.19) we get

$$\int_{c_5}^{v(t)} \frac{ds}{\psi(s)} \leq \widehat{M}\, n\, \mathcal{S}_n \int_0^t p(s)\, ds$$

$$\leq \widehat{M}\, n\, \mathcal{S}_n \int_0^n p(s)\, ds$$

$$< \int_{c_5}^{+\infty} \frac{ds}{\psi(s)}.$$

Thus, there exists a constant Λ_n such that $v(t) \leq \Lambda_n$, $t \in [0, n]$ and hence $\mu(t) \leq \Lambda_n$, $t \in [0, n]$. Since for every $t \in [0, n]$, $\|y_t\| \leq \mu(t)$, we have $\|y\|_n \leq \max\{\|\varphi\|, \Lambda_n\} := \Delta_n$. Set

$$Y = \{ y \in \mathcal{C}([-r, +\infty), E) : \sup\{|y(t)| : 0 \leq t \leq n\} \leq \Delta_n + 1 \text{ for all } n \in \mathbb{N} \}.$$

Clearly, Y is a closed subset of $\mathcal{C}([-r, +\infty), E)$.

We shall show that $N_5 : Y \to \mathcal{C}([-r, +\infty), E)$ is a contraction operator. Indeed, consider $y, \bar{y} \in \mathcal{C}([-r, +\infty), E)$, thus using (2.1.1), (2.1.3), and (2.9.1) for each $t \in [0, n]$ and $n \in \mathbb{N}$ we get

$$|(N_5 y)(t) - (N_5 \bar{y})(t)| = \left| \int_0^t U(t, s) \int_0^s \mathcal{K}(s, \tau) [f(\tau, y_\tau) - f(\tau, \bar{y}_\tau)] \, d\tau \, ds \right|$$

$$\leq \int_0^t \|U(t, s)\|_{B(E)} \int_0^s |\mathcal{K}(s, \tau)| |f(\tau, y_\tau) - f(\tau, \bar{y}_\tau)| \, d\tau \, ds$$

$$\leq \int_0^t \widehat{M} \int_0^s |\mathcal{K}(s, \tau)| |f(\tau, y_\tau) - f(\tau, \bar{y}_\tau)| \, d\tau \, ds$$

$$\leq \int_0^t \widehat{M} \, n \, \mathcal{S}_n \, l_n(s) \, \|y_s - \bar{y}_s\| \, ds$$

$$\leq \int_0^t [\bar{l}_n(s) \, e^{\tau \, L_n^*(s)}] \, [e^{-\tau \, L_n^*(s)} \, \|y_s - \bar{y}_s\|] \, ds$$

$$\leq \int_0^t \left[\frac{e^{\tau \, L_n^*(s)}}{\tau} \right]' \, ds \, \|y - \bar{y}\|_n$$

$$\leq \frac{1}{\tau} \, e^{\tau \, L_n^*(t)} \, \|y - \bar{y}\|_n.$$

Therefore,

$$\|N_5(y) - N_5(\bar{y})\|_n \leq \frac{1}{\tau} \, \|y - \bar{y}\|_n.$$

So, for $\tau > 1$, the operator N_5 is a contraction for all $n \in \mathbb{N}$. From the choice of Y there is no $y \in \partial Y^n$ such that $y = \lambda \, N_5(y)$ for some $\lambda \in (0, 1)$. Then the statement (S2) in Theorem 1.29 does not hold. A consequence of the nonlinear alternative of Frigon and Granas [116] that (S1) holds, we deduce that the operator N_5 has a unique fixed point y^* in $\overline{Y}$, which is the unique mild solution of the problem (2.15)–(2.16). $\square$

2.4.3 An Example

As an application of our results we consider the following partial functional integro-differential equation

$$
\begin{cases}
\dfrac{\partial z(t,x)}{\partial t} = a(t,x)\dfrac{\partial^2 z(t,x)}{\partial x^2} \\
\qquad + \displaystyle\int_{-r}^{t} \alpha(t,s)Q(s,z(s-r,x))ds, \; t \geq 0, x \in [0,\pi] \\[2mm]
z(t,0) = z(t,\pi) = 0 \qquad\qquad\qquad t \geq 0 \\[2mm]
z(t,x) = \Phi(t,x) \qquad\qquad\qquad\quad t \in H, \; x \in [0,\pi]
\end{cases}
\tag{2.20}
$$

where $a(t,x)$ is a continuous function and is uniformly Hölder continuous in t, $\alpha :$ $[0,+\infty) \times [0,+\infty) \to \mathbb{R}$, $Q : [0,+\infty) \times \mathbb{R} \to \mathbb{R}$ and $\Phi : H \times [0,\pi] \to \mathbb{R}$ are continuous functions.

Consider $E = L^2([0,\pi],\mathbb{R})$ and define $A(t)$ by $A(t)w = a(t,x)w''$ with domain

$$D(A) = \{w \in E : w, \; w' \text{ are absolutely continuous, } w'' \in E, \; w(0) = w(\pi) = 0 \}$$

Then $A(t)$ generates an evolution system $U(t,s)$ satisfying assumption (2.1.1) (see [112, 149]).

For $x \in [0,\pi]$, we have

$$
\begin{aligned}
y(t)(x) &= z(t,x) \quad t \in \mathbb{R}_+, \\
\mathcal{K}(t,s) &= \alpha(t,s) \quad t,s \in \mathbb{R}_+, \\
f(t,y_t)(x) &= Q(t,z(t-r,x)) \quad t \in \mathbb{R}_+
\end{aligned}
$$

and

$$\varphi(t)(x) = \Phi(t,x) \quad -r \leq t \leq 0.$$

Thus, under the above definitions of f, $\mathcal{K}$, and $A(\cdot)$, the system (2.20) can be represented by the abstract partial functional integro-differential evolution problem (2.15)–(2.16). Furthermore, more appropriate conditions on Q ensure the existence of unique mild solution for (2.20) by Theorems 2.10 and 1.29.

2.4.4 Nonlocal Case

In this section, we extend the above results of existence and uniqueness of mild solution to the partial functional integro-differential evolution equations with nonlocal conditions (2.17)–(2.18). Nonlocal conditions were initiated by Byszewski [89]. First, we define the mild solution.

Definition 2.11. A function $y \in C([-r, +\infty), E)$ is said to be a mild solution of (2.17)–(2.18) if $y(t) = \varphi(t) - h_t(y)$ for all $t \in H$ and y satisfies the following integral equation

$$y(t) = U(t, 0) [\varphi(0) - h_0(y)] + \int_0^t U(t, s) \int_0^s \mathcal{K}(s, \tau) f(\tau, y_\tau) \, d\tau \, ds, \quad \text{for } t \in \mathbb{R}_+.$$

Under the same assumptions in Sect. 2.2.4 for the function $h_t(\cdot)$, we establish that:

Theorem 2.12. *Assume that the hypotheses* (2.1.1)–(2.1.3), (2.9.1), (D1), *and* (D2) *hold and moreover*

$$\int_{c_6}^{+\infty} \frac{ds}{\psi(s)} > \widehat{M} \, n \, S_n \int_0^n p(s) \, ds, \quad \text{for each } n > 0 \qquad (2.21)$$

where $c_6 = \widehat{M}(\|\varphi\| + \sigma)$. *Then the nonlocal integro-differential evolution problem* (2.17)–(2.18) *has a unique mild solution.*

Proof. Transform the problem (2.17)–(2.18) into a fixed point problem. Consider the operator $N_6 : C([-r, +\infty), E) \to C([-r, +\infty), E)$ defined by:

$$(N_6 y)(t) = \begin{cases} \varphi(t) - h_t(y), & \text{if } t \in H; \\ U(t, 0) [\varphi(0) - h_0(y)] \\ \quad + \int_0^t U(t, s) \int_0^s \mathcal{K}(s, \tau) f(\tau, y_\tau) \, d\tau \, ds, & \text{if } t \geq 0. \end{cases}$$

Clearly, the fixed points of the operator N_6 are mild solutions of the problem (2.17)–(2.18).

Then, by parallel steps of the Theorem 2.10's proof, we can easily show that the operator N_6 is a contraction which have a unique fixed point by statement (S1) in Theorem 1.29. The details are left to the reader. $\qquad \square$

2.5   Neutral Functional Integro-Differential Evolution Equations

2.5.1   Introduction

In this section, we investigate neutral functional integro-differential evolution equations with local and nonlocal conditions where the existence of the unique mild solution is assured. Firstly, we study in Sect. 2.5.2 the neutral functional integro-differential evolution equations of the form

$$\frac{d}{dt}[y(t) - g(t, y_t)] = A(t)y(t) + \int_0^t \mathcal{K}(t, s) f(s, y_s)\, ds, \text{ a.e. } t \in J = \mathbb{R}_+ \quad (2.22)$$

$$y(t) = \varphi(t), \ t \in H, \quad (2.23)$$

where $\mathcal{K} : J \times J \to E$, $f, g : J \times C(H, E) \to E$ and $\varphi \in C(H, E)$ are given functions and $\{A(t)\}_{t \geq 0}$ is a family of linear closed (not necessarily bounded) operators from E into E that generate an evolution system of operators $\{U(t, s)\}_{(t,s) \in J \times J}$.

An extension of these existence results, we consider the following neutral functional evolution equation with nonlocal conditions

$$\frac{d}{dt}[y(t) - g(t, y_t)] = A(t)y(t) + \int_0^t \mathcal{K}(t, s) f(s, y_s)\, ds, \text{ a.e. } t \in J = \mathbb{R}_+ \quad (2.24)$$

$$y(t) + h_t(y) = \varphi(t), \ t \in H, \quad (2.25)$$

where $A(\cdot)$, $\mathcal{K}, f, g$, and φ are as in problem (2.22)–(2.23) and $h_t : C([-r, \infty), E) \to E$ is a given function.

The problem of proving the existence of mild solutions for integro-differential equations and inclusions in abstract spaces has been studied by several authors; see Balachandran and Anandhi [41, 42], Balachandran and Leelamani [45], Benchohra et al. [72, 77], Benchohra and Ntouyas [60, 61, 66], Ntouyas [165]. We are motivated by the mixed problems in [42, 64, 66].

Here we are interested to study of the existence and uniqueness of the mild solution for the neutral functional integro-differential evolution equations (2.22)–(2.23) and the corresponding nonlocal problem (2.24)–(2.25). These results are an extension of [37] for the neutral case.

2.5.2   Main Result

We give first the definition of the mild solution of the neutral functional integro-differential evolution problem (2.22)–(2.23).

Definition 2.13. We say that the continuous function $y(\cdot) : [-r, +\infty) \to E$ is a mild solution of (2.22)–(2.23) if $y(t) = \varphi(t)$ for all $t \in H$ and y satisfies the following integral equation

$$y(t) = U(t, 0) \left[\varphi(0) - g(0, \varphi)\right] + g(t, y_t) + \int_0^t U(t, s) A(s) g(s, y_s) \, ds$$
$$+ \int_0^t U(t, s) \int_0^s \mathcal{K}(s, \tau) f(\tau, y_\tau) \, d\tau \, ds \qquad \text{for each } t \in \mathbb{R}_+.$$

For every $n \in \mathbb{N}$, we define in $\mathcal{C}([-r, +\infty), E)$ the semi-norms by:

$$\|y\|_n := \sup \left\{ e^{-\tau \, L_n^*(t)} \, |y(t)| : t \in [0, n] \right\}$$

where $L_n^*(t) = \int_0^t \bar{l}_n(s) \, ds$, $\bar{l}_n(t) = \widehat{M} \left(L_* + n\mathcal{S}_n l_n(t)\right)$ and l_n is the function from (2.1.3).

Then $\mathcal{C}([-r, +\infty), E)$ is a Fréchet space with the family of semi-norms $\{\|\cdot\|_n\}_{n \in \mathbb{N}}$. Let us fix $\tau > 0$ and assume $\left[\overline{M}_0 \, L_* + \dfrac{1}{\tau}\right] < 1$.

Theorem 2.14. *Suppose that hypotheses (2.1.1)–(2.1.3), (2.9.1), and (G1)–(G3) are satisfied. If*

$$\int_{c_{7,n}}^{+\infty} \frac{ds}{s + \psi(s)} > \frac{\widehat{M}}{1 - \overline{M}_0 L} \int_0^n \max(L, n\mathcal{S}_n p(s)) ds, \qquad \text{for each } n > 0 \qquad (2.26)$$

with

$$c_{7,n} = \frac{\widehat{M} \|\varphi\| (1 + \overline{M}_0 L) + \overline{M}_0 L(\widehat{M} + 1) + \widehat{M} Ln}{1 - \overline{M}_0 L};$$

Then the neutral functional integro-differential evolution problem (2.22)–(2.23) has a unique mild solution.

Proof. Transform the problem (2.22)–(2.23) into a fixed point problem. Consider the operator $N_7 : \mathcal{C}([-r, +\infty), E) \to \mathcal{C}([-r, +\infty), E)$ defined by:

$$(N_7 y)(t) = \begin{cases} \varphi(t), & \text{if } t \in H \\[2ex] U(t, 0) \left[\varphi(0) - g(0, \varphi)\right] + g(t, y_t) + \displaystyle\int_0^t U(t, s) A(s) g(s, y_s) ds \\[2ex] + \displaystyle\int_0^t U(t, s) \int_0^s \mathcal{K}(s, \tau) f(\tau, y_\tau) \, d\tau \, ds, & \text{if } t \in \mathbb{R}_+. \end{cases}$$

Clearly, the fixed points of the operator N_7 are mild solutions of the problem (2.22)–(2.23).

Let y be a possible solution of the problem (2.22)–(2.23). Given $n \in \mathbb{N}$ and $t \leq n$, then from (2.1.1), (2.9.1), $(G1)$, and $(G2)$, we have

$$|y(t)| \leq |U(t,0)| \, |\varphi(0) - g(0,\varphi)| + |g(t,y_t)| + \int_0^t \|U(t,s)\|_{B(E)} \, |A(s) \, g(s,y_s)| \, ds$$

$$+ \int_0^t \|U(t,s)\|_{B(E)} \left| \int_0^s \mathcal{K}(s,\tau) f(\tau,y_\tau) \, d\tau \right| ds$$

$$\leq \widehat{M} \, \|\varphi\| + \widehat{M} \, \|A^{-1}(0)\| \, \|A(0) \, g(0,\varphi)\| + \|A^{-1}(t)\| \, \|A(t) \, g(t,y_t)\|$$

$$+ \widehat{M} \int_0^t \|A(s) \, g(s,y_s)\| \, ds + \widehat{M} \int_0^t \int_0^s |\mathcal{K}(s,\tau)| \, |f(\tau,y_\tau)| \, d\tau \, ds$$

$$\leq \widehat{M} \, \|\varphi\| + \widehat{M} \, \overline{M}_0 \, L \, (\|\varphi\| + 1) + \overline{M}_0 \, L \, (\|y_t\|) + 1)$$

$$+ \widehat{M} \, L \int_0^t (\|y_s\| + 1) \, ds + \widehat{M} \int_0^t \int_0^s |\mathcal{K}(s,\tau)| \, p(\tau) \psi (\|y_\tau\|) \, d\tau \, ds$$

$$\leq \widehat{M} \, \|\varphi\| (1 + \overline{M}_0 \, L) + \overline{M}_0 \, L(\widehat{M} + 1) + \widehat{M} \, L \, n$$

$$+ \overline{M}_0 \, L \|y_t\| + \widehat{M} \, L \int_0^t \|y_s\| \, ds + \widehat{M} \, n \, \mathcal{S}_n \int_0^t p(s) \psi (\|y_s\|) \, ds.$$

We consider the function μ defined by

$$\mu(t) := \sup\{ |y(s)| \; : \; 0 \leq s \leq t \}, \quad 0 \leq t < +\infty.$$

Let $t^* \in [-r, t]$ be such that $\mu(t) = |y(t^*)|$. If $t^* \in [0, n]$, by the previous inequality we get for $t \in [0, n]$

$$\mu(t) \leq \widehat{M} \|\varphi\| (1 + \overline{M}_0 L) + \overline{M}_0 L(\widehat{M} + 1) + \widehat{M} L n$$

$$+ \overline{M}_0 L \mu(t) + \widehat{M} L \int_0^t \mu(s) ds + \widehat{M} n S_n \int_0^t p(s) \psi (\mu(s)) ds.$$

Then

$$(1 - \overline{M}_0 L)\mu(t) \leq \widehat{M} \|\varphi\| (1 + \overline{M}_0 L) + \overline{M}_0 L(\widehat{M} + 1) + \widehat{M} L n$$

$$+ \widehat{M} L \int_0^t \mu(s) ds + \widehat{M} n S_n \int_0^t p(s) \psi (\mu(s)) ds.$$

Then

$$\mu(t) \leq \frac{1}{1 - \overline{M}_0 L} \left[\widehat{M} \|\varphi\| (1 + \overline{M}_0 L) + \overline{M}_0 L(\widehat{M} + 1) + \widehat{M} L n \right]$$

$$+ \frac{\widehat{M} L}{1 - \overline{M}_0 L} \int_0^t \mu(s) ds + \frac{\widehat{M} n S_n}{1 - \overline{M}_0 L} \int_0^t p(s) \psi (\mu(s)) ds$$

$$= c_{7,n} + \frac{\widehat{M}}{1 - \overline{M}_0 L} \left[\int_0^t L\mu(s)ds + \int_0^t nS_n p(s)\psi(\mu(s))ds \right].$$

If $t^* \in H$, then $\mu(t) = \|\varphi\|$ and the previous inequality holds.

Let us take the right-hand side of the above inequality as $v(t)$. Then we have $\mu(t) \le v(t)$ for all $t \in [0, n]$. From the definition of v, we get

$$c := v(0) = c_{7,n} \quad \text{and} \quad v'(t) = \frac{\widehat{M}}{1 - \overline{M}_0 L} [L\mu(t) + nS_n p(t)\psi(\mu(t))] \quad \text{a.e. } t \in [0, n].$$

Using the nondecreasing character of ψ, we have for a.e. $t \in [0, n]$

$$v'(t) \le \frac{\widehat{M}}{1 - \overline{M}_0 L} [Lv(t) + nS_n p(t)\psi(v(t))].$$

This implies that for each $t \in [0, n]$ and using (2.26) we get

$$\int_{c_{7,n}}^{v(t)} \frac{ds}{s + \psi(s)} \le \frac{\widehat{M}}{1 - \overline{M}_0 L} \int_0^t \max(L, nS_n p(s))ds$$

$$\le \frac{\widehat{M}}{1 - \overline{M}_0 L} \int_0^n \max(L, nS_n p(s))ds$$

$$< \int_{c_{7,n}}^{+\infty} \frac{ds}{s + \psi(s)}.$$

Thus, there exists a constant Λ_n such that $v(t) \le \Lambda_n$, $t \in [0, n]$ and hence $\mu(t) \le \Lambda_n$, $t \in [0, n]$. Since for every $t \in [0, n]$, $\|y_t\| \le \mu(t)$, we have $\|y\|_n \le \max\{\|\varphi\|, \Lambda_n\} := \Delta_n$. Set

$$Y = \{ y \in C([-r, +\infty), E) : \|y\|_\infty \le \Delta_n + 1 \text{ for all } n \in \mathbb{N} \}.$$

Clearly, Y is a closed subset of $C([-r, +\infty), E)$.

We shall show that $N_7 : \overline{Y} \to C([-r, +\infty), E)$ is a contraction operator. Indeed, consider $y, \overline{y} \in C([-r, +\infty), E)$, thus using (2.1.1), (2.1.3), (2.9.1), (G1) and (G3) for each $t \in [0, n]$ and $n \in \mathbb{N}$, we get

$$|(N_7 y)(t) - (N_7 \overline{y})(t)| \le |g(t, y_t) - g(t, \overline{y}_t)|$$

$$+ \int_0^t \|U(t,s)\|_{B(E)} |A(s)(g(s, y_s) - g(s, \overline{y}_s))| ds$$

$$+ \left| \int_0^t U(t,s) \int_0^s K(s, \tau)[f(\tau, y_\tau) - f(\tau, \overline{y}_\tau)] d\tau ds \right|$$

$$\le \|A^{-1}(t)\| \|A(t)g(t, y_t) - A(t)g(t, \overline{y}_t)\|$$

$$+\widehat{M} \int_0^t \|A(s)g(s,y_s) - A(s)g(s,\bar{y}_s)\| ds$$

$$+ \int_0^t \|U(t,s)\|_{B(E)} \int_0^s |\mathcal{K}(s,\tau)||f(\tau,y_\tau) - f(\tau,\bar{y}_\tau)| d\tau ds$$

$$\leq \widehat{M}_0 L_* \|y_t - \bar{y}_t\| + \int_0^t \widehat{M} L_* \|y_s - \bar{y}_s\| ds$$

$$+ \int_0^t \widehat{M} n \mathcal{S}_n l_n(s) \|y_s - \bar{y}_s\| ds$$

$$\leq \widehat{M}_0 L_* \|y_t - \bar{y}_t\| + \int_0^t \left[\widehat{M} L_* + \widehat{M} n \mathcal{S}_n l_n(s) \right] \|y_s - \bar{y}_s\| ds$$

$$\leq \overline{M}_0 L_* e^{\tau L_n^*(t)} [e^{-\tau L_n^*(t)} \|y_t - \bar{y}_t\|]$$

$$+ \int_0^t \bar{l}_n(s) e^{\tau L_n^*(s)} [e^{-\tau L_n^*(s)} \|y_s - \bar{y}_s\|] ds$$

$$\leq \overline{M}_0 L_* e^{\tau L_n^*(t)} \|y - \bar{y}\|_n + \int_0^t \left[\frac{e^{\tau L_n^*(s)}}{\tau} \right]' ds \|y - \bar{y}\|_n$$

$$\leq \overline{M}_0 L_* e^{\tau L_n^*(t)} \|y - \bar{y}\|_n + \frac{1}{\tau} e^{\tau L_n^*(t)} \|y - \bar{y}\|_n$$

$$\leq \left[\overline{M}_0 L_* + \frac{1}{\tau} \right] e^{\tau L_n^*(t)} \|y - \bar{y}\|_n.$$

Therefore,

$$\|N_7(y) - N_7(\bar{y})\|_n \leq \left[\overline{M}_0 L_* + \frac{1}{\tau} \right] \|y - \bar{y}\|_n.$$

So, for $\left[\overline{M}_0 L_* + \frac{1}{\tau} \right] < 1$, the operator N_7 is a contraction for all $n \in \mathbb{N}$. From the choice of Y there is no $y \in \partial Y^n$ such that $y = \lambda N_7(y)$ for some $\lambda \in (0,1)$. Then the statement (S2) in Theorem 1.29 does not hold. A consequence of the nonlinear alternative of Frigon and Granas [116] that (S1) holds, we deduce that the operator N_7 has a unique fixed point y^* in $\overline{Y}$, which is the unique mild solution of the problem (2.22)–(2.23). □

2.5.3 An Example

As an application of our results we consider the following neutral functional integro-differential evolution equation

$$\begin{cases} llc\frac{\partial}{\partial t}\left[z(t,x) - \int_{-r}^{t}\int_{0}^{\pi} b(s-t,u,x)\,z(s,u)\,du\,ds\right] = & a(t,x)\frac{\partial^2 z}{\partial x^2}(t,x) \\ \quad + \int_{0}^{t}\alpha(t,s)Q\left(s,z(s-r,x),\frac{\partial z}{\partial x}(s-r,x)\right)ds, & t \geq 0, x \in [0,\pi] \\ \\ z(t,0) = z(t,\pi) = 0, & t \geq 0 \\ \\ z(t,x) = \Phi(t,x), & t \in H,\ x \in [0,\pi] \end{cases}$$

$$(2.27)$$

where $a(t,x)$ is a continuous function and is uniformly Hölder continuous in t, α : $[0,+\infty) \times [0,+\infty) \to \mathbb{R}$, $Q : [0,+\infty) \times \mathbb{R} \times \mathbb{R} \to \mathbb{R}$ and $\Phi : H \times [0,\pi] \to \mathbb{R}$ are continuous functions.

Consider $E = L^2([0,\pi], \mathbb{R})$ and define $A(t)$ by $A(t)w = a(t,x)w''$ with domain

$$D(A) = \{w \in E : w, w' \text{ are absolutely continuous}, w'' \in E,\ w(0) = w(\pi) = 0\}$$

Then $A(t)$ generates an evolution system $U(t,s)$ satisfying assumptions (2.1.1) and (G1) (see [112, 149]).

For $x \in [0,\pi]$, we have

$$y(t)(x) = z(t,x) \quad t \in \mathbb{R}_+,$$

$$\mathcal{K}(t,s) = \alpha(t,s) \quad t,s \in \mathbb{R}_+,$$

$$f(t,y_t)(x) = Q(t,z(t-r,x)) \quad t \in \mathbb{R}_+,$$

$$g(t,y_t)(x) = \int_{-r}^{t}\int_{0}^{\pi} b(s-t,u,x)z(s,u)du ds, \quad x \in [0,\pi]$$

and

$$\varphi(t)(x) = \Phi(t,x) \quad t \in H.$$

Here we consider that $\varphi : H \to E$ is Lebesgue measurable and $h \|\varphi\|^2$ is Lebesgue integrable on H where $h : H \to \mathbb{R}$ is a positive integrable function. The norm is defined here by:

$$\|\varphi\| = \|\Phi(0)\| + \left(\int_{-r}^{0} h(s) \|\varphi\|^2\,ds\right)^{\frac{1}{2}}.$$

(i) The function b is measurable and

$$\int_{0}^{\pi}\int_{-r}^{t}\int_{0}^{\pi} \frac{b^2(s,u,x)}{h(s)}dsdudx < \infty.$$

(ii) The function $(\frac{\partial b}{\partial x}(s, u, x))$ and $(\frac{\partial^2 b}{\partial x^2}(s, u, x))$ are measurable, $b(s, u, 0) = b(s, u, \pi)) = 0$ and $\sup_{t \in [0,b]} \mathcal{N}(t) < \infty$, where

$$\mathcal{N}(t) = \int_0^\pi \int_{-r}^t \int_0^\pi \frac{1}{h(s)} \left(a(s, x) \frac{\partial^2}{\partial x^2} b(s, u, x) \right)^2 dsdudx.$$

Thus, under the above definitions of f, g, $\mathcal{K}$, and $A(\cdot)$, the system (2.27) can be represented by the abstract neutral functional integro-differential evolution problem (2.22)–(2.23). Furthermore, more appropriate conditions on Q ensure the existence of a least one mild solution for (2.27) by Theorems (2.14) and (1.29).

2.5.4   Nonlocal Case

An extension of these results is given here for the neutral functional integro-differential evolution equation with nonlocal conditions (2.24)–(2.25). Nonlocal conditions were initiated by Byszewski [89]. Before giving the main result, we give first the definition of the mild solution.

Definition 2.15. A function $y \in \mathcal{C}([-r, +\infty), E)$ is said to be a mild solution of (2.24)–(2.25) if $y(t) = \varphi(t) - h_t(y)$ for all $t \in H$ and y satisfies the following integral equation

$$y(t) = U(t, 0) \left[\varphi(0) - h_0(y) - g(0, \varphi) \right] + g(t, y_t) + \int_0^t U(t, s) A(s) g(s, y_s) ds$$

$$+ \int_0^t U(t, s) \int_0^s \mathcal{K}(s, \tau) f(\tau, y_\tau) d\tau ds, \quad \text{for each } t \in \mathbb{R}_+.$$

Under the same assumptions in Sect. 2.2.4 for the function $h_t(\cdot)$, we establish that

Theorem 2.16. *Assume that the hypotheses* (2.1.1)–(2.1.3), (2.9.1), (G1)–(G3), (D1), *and* (D2) *hold. If*

$$\int_{c_{8,n}}^{+\infty} \frac{ds}{\psi(s)} > \frac{\widehat{M}}{1 - \overline{M}_0 L} \int_0^n \max(L, n\mathcal{S}_n p(s)) \, ds, \quad \text{for each } n > 0 \qquad (2.28)$$

with

$$c_{8,n} = \frac{\widehat{M} \left[\|\varphi\|(1 + \overline{M}_0 L) + \sigma \right] + \overline{M}_0 L (\widehat{M} + 1) + \widehat{M} L n}{1 - \overline{M}_0 L}.$$

Then the nonlocal neutral functional integro-differential evolution problem (2.24)–(2.25) *has a unique mild solution.*

Proof. Transform the problem (2.24)–(2.25) into a fixed point problem. Consider the operator $N_8 : C([-r, +\infty), E) \rightarrow C([-r, +\infty), E)$ defined by:

$$(N_8 y)(t) = \begin{cases} \varphi(t) - h_t(y), & \text{if } t \in H; \\ U(t,0)\left[\varphi(0) - h_0(y) - g(0,\varphi)\right] + \displaystyle\int_0^t U(t,s) A(s) g(s, y_s) \, ds \\ \quad + \displaystyle\int_0^t U(t,s) \int_0^s \mathcal{K}(s,\tau) f(\tau, y_\tau) \, d\tau \, ds, & \text{for each } t \geq 0. \end{cases}$$

Clearly, the fixed points of the operator N_8 are mild solutions of the problem (2.24)–(2.25).

Then, by parallel steps of the Theorem 2.14's proof, we can prove that the operator N_8 is a contraction which have a fixed point by statement (S1) in Theorem 1.29. The details are left to the reader.   □

2.6   Notes and Remarks

The results of Chap. 2 are taken from Baghli and Benchohra [33, 33, 37]. Other results may be found in [41, 42, 44].

Chapter 3
Partial Functional Evolution Equations with Infinite Delay

3.1 Introduction

In this chapter, we provide sufficient conditions for the existence of the unique mild solution on the positive half-line $\mathbb{R}_+$ for some classes of first order partial functional and neutral functional differential evolution equations with infinite delay.

3.2 Partial Functional Evolution Equations

3.2.1 Introduction

In this section, we consider the following partial functional evolution equations with infinite delay

$$y'(t) = A(t)y(t) + f(t, y_t), \quad \text{a.e.} \quad t \in J = \mathbb{R}_+ \tag{3.1}$$

$$y_0 = \phi \in \mathcal{B}, \tag{3.2}$$

where $f : J \times \mathcal{B} \to E$ and $\phi \in \mathcal{B}$ are given functions and $\{A(t)\}_{0 \le t < +\infty}$ is a family of linear closed (not necessarily bounded) operators from E into E that generate an evolution system of operators $\{U(t, s)\}_{(t,s) \in J \times J}$ for $0 \le s \le t < +\infty$. Here $y_t(\cdot)$ represents the history of the state from time $t - r$ up to the present time t defined by $y_t(\theta) = y(t + \theta)$ for $\theta \in (-\infty, 0]$. We assume that the histories y_t belongs to some abstract *phase space* $\mathcal{B}$.

© Springer International Publishing Switzerland 2015
S. Abbas, M. Benchohra, *Advanced Functional Evolution Equations and Inclusions*, Developments in Mathematics 39,
DOI 10.1007/978-3-319-17768-7_3

Here we are interested to give existence and uniqueness results of the mild solution for the partial functional evolution equations (3.1)–(3.2) by using Theorem 1.29 due to Frigon and Granas [116].

3.2.2 Existence and Uniqueness of Mild Solution

Before stating and proving the main result, we give first the definition of mild solution of the partial functional evolution problem (3.1)–(3.2).

Definition 3.1. We say that the continuous function $y(\cdot) : \mathbb{R} \to E$ is a mild solution of (3.1)–(3.2) if $y(t) = \phi(t)$ for all $t \in (-\infty, 0]$ and y satisfies the following integral equation

$$y(t) = U(t, 0)\, \phi(0) + \int_0^t U(t, s) f(s, y_s)\, ds, \qquad \text{for each } t \in \mathbb{R}_+.$$

We will need to introduce the following hypotheses which are assumed hereafter:

(3.1.1) There exists a constant $\widehat{M} \geq 1$ such that:

$$\|U(t, s)\|_{B(E)} \leq \widehat{M} \quad \text{for every } (t, s) \in \Delta.$$

(3.1.2) There exist a function $p \in L^1_{loc}(J, \mathbb{R}_+)$ and a continuous nondecreasing function $\psi : \mathbb{R}_+ \to (0, \infty)$ such that :

$$|f(t, u)| \leq p(t)\, \psi(\|u\|_B) \text{ for a.e. } t \in J \text{ and each } u \in B.$$

(3.1.3) For all $R > 0$, there exists $l_R \in L^1_{loc}(\mathbb{R}, \mathbb{R}_+)$ such that:

$$|f(t, u) - f(t, v)| \leq l_R(t)\, \|u - v\|_B$$

for all $u, v \in B$ with $\|u\|_B \leq R$ and $\|v\|_B \leq R$.

Consider the following space

$$B_{+\infty} = \left\{ y : \mathbb{R} \to E : y|_{[0,T]} \in C([0, T], E), \ y_0 \in B \right\},$$

where $y|_{[0,T]}$ is the restriction of y to any real compact interval $[0, T]$.

For every $n \in \mathbb{N}$, we define in $B_{+\infty}$ the semi-norms by:

$$\|y\|_n := \sup \left\{ e^{-\tau\, L_n^*(t)}\, |y(t)| : t \in [0, n] \right\}$$

where $L_n^*(t) = \int_0^t \bar{l}_n(s)\, ds$, $\bar{l}_n(t) = K_n \widehat{M} l_n(t)$ and l_n is the function from (3.1.3).

Then $B_{+\infty}$ is a Fréchet space with the family of semi-norms $\| \cdot \|_{n \in \mathbb{N}}$. In what follows let us fix $\tau > 1$.

Theorem 3.2 ([33]). *Suppose that hypotheses* (3.1.1)–(3.1.3) *are satisfied and moreover*

$$\int_{c_{9,n}}^{+\infty} \frac{ds}{\psi(s)} > K_n \widehat{M} \int_0^n p(s)\, ds, \quad \text{for each } n > 0 \tag{3.3}$$

with $c_{9,n} = (K_n \widehat{M} H + M_n)\|\phi\|_{\mathcal{B}}$. *Then the problem* (3.1)–(3.2) *has a unique mild solution.*

Proof. Consider the operator $N_9 : B_{+\infty} \to B_{+\infty}$ defined by:

$$(N_9 y)(t) = \begin{cases} \phi(t), & \text{if } t \leq 0; \\[2mm] U(t,0)\,\phi(0) + \displaystyle\int_0^t U(t,s) f(s, y_s)\, ds, & \text{if } t \geq 0. \end{cases}$$

Clearly, fixed points of the operator N_9 are mild solutions of the problem (3.1)–(3.2).
   For $\phi \in \mathcal{B}$, we will define the function $x(.) : \mathbb{R} \to E$ by

$$x(t) = \begin{cases} \phi(t), & \text{if } t \in (-\infty, 0]; \\[2mm] U(t,0)\,\phi(0), & \text{if } t \in J. \end{cases}$$

Then $x_0 = \phi$. For each function $z \in C(J, E)$, set

$$y(t) = z(t) + x(t).$$

It is obvious that y satisfies Definition 3.1 if and only if z satisfies $z_0 = 0$ and

$$z(t) = \int_0^t U(t,s) f(s, z_s + x_s)\, ds, \quad \text{for } t \in J.$$

Let

$$B_{+\infty}^0 = \{z \in B_{+\infty} : z_0 = 0\}.$$

Define the operator $F : B_{+\infty}^0 \to B_{+\infty}^0$ by:

$$F(z)(t) = \int_0^t U(t,s) f(s, z_s + x_s)\, ds, \quad \text{for } t \in J.$$

Obviously the operator N_9 has a fixed point is equivalent to F has one, so it turns to prove that F has a fixed point.

Let $z \in B_{+\infty}^0$ be a possible fixed point of the operator F. By the hypotheses (3.1.1) and (3.1.2), we have for each $t \in [0, n]$

$$|z(t)| \leq \int_0^t \|U(t, s)\|_{B(E)} \, |f(s, z_s + x_s)| \, ds$$

$$\leq \widehat{M} \int_0^t p(s) \, \psi \, (\|z_s + x_s\|_B) \, ds.$$

Assumption (A_1) gives

$$\|z_s + x_s\|_B \leq \|z_s\|_B + \|x_s\|_B$$
$$\leq K(s)|z(s)| + M(s)\|z_0\|_B + K(s)|x(s)| + M(s)\|x_0\|_B$$
$$\leq K_n|z(s)| + K_n\|U(s, 0)\|_{B(E)}|\phi(0)| + M_n\|\phi\|_B$$
$$\leq K_n|z(s)| + K_n\widehat{M}|\phi(0)| + M_n\|\phi\|_B$$
$$\leq K_n|z(s)| + K_n\widehat{M}H\|\phi\|_B + M_n\|\phi\|_B$$
$$\leq K_n|z(s)| + (K_n\widehat{M}H + M_n)\|\phi\|_B.$$

Set $\alpha_n := c_{9,n} = (K_n\widehat{M}H + M_n)\|\phi\|_B$, then we have

$$\|z_s + x_s\|_B \leq K_n|z(s)| + \alpha_n \tag{3.4}$$

Using the nondecreasing character of ψ, we get

$$|z(t)| \leq \widehat{M} \int_0^t p(s) \, \psi \, (K_n|z(s)| + \alpha_n) \, ds.$$

Then

$$K_n|z(t)| + \alpha_n \leq K_n\widehat{M} \int_0^t p(s)\psi \, (K_n|z(s)| + \alpha_n)ds + c_{9,n}.$$

Consider the function μ defined by

$$\mu(t) := \sup \{ K_n|z(s)| + \alpha_n \, : \, 0 \leq s \leq t \}, \quad 0 \leq t < +\infty.$$

Let $t^* \in [0, t]$ be such that $\mu(t) = K_n|z(t^*)| + \alpha_n$. By the previous inequality, we have

$$\mu(t) \leq K_n\widehat{M} \int_0^t p(s) \, \psi(\mu(s)) \, ds + c_{9,n}, \quad \text{for} \quad t \in [0, n].$$

Let us take the right-hand side of the above inequality as $v(t)$. Then, we have

$$\mu(t) \le v(t) \text{ for all } t \in [0, n].$$

From the definition of v, we have

$$v(0) = c_{9,n} \text{ and } v'(t) = K_n \widehat{M} p(t) \psi(\mu(t)) \text{ a.e. } t \in [0, n].$$

Using the nondecreasing character of ψ, we get

$$v'(t) \le K_n \widehat{M} p(t) \psi(v(t)) \text{ a.e. } t \in [0, n].$$

This implies that for each $t \in [0, n]$ and using the condition (3.3), we get

$$\int_{c_{9,n}}^{v(t)} \frac{ds}{\psi(s)} \le K_n \widehat{M} \int_0^t p(s) \, ds$$

$$\le K_n \widehat{M} \int_0^n p(s) \, ds$$

$$< \int_{c_{9,n}}^{+\infty} \frac{ds}{\psi(s)}.$$

Thus, for every $t \in [0, n]$, there exists a constant Λ_n such that $v(t) \le \Lambda_n$ and hence $\mu(t) \le \Lambda_n$. Since $\|z\|_n \le \mu(t)$, we have $\|z\|_n \le \Lambda_n$. Set

$$Z = \{ z \in B^0_{+\infty} : \sup\{ |z(t)| : 0 \le t \le n \} \le \Lambda_n + 1 \text{ for all } n \in \mathbb{N} \}.$$

Clearly, Z is a closed subset of $B^0_{+\infty}$.

We shall show that $F : Z \to B^0_{+\infty}$ is a contraction operator. Indeed, consider $z, \overline{z} \in B^0_{+\infty}$, thus using (3.1.1) and (3.1.3) for each $t \in [0, n]$ and $n \in \mathbb{N}$

$$|(Fz)(t) - (F\overline{z})(t)| = \left| \int_0^t U(t, s) [f(s, z_s + x_s) - f(s, \overline{z}_s + x_s)] \, ds \right|$$

$$\le \int_0^t \|U(t, s)\|_{B(E)} |f(s, z_s + x_s) - f(s, \overline{z}_s + x_s)| \, ds$$

$$\le \int_0^t \widehat{M} l_n(s) \|z_s + x_s - \overline{z}_s - x_s\|_B \, ds$$

$$\le \int_0^t \widehat{M} l_n(s) \|z_s - \overline{z}_s\|_B \, ds.$$

Using (A_1), we obtain

$$|(Fz)(t) - (F\bar{z})(t)| \leq \int_0^t \widehat{M}\, l_n(s)\ (K(s)\, |z(s) - \bar{z}(s)| + M(s)\, \|z_0 - \bar{z}_0\|_{\mathcal{B}})\ ds$$

$$\leq \int_0^t \widehat{M}\, K_n\, l_n(s)\, |z(s) - \bar{z}(s)|\ ds$$

$$\leq \int_0^t [\bar{l}_n(s)\, e^{\tau\, L_n^*(s)}\,]\, [e^{-\tau\, L_n^*(s)}\, |z(s) - \bar{z}(s)|]\ ds$$

$$\leq \int_0^t \left[\frac{e^{\tau\, L_n^*(s)}}{\tau}\right]'\ ds\, \|z - \bar{z}\|_n$$

$$\leq \frac{1}{\tau}\, e^{\tau\, L_n^*(t)}\, \|z - \bar{z}\|_n.$$

Therefore,

$$\|F(z) - F(\bar{z})\|_n \leq \frac{1}{\tau}\, \|z - \bar{z}\|_n.$$

So, for $\tau > 1$, the operator F is a contraction for all $n \in \mathbb{N}$. From the choice of Z there is no $z \in \partial Z^n$ such that $z = \lambda\, F(z)$ for some $\lambda \in (0, 1)$. Then the statement $(S2)$ in Theorem 1.29 does not hold. A consequence of the nonlinear alternative of Frigon and Granas that $(S1)$ holds, we deduce that the operator F has a unique fixed point z^*. Then $y^*(t) = z^*(t) + x(t)$, $t \in \mathbb{R}$ is a fixed point of the operator N_9, which is the unique mild solution of the problem (3.1)–(3.2). $\square$

3.2.3 An Example

Consider the following partial functional differential equation

$$\begin{cases} \dfrac{\partial z}{\partial t}(t, x) = a(t, x)\dfrac{\partial^2 z}{\partial x^2}(t, x) + Q(t, z(t - r, x))\ t \geq 0,\ x \in [0, \pi] \\[2ex] z(t, 0) = z(t, \pi) = 0 \qquad\qquad\qquad\qquad\qquad t \geq 0 \\[2ex] z(t, x) = \Phi(t, x) \qquad\qquad\qquad\qquad\qquad t \leq 0,\ x \in [0, \pi], \end{cases} \qquad (3.5)$$

where $a(t, x)$ is a continuous function and is uniformly Hölder continuous in t, $Q : \mathbb{R}_+ \times \mathbb{R} \to \mathbb{R}$ and $\Phi : \mathcal{B} \times [0, \pi] \to \mathbb{R}$ are continuous functions.

Let

$$y(t)(x) = z(t, x), \ t \in [0, \infty), \ x \in [0, \pi],$$

$$\phi(\theta)(x) = \Phi(\theta, x), \ \theta \leq 0, \ x \in [0, \pi]$$

and

$$f(t, \phi)(x) = Q(t, \phi(\theta, x)), \ \theta \leq 0, \ x \in [0, \pi].$$

Consider $E = L^2([0, \pi], \mathbb{R})$ and define $A(t)$ by $A(t)w = a(t, x)w''$ with domain

$D(A) = \{ w \in E : w, w' \text{ are absolutely continuous, } w'' \in E, \ w(0) = w(\pi) = 0 \}.$

Then $A(t)$ generates an evolution system $U(t, s)$ satisfying assumption (3.1.1) (see [112, 149]).

Thus, under the above definitions of f and $A(\cdot)$, the system (3.5) can be represented by the abstract evolution problem (3.1)–(3.2). Furthermore, more appropriate conditions on Q ensure the existence of the unique mild solution of (3.5) by Theorems 3.2 and 1.29.

3.3   Controllability on Finite Interval for Partial Evolution Equations

3.3.1   Introduction

In this section, we give sufficient conditions to ensure the controllability of mild solutions on a bounded interval $J_T := [0, T]$ for $T > 0$ for the partial functional evolution equations with infinite delay of the form

$$y'(t) = A(t)y(t) + Cu(t) + f(t, y_t), \quad \text{a.e.} \ t \in J_T \tag{3.6}$$

$$y_0 = \phi \in \mathcal{B}, \tag{3.7}$$

where $f : J \times \mathcal{B} \to E$ and $\phi \in \mathcal{B}$ are given functions, the control function $u(.)$ is given in $L^2([0, T], E)$, the Banach space of admissible control functions with E be a real separable Banach space with the norm $|\cdot|$, C is a bounded linear operator from E into E, and $\{A(t)\}_{0 \leq t \leq T}$ is a family of linear closed (not necessarily bounded) operators from E into E that generate an evolution system of operators $\{U(t, s)\}_{(t,s) \in J \times J}$ for $0 \leq s \leq t \leq T$.

3.3.2 Controllability of Mild Solutions

Before stating and proving the main result, we give first the definition of mild solution of problem (3.6)–(3.7) and the definition of controllability of the mild solution.

Definition 3.3. We say that the continuous function $y(\cdot) : \mathbb{R} \to E$ is a mild solution of (3.6)–(3.7) if $y(t) = \phi(t)$ for all $t \in (-\infty, 0]$ and y satisfies the following integral equation

$$y(t) = U(t,0)\phi(0) + \int_0^t U(t,s)Cu(s)ds + \int_0^t U(t,s)f(s,y_s)ds, \quad \text{for each } t \in [0,T].$$

Definition 3.4. The problem (3.6)–(3.7) is said to be controllable on the interval $[0,T]$ if for every initial function $\phi \in \mathcal{B}$ and $\tilde{y} \in E$ there exists a control $u \in L^2([0,T],E)$ such that the mild solution $y(\cdot)$ of (3.6)–(3.7) satisfies $y(T) = \tilde{y}$.

We will need to introduce the following hypotheses which are assumed hereafter:

(3.4.1) $U(t,s)$ is compact for $t - s > 0$ and there exists a constant $\widehat{M} \geq 1$ such that:

$$\|U(t,s)\|_{B(E)} \leq \widehat{M} \quad \text{for every } 0 \leq s \leq t \leq T.$$

(3.4.2) There exists a function $p \in L^1(J_T, \mathbb{R}_+)$ and a continuous nondecreasing function $\psi : \mathbb{R}_+ \to (0, \infty)$ such that:

$$|f(t,u)| \leq p(t)\, \psi(\|u\|_{\mathcal{B}}) \text{ for a.e. } t \in J_T \text{ and each } u \in \mathcal{B}.$$

(3.4.3) The linear operator $W : L^2([0,T],E) \to \mathcal{C}([0,T],E)$ is defined by

$$Wu = \int_0^T U(T,s)Cu(s)ds,$$

has a bounded inverse operator W^{-1} which takes values in $L^2([0,T],E)/\ker W$ and there exists positive constants $\tilde{M}$ and $\tilde{M}_1$ such that:

$$\|C\| \leq \tilde{M} \quad \text{and} \quad \|W^{-1}\| \leq \tilde{M}_1.$$

Remark 3.5. For the construction of W see the book of Carmichael and Quinn [93].

Consider the following space

$$B_T = \{y : (-\infty, T] \to E : y|_J \in C(J,E), \ y_0 \in \mathcal{B}\},$$

where $y|_J$ is the restriction of y to J.

Theorem 3.6. *Suppose that hypotheses (3.4.1)–(3.4.3) are satisfied and moreover there exists a constant $M_* > 0$ such that*

$$\frac{M_*}{c_{10,T} + K_T \widehat{M} \left(\widehat{M} \widetilde{M} \widetilde{M}_1 T + 1 \right) \psi(M_*) \|p\|_{L^1}} > 1, \qquad (3.8)$$

with

$$c_{10,T} = c_{10}(\phi, \tilde{y}, T) = \left[K_T \widehat{M} H \left(\widehat{M} \widetilde{M} \widetilde{M}_1 T + 1 \right) + M_T \right] \|\phi\|_{\mathcal{B}} + K_T \widehat{M} \widehat{M} \widetilde{M}_1 T |\tilde{y}| .$$

Then the problem (3.6)–(3.7) is controllable on $(-\infty, T]$.

Proof. Transform the problem (3.6)–(3.7) into a fixed point problem. Consider the operator $N_{10} : B_T \to B_T$ defined by:

$$(N_{10}y)(t) = \begin{cases} \phi(t), & \text{if } t \in (-\infty, 0]; \\[2mm] U(t,0)\,\phi(0) + \displaystyle\int_0^t U(t,s)\,C\,u_y(s)\,ds \\[2mm] \quad + \displaystyle\int_0^t U(t,s)\,f(s,y_s)\,ds, & \text{if } t \in [0,T]. \end{cases}$$

Clearly, fixed points of the operator N_{10} are mild solutions of the problem (3.6)–(3.7).

Using assumption (3.4.3), for arbitrary function $y(\cdot)$, we define the control

$$u_y(t) = W^{-1} \left[\tilde{y} - U(T,0)\,\phi(0) - \int_0^T U(T,s)\,f(s,y_s)\,ds \right](t).$$

Noting that, we have

$$|u_y(t)| \leq \|W^{-1}\| \left[|\tilde{y}| + \|U(T,0)\|_{B(E)} |\phi(0)| + \int_0^T \|U(T,\tau)\|_{B(E)} |f(\tau, y_\tau)| d\tau \right]$$

$$\leq \tilde{M}_1 \left[|\tilde{y}| + \widehat{M} H \|\phi\|_{\mathcal{B}} + \widehat{M} \int_0^T |f(\tau, y_\tau)| d\tau \right]$$

$$\leq \tilde{M}_1 \left[|\tilde{y}| + \widehat{M} H \|\phi\|_{\mathcal{B}} + \widehat{M} \int_0^T p(\tau)\,\psi(\|y_\tau\|_{\mathcal{B}})\,d\tau \right].$$

For $\phi \in \mathcal{B}$, we will define the function $x(.) : \mathbb{R} \to E$ by

$$x(t) = \begin{cases} \phi(t), & \text{if } t \in (-\infty, 0]; \\[2mm] U(t,0)\,\phi(0), & \text{if } t \in J_T. \end{cases}$$

Then $x_0 = \phi$. For each function $z \in B_T$, set $y(t) = z(t) + x(t)$. It is obvious that y satisfies Definition 3.3 if and only if z satisfies $z_0 = 0$ and

$$z(t) = \int_0^t U(t, s)\, C\, u_z(s)\, ds + \int_0^t U(t, s)\, f(s, z_s + x_s)\, ds, \quad \text{for } t \in J_T.$$

Let $B_T^0 = \{z \in B_T : z_0 = 0\}$. For any $z \in B_T^0$ we have

$$\|z\|_T = \sup\{ |z(t)| \; : \; t \in [0, T] \} + \|z_0\|_B = \sup\{ |z(t)| \; : \; t \in [0, T] \}.$$

Thus $(B_T^0, \| \cdot \|_T)$ is a Banach space.

Define the operator $F : B_T^0 \to B_T^0$ by:

$$(Fz)(t) = \int_0^t U(t, s)\, C\, u_z(s)\, ds + \int_0^t U(t, s)\, f(s, z_s + x_s)\, ds, \quad \text{for } t \in J_T.$$

Obviously the operator N_{10} has a fixed point is equivalent to F has one, so it turns to prove that F has a fixed point. The proof will be given in several steps.

Let us first show that the operator F is continuous and compact.

Step 1: F is continuous. Let $(z_n)_n$ be a sequence in B_T^0 such that $z_n \to z$ in B_T^0. Then, we get

$$|(Fz_n)(t) - (Fz)(t)| \leq \int_0^t \|U(t, s)\|_{B(E)}\, \|C\|\, |u_{z_n}(s) - u_z(s)|\, ds$$

$$+ \int_0^t \|U(t, s)\|_{B(E)}\, |f(s, z_{ns} + x_s) - f(s, z_s + x_s)|\, ds$$

$$\leq \widehat{M}\widetilde{M} \int_0^t \widetilde{M}_1 \widehat{M} \int_0^T |f(\tau, z_{n\tau} + x_\tau) - f(\tau, z_\tau + x_\tau)|\, d\tau\, ds$$

$$+ \widehat{M} \int_0^t |f(s, z_{ns} + x_s) - f(s, z_s + x_s)|\, ds$$

$$\leq \widehat{M}^2 \widetilde{M}\widetilde{M}_1 T \int_0^T |f(s, z_{ns} + x_s) - f(s, z_s + x_s)|\, ds$$

$$+ \widehat{M} \int_0^T |f(s, z_{ns} + x_s) - f(s, z_s + x_s)|\, ds$$

$$\leq \widehat{M} \left(\widehat{M}\widetilde{M}\widetilde{M}_1 T + 1 \right) \|f(\cdot, z_n. + x.) - f(\cdot, z. + x.)\|_{L^1}.$$

Since f is continuous, we obtain by the Lebesgue dominated convergence theorem

$$|F(z_n)(t) - F(z)(t)| \to 0 \quad \text{as } n \to +\infty.$$

Thus F is continuous.

Step 2: F maps bounded sets of B_T^0 into bounded sets. For any $d > 0$, there exists a positive constant ℓ such that for each $z \in B_d = \{z \in B_T^0 : \|z\|_T \le d\}$ we have $\|F(z)\|_T \le \ell$. Let $z \in B_d$, for each $t \in [0, T]$, we have

$$
\begin{aligned}
|(Fz)(t)| &\le \int_0^t \|U(t,s)\|_{B(E)} \, \|C\| \, |u_z(s)| \, ds + \int_0^t \|U(t,s)\|_{B(E)} \, |f(s, z_s + x_s)| \, ds \\
&\le \widehat{M}\widetilde{M} \int_0^t |u_z(s)| \, ds + \widehat{M} \int_0^t |f(s, z_s + x_s)| \, ds \\
&\le \widehat{M}\widetilde{M} \int_0^t \widetilde{M}_1 \left[|\tilde{y}| + \widehat{M}H\|\phi\|_{\mathcal{B}} + \widehat{M} \int_0^T p(\tau)\, \psi(\|z_\tau + x_\tau\|_{\mathcal{B}})\, d\tau \right] ds \\
&\quad + \widehat{M} \int_0^t |f(s, z_s + x_s)| \, ds \\
&\le \widehat{M}\widetilde{M}\widetilde{M}_1 T \left[|\tilde{y}| + \widehat{M}H\|\phi\|_{\mathcal{B}} + \widehat{M} \int_0^T p(s)\, \psi(\|z_s + x_s\|_{\mathcal{B}})\, ds \right] \\
&\quad + \widehat{M} \int_0^t p(s)\, \psi(\|z_s + x_s\|_{\mathcal{B}})\, ds \\
&\le \widehat{M}\widetilde{M}\widetilde{M}_1 T \left[|\tilde{y}| + \widehat{M}H\|\phi\|_{\mathcal{B}} \right] \\
&\quad + \widehat{M} \left(\widehat{M}\widetilde{M}\widetilde{M}_1 T + 1 \right) \int_0^T p(s)\, \psi(\|z_s + x_s)\|_{\mathcal{B}})\, ds.
\end{aligned}
$$

By (3.4) on J_T, we get for each $z \in B_d$

$$
\|z_s + x_s\|_{\mathcal{B}} \le K_T d + \alpha_T := \delta_T. \tag{3.9}
$$

Then, using the nondecreasing character of ψ, we get for each $t \in [0, T]$

$$
|(Fz)(t)| \le \widehat{M}\widetilde{M}\widetilde{M}_1 T \left[|\tilde{y}| + \widehat{M}H\|\phi\|_{\mathcal{B}} \right] + \widehat{M} \left(\widehat{M}\widetilde{M}\widetilde{M}_1 T + 1 \right) \psi(\delta_T)\, \|p\|_{L^1} := \ell.
$$

Thus there exists a positive number ℓ such that $\|F(z)\|_T \le \ell$. Hence $F(B_d) \subset B_d$.

Step 3: F maps bounded sets into equi-continuous sets of B_T^0. We consider B_d as in Step 2 and we show that $F(B_d)$ is equi-continuous. Let $\tau_1, \tau_2 \in J_T$ with $\tau_2 > \tau_1$ and $z \in B_d$. Then

$$
\begin{aligned}
|(Fz)(\tau_2) - (Fz)(\tau_1)| &\le \int_0^{\tau_1} \|U(\tau_2, s) - U(\tau_1, s)\|_{B(E)} \, \|C\| \, |u_z(s)| \, ds \\
&\quad + \int_0^{\tau_1} \|U(\tau_2, s) - U(\tau_1, s)\|_{B(E)} \, |f(s, z_s + x_s)| \, ds \\
&\quad + \int_{\tau_1}^{\tau_2} \|U(\tau_2, s)\|_{B(E)} \, \|C\| \, |u_z(s)| \, ds \\
&\quad + \int_{\tau_1}^{\tau_2} \|U(\tau_2, s)\|_{B(E)} \, |f(s, z_s + x_s)| \, ds.
\end{aligned}
$$

In the property of u_y, we use (3.9) and the nondecreasing character of ψ to get

$$|u_z(t)| \leq \tilde{M}_1 \left[|\bar{y}| + \widehat{M}H\|\phi\|_B + \widehat{M} \, \psi(\delta_T) \, \|p\|_{L^1} \right] := \omega.$$

Then

$$|(Fz)(\tau_2) - (Fz)(\tau_1)| \leq \|C\|_{B(E)} \, \omega \int_0^{\tau_1} \|U(\tau_2, s) - U(\tau_1, s)\|_{B(E)} \, ds$$

$$+ \psi(\delta_T) \int_0^{\tau_1} \|U(\tau_2, s) - U(\tau_1, s)\|_{B(E)} \, p(s) \, ds$$

$$+ \|C\|_{B(E)} \, \omega \int_{\tau_1}^{\tau_2} \|U(\tau_2, s)\|_{B(E)} \, ds$$

$$+ \psi(\delta_T) \int_{\tau_1}^{\tau_2} \|U(\tau_2, s)\|_{B(E)} \, p(s) \, ds.$$

Noting that $|(Fz)(\tau_2) - (Fz)(\tau_1)|$ tends to zero as $\tau_2 - \tau_1 \to 0$ independently of $z \in B_d$. The right-hand side of the above inequality tends to zero as $\tau_2 - \tau_1 \to 0$. Since $U(t, s)$ is a strongly continuous operator and the compactness of $U(t, s)$ for $t > s$ implies the continuity in the uniform operator topology (see [16, 168]). As a consequence of Steps 1 to 3 together with the Arzelá–Ascoli theorem it suffices to show that the operator F maps B_d into a precompact set in E.

Let $t \in J_T$ be fixed and let ϵ be a real number satisfying $0 < \epsilon < t$. For $z \in B_d$ we define

$$(F_\epsilon z)(t) = U(t, t - \epsilon) \int_0^{t-\epsilon} U(t - \epsilon, s) \, C \, u_z(s) \, ds$$

$$+ U(t, t - \epsilon) \int_0^{t-\epsilon} U(t - \epsilon, s) f(s, z_s + x_s) \, ds.$$

Since $U(t, s)$ is a compact operator, the set $Z_\epsilon(t) = \{F_\epsilon(z)(t) : z \in B_d\}$ is precompact in E for every ϵ sufficiently small, $0 < \epsilon < t$. Moreover using the definition of ω, we have

$$|(Fz)(t) - (F_\epsilon z)(t)| \leq \int_{t-\epsilon}^t \|U(t, s)\|_{B(E)} \|C\| \, |u_z(s)| \, ds$$

$$+ \int_{t-\epsilon}^t \|U(t, s)\|_{B(E)} \, |f(s, z_s + x_s)| \, ds$$

$$\leq \|C\|_{B(E)} \, \omega \int_{t-\epsilon}^t \|U(t, s)\|_{B(E)} \, ds$$

$$+ \psi(\delta_T) \int_{t-\epsilon}^t \|U(t, s)\|_{B(E)} \, p(s) \, ds.$$

Therefore there are precompact sets arbitrary close to the set $\{F(z)(t) : z \in B_d\}$. Hence the set $\{F(z)(t) : z \in B_d\}$ is precompact in E. So we deduce from Steps 1, 2, and 3 that F is a compact operator.

Step 4: For applying Theorem 1.27, we must check (S2): i.e., it remains to show that the set

$$\mathcal{E} = \{z \in B_T^0 : z = \lambda \, F(z) \text{ for some } 0 < \lambda < 1\}$$

is bounded.

Let $z \in \mathcal{E}$, for each $t \in [0, T]$ we have

$$|z(t)| \leq \int_0^t \|U(t,s)\|_{B(E)} \, \|C\| \, |u_z(s)| \, ds + \int_0^t \|U(t,s)\|_{B(E)} \, |f(s, z_s + x_s)| \, ds$$

$$\leq \widehat{M}\tilde{M} \int_0^t \tilde{M}_1 \left[|\tilde{y}| + \widehat{M}H\|\phi\|_B + \widehat{M} \int_0^T p(\tau) \, \psi(\|z_\tau + x_\tau\|_B) \, d\tau \right] ds$$

$$+ \widehat{M} \int_0^t p(s) \, \psi(\|z_s + x_s\|_B) \, ds$$

$$\leq \widehat{M}\tilde{M}\tilde{M}_1 T \left[|\tilde{y}| + \widehat{M}H\|\phi\|_B + \widehat{M} \int_0^T p(s) \, \psi(\|z_s + x_s\|_B) \, ds \right]$$

$$+ \widehat{M} \int_0^t p(s) \, \psi(\|z_s + x_s\|_B) \, ds.$$

Using the inequality (3.4) over J_T and the nondecreasing character of ψ, we get

$$|z(t)| \leq \widehat{M}\tilde{M}\tilde{M}_1 T \left[|\tilde{y}| + \widehat{M}H\|\phi\|_B + \widehat{M} \int_0^T p(s) \, \psi(K_T|z(s)| + \alpha_T) \, ds \right]$$

$$+ \widehat{M} \int_0^t p(s) \, \psi(K_T|z(s)| + \alpha_T) \, ds.$$

Then

$$K_T|z(t)| + \alpha_T \leq \alpha_T + K_T\widehat{M}\tilde{M}\tilde{M}_1 T$$

$$\left[|\tilde{y}| + \widehat{M}H\|\phi\|_B\widehat{M} \int_0^T p(s) \, \psi(K_T|z(s)| + \alpha_T) \, ds \right]$$

$$+ K_T\widehat{M} \int_0^t p(s) \, \psi(K_T|z(s)| + \alpha_T) \, ds.$$

Set $c_{10,T} := \alpha_T + K_T \widehat{M} \tilde{M} \tilde{M}_1 T \left[|\bar{y}| + \widehat{M} H \|\phi\|_{\mathcal{B}} \right]$, thus

$$K_T |z(t)| + \alpha_T \leq c_{10,T} + \widehat{M}^2 K_T \tilde{M} \tilde{M}_1 T \int_0^T p(s) \, \psi(K_T |z(s)| + \alpha_T) \, ds$$

$$+ K_T \widehat{M} \int_0^t p(s) \, \psi(K_T |z(s)| + \alpha_T) \, ds.$$

We consider the function μ defined by

$$\mu(t) := \sup \{ K_T |z(s)| + \alpha_T : 0 \leq s \leq t \}, \quad 0 \leq t \leq T.$$

Let $t^* \in [0, t]$ be such that $\mu(t) = K_T |z(t^*)| + \alpha_T$. If $t^* \in [0, T]$, by the previous inequality, we have for $t \in [0, T]$

$$\mu(t) \leq c_{10,T} + \widehat{M}^2 K_T \tilde{M} \tilde{M}_1 T \int_0^T p(s) \, \psi(\mu(s)) \, ds + K_T \widehat{M} \int_0^t p(s) \, \psi(\mu(s)) \, ds.$$

Then, we have

$$\mu(t) \leq c_{10,T} + K_T \widehat{M} \left(\widehat{M} \tilde{M} \tilde{M}_1 T + 1 \right) \int_0^T p(s) \, \psi(\mu(s)) \, ds.$$

Consequently,

$$\frac{\|z\|_T}{c_{10,T} + K_T \widehat{M} \left(\widehat{M} \tilde{M} \tilde{M}_1 T + 1 \right) \psi(\|z\|_T) \|p\|_{L^1}} \leq 1.$$

Then by (3.8), there exists a constant M_* such that $\|z\|_T \neq M_*$. Set

$$Z = \{ z \in B_T^0 : \|z\|_T \leq M_* + 1 \}.$$

Clearly, Z is a closed subset of B_T^0. From the choice of Z there is no $z \in \partial Z$ such that $z = \lambda \, F(z)$ for some $\lambda \in (0, 1)$. Then the statement $(S2)$ in Theorem 1.27 does not hold. As a consequence of the nonlinear alternative of Leray–Schauder type [128], we deduce that $(S1)$ holds: i.e., the operator F has a fixed point z^*. Then $y^*(t) = z^*(t) + x(t), t \in (-\infty, T]$ is a fixed point of the operator N_{10}, which is a mild solution of the problem (3.6)–(3.7). Thus the evolution system (3.6)–(3.7) is controllable on $(-\infty, T]$. $\qquad \square$

3.3.3 An Example

As an application of Theorem 3.6, we present the following control problem

$$
\begin{cases}
\dfrac{\partial v}{\partial t}(t,\xi) = a(t,\xi)\dfrac{\partial^2 v}{\partial \xi^2}(t,\xi) + d(\xi)u(t) \\
\qquad + \displaystyle\int_{-\infty}^{0} P(\theta)r(t,v(t+\theta,\xi))d\theta \quad t \in [0,T] \quad \xi \in [0,\pi] \\
v(t,0) \ = \ v(t,\pi) \ = 0 \qquad\qquad\qquad\quad t \in [0,T] \\
v(\theta,\xi) \ = \ v_0(\theta,\xi) \qquad\qquad\qquad\qquad -\infty < \theta \le 0,\ \xi \in [0,\pi],
\end{cases}
\tag{3.10}
$$

where $a(t,\xi)$ is a continuous function and is uniformly Hölder continuous in t ; $P : (-\infty,0] \to \mathbb{R}$; $r : [0,T] \times \mathbb{R} \to \mathbb{R}$; $v_0 : (-\infty,0] \times [0,\pi] \to \mathbb{R}$ and $d : [0,\pi] \to E$ are continuous functions. $u(\cdot) : [0,T] \to E$ is a given control.

Consider $E = L^2([0,\pi],\mathbb{R})$ and define $A(t)$ by $A(t)w = a(t,\xi)w''$ with domain

$$D(A) = \{\, w \in E \ : \ w,\ w' \text{ are absolutely continuous, } w'' \in E,\ w(0) = w(\pi) = 0 \,\}$$

Then $A(t)$ generates an evolution system $U(t,s)$ satisfying assumption (3.5.1) (see [112, 149]).

For the phase space $\mathcal{B}$, we choose the well-known space $BUC(\mathbb{R}^-, E)$: the space of uniformly bounded continuous functions endowed with the following norm

$$\|\varphi\| = \sup_{\theta \le 0}|\varphi(\theta)| \quad \text{for} \quad \varphi \in \mathcal{B}.$$

If we put for $\varphi \in BUC(\mathbb{R}^-, E)$ and $\xi \in [0,\pi]$

$$y(t)(\xi) = v(t,\xi),\ t \in [0,T],\ \xi \in [0,\pi],$$

$$\phi(\theta)(\xi) = v_0(\theta,\xi),\ -\infty < \theta \le 0,\ \xi \in [0,\pi],$$

and

$$f(t,\varphi)(\xi) = \int_{-\infty}^{0} P(\theta)r(t,\varphi(\theta)(\xi))d\theta,\ -\infty < \theta \le 0,\ \xi \in [0,\pi]$$

Finally let $C \in B(\mathbb{R}, E)$ be defined as

$$Cu(t)(\xi) = d(\xi)u(t),\ t \in [0,T],\ \xi \in [0,\pi],\ u \in \mathbb{R},\ d(\xi) \in E.$$

Then, problem (3.10) takes the abstract evolution form (3.6)–(3.7). In order to show the controllability of mild solutions of system (3.10), we suppose the following assumptions:

- There exists a continuous function $p \in L^1(J_T, \mathbb{R}^+)$ and a nondecreasing continuous function $\psi : [0, \infty) \to [0, \infty)$ such that

$$|r(t, u)| \leq p(t)\psi(|u|), \text{ for } t \in J_T, \text{ and } u \in \mathbb{R}.$$

- P is integrable on $(-\infty, 0]$.

By the dominated convergence theorem, one can show that f is a continuous function mapping B into E. In fact, we have for $\varphi \in B$ and $\xi \in [0, \pi]$

$$|f(t, \varphi)(\xi)| \leq \int_{-\infty}^0 |p(t)P(\theta)| \psi(|(\varphi(\theta))(\xi)|)d\theta.$$

Since the function ψ is nondecreasing, it follows that

$$|f(t, \varphi)| \leq p(t) \int_{-\infty}^0 |P(\theta)| d\theta \psi(|\varphi|), \text{ for } \varphi \in B.$$

Proposition 3.7. *Under the above assumptions, if we assume that condition (3.8) in Theorem 3.6 is true, $\varphi \in B$, then the problem (3.10) is controllable on $(-\infty, T]$.*

3.4 Controllability on Semi-infinite Interval for Partial Evolution Equations

3.4.1 Introduction

We obtain in this section the controllability of mild solutions on the semi-infinite interval $J = \mathbb{R}_+$ for the partial functional evolution equations with infinite delay of the form

$$y'(t) = A(t)y(t) + Cu(t) + f(t, y_t), \quad \text{a.e. } t \in J = \mathbb{R}_+ \tag{3.11}$$

$$y_0 = \phi \in (-\infty, 0], \tag{3.12}$$

where $f : J \times B \to E$ and $\phi \in B$ are given functions, the control function $u(.)$ is given in $L^2([0, \infty), E)$, the Banach space of admissible control function with E is a real separable Banach space with the norm $|\cdot|$ for some $n > 0$, C is a bounded linear operator from E into E, and $\{A(t)\}_{0 \leq t < +\infty}$ is a family of linear closed (not necessarily bounded) operators from E into E that generate an evolution system of operators $\{U(t, s)\}_{(t,s) \in J \times J}$ for $0 \leq s \leq t < +\infty$.

3.4.2   Controllability of Mild Solutions

In this section, we give controllability result for the system (3.11)–(3.12). Before this, we introduce the the following type of solutions for the problem (3.11)–(3.12).

Definition 3.8. We say that the continuous function $y(\cdot) : \mathbb{R} \to E$ is a mild solution of (3.11)–(3.12) if $y(t) = \phi(t)$ for all $t \in (-\infty, 0]$ and y satisfies the following integral equation

$$y(t) = U(t, 0)\phi(0) + \int_0^t U(t, s)Cu(s)ds + \int_0^t U(t, s)f(s, y_s)ds, \quad \text{for each } t \in \mathbb{R}_+$$

Definition 3.9. The evolution problem (3.11)–(3.12) is said to be controllable if for every initial function $\phi \in \mathcal{B}$ and $\hat{y} \in E$, there is some control $u \in L^2([0, n], E)$ such that the mild solution $y(\cdot)$ of (3.11)–(3.12) satisfies the terminal condition $y(n) = \hat{y}$.

We will consider the hypotheses (3.1.1)–(3.1.3) and we will need to introduce the following one which is assumed hereafter:

(3.9.1) For each $n \in \mathbb{N}$, the linear operator $W : L^2([0, n], E) \to E$ is defined by

$$Wu = \int_0^n U(n, s)Cu(s)ds,$$

has a bounded inverse operator W^{-1} which takes values in $L^2([0, n], E)/\ker W$ and there exists positive constants $\tilde{M}$ and $\tilde{M}_1$ such that:

$$\|C\| \leq \tilde{M} \quad \text{and} \quad \|W^{-1}\| \leq \tilde{M}_1.$$

Remark 3.10. For the construction of W see [93].

Consider the following space

$$B_{+\infty} = \left\{ y : \mathbb{R} \to E : y|_{[0,T]} \in C([0, T], E), \ y_0 \in \mathcal{B} \right\},$$

where $y|_{[0,T]}$ is the restriction of y to any real compact interval $[0, T]$.

For every $n \in \mathbb{N}$, we define in $B_{+\infty}$ the semi-norms by

$$\|y\|_n := \sup \left\{ e^{-\tau L_n^*(t)} |y(t)| : t \in [0, n] \right\}$$

where $L_n^*(t) = \int_0^t \bar{l}_n(s) \, ds$, $\bar{l}_n(t) = K_n \widehat{M} l_n(t)$ and l_n is the function from (3.1.3).

Then $B_{+\infty}$ is a Fréchet space with the family of semi-norms $\{\|\cdot\|_n\}_{n \in \mathbb{N}}$. In what follows let us fix $\tau > 1$.

Theorem 3.11. *Suppose that hypotheses* (3.1.1)–(3.1.3), (3.9.1) *are satisfied and moreover there exists a constant* $M^* > 0$

$$\frac{M^*}{c_{11,n} + K_n\widehat{M}(\widehat{M}\tilde{M}\tilde{M}_1 n + 1)\,\psi(M^*)\,\|p\|_{L^1}} > 1, \tag{3.13}$$

with

$$c_{11,n} = c_{11}(\hat{y}, \phi, n) := \left[K_n\widehat{M}H\left(\widehat{M}\tilde{M}\tilde{M}_1 n + 1\right) + M_n\right]\|\phi\|_{\mathcal{B}} + K_n\widehat{M}\tilde{M}\tilde{M}_1 n\,|\hat{y}|.$$

Then the evolution problem (3.11)–(3.12) *is controllable on* $\mathbb{R}$.

Proof. Consider the operator $N_{11} : B_{+\infty} \to B_{+\infty}$ defined by:

$$(N_{11}y)(t) = \begin{cases} \phi(t), & \text{if } t \le 0; \\[2mm] U(t,0)\,\phi(0) + \displaystyle\int_0^t U(t,s)\,C\,u_y(s)\,ds \\[2mm] \quad + \displaystyle\int_0^t U(t,s)\,f(s,y_s)\,ds, & \text{if } t \ge 0. \end{cases}$$

Using assumption (3.9.1), for arbitrary function $y(\cdot)$, we define the control

$$u_y(t) = W^{-1}\left[\hat{y} - U(n,0)\,\phi(0) - \int_0^n U(n,s)\,f(s,y_s)\,ds\right](t).$$

Noting that, we have

$$|u_y(t)| \le \|W^{-1}\|\left[|\hat{y}| + \|U(t,0)\|_{B(E)}|\phi(0)| + \int_0^n \|U(n,\tau)\|_{B(E)}|f(\tau,y_\tau)|d\tau\right]$$

$$\le \tilde{M}_1\left[|\hat{y}| + \widehat{M}H\|\phi\|_{\mathcal{B}} + \widehat{M}\int_0^n |f(\tau,y_\tau)|d\tau\right]$$

$$\le \tilde{M}_1\left[|\hat{y}| + \widehat{M}H\|\phi\|_{\mathcal{B}} + \widehat{M}\int_0^n p(\tau)\,\psi(\|y_\tau\|_{\mathcal{B}})\,d\tau\right].$$

We shall show that using this control the operator N_{11} has a fixed point $y(\cdot)$. Then $y(\cdot)$ is a mild solution of the evolution system (3.11)–(3.12).

For $\phi \in \mathcal{B}$, we will define the function $x(.) : \mathbb{R} \to E$ by

$$x(t) = \begin{cases} \phi(t), & \text{if } t \le 0; \\[2mm] U(t,0)\,\phi(0), & \text{if } t \ge 0. \end{cases}$$

Then $x_0 = \phi$. For each function $z \in B_{+\infty}$, set $y(t) = z(t) + x(t)$. Then z satisfies $z_0 = 0$ and

$$z(t) = \int_0^t U(t,s)\, C\, u_{z+x}(s)\, ds + \int_0^t U(t,s)\, f(s, z_s + x_s)\, ds, \quad \text{for } t \geq 0.$$

Let $B_{+\infty}^0 = \{z \in B_{+\infty} : z_0 = 0\}$. Define the operators $F, G : B_{+\infty}^0 \to B_{+\infty}^0$ by:

$$F(z)(t) = \int_0^t U(t,s)\, C\, u_{z+x}(s)\, ds, \quad \text{for } t \geq 0.$$

and

$$G(z)(t) = \int_0^t U(t,s)\, f(s, z_s + x_s)\, ds, \quad \text{for } t \geq 0.$$

Obviously the operator N_{11} has a fixed point is equivalent to $F + G$ has one, so it turns to prove that $F + G$ has a fixed point. The proof will be given in several steps.

We can show as in Sect. 3.3.2 that the operator F is continuous and compact.

We can prove also that the operator G is a contraction as in the proof of Theorem 2.2).

For applying Theorem 1.30, it remains to show that (S2) doesn't hold: i.e., we will prove that the following set is bounded

$$\mathcal{E} = \left\{z \in B_{+\infty}^0 : z = \lambda\, F(z) + \lambda\, G\left(\frac{z}{\lambda}\right) \text{ for some } 0 < \lambda < 1\right\}.$$

Let $z \in \mathcal{E}$, for each $t \in [0, n]$, we have

$$|z(t)| \leq \lambda \int_0^t \|U(t,s)\|_{B(E)}\, \|C\|\, |u_{z+x}(s)|\, ds$$

$$+ \lambda \int_0^t \|U(t,s)\|_{B(E)}\, \left|f\left(s, \frac{z_s}{\lambda} + x_s\right)\right|\, ds.$$

Then

$$\frac{1}{\lambda}|z(t)| \leq \widehat{M}\tilde{M} \int_0^t \tilde{M}_1\left[|\hat{y}| + \widehat{M}H\|\phi\|_B + \widehat{M}\int_0^n p(\tau)\, \psi(\|z_\tau + x_\tau\|_B)\, d\tau\right] ds$$

$$+ \widehat{M}\int_0^t p(s)\, \psi\left(\left\|\frac{z_s}{\lambda} + x_s\right\|_B\right) ds$$

$$\leq \widehat{M}\tilde{M}\tilde{M}_1 t\left[|\hat{y}| + \widehat{M}H\|\phi\|_B + \widehat{M}\int_0^n p(s)\, \psi(\|z_s + x_s\|_B)\, ds\right]$$

$$+ \widehat{M}\int_0^t p(s)\, \psi\left(\left\|\frac{z_s}{\lambda} + x_s\right\|_B\right) ds$$

$$\leq \widehat{M}\tilde{M}\tilde{M}_1 n \left[|\hat{y}| + \widehat{M}H\|\phi\|_{\mathcal{B}} \right] + \widehat{M}^2 \tilde{M}\tilde{M}_1 n \int_0^n p(s)\, \psi\,(\|z_s + x_s\|_{\mathcal{B}})\, ds$$

$$+\widehat{M} \int_0^t p(s)\, \psi\left(\left\| \frac{z_s}{\lambda} + x_s \right\|_{\mathcal{B}} \right) ds.$$

Using the inequality (3.4) and the nondecreasing character of ψ, we get

$$\frac{1}{\lambda}|z(t)| \leq \widehat{M}\tilde{M}\tilde{M}_1 n \left[|\hat{y}| + \widehat{M}H\|\phi\|_{\mathcal{B}} \right] + \widehat{M}^2 \tilde{M}\tilde{M}_1 n \int_0^n p(s)\, \psi\,(K_n|z(s)| + \alpha_n)\, ds$$

$$+\widehat{M} \int_0^t p(s)\, \psi\left(\frac{K_n}{\lambda}|z(s)| + \alpha_n \right) ds.$$

Then, we get

$$\frac{K_n}{\lambda}|z(t)| + \alpha_n \leq \alpha_n + K_n\widehat{M}\tilde{M}\tilde{M}_1 n \left[|\hat{y}| + \widehat{M}H\|\phi\|_{\mathcal{B}} \right]$$

$$+K_n\widehat{M}^2 \tilde{M}\tilde{M}_1 n \int_0^n p(s)\, \psi\,(K_n|z(s)| + \alpha_n)\, ds$$

$$+K_n\widehat{M} \int_0^t p(s)\, \psi\left(\frac{K_n}{\lambda}|z(s)| + \alpha_n \right) ds.$$

Set $c_{11,n} := \alpha_n + K_n\widehat{M}\tilde{M}\tilde{M}_1 n \left[|\hat{y}| + \widehat{M}H\|\phi\|_{\mathcal{B}} \right]$. By the nondecreasing character of ψ and for $\lambda < 1$, we obtain

$$\frac{K_n}{\lambda}|z(t)| + \alpha_n \leq c_{11,n} + K_n\widehat{M}^2 \tilde{M}\tilde{M}_1 n \int_0^n p(s)\, \psi\left(\frac{K_n}{\lambda}|z(s)| + \alpha_n \right) ds$$

$$+ K_n\widehat{M} \int_0^t p(s)\, \psi\left(\frac{K_n}{\lambda}|z(s)| + \alpha_n \right) ds.$$

We consider the function μ defined by

$$\mu(t) := \sup\left\{ \frac{K_n}{\lambda}|z(s)| + \alpha_n \; : \; 0 \leq s \leq t \right\}, \quad 0 \leq t \leq n.$$

Let $t^* \in [0, t]$ be such that $\mu(t) = \frac{K_n}{\lambda}|z(t^*)| + \alpha_n$. If $t^* \in [0, n]$, by the previous inequality, we have for $t \in [0, n]$

$$\mu(t) \leq c_{11,n} + K_n\widehat{M}^2 \tilde{M}\tilde{M}_1 n \int_0^n p(s)\, \psi(\mu(s))\, ds + K_n\widehat{M} \int_0^t p(s)\, \psi(\mu(s))\, ds.$$

Then, we have

$$\mu(t) \le c_{11,n} + K_n \widehat{M}(\widehat{M}\tilde{M}\tilde{M}_1 n + 1) \int_0^n p(s)\, \psi(\mu(s))\, ds.$$

Consequently,

$$\frac{\|z\|_n}{c_{11,n} + K_n \widehat{M}(\widehat{M}\tilde{M}\tilde{M}_1 n + 1)\, \psi(\|z\|_n)\, \|p\|_{L^1}} \le 1.$$

Then by the condition (3.13), there exists a constant M^* such that $\mu(t) \le M^*$. Since $\|z\|_n \le \mu(t)$, we have $\|z\|_n \le M^*$. This shows that the set $\mathcal{E}$ is bounded, i.e., the statement (S2) in Theorem 1.30 does not hold. Then the nonlinear alternative of Avramescu [32] implies that (S1) holds, i.e., the operator $F + G$ has a fixed-point z^*. Then $y^*(t) = z^*(t) + x(t)$, $t \in \mathbb{R}$ is a fixed point of the operator N_{11}, which is a mild solution of the problem (3.11)–(3.12). Thus the evolution system (3.11)–(3.12) is controllable. □

3.4.3 An Example

As an application of Theorem 3.11, we present the following control problem

$$\begin{cases} \dfrac{\partial v}{\partial t}(t,\xi) = a(t,\xi)\dfrac{\partial^2 v}{\partial \xi^2}(t,\xi) + d(\xi)u(t) \\[2mm] \qquad + \displaystyle\int_{-\infty}^0 P(\theta) r(t, v(t+\theta,\xi)) d\theta \qquad t \ge 0 \qquad \xi \in [0,\pi] \\[4mm] v(t,0) = v(t,\pi) = 0 \qquad\qquad\qquad t \ge 0 \\[2mm] v(\theta,\xi) = v_0(\theta,\xi) \qquad\qquad\qquad -\infty < \theta \le 0,\ \xi \in [0,\pi], \end{cases} \qquad (3.14)$$

where $a(t,\xi)$ is a continuous function and is uniformly Hölder continuous in t; $P : (-\infty, 0] \to \mathbb{R}$; $r : \mathbb{R}_+ \times \mathbb{R} \to \mathbb{R}$; $v_0 : (-\infty, 0] \times [0,\pi] \to \mathbb{R}$ and $d : [0,\pi] \to E$ are continuous functions. $u(\cdot) : \mathbb{R}_+ \to E$ is a given control.

Consider $E = L^2([0,\pi], \mathbb{R})$ and define $A(t)$ by $A(t)w = a(t,\xi)w''$ with domain

$$D(A) = \{\, w \in E \ : \ w, w' \text{ are absolutely continuous, } w'' \in E,\ w(0) = w(\pi) = 0 \,\}$$

Then $A(t)$ generates an evolution system $U(t,s)$ satisfying assumption (3.1.1) (see [112, 149]).

For the phase space $\mathcal{B}$, we choose the well-known space $BUC(\mathbb{R}^-, E)$: the space of uniformly bounded continuous functions endowed with the following norm

$$\|\varphi\| = \sup_{\theta \leq 0} |\varphi(\theta)| \quad \text{for} \quad \varphi \in \mathcal{B}.$$

If we put for $\varphi \in BUC(\mathbb{R}^-, E)$ and $\xi \in [0, \pi]$

$$y(t)(\xi) = v(t, \xi), \ t \geq 0, \ \xi \in [0, \pi],$$

$$\phi(\theta)(\xi) = v_0(\theta, \xi), \ -\infty < \theta \leq 0, \ \xi \in [0, \pi],$$

and

$$f(t, \varphi)(\xi) = \int_{-\infty}^{0} P(\theta) r(t, \varphi(\theta)(\xi)) d\theta, \ -\infty < \theta \leq 0, \ \xi \in [0, \pi].$$

Finally let $C \in B(\mathbb{R}, E)$ be defined as

$$Cu(t)(\xi) = d(\xi) u(t), \ t \geq 0, \ \xi \in [0, \pi], \ u \in \mathbb{R}, \ d(\xi) \in E.$$

Then, problem (3.14) takes the abstract evolution form (3.11)–(3.12). Furthermore, more appropriate conditions on P and r ensure the controllability of mild solutions on $(-\infty, +\infty)$ of the system (3.14) by Theorems 3.11 and 1.30.

3.5 Neutral Functional Evolution Equations

3.5.1 Introduction

In this section, we investigate the following neutral functional differential evolution equation with infinite delay

$$\frac{d}{dt}[y(t) - g(t, y_t)] = A(t)y(t) + f(t, y_t), \quad \text{a.e.} \ \ t \in J = \mathbb{R}_+ \tag{3.15}$$

$$y_0 = \phi \in \mathcal{B}, \tag{3.16}$$

where $A(\cdot), f$, and ϕ are as in problem (3.1)–(3.2) and $g : J \times \mathcal{B} \rightarrow E$ is a given function.

3.5.2   Existence and Uniqueness of Mild Solution

We give first the definition of the mild solution of our neutral functional evolution problem (3.15)–(3.16) before stating our main result and proving it.

Definition 3.12. We say that the continuous function $y(\cdot) : \mathbb{R} \to E$ is a mild solution of (3.15)–(3.16) if $y(t) = \phi(t)$ for all $t \in (-\infty, 0]$ and y satisfies the following integral equation

$$y(t) = U(t, 0)[\phi(0) - g(0, \phi)] + g(t, y_t) + \int_0^t U(t, s)A(s)g(s, y_s)ds$$

$$+ \int_0^t U(t, s)f(s, y_s)\, ds, \quad \text{for each } t \in \mathbb{R}_+.$$

We will need to introduce the following assumptions which are assumed hereafter:

(G1)   There exists a constant $\overline{M}_0 > 0$ such that:

$$\|A^{-1}(t)\|_{B(E)} \le \overline{M}_0 \quad \text{for all } t \in J.$$

(G2)   There exists a constant $0 < L < \dfrac{1}{\overline{M}_0 K_n}$ such that:

$$|A(t)\, g(t, \phi)| \le L\,(\|\phi\|_B + 1) \text{ for all } t \in J \text{ and } \phi \in \mathcal{B}.$$

(G3)   There exists a constant $L_* > 0$ such that:

$$|A(s)\, g(s, \phi) - A(\bar{s})\, g(\bar{s}, \bar{\phi})| \le L_*\,(|s - \bar{s}| + \|\phi - \bar{\phi}\|_\mathcal{B})$$

for all $s, \bar{s} \in J$ and $\phi, \bar{\phi} \in \mathcal{B}$.

Consider the following space

$$B_{+\infty} = \{y : \mathbb{R} \to E : y|_{[0,T]} \in C([0, T], E), \ y_0 \in \mathcal{B}\},$$

where $y|_{[0,T]}$ is the restriction of y to any real compact interval $[0, T]$.

For every $n \in \mathbb{N}$, we define in $B_{+\infty}$ the semi-norms by:

$$\|y\|_n := \sup\{ e^{-\tau\, L_n^*(t)}\, |y(t)| : t \in [0, n]\}$$

where $L_n^*(t) = \displaystyle\int_0^t \bar{l}_n(s)\, ds$, $\bar{l}_n(t) = K_n \widehat{M}[L_* + l_n(t)]$ and l_n is the function from (3.1.3).

Then $B_{+\infty}$ is a Fréchet space with the family of semi-norms $\|\cdot\|_{n \in \mathbb{N}}$. Let us fix $\tau > 0$ and assume that $\left[\overline{M}_0 L_* K_n + \dfrac{1}{\tau}\right] < 1$.

Theorem 3.13. *Suppose that hypotheses* (3.1.1)–(3.1.3) *and assumptions* (G1)–(G3) *are satisfied and moreover*

$$\int_{c_{12,n}}^{+\infty} \frac{ds}{s + \psi(s)} > \frac{K_n \widehat{M}}{1 - \overline{M}_0 L K_n} \int_0^n \max(L, p(s)) ds, \quad \text{for each } n > 0 \quad (3.17)$$

with

$$c_{12,n} = \left[\frac{\overline{M}_0 L K_n \widehat{M}}{1 - \overline{M}_0 L K_n} + (K_n \widehat{M} H + M_n) \left(1 + \frac{\overline{M}_0 L K_n}{1 - \overline{M}_0 L K_n} \right) \right] \|\phi\|_{\mathcal{B}}$$

$$+ \frac{K_n}{1 - \overline{M}_0 L K_n} \left[\overline{M}_0 L (\widehat{M} + 1) + \widehat{M} L n \right],$$

then the problem (3.15)–(3.16) *has a unique mild solution.*

Proof. Let the operator $N_{12} : B_{+\infty} \to B_{+\infty}$ be defined by:

$$(N_{12} y)(t) = \begin{cases} \phi(t), & \text{if } t \leq 0; \\ U(t, 0) \left[\phi(0) - g(0, \phi) \right] + g(t, y_t) \\ \quad + \int_0^t U(t, s) A(s) g(s, y_s) ds + \int_0^t U(t, s) f(s, y_s) ds, & \text{if } t \geq 0. \end{cases}$$

Then, fixed points of the operator N_{12} are mild solutions of the problem (3.15)–(3.16).

For $\phi \in \mathcal{B}$, we will define the function $x(.) : \mathbb{R} \to E$ by

$$x(t) = \begin{cases} \phi(t), & \text{if } t \in (-\infty, 0]; \\ U(t, 0) \phi(0), & \text{if } t \in J. \end{cases}$$

Then $x_0 = \phi$. For each function $z \in B_{+\infty}$, set $y(t) = z(t) + x(t)$. It is obvious that y satisfies Definition 3.12 if and only if z satisfies $z_0 = 0$ and for $t \in J$, we get

$$z(t) = g(t, z_t + x_t) - U(t, 0) g(0, \phi)$$

$$+ \int_0^t U(t, s) A(s) g(s, z_s + x_s) ds + \int_0^t U(t, s) f(s, z_s + x_s) ds.$$

Let $B_{+\infty}^0 = \{ z \in B_{+\infty} : z_0 = 0 \}$. Define the operator $F : B_{+\infty}^0 \to B_{+\infty}^0$ by:

$$(Fz)(t) = g(t, z_t + x_t) - U(t, 0) g(0, \phi)$$

$$+ \int_0^t U(t, s) A(s) g(s, z_s + x_s) ds + \int_0^t U(t, s) f(s, z_s + x_s) ds.$$

Obviously the operator N_{12} has a fixed point is equivalent to F has one, so it turns to prove that F has a fixed point.

Let $z \in B^0_{+\infty}$ be a possible F fixed point of the operator . Then, using (3.1.1), (3.1.2), $(G1)$ and $(G2)$, we have for each $t \in [0, n]$

$$|z(t)| \leq |g(t, z_t + x_t)| + |U(t, 0)g(0, \phi)| + |\int_0^t U(t, s)A(s)g(s, z_s + x_s)ds|$$

$$+ |\int_0^t U(t, s)f(s, z_s + x_s)ds|$$

$$\leq \|A^{-1}(t)\|_{B(E)}\|A(t)g(t, z_t + x_t)\| + \|U(t, 0)\|_{B(E)}\|A^{-1}(0)\|\|A(0) g(0, \phi)\|$$

$$+ \int_0^t \|U(t, s)\|_{B(E)}\|A(s)g(s, z_s + x_s)\|ds + \int_0^t \|U(t, s)\|_{B(E)}|f(s, z_s + x_s)|ds$$

$$\leq \overline{M}_0 L(\|z_t + x_t\|_B + 1) + \widehat{M}\overline{M}_0 L(\|\phi\|_B + 1)$$

$$+ \widehat{M} \int_0^t L(\|z_s + x_s\|_B + 1)ds + \widehat{M} \int_0^t p(s)\psi(\|z_s + x_s\|_B)ds$$

$$\leq \overline{M}_0 L\|z_t + x_t\|_B + \overline{M}_0 L(\widehat{M} + 1) + \widehat{M}Ln + \widehat{M}\overline{M}_0 L\|\phi\|_B$$

$$+ \widehat{M} \int_0^t L\|z_s + x_s\|_B ds + \widehat{M} \int_0^t p(s)\psi(\|z_s + x_s\|_B)ds.$$

Using the inequality (3.4) and the nondecreasing character of ψ, we get

$$|z(t)| \leq \overline{M}_0 L(K_n|z(t)| + \alpha_n) + \overline{M}_0 L(\widehat{M} + 1) + \widehat{M}Ln + \widehat{M}\overline{M}_0 L\|\phi\|_B$$

$$+ \widehat{M} \int_0^t L(K_n|z(s)| + \alpha_n)ds + \widehat{M} \int_0^t p(s)\psi(K_n|z(s)| + \alpha_n)ds$$

$$\leq \overline{M}_0 L K_n|z(t)| + \overline{M}_0 L\alpha_n + \widehat{M}\overline{M}_0 L\|\phi\|_B + \overline{M}_0 L(\widehat{M} + 1) + \widehat{M}Ln$$

$$+ \widehat{M} \int_0^t L(K_n|z(s)| + \alpha_n)ds + \widehat{M} \int_0^t p(s)\psi(K_n|z(s)| + \alpha_n)ds.$$

Then

$$(1 - \overline{M}_0 L K_n)|z(t)| \leq \overline{M}_0 L\alpha_n + \widehat{M}\overline{M}_0 L\|\phi\|_B + \overline{M}_0 L(\widehat{M} + 1) + \widehat{M}Ln$$

$$+ \widehat{M} \int_0^t L(K_n|z(s)| + \alpha_n)ds + \widehat{M} \int_0^t p(s)\psi(K_n|z(s)| + \alpha_n)ds.$$

Set $c_{12,n} := \alpha_n + \dfrac{K_n}{1 - \overline{M}_0 L K_n} \left[\overline{M}_0 L \left(\alpha_n + \widehat{M} \|\phi\|_B \right) + \overline{M}_0 L (\widehat{M} + 1) + \widehat{M} L n \right]$. Thus

$$K_n |z(t)| + \alpha_n \leq c_{12,n} + \frac{K_n \widehat{M}}{1 - \overline{M}_0 L K_n} \int_0^t L(K_n |z(s)| + \alpha_n) ds$$

$$+ \frac{K_n \widehat{M}}{1 - \overline{M}_0 L K_n} \int_0^t p(s) \psi (K_n |z(s)| + \alpha_n) ds.$$

We consider the function μ defined by

$$\mu(t) := \sup \{ K_n |z(s)| + \alpha_n \; : \; 0 \leq s \leq t \}, \quad 0 \leq t < +\infty.$$

Let $t^* \in [0, t]$ be such that $\mu(t) = K_n |z(t^*)| + \alpha_n$. By the previous inequality, we have

$$\mu(t) \leq c_{12,n} + \frac{K_n \widehat{M}}{1 - \overline{M}_0 L K_n} \left[\int_0^t L \mu(s) ds + \int_0^t p(s) \psi (\mu(s)) ds, \right] \quad \text{for} \; t \in [0, n].$$

Let us take the right-hand side of the above inequality as $v(t)$. Then, we have

$$\mu(t) \leq v(t) \; \text{for all} \; t \in [0, n].$$

From the definition of v, we have $v(0) = c_{12,n}$ and

$$v'(t) = \frac{K_n \widehat{M}}{1 - \overline{M}_0 L K_n} [L \mu(t) + p(t) \psi (\mu(t))] \quad \text{a.e.} \; t \in [0, n].$$

Using the nondecreasing character of ψ, we get

$$v'(t) \leq \frac{K_n \widehat{M}}{1 - \overline{M}_0 L K_n} [L v(t) + p(t) \psi (v(t))] \quad \text{a.e.} \; t \in [0, n].$$

This implies that for each $t \in [0, n]$ and using the condition (3.17), we get

$$\int_{c_{12,n}}^{v(t)} \frac{ds}{s + \psi(s)} \leq \frac{K_n \widehat{M}}{1 - \overline{M}_0 L K_n} \int_0^t \max(L, p(s)) ds$$

$$\leq \frac{K_n \widehat{M}}{1 - \overline{M}_0 L K_n} \int_0^n \max(L, p(s)) ds$$

$$< \int_{c_{12,n}}^{+\infty} \frac{ds}{s + \psi(s)}.$$

Thus, for every $t \in [0, n]$, there exists a constant Λ_n such that $v(t) \leq \Lambda_n$ and hence $\mu(t) \leq \Lambda_n$. Since $\|z\|_n \leq \mu(t)$, we have $\|z\|_n \leq \Lambda_n$. Set

$$Z = \{z \in B^0_{+\infty} : \sup\{|z(t)| \; 0 \leq t \leq n\} \leq \Lambda_n + 1 \quad \text{for all} \;\; n \in \mathbb{N}\}.$$

Clearly, Z is a closed subset of $B^0_{+\infty}$.

Now, we shall show that $F : Z \to B^0_{+\infty}$ is a contraction operator. Indeed, consider $z, \bar{z} \in Z$, thus for each $t \in [0, n]$ and $n \in \mathbb{N}$ and using (3.1.1), (3.1.3), $G1$ and $(G3)$, we get

$$
\begin{aligned}
|F(z)(t) - F(\bar{z})(t)| &\leq |g(t, z_t + x_t) - g(t, \bar{z}_t + x_t)| \\
&+ \int_0^t \|U(t, s)\|_{B(E)} |A(s)[g(s, z_s + x_s) - g(s, \bar{z}_s + x_s)]| ds \\
&+ \int_0^t \|U(t, s)\|_{B(E)} |f(s, z_s + x_s) - f(s, \bar{z}_s + x_s)| ds \\
&\leq \|A^{-1}(t)\|_{B(E)} |A(t)g(t, z_t + x_t) - A(t)g(t, \bar{z}_t + x_t)| \\
&+ \int_0^t \|U(t, s)\|_{B(E)} |A(s)g(s, z_s + x_s) - A(s)g(s, \bar{z}_s + x_s)| ds \\
&+ \int_0^t \|U(t, s)\|_{B(E)} |f(s, z_s + x_s) - f(s, \bar{z}_s + x_s)| ds \\
&\leq \overline{M}_0 L_* \|z_t - \bar{z}_t\|_B + \int_0^t \widehat{M} L_* \|z_s - \bar{z}_s\|_B ds \\
&+ \int_0^t \widehat{M} l_n(s) \|z_s - \bar{z}_s\|_B ds.
\end{aligned}
$$

Using (A_1), we obtain

$$
\begin{aligned}
|F(z)(t) - F(\bar{z})(t)| &\leq \overline{M}_0 L_* (K(t) |z(t) - \bar{z}(t)| + M(t) \|z_0 - \bar{z}_0\|_B) \\
&+ \int_0^t \widehat{M} [L_* + l_n(s)] (K(s) |z(s) - \bar{z}(s)| + M(s) \|z_0 - \bar{z}_0\|_B) ds \\
&\leq \overline{M}_0 L_* K_n |z(t) - \bar{z}(t)| + \int_0^t K_n \widehat{M} [L_* + l_n(s)] |z(s) - \bar{z}(s)| ds \\
&\leq \overline{M}_0 L_* K_n |z(t) - \bar{z}(t)| + \int_0^t \bar{l}_n(s) |z(s) - \bar{z}(s)| ds \\
&\leq \left[\overline{M}_0 L_* K_n e^{\tau L_n^*(t)} \right] \left[e^{-\tau L_n^*(t)} |z(t) - \bar{z}(t)| \right] \\
&+ \int_0^t \left[\bar{l}_n(s) e^{\tau L_n^*(s)} \right] \left[e^{-\tau L_n^*(s)} |z(s) - \bar{z}(s)| \right] ds
\end{aligned}
$$

$$\leq \overline{M}_0 L_* K_n \, e^{\tau \, L_n^*(t)} \, \|z - \bar{z}\|_n + \int_0^t \left[\frac{e^{\tau \, L_n^*(s)}}{\tau} \right]' \, ds \, \|z - \bar{z}\|_n$$

$$\leq \overline{M}_0 L_* K_n \, e^{\tau \, L_n^*(t)} \, \|z - \bar{z}\|_n + \frac{1}{\tau} \, e^{\tau \, L_n^*(t)} \, \|z - \bar{z}\|_n$$

$$\leq \left[\overline{M}_0 L_* K_n + \frac{1}{\tau} \right] e^{\tau \, L_n^*(t)} \, \|z - \bar{z}\|_n.$$

Therefore,

$$\|F(z) - F(\bar{z})\|_n \leq \left[\overline{M}_0 L_* K_n + \frac{1}{\tau} \right] \|z - \bar{z}\|_n.$$

So, for $\left[\overline{M}_0 L_* K_n + \dfrac{1}{\tau} \right] < 1$, the operator F is a contraction for all $n \in \mathbb{N}$. From the choice of Z there is no $z \in \partial Z^n$ such that $z = \lambda \, F(z)$ for some $\lambda \in (0, 1)$. Then the statement $(S2)$ in Theorem 1.29 does not hold. We deduce that the operator F has a unique fixed point z^*. Then $y^*(t) = z^*(t) + x(t)$, $t \in \mathbb{R}$ is a fixed point of the operator N_{12}, which is the unique mild solution of the problem (3.15)–(3.16). $\square$

3.5.3 An Example

As an application we consider the following neutral functional evolution equation

$$\begin{cases} \dfrac{\partial}{\partial t} \left[z(t, x) - \displaystyle\int_{-\infty}^t \int_0^\pi b(s - t, u, x) \, z(s, u) \, du \, ds \right] \\[2mm] \quad = a(t, x) \dfrac{\partial^2 z}{\partial x^2}(t, x) + Q(t, z(t - r, x), \dfrac{\partial z}{\partial x}(t - r, x)), \, t \geq 0, \, x \in [0, \pi] \\[2mm] z(t, 0) = z(t, \pi) = 0, \hspace{4cm} t \geq 0 \\[2mm] z(t, x) = \Phi(t, x), \hspace{3.5cm} t \leq 0, \, x \in [0, \pi] \end{cases} \tag{3.18}$$

where $r > 0$, $a(t, x)$ is a continuous function and is uniformly Hölder continuous in t, $Q : \mathbb{R}_+ \times \mathbb{R} \times \mathbb{R} \to \mathbb{R}$ and $\Phi : \mathcal{B} \times [0, \pi] \to \mathbb{R}$ are continuous functions.

Let

$$y(t)(x) = z(t, x), \, t \in [0, \infty), \, x \in [0, \pi],$$

$$\phi(\theta)(x) = \Phi(\theta, x), \, \theta \leq 0, \, x \in [0, \pi],$$

$$g(t, \phi)(x) = \int_{-\infty}^t \int_0^\pi b(s - t, u, x) \phi(s, u) \, du \, ds, \, x \in [0, \pi]$$

and

$$f(t, \phi)(x) = Q\left(t, \phi(\theta, x), \frac{\partial \phi}{\partial x}(\theta, x)\right), \ \theta \le 0, \ x \in [0, \pi].$$

Consider $E = L^2[0, \pi]$ and define $A(t)$ by $A(t)w = a(t, x)w''$ with domain

$D(A) = \{ w \in E \mid w, w' \text{ are absolutely continuous, } w'' \in E, \ w(0) = w(\pi) = 0 \}.$

Then $A(t)$ generates an evolution system $U(t, s)$ satisfying assumptions (3.1.1) and (G1) (see [112, 149]).

Here we consider that $\varphi : (-\infty, 0] \to E$ such that φ is Lebesgue measurable and $h(s)|\varphi(s)|^2$ is Lebesgue integrable on H where $h : (-\infty, 0] \to \mathbb{R}$ is a positive integrable function. The norm is defined here by:

$$\|\varphi\| = |\Phi(0)| + \left(\int_{-\infty}^{0} h(s) \, |\varphi(s)|^2 \, ds\right)^{\frac{1}{2}}.$$

The function b is measurable on $\mathbb{R}_+ \times [0, \pi] \times [0, \pi]$,

$$b(s, u, 0) = b(s, u, \pi) = 0, \ (s, u) \in \mathbb{R}_+ \times [0, \pi],$$

$$\int_0^\pi \int_{-\infty}^t \int_0^\pi \frac{b^2(s, u, x)}{h(s)} ds du dx < \infty,$$

and $\sup_{t \in \mathbb{R}_+} \mathcal{N}(t) < \infty$, where

$$\mathcal{N}(t) = \int_0^\pi \int_{-\infty}^t \int_0^\pi \frac{1}{h(s)} \left(a(s, x)\frac{\partial^2}{\partial x^2}b(s, u, x)\right)^2 ds du dx.$$

Thus, under the above definitions of f, g, and $A(\cdot)$, the system (3.18) can be represented by the abstract neutral functional evolution problem (3.15)–(3.16). Furthermore, more appropriate conditions on Q ensure the existence of the unique mild solution of (3.18) by Theorem 3.13 and 1.29.

3.6 Controllability on Finite Interval for Neutral Evolution Equations

3.6.1 Introduction

In this section, we give sufficient conditions ensuring the controllability of mild solutions on a bounded interval $J_T := [0, T]$ for $T > 0$ for the neutral functional differential evolution equation with infinite delay of the form

$$\frac{d}{dt}[y(t) - g(t, y_t)] = A(t)y(t) + Cu(t) + f(t, y_t), \quad \text{a.e.} \quad t \in J_T = [0, T] \qquad (3.19)$$

$$y_0 = \phi \in \mathcal{B}, \qquad (3.20)$$

where $A(\cdot), f, u, C$, and ϕ are as in problem (3.6)–(3.7) and $g : J_T \times \mathcal{B} \to E$ is a given function.

3.6.2   Controllability of Mild Solutions

Before stating and proving the controllability result, we give first the definition of mild solution of our evolution problem (3.19)–(3.20).

Definition 3.14. We say that the continuous function $y(\cdot) : (-\infty, T] \to E$ is a mild solution of (3.19)–(3.20) if $y(t) = \phi(t)$ for all $t \in (-\infty, 0]$ and y satisfies the following integral equation

$$y(t) = U(t, 0)[\phi(0) - g(0, \phi)] + g(t, y_t) + \int_0^t U(t, s)A(s)g(s, y_s)ds$$

$$+ \int_0^t U(t, s)Cu(s)ds + \int_0^t U(t, s)f(s, y_s) \, ds, \quad \text{for each } t \in [0, T].$$

Definition 3.15. The neutral functional evolution problem (3.19)–(3.20) is said to be controllable on the interval $[0, T]$ if for every initial function $\phi \in \mathcal{B}$ and $\tilde{y} \in E$ there exists a control $u \in L^2([0, T], E)$ such that the mild solution $y(\cdot)$ of (3.19)–(3.20) satisfies $y(T) = \tilde{y}$.

We consider the hypotheses (3.4.1)–(3.4.3) and we will need to introduce the following assumptions which are assumed hereafter:

$(\widetilde{G1})$ There exists a constant $\overline{M}_0 > 0$ such that:

$$\|A^{-1}(t)\|_{B(E)} \leq \overline{M}_0 \quad \text{for all } t \in J_T.$$

$(\widetilde{G2})$ There exists a constant $0 < L < \dfrac{1}{\overline{M}_0 K_T}$ such that:

$$|A(t) \, g(t, \phi)| \leq L \, (\|\phi\|_\mathcal{B} + 1) \text{ for all } t \in J_T \text{ and } \phi \in \mathcal{B}.$$

$(\widetilde{G3})$ There exists a constant $L_* > 0$ such that:

$$|A(t) \, g(s, \phi) - A(t) \, g(\overline{s}, \overline{\phi})| \leq L_* \, (|s - \overline{s}| + \|\phi - \overline{\phi}\|_\mathcal{B})$$

for all $0 \leq t, s, \overline{s} \leq T$ and $\phi, \overline{\phi} \in \mathcal{B}$.

(G̃4) The function g is completely continuous and for any bounded set $Q \subseteq B_T$ the set $\{t \to g(t, x_t) : x \in Q\}$ is equi-continuous in $C([0, T], E)$.

Consider the following space

$$B_T = \{y : (-\infty, T] \to E : y|_J \in C(J, E), \ y_0 \in B\},$$

where $y|_J$ is the restriction of y to J.

Theorem 3.16. *Suppose that hypotheses* (3.4.1)–(3.4.3) *and assumptions* (G̃1)–(G̃4) *are satisfied and moreover there exists a constant $M_* > 0$ with*

$$\frac{M_*}{c_{13,T} + K_T \widehat{M} \dfrac{\widehat{M} \tilde{M} \tilde{M}_1 T + 1}{1 - \overline{M}_0 L K_T} \ [M_* + \psi(M_*)] \ \|\zeta\|_{L^1}} > 1, \tag{3.21}$$

where $\zeta(t) = \max(L, p(t))$ and

$$c_{13,T} = c_{13}(\phi, \tilde{y}, T) = \frac{K_T(\widehat{M} \tilde{M} \tilde{M}_1 T + 1)}{1 - \overline{M}_0 L K_T} \left[\overline{M}_0 L(\widehat{M} + 1) + \widehat{M} L T \right]$$

$$+ \left[\frac{K_T \widehat{M}}{1 - \overline{M}_0 L K_T} \left(\overline{M}_0 L(\widehat{M} \tilde{M} \tilde{M}_1 T + 1) + \tilde{M} \tilde{M}_1 T(\widehat{M} H + \overline{M}_0 L M_T) \right) \right.$$

$$\left. + (K_T \widehat{M} H + M_T) \left(1 + \frac{\overline{M}_0 L K_T}{1 - \overline{M}_0 L K_T} \right) \right] \|\phi\|_B$$

$$+ K_T \widehat{M} \tilde{M} \tilde{M}_1 T \frac{1 + \overline{M}_0 L K_T}{1 - \overline{M}_0 L K_T} |\tilde{y}|$$

then the neutral functional evolution problem (3.19)–(3.20) *is controllable on* $(-\infty, T]$.

Proof. Consider the operator $N_{13} : B_T \to B_T$ defined by:

$$N_{13}(y)(t) = \begin{cases} \phi(t), & \text{if } t \in (-\infty, 0]; \\[2ex] U(t, 0)\,[\phi(0) - g(0, \phi)] + g(t, y_t) \\[1ex] \quad + \displaystyle\int_0^t U(t, s)A(s)g(s, y_s)ds \\[1ex] \quad + \displaystyle\int_0^t U(t, s)Cu_y(s)ds + \int_0^t U(t, s)f(s, y_s)ds, & \text{if } t \in J_T. \end{cases}$$

Using assumption (3.4.3), for arbitrary function $y(\cdot)$, we define the control

$$u_y(t) = W^{-1}\Big[\tilde{y} - U(T,0)\,(\phi(0) - g(0,\phi)) - g(T, y_T)$$

$$- \int_0^T U(T,s)A(s)g(s, y_s)ds - \int_0^T U(T,s)f(s, y_s)ds\Big](t).$$

Noting that

$$|u_y(t)| \le \|W^{-1}\|\Big[\,|\tilde{y}| + \|U(t,0)\|_{B(E)}\,(|\phi(0)| + \|A^{-1}(0)\|\,|A(0)g(0,\phi)|)$$

$$+\|A^{-1}(T)\|\,|A(T)g(T, y_T)| + \int_0^T \|U(T,\tau)\|_{B(E)}|A(\tau)g(\tau, y_\tau)|d\tau$$

$$+ \int_0^T \|U(T,\tau)\|_{B(E)}|f(\tau, y_\tau)|d\tau\Big]$$

$$\le \tilde{M}_1\Big[\,|\tilde{y}| + \widehat{M}H\|\phi\|_B + \widehat{M}\overline{M}_0L(\|\phi\|_B + 1) + \overline{M}_0L(\|y_T\|_B + 1)\Big]$$

$$+\tilde{M}_1\widehat{M}L\int_0^T (\|y_\tau\|_B + 1)d\tau + \tilde{M}_1\widehat{M}\int_0^T |f(\tau, y_\tau)|d\tau.$$

From (3.4.2), we get

$$|u_y(t)| \le \tilde{M}_1\Big[\,|\tilde{y}| + \widehat{M}(H + \overline{M}_0L)\|\phi\|_B + \overline{M}_0L(\widehat{M} + 1) + \widehat{M}LT\Big]$$

$$+\tilde{M}_1\overline{M}_0L\|y_T\|_B + \tilde{M}_1\widehat{M}L\int_0^T \|y_\tau\|_B\,d\tau + \tilde{M}_1\widehat{M}\int_0^T |f(\tau, y_\tau)|\,d\tau$$

$$\le \tilde{M}_1\Big[\,|\tilde{y}| + \widehat{M}(H + \overline{M}_0L)\|\phi\|_B + \overline{M}_0L(\widehat{M} + 1) + \widehat{M}LT\Big]$$

$$+\tilde{M}_1\overline{M}_0L\|y_T\|_B + \tilde{M}_1\widehat{M}L\int_0^T \|y_\tau\|_Bd\tau + \tilde{M}_1\widehat{M}\int_0^T p(\tau)\psi(\|y_\tau\|_B)d\tau.$$

It shall be shown that using this control the operator N_{13} has a fixed point $y(\cdot)$. Then $y(\cdot)$ is a mild solution of the neutral functional evolution system (3.19)–(3.20).

For $\phi \in B$, we will define the function $x(.) : \mathbb{R} \to E$ by

$$x(t) = \begin{cases} \phi(t), & \text{if } t \in (-\infty, 0]; \\[2mm] U(t,0)\,\phi(0), & \text{if } t \in J_T. \end{cases}$$

Then $x_0 = \phi$. For each function $z \in B_T$, set $y(t) = z(t) + x(t)$. It is obvious that y satisfies Definition 3.15 if and only if z satisfies $z_0 = 0$ and for $t \in J_T$, we get

$$z(t) = g(t, z_t + x_t) - U(t,0)g(0,\phi) + \int_0^t U(t,s)A(s)g(s, z_s + x_s)ds$$

$$+ \int_0^t U(t,s)Cu_z(s)ds + \int_0^t U(t,s)f(s, z_s + x_s)ds.$$

Define the operator $F : B_T^0 \to B_T^0$ by:

$$F(z)(t) = g(t, z_t + x_t) - U(t, 0)\, g(0, \phi) + \int_0^t U(t, s)\, A(s)\, g(s, z_s + x_s)\, ds$$
$$+ \int_0^t U(t, s)\, C\, u_z(s)\, ds + \int_0^t U(t, s)\, f(s, z_s + x_s)\, ds.$$

Obviously the operator N_{13} has a fixed point is equivalent to F has one, so it turns to prove that F has a fixed point. The proof will be given in several steps.

We can show that the operator F is continuous and compact. For applying Theorem 1.27, we must check (S2): i.e., it remains to show that the set

$$\mathcal{E} = \{ z \in B_T^0 : z = \lambda\, F(z) \text{ for some } 0 < \lambda < 1 \}$$

is bounded.

Let $z \in \mathcal{E}$. By (3.4.1)–(3.4.3), $(\widetilde{G1})$ and $(\widetilde{G2})$, we have for each $t \in [0, T]$

$$|z(t)| \leq \|A^{-1}(t)\|\, |A(t)g(t, z_t + x_t)| + \|U(t, 0)\|_{B(E)}\, \|A^{-1}(0)\|\, |A(0)g(0, \phi)|$$
$$+ \int_0^t \|U(t, s)\|_{B(E)} |A(s)g(s, z_s + x_s)|\, ds + \int_0^t \|U(t, s)\|_{B(E)}\, \|C\|\, |u_z(s)|\, ds$$
$$+ \int_0^t \|U(t, s)\|_{B(E)}\, |f(s, z_s + x_s)|\, ds$$

$$\leq \overline{M}_0 L\, (\|z_t + x_t\|_{\mathcal{B}} + 1) + \widehat{M}\overline{M}_0 L\, (\|\phi\|_{\mathcal{B}} + 1) + \widehat{M} L \int_0^t (\|z_s + x_s\|_{\mathcal{B}} + 1)\, ds$$

$$+ \widehat{M}\tilde{M} \int_0^t \tilde{M}_1 \Big[|\tilde{y}| + \widehat{M}(H + \overline{M}_0 L)\|\phi\|_{\mathcal{B}} + \overline{M}_0 L(\widehat{M} + 1) + \widehat{M} L T$$

$$+ \overline{M}_0 L\|z_T + x_T\|_{\mathcal{B}} + \widehat{M} L \int_0^T \|z_\tau + x_\tau\|_{\mathcal{B}} d\tau + \widehat{M} \int_0^T p(\tau)\psi(\|z_\tau + x_\tau\|_{\mathcal{B}}) d\tau \Big]\, ds$$

$$+ \widehat{M} \int_0^t p(s)\, \psi(\|z_s + x_s\|_{\mathcal{B}})\, ds$$

$$\leq \Big[\overline{M}_0 L(\widehat{M} + 1) + \widehat{M} L T \Big] (\widehat{M}\tilde{M}\tilde{M}_1 T + 1) + \widehat{M}\tilde{M}\tilde{M}_1 T\, |\tilde{y}|$$

$$+ \widehat{M} \Big[\overline{M}_0 L(\widehat{M}\tilde{M}\tilde{M}_1 T + 1) + \widehat{M}\tilde{M}\tilde{M}_1 T H \Big] \|\phi\|_{\mathcal{B}} + \widehat{M}\tilde{M}\tilde{M}_1 \overline{M}_0 L T \|z_T + x_T\|_{\mathcal{B}}$$

$$+ \overline{M}_0 L\|z_t + x_t\|_{\mathcal{B}} + \widehat{M} L \int_0^t \|z_s + x_s\|_{\mathcal{B}}\, ds + \widehat{M}^2 \tilde{M}\tilde{M}_1 L T \int_0^T \|z_s + x_s\|_{\mathcal{B}}\, ds$$

$$+ \widehat{M}^2 \tilde{M}\tilde{M}_1 T \int_0^T p(s)\psi(\|z_s + x_s\|_{\mathcal{B}})\, ds + \widehat{M} \int_0^t p(s)\psi(\|z_s + x_s\|_{\mathcal{B}})\, ds.$$

Noting that we have $\|z_T + x_T\|_B \leq K_T |\bar{y}| + M_T\|\phi\|_B$ and using (3.4) and by the nondecreasing character of ψ, we obtain

$$|z(t)| \leq \left[\overline{M}_0 L(\widehat{M}+1) + \widehat{M}LT\right](\widehat{M}\tilde{M}\tilde{M}_1 T + 1) + \widehat{M}\tilde{M}\tilde{M}_1 T\left(1 + \overline{M}_0 LK_T\right)|\bar{y}|$$

$$+\widehat{M}\left[\overline{M}_0 L(\widehat{M}\tilde{M}\tilde{M}_1 T + 1) + \tilde{M}\tilde{M}_1 T\left(\widehat{M}H + \overline{M}_0 LM_T\right)\right]\|\phi\|_B$$

$$+\overline{M}_0 L\left(K_T|z(t)| + \alpha_T\right)$$

$$+\widehat{M}L\int_0^t \left(K_T|z(s)| + \alpha_T\right)\,ds + \widehat{M}^2\tilde{M}\tilde{M}_1 LT\int_0^T \left(K_T|z(s)| + \alpha_T\right)\,ds$$

$$+\widehat{M}^2\tilde{M}\tilde{M}_1 T\int_0^T p(s)\psi\left(K_T|z(s)| + \alpha_T\right)\,ds$$

$$+\widehat{M}\int_0^t p(s)\psi\left(K_T|z(s)| + \alpha_T\right)\,ds.$$

Then

$$\left(1 - \overline{M}_0 LK_T\right)|z(t)| \leq \left[\overline{M}_0 L(\widehat{M}+1) + \widehat{M}LT\right](\widehat{M}\tilde{M}\tilde{M}_1 T + 1)$$

$$+\widehat{M}\tilde{M}\tilde{M}_1 T\left(1 + \overline{M}_0 LK_T\right)|\bar{y}| + \overline{M}_0 L\alpha_T$$

$$+\widehat{M}\left[\overline{M}_0 L(\widehat{M}\tilde{M}\tilde{M}_1 T + 1)\right.$$

$$\left. + \tilde{M}\tilde{M}_1 T\left(\widehat{M}H + \overline{M}_0 LM_T\right)\right]\|\phi\|_B$$

$$+\widehat{M}L\int_0^t \left(K_T|z(s)| + \alpha_T\right)\,ds$$

$$+\widehat{M}^2\tilde{M}\tilde{M}_1 LT\int_0^T \left(K_T|z(s)| + \alpha_T\right)\,ds$$

$$+\widehat{M}^2\tilde{M}\tilde{M}_1 T\int_0^T p(s)\psi\left(K_T|z(s)| + \alpha_T\right)\,ds$$

$$+\widehat{M}\int_0^t p(s)\psi\left(K_T|z(s)| + \alpha_T\right)\,ds.$$

Set

$$c_{13,T} := \alpha_T + \frac{K_T}{1 - \overline{M}_0 LK_T}$$

$$\times \left\{\left[\overline{M}_0 L(\widehat{M}+1) + \widehat{M}LT\right](\widehat{M}\tilde{M}\tilde{M}_1 T + 1) + \widehat{M}\tilde{M}\tilde{M}_1 T\left(1 + \overline{M}_0 LK_T\right)|\bar{y}|\right.$$

$$\left. + \overline{M}_0 L\alpha_T + \widehat{M}\left[\overline{M}_0 L(\widehat{M}\tilde{M}\tilde{M}_1 T + 1) + \tilde{M}\tilde{M}_1 T\left(\widehat{M}H + \overline{M}_0 LM_T\right)\right]\|\phi\|_B\right\}$$

thus

$$
K_T|z(t)| + \alpha_T \le c_{13,T} + \frac{K_T\widehat{M}}{1 - \overline{M}_0 L K_T}
$$

$$
\times \left[L \int_0^t \left(K_T|z(s)| + \alpha_T \right) \, ds + \widehat{M}\tilde{M}\tilde{M}_1 L T \int_0^T \left(K_T|z(s)| + \alpha_T \right) \, ds \right.
$$

$$
\left. + \widehat{M}\tilde{M}\tilde{M}_1 T \int_0^T p(s)\psi \left(K_T|z(s)| + \alpha_T \right) \, ds + \int_0^t p(s)\psi \left(K_T|z(s)| + \alpha_T \right) \, ds \right].
$$

We consider the function μ defined by

$$
\mu(t) := \sup \{ K_T|z(s)| + \alpha_T \; : \; 0 \le s \le t \}, \quad 0 \le t \le T.
$$

Let $t^* \in [0, t]$ be such that $\mu(t) = K_T|z(t^*)| + \alpha_T$. If $t^* \in [0, T]$, by the previous inequality, we have

$$
\mu(t) \le c_{13,T} + \frac{K_T\widehat{M}}{1 - \overline{M}_0 L K_T} \left[L \int_0^t \mu(s) \, ds + \widehat{M}\tilde{M}\tilde{M}_1 L T \int_0^T \mu(s) \, ds \right.
$$

$$
\left. + \widehat{M}\tilde{M}\tilde{M}_1 T \int_0^T p(s)\psi(\mu(s)) \, ds + \int_0^t p(s)\psi(\mu(s)) \, ds \right].
$$

Then, we have

$$
\mu(t) \le c_{13,T} + K_T\widehat{M}\frac{\widehat{M}\tilde{M}\tilde{M}_1 T + 1}{1 - \overline{M}_0 L K_T} \left[L \int_0^T \mu(s) \, ds + \int_0^T p(s)\psi(\mu(s)) \, ds \right].
$$

Set $\zeta(t) := \max(L, p(t))$ for $t \in [0, T]$

$$
\mu(t) \le c_{13,T} + K_T\widehat{M}\frac{\widehat{M}\tilde{M}\tilde{M}_1 T + 1}{1 - \overline{M}_0 L K_T} \int_0^T \zeta(s) \left[\mu(s) + \psi(\mu(s)) \right] \, ds.
$$

Consequently,

$$
\frac{\|z\|_T}{c_{13,T} + K_T\widehat{M}\dfrac{\widehat{M}\tilde{M}\tilde{M}_1 T + 1}{1 - \overline{M}_0 L K_T} \left[\|z\|_T + \psi(\|z\|_T) \right] \|\zeta\|_{L^1}} \le 1.
$$

Then by (3.21), there exists a constant M_* such that $\|z\|_T \neq M_*$. Set

$$
\tilde{Z} = \{ z \in B_T^0 \; : \; \|z\|_T \le M_* + 1 \}.
$$

Clearly, $\tilde{Z}$ is a closed subset of B_T^0. From the choice of $\tilde{Z}$ there is no $z \in \partial\tilde{Z}$ such that $z = \lambda\, F(z)$ for some $\lambda \in (0,1)$. Then the statement $(S2)$ in Theorem 1.27 does not hold. As a consequence of the nonlinear alternative of Leray–Schauder type [128], we deduce that $(S1)$ holds: i.e., the operator F has a fixed point z^*. Then $y^*(t) = z^*(t) + x(t)$, $t \in (-\infty, T]$ is a fixed point of the operator N_{13}, which is a mild solution of the problem (3.19)–(3.20). Thus the evolution system (3.19)–(3.20) is controllable on $(-\infty, T]$. $\qquad\qquad\qquad\qquad\qquad\qquad\qquad\qquad\qquad\qquad\square$

3.6.3   An Example

As an application of Theorem 3.16, we present the following control problem

$$
\begin{cases}
\dfrac{\partial}{\partial t}\left[v(t,\xi) - \displaystyle\int_{-\infty}^{0} T(\theta)w(t, v(t+\theta,\xi))d\theta \right] \\[2mm]
\quad = a(t,\xi)\dfrac{\partial^2 v}{\partial\xi^2}(t,\xi) + d(\xi)u(t) \\[2mm]
\qquad + \displaystyle\int_{-\infty}^{0} P(\theta)r(t, v(t+\theta,\xi))d\theta & t \in [0,T] \quad \xi \in [0,\pi] \\[4mm]
v(t,0) \;=\; v(t,\pi) = 0 & t \in [0,T] \\[2mm]
v(\theta,\xi) \;=\; v_0(\theta,\xi) & -\infty < \theta \le 0,\ \xi \in [0,\pi],
\end{cases}
\tag{3.22}
$$

where $a(t,\xi)$ is a continuous function and is uniformly Hölder continuous in t ; $T, P : (-\infty, 0] \to \mathbb{R}$; $w, r : [0,T] \times \mathbb{R} \to \mathbb{R}$; $v_0 : (-\infty, 0] \times [0,\pi] \to \mathbb{R}$ and $d : [0,\pi] \to E$ are continuous functions. $u(\cdot) : [0,T] \to E$ is a given control.

Consider $E = L^2([0,\pi], \mathbb{R})$ and define $A(t)$ by $A(t)w = a(t,\xi)w''$ with domain

$$D(A) = \{\, w \in E \,:\, w,\ w'\ \text{are absolutely continuous},\ w'' \in E,\ w(0) = w(\pi) = 0 \,\}$$

Then $A(t)$ generates an evolution system $U(t,s)$ satisfying assumptions (3.6.1) and $(\widetilde{G1})$ (see [112, 149]).

For the phase space $\mathcal{B}$, we choose the well-known space $BUC(\mathbb{R}^-, E)$: the space of uniformly bounded continuous functions endowed with the following norm

$$\|\varphi\| = \sup_{\theta \le 0} |\varphi(\theta)| \quad \text{for} \quad \varphi \in \mathcal{B}.$$

If we put for $\varphi \in BUC(\mathbb{R}^-, E)$ and $\xi \in [0,\pi]$

$$y(t)(\xi) = v(t,\xi),\ t \in [0,T],\ \xi \in [0,\pi],$$

$$\phi(\theta)(\xi) = v_0(\theta, \xi), \quad -\infty < \theta \le 0, \ \xi \in [0, \pi],$$

$$g(t, \varphi)(\xi) = \int_{-\infty}^{0} T(\theta)w(t, \varphi(\theta)(\xi))d\theta, \quad -\infty < \theta \le 0, \ \xi \in [0, \pi],$$

and

$$f(t, \varphi)(\xi) = \int_{-\infty}^{0} P(\theta)r(t, \varphi(\theta)(\xi))d\theta, \quad -\infty < \theta \le 0, \ \xi \in [0, \pi]$$

Finally let $C \in B(\mathbb{R}, E)$ be defined as

$$Cu(t)(\xi) = d(\xi)u(t), \ t \in [0, T], \ \xi \in [0, \pi], \ u \in \mathbb{R}, \ d(\xi) \in E.$$

Then, problem (3.22) takes the abstract neutral functional evolution form (3.19)–(3.20). In order to show the controllability of mild solutions of system (3.22), we suppose the following assumptions:

- w is Lipschitz with respect to its second argument. Let $lip(w)$ denotes the Lipschitz constant of w.
- There exist a function $p \in L^1(J_T, \mathbb{R}^+)$ and a nondecreasing continuous function $\psi : [0, \infty) \to [0, \infty)$ such that

$$|r(t, u)| \le p(t)\psi(|u|), \ \text{for } t \in J_T, \ \text{and } u \in \mathbb{R}.$$

- T and P are integrable on $(-\infty, 0]$.

By the dominated convergence theorem, one can show that f is a continuous function from B to E. Moreover the mapping g is Lipschitz continuous in its second argument, in fact, we have

$$|g(t, \varphi_1) - g(t, \varphi_2)| \le \overline{M}_0 L_* lip(w) \int_{-\infty}^{0} |T(\theta)| \, d\theta \, |\varphi_1 - \varphi_2|, \ \text{for } \varphi_1, \varphi_2 \in B.$$

On the other hand, we have for $\varphi \in B$ and $\xi \in [0, \pi]$

$$|f(t, \varphi)(\xi)| \le \int_{-\infty}^{0} |p(t)P(\theta)| \, \psi(|(\varphi(\theta))(\xi)|)d\theta.$$

Since the function ψ is nondecreasing, it follows that

$$|f(t, \varphi)| \le p(t) \int_{-\infty}^{0} |P(\theta)| \, d\theta \psi(|\varphi|), \ \text{for } \varphi \in B.$$

Proposition 3.17. *Under the above assumptions, if we assume that condition* (3.21) *in Theorem 3.16 is true, $\varphi \in B$, then the problem* (3.22) *is controllable on* $(-\infty, T]$.

3.7   Controllability on Semi-infinite Interval for Neutral Evolution Equations

3.7.1   Introduction

We investigate in this section the controllability of mild solutions on the semi-infinite interval $J = \mathbb{R}_+$ for the following neutral functional evolution equations with infinite delay

$$\frac{d}{dt}[y(t) - g(t, y_t)] = A(t)y(t) + Cu(t) + f(t, y_t), \quad \text{a.e.} \quad t \in J = \mathbb{R}_+ \tag{3.23}$$

$$y_0 = \phi \in (-\infty, 0], \tag{3.24}$$

where $A(\cdot), f, u, C$, and ϕ are as in problem (3.11)–(3.12) (Sect. 3.4) and $g : J \times \mathcal{B} \to E$ is a given function.

Here we are interested to give an application of (3.15) in [33] to control theory on the semi-infinite interval $J = \mathbb{R}_+$ for the partial functional evolution equations (3.23)–(3.24) by Theorem 1.30 due to Avramescu in [32] for sum of compact and contraction operators in Fréchet spaces, combined with the semigroup theory [16, 168].

3.7.2   Controllability of Mild Solutions

Before stating and proving the controllability result, we give first the definition of mild solution of the evolution problem (3.23)–(3.24).

Definition 3.18. We say that the function $y(\cdot) : \mathbb{R} \to E$ is a mild solution of (3.23)–(3.24) if $y(t) = \phi(t)$ for all $t \in (-\infty, 0]$ and y satisfies the following integral equation

$$y(t) = U(t, 0)[\phi(0) - g(0, \phi)] + g(t, y_t) + \int_0^t U(t, s)A(s)g(s, y_s)ds$$

$$+ \int_0^t U(t, s)Cu(s)ds + \int_0^t U(t, s)f(s, y_s)\, ds, \quad \text{for each } t \in \mathbb{R}_+.$$

Definition 3.19. The neutral functional evolution problem (3.23)–(3.24) is said to be controllable if for every initial function $\phi \in \mathcal{B}$ and $\hat{y} \in E$, there is some control $u \in L^2([0, n], E)$ such that the mild solution $y(\cdot)$ of (3.23)–(3.24) satisfies $y(n) = \hat{y}$.

We consider the hypotheses (3.3.1)–(3.1.3) given in Sect. 3.2.2 and the assumption (3.9.1) of Sect. 3.4.2 and we will need to introduce the following one which is assumed hereafter:

(G4) The function g is completely continuous and for any bounded set $Q \subseteq B_T$ the set $\{t \to g(t, x_t) : x \in Q\}$ is equi-continuous in $C(\mathbb{R}_+, E)$.

Consider the following space

$$B_{+\infty} = \{y : \mathbb{R} \to E : y|_{[0,T]} \in C([0,T], E), \ y_0 \in \mathcal{B}\},$$

where $y|_{[0,T]}$ is the restriction of y to any real compact interval $[0, T]$.

For every $n \in \mathbb{N}$, we define in $B_{+\infty}$ the semi-norms by:

$$\|y\|_n := \sup\{e^{-\tau L_n^*(t)} |y(t)| : t \in [0, n]\}$$

where $L_n^*(t) = \int_0^t \bar{l}_n(s)\, ds$, $\bar{l}_n(t) = K_n \widehat{M}[L_* + l_n(t)]$, and l_n is the function from (3.1.3).

Then $B_{+\infty}$ is a Fréchet space with the family of semi-norms $\| \cdot \|_{n \in \mathbb{N}}$. Let us fix $\tau > 0$ and assume that $\left[\overline{M}_0 L_* K_n + \dfrac{1}{\tau} \right] < 1$.

Theorem 3.20. *Suppose that hypotheses* (3.1.1)–(3.1.3), (3.9.1) *and the assumptions* (G1)–(G3) *are satisfied and moreover there exists a constant* $M^* > 0$ *with*

$$\frac{M^*}{c_{14,n} + K_n \widehat{M} \dfrac{\widehat{M}\tilde{M}\tilde{M}_1 n + 1}{1 - \overline{M}_0 L K_n} [M^* + \psi(M^*)] \|\zeta\|_{L^1}} > 1, \tag{3.25}$$

with $\zeta(t) = \max(L, p(t))$ *and*

$$c_{14,n} = c_{14}(\Phi, \hat{y}, n) = \frac{K_n(\widehat{M}\tilde{M}\tilde{M}_1 n + 1)}{1 - \overline{M}_0 L K_n} \left[\overline{M}_0 L(\widehat{M} + 1) + \widehat{M} L n \right]$$

$$+ \left[\frac{K_n \widehat{M}}{1 - \overline{M}_0 L K_n} \left(\overline{M}_0 L(\widehat{M}\tilde{M}\tilde{M}_1 n + 1) + \tilde{M}\tilde{M}_1 n(\widehat{M} H + \overline{M}_0 L M_n) \right) \right.$$

$$\left. + (K_n \widehat{M} H + M_n) \left(1 + \frac{\overline{M}_0 L K_n}{1 - \overline{M}_0 L K_n} \right) \right] \|\phi\|_{\mathcal{B}}$$

$$+ K_n \widehat{M}\tilde{M}\tilde{M}_1 n \frac{1 + \overline{M}_0 L K_n}{1 - \overline{M}_0 L K_n} |\hat{y}|$$

Then the neutral functional evolution problem (3.23)–(3.24) *is controllable on* $\mathbb{R}$.

Proof. Consider the operator $N_{14} : B_{+\infty} \to B_{+\infty}$ defined by:

$$
(N_{14}y)(t) = \begin{cases}
\phi(t), & \text{if } t \le 0; \\[2mm]
U(t,0)\,[\phi(0) - g(0,\phi)] + g(t,y_t) \\[2mm]
\quad + \displaystyle\int_0^t U(t,s)A(s)g(s,y_s)ds \\[2mm]
\quad + \displaystyle\int_0^t U(t,s)Cu_y(s)ds + \int_0^t U(t,s)f(s,y_s)ds, & \text{if } t \ge 0.
\end{cases}
$$

Using assumption (3.9.1), for arbitrary function $y(\cdot)$, we define the control

$$
u_y(t) = W^{-1}\left[\hat{y} - U(n,0)\,(\phi(0) - g(0,\phi)) - g(n,y_n) - \int_0^n U(n,s)A(s)g(s,y_s)ds \right.
$$
$$
\left. - \int_0^n U(n,s)f(s,y_s)ds \right](t).
$$

Noting that

$$
|u_y(t)| \le \|W^{-1}\| \Big[|\hat{y}| + \|U(t,0)\|_{B(E)} \left(|\phi(0)| + \|A^{-1}(0)\| \|A(0)g(0,\phi)| \right)
$$
$$
+ \|A^{-1}(n)\| |A(n)g(n,y_n)| + \int_0^n \|U(n,\tau)\|_{B(E)} |A(\tau)g(\tau,y_\tau)| d\tau
$$
$$
+ \int_0^n \|U(n,\tau)\|_{B(E)} |f(\tau,y_\tau)| d\tau \Big]
$$
$$
\le \tilde{M}_1 \Big[|\hat{y}| + \widehat{M}H\|\phi\|_B + \widehat{M}\overline{M}_0 L(\|\phi\|_B + 1) + \overline{M}_0 L(\|y_n\|_B + 1) \Big]
$$
$$
+ \tilde{M}_1 \widehat{M} L \int_0^n (\|y_\tau\|_B + 1)d\tau + \tilde{M}_1 \widehat{M} \int_0^n |f(\tau,y_\tau)| d\tau.
$$

Applying (3.1.2), we get

$$
|u_y(t)| \le \tilde{M}_1 \Big[|\hat{y}| + \widehat{M}\left(H + \overline{M}_0 L\right)\|\phi\|_B + \overline{M}_0 L(\widehat{M} + 1) + \widehat{M}Ln \Big]
$$
$$
+ \tilde{M}_1 \overline{M}_0 L\|y_n\|_B + \tilde{M}_1 \widehat{M} L \int_0^n \|y_\tau\|_B \, d\tau + \tilde{M}_1 \widehat{M} \int_0^n |f(\tau,y_\tau)| \, d\tau
$$
$$
\le \tilde{M}_1 \Big[|\hat{y}| + \widehat{M}\left(H + \overline{M}_0 L\right)\|\phi\|_B + \overline{M}_0 L(\widehat{M} + 1) + \widehat{M}Ln \Big]
$$
$$
+ \tilde{M}_1 \overline{M}_0 L\|y_n\|_B + \tilde{M}_1 \widehat{M} L \int_0^n \|y_\tau\|_B d\tau + \tilde{M}_1 \widehat{M} \int_0^n p(\tau)\psi(\|y_\tau\|_B)d\tau.
$$

Using this control the operator N_{14} has a fixed point $y(\cdot)$. Then $y(\cdot)$ is a mild solution of the neutral functional evolution system (3.23)–(3.24).

For $\phi \in \mathcal{B}$, we will define the function $x(.) : \mathbb{R} \to E$ by

$$x(t) = \begin{cases} \phi(t), & \text{if } t \leq 0; \\ U(t,0)\,\phi(0), & \text{if } t \geq 0. \end{cases}$$

Then $x_0 = \phi$. For each function $z \in B_{+\infty}$, set $y(t) = z(t) + x(t)$. It is obvious that y satisfies Definition 3.19 if and only if z satisfies $z_0 = 0$ and for $t \geq 0$, we get

$$z(t) = g(t, z_t + x_t) - U(t,0)\,g(0,\phi) + \int_0^t U(t,s)\,A(s)\,g(s, z_s + x_s)ds$$

$$+ \int_0^t U(t,s)\,C\,u_{z+x}(s)\,ds + \int_0^t U(t,s)\,f(s, z_s + x_s)\,ds.$$

Let $B_{+\infty}^0 = \{z \in B_{+\infty} : z_0 = 0\}$. Define the operators $F, G : B_{+\infty}^0 \to B_{+\infty}^0$ by:

$$F(z)(t) = g(t, z_t + x_t) - U(t,0)\,g(0,\phi) + \int_0^t U(t,s)\,A(s)\,g(s, z_s + x_s)\,ds$$

$$+ \int_0^t U(t,s)\,C\,u_{z+x}(s)\,ds.$$

and

$$G(z)(t) = \int_0^t U(t,s)\,f(s, z_s + x_s)\,ds.$$

Obviously the operator N_{14} has a fixed point is equivalent to the operator sum $F + G$ has one, so it turns to prove that $F + G$ has a fixed point. The proof will be given in several steps.

We can show as in above sections that the operator F is continuous and compact and we have shown in Sect. 3.4.2 (Step 4) that the operator G is a contraction.

For applying Avramescu nonlinear alternative, we must check $(S2)$ in Theorem 1.30: i.e., it remains to show that the following set

$$\mathcal{E} = \left\{ z \in B_{+\infty}^0 : z = \lambda F(z) + \lambda G\left(\frac{z}{\lambda}\right) \text{ for some } 0 < \lambda < 1 \right\}$$

is bounded.

Let $z \in \mathcal{E}$. Then, by (3.1.1)–(3.1.3), (3.9.1), (G1), and (G2), we have for each $t \in [0, n]$

$$|z(t)| \leq \lambda \left\{ \|A^{-1}(t)\| \, |A(t)g(t, z_t + x_t)| + \|U(t, 0)\|_{B(E)} \, \|A^{-1}(0)\| \, |A(0)g(0, \phi)| \right.$$

$$+ \int_0^t \|U(t, s)\|_{B(E)} |A(s)g(s, z_s + x_s)| \, ds$$

$$\left. + \int_0^t \|U(t, s)\|_{B(E)} \, \|C\| \, |u_{z+x}(s)| \, ds \right\}$$

$$+ \lambda \int_0^t \|U(t, s)\|_{B(E)} \left| f\left(s, \frac{z_s}{\lambda} + x_s\right) \right| \, ds$$

$$\leq \lambda \left\{ \overline{M}_0 L (\|z_t + x_t\|_{\mathcal{B}} + 1) + \widehat{M} \overline{M}_0 L (\|\phi\|_{\mathcal{B}} + 1) \right.$$

$$+ \widehat{M} L \int_0^t (\|z_s + x_s\|_{\mathcal{B}} + 1) \, ds$$

$$+ \widehat{M} \tilde{M} \int_0^t \tilde{M}_1 \left[|\hat{y}| + \widehat{M} (H + \overline{M}_0 L) \|\phi\|_{\mathcal{B}} + \overline{M}_0 L (\widehat{M} + 1) + \widehat{M} L n \right.$$

$$+ \overline{M}_0 L \|z_n + x_n\|_{\mathcal{B}} + \widehat{M} L \int_0^n \|z_\tau + x_\tau\|_{\mathcal{B}} d\tau$$

$$\left. + \widehat{M} \int_0^n p(\tau) \, \psi(\|z_\tau + x_\tau\|_{\mathcal{B}}) d\tau \right] ds$$

$$\left. + \widehat{M} \int_0^t p(s) \, \psi\left(\left\| \frac{z_s}{\lambda} + x_s \right\|_{\mathcal{B}} \right) ds \right\} .$$

Then

$$|z(t)| \leq \lambda \left\{ \left[\overline{M}_0 L (\widehat{M} + 1) + \widehat{M} L n \right] (\widehat{M} \tilde{M} \tilde{M}_1 n + 1) + \widehat{M} \tilde{M} \tilde{M}_1 n |\hat{y}| \right.$$

$$+ \widehat{M} \left[\overline{M}_0 L (\widehat{M} \tilde{M} \tilde{M}_1 n + 1) + \widehat{M} \tilde{M} \tilde{M}_1 n H \right] \|\phi\|_{\mathcal{B}}$$

$$+ \widehat{M} \tilde{M} \tilde{M}_1 \overline{M}_0 L n \|z_n + x_n\|_{\mathcal{B}} + \overline{M}_0 L \|z_t + x_t\|_{\mathcal{B}} + \widehat{M} L$$

$$\times \int_0^t \|z_s + x_s\|_{\mathcal{B}} \, ds + \widehat{M}^2 \tilde{M} \tilde{M}_1 L n \int_0^n \|z_s + x_s\|_{\mathcal{B}} \, ds$$

$$+ \widehat{M}^2 \tilde{M} \tilde{M}_1 n \int_0^n p(s) \, \psi(\|z_s + x_s\|_{\mathcal{B}}) \, ds$$

$$\left. + \widehat{M} \int_0^t p(s) \, \psi\left(\left\| \frac{z_s}{\lambda} + x_s \right\|_{\mathcal{B}} \right) ds \right\} .$$

Noting that we have $\|z_n + x_n\|_\mathcal{B} \leq K_n |\hat{y}| + M_n \|\phi\|_\mathcal{B}$ and using the inequality (3.4), then by the nondecreasing character of ψ, we obtain

$$|z(t)| \leq \lambda \left\{ \left[\overline{M}_0 L(\widehat{M} + 1) + \widehat{M} L n \right] (\widehat{M} \tilde{M} \tilde{M}_1 n + 1) + \widehat{M} \tilde{M} \tilde{M}_1 n \left(1 + K_n \overline{M}_0 L \right) |\hat{y}| \right.$$

$$+ \widehat{M} \left[\overline{M}_0 L(\widehat{M} \tilde{M} \tilde{M}_1 n + 1) + \tilde{M} \tilde{M}_1 n \left(\widehat{M} H + \overline{M}_0 L M_n \right) \right] \|\phi\|_\mathcal{B}$$

$$+ \overline{M}_0 L \left(K_n |z(t)| + \alpha_n \right)$$

$$+ \widehat{M} L \left[\int_0^t (K_n |z(s)| + \alpha_n) \, ds + \widehat{M} \tilde{M} \tilde{M}_1 n \int_0^n (K_n |z(s)| + \alpha_n) \, ds \right]$$

$$\left. + \widehat{M} \left[\widehat{M} \tilde{M} \tilde{M}_1 n \int_0^n p(s) \psi \left(K_n |z(s)| + \alpha_n \right) ds + \int_0^t p(s) \psi \left(\frac{K_n |z(s)|}{\lambda} + \alpha_n \right) ds \right] \right\}$$

$$\leq \overline{M}_0 L K_n |z(t)| + \lambda \overline{M}_0 L \alpha_n$$

$$+ \lambda \left\{ \left[\overline{M}_0 L(\widehat{M} + 1) + \widehat{M} L n \right] (\widehat{M} \tilde{M} \tilde{M}_1 n + 1) + \widehat{M} \tilde{M} \tilde{M}_1 n \left(1 + K_n \overline{M}_0 L \right) |\hat{y}| \right.$$

$$+ \widehat{M} \left[\overline{M}_0 L(\widehat{M} \tilde{M} \tilde{M}_1 n + 1) + \tilde{M} \tilde{M}_1 n \left(\widehat{M} H + \overline{M}_0 L M_n \right) \right] \|\phi\|_\mathcal{B}$$

$$+ \widehat{M} L \left[\int_0^t (K_n |z(s)| + \alpha_n) \, ds + \widehat{M} \tilde{M} \tilde{M}_1 n \int_0^n (K_n |z(s)| + \alpha_n) \, ds \right]$$

$$\left. + \widehat{M} \left[\widehat{M} \tilde{M} \tilde{M}_1 n \int_0^n p(s) \psi \left(K_n |z(s)| + \alpha_n \right) ds + \int_0^t p(s) \psi \left(\frac{K_n |z(s)|}{\lambda} + \alpha_n \right) ds \right] \right\}.$$

Then,

$$\frac{|z(t)|}{\lambda} \left(1 - \overline{M}_0 L K_n \right)$$

$$\leq \overline{M}_0 L \alpha_n + \left[\overline{M}_0 L(\widehat{M} + 1) + \widehat{M} L n \right] (\widehat{M} \tilde{M} \tilde{M}_1 n + 1) + \widehat{M} \tilde{M} \tilde{M}_1 n \left(1 + K_n \overline{M}_0 L \right) |\hat{y}|$$

$$+ \widehat{M} \left[\overline{M}_0 L(\widehat{M} \tilde{M} \tilde{M}_1 n + 1) + \tilde{M} \tilde{M}_1 n \left(\widehat{M} H + \overline{M}_0 L M_n \right) \right] \|\phi\|_\mathcal{B}$$

$$+ \widehat{M} L \left[\int_0^t (K_n |z(s)| + \alpha_n) \, ds + \widehat{M} \tilde{M} \tilde{M}_1 n \int_0^n (K_n |z(s)| + \alpha_n) \, ds \right]$$

$$+ \widehat{M} \left[\widehat{M} \tilde{M} \tilde{M}_1 n \int_0^n p(s) \psi \left(K_n |z(s)| + \alpha_n \right) ds + \int_0^t p(s) \psi \left(\frac{K_n}{\lambda} |z(s)| + \alpha_n \right) ds \right].$$

Set

$$c_{14,n} := \alpha_n + \frac{K_n}{1 - \overline{M}_0 L K_n}$$

$$\times \left\{ \overline{M}_0 L \alpha_n \right.$$

$$+ \left[\overline{M}_0 L(\widehat{M} + 1) + \widehat{M} L n \right] (\widehat{M} \tilde{M} \tilde{M}_1 n + 1) + \widehat{M} \tilde{M} \tilde{M}_1 n \left(1 + K_n \overline{M}_0 L \right) |\hat{y}|$$

$$\left. + \widehat{M} \left[\overline{M}_0 L(\widehat{M} \tilde{M} \tilde{M}_1 n + 1) + \tilde{M} \tilde{M}_1 n \left(\widehat{M} H + \overline{M}_0 L M_n \right) \right] \|\phi\|_\mathcal{B} \right\}.$$

Thus

$$\frac{K_n|z(t)|}{\lambda} + \alpha_n \le c_{14,n} + \frac{K_n}{1 - \overline{M}_0 L K_n}$$

$$\times \left\{ \widehat{M} L \left[\int_0^t (K_n|z(s)| + \alpha_n) \, ds + \widehat{M} \tilde{M} \tilde{M}_1 n \int_0^n (K_n|z(s)| + \alpha_n) \, ds \right] \right.$$

$$\left. + \widehat{M} \left[\widehat{M} \tilde{M} \tilde{M}_1 n \int_0^n p(s) \psi \, (K_n|z(s)| + \alpha_n) \, ds + \int_0^t p(s) \psi \left(\frac{K_n|z(s)|}{\lambda} + \alpha_n \right) ds \right] \right\}.$$

By the nondecreasing character of ψ, we get for $\lambda < 1$

$$\frac{K_n|z(t)|}{\lambda} + \alpha_n \le c_{14,n} + \frac{K_n \widehat{M}}{1 - \overline{M}_0 L K_n}$$

$$\times \left\{ L \left[\int_0^t \left(\frac{K_n|z(s)|}{\lambda} + \alpha_n \right) ds + \widehat{M} \tilde{M} \tilde{M}_1 n \int_0^n \left(\frac{K_n}{\lambda}|z(s)| + \alpha_n \right) ds \right] \right.$$

$$\left. + \left[\widehat{M} \tilde{M} \tilde{M}_1 n \int_0^n p(s) \psi \left(\frac{K_n|z(s)|}{\lambda} + \alpha_n \right) ds + \int_0^t p(s) \psi \left(\frac{K_n|z(s)|}{\lambda} + \alpha_n \right) ds \right] \right\}.$$

We consider the function μ defined by

$$\mu(t) := \sup \left\{ \frac{K_n}{\lambda}|z(s)| + \alpha_n : 0 \le s \le t \right\}, \quad 0 \le t < +\infty.$$

Let $t^* \in [0, t]$ be such that $\mu(t) = \frac{K_n}{\lambda}|z(t^*)| + \alpha_n$. If $t^* \in [0, n]$, by the previous inequality, we have for $t \in [0, n]$

$$\mu(t) \le c_{14,n} + \frac{K_n \widehat{M}}{1 - \overline{M}_0 L K_n} \left\{ L \left[\int_0^t \mu(s) ds + \widehat{M} \tilde{M} \tilde{M}_1 n \int_0^n \mu(s) ds \right] \right.$$

$$\left. + \left[\widehat{M} \tilde{M} \tilde{M}_1 n \int_0^n p(s) \psi(\mu(s)) ds + \int_0^t p(s) \psi(\mu(s)) ds \right] \right\}.$$

Then, we have

$$\mu(t) \le c_{14,n} + K_n \widehat{M} \frac{\widehat{M} \tilde{M} \tilde{M}_1 n + 1}{1 - \overline{M}_0 L K_n} \left[L \int_0^n \mu(s) ds + \int_0^n p(s) \psi(\mu(s)) ds \right].$$

Set $\zeta(t) := \max(L, p(t))$ for $t \in [0, n]$, then

$$\mu(t) \le c_{14,n} + K_n \widehat{M} \frac{\widehat{M} \tilde{M} \tilde{M}_1 n + 1}{1 - \overline{M}_0 L K_n} \int_0^n \zeta(s) \left[\mu(s) + \psi(\mu(s)) \right] ds.$$

Consequently,

$$\frac{\|z\|_n}{c_{14,n} + K_n \widehat{M} \dfrac{\widehat{M}\widetilde{M}\widetilde{M}_1 n + 1}{1 - \overline{M}_0 L K_n} [\|z\|_n + \psi(\|z\|_n)]] \|\zeta\|_{L^1}} \leq 1.$$

Then by the condition (3.25), there exists a constant M^* such that $\mu(t) \leq M^*$. Since $\|z\|_n \leq \mu(t)$, we have $\|z\|_n \leq M^*$. This shows that the set $\mathcal{E}$ is bounded, i.e., the statement (S2) in Theorem 1.30 does not hold. Then the nonlinear alternative of Avramescu [32] implies that (S1) holds: i.e., the operator $F + G$ has a fixed-point z^*. Then $y^*(t) = z^*(t) + x(t)$, $t \in \mathbb{R}$ is a fixed point of the operator N_{14}, which is a mild solution of the problem (3.23)–(3.24). Thus the evolution system (3.23)–(3.24) is controllable on $\mathbb{R}$. □

3.7.3 An Example

To illustrate the previous results, we consider the following model

$$\begin{cases} \dfrac{\partial}{\partial t}\left[v(t,\xi) - \displaystyle\int_{-\infty}^{0} T(\theta)w(t, v(t+\theta, \xi))d\theta\right] \\ \qquad = a(t,\xi)\dfrac{\partial^2 v}{\partial \xi^2}(t,\xi) + d(\xi)u(t) \\ \qquad\quad + \displaystyle\int_{-\infty}^{0} P(\theta)r(t, v(t+\theta, \xi))d\theta & t \geq 0 \;\; \xi \geq 0 \qquad (3.26) \\ v(t,0) = v(t,+\infty) = 0 & t \geq 0 \\ v(\theta,\xi) = v_0(\theta,\xi) & \theta \leq 0, \xi \geq 0, \end{cases}$$

where $a(t,\xi)$ is a continuous function and is uniformly Hölder continuous in t ; $T, P : (-\infty, 0] \to \mathbb{R}$; $w, r : \mathbb{R}^+ \times \mathbb{R} \to \mathbb{R}$; $v_0 : (-\infty, 0] \times \mathbb{R}_+ \to \mathbb{R}$ and $d : [0, \pi] \to E$ are continuous functions. $u(\cdot) : \mathbb{R}_+ \to E$ is a given control.

Consider $E = L^2(\mathbb{R}_+, \mathbb{R})$ and define $A(t)$ by $A(t)w = a(t, \xi)w''$ with domain

$$D(A) = \{ w \in E : w, w' \text{ are absolutely continuous, } w'' \in E, w(0) = w(+\infty) = 0 \}$$

Then $A(t)$ generates an evolution system $U(t, s)$ satisfying assumptions (3.1.1) and (G1) (see [112, 149]).

For the phase space $\mathcal{B}$, we choose the well-known space $BUC(\mathbb{R}^-, E)$: the space of uniformly bounded continuous functions endowed with the following norm

$$\|\varphi\| = \sup_{\theta \leq 0} |\varphi(\theta)| \quad \text{for} \quad \varphi \in \mathcal{B}.$$

If we put for $\varphi \in BUC(\mathbb{R}^-, E)$ and $\xi \geq 0$

$$y(t)(\xi) = v(t, \xi), \ t \geq 0, \ \xi \geq 0,$$

$$\phi(\theta)(\xi) = v_0(\theta, \xi), \ \theta \leq 0, \ \xi \geq 0,$$

$$g(t, \varphi)(\xi) = \int_{-\infty}^{0} T(\theta) w(t, \varphi(\theta)(\xi)) d\theta, \ \theta \leq 0, \ \xi \geq 0,$$

and

$$f(t, \varphi)(\xi) = \int_{-\infty}^{0} P(\theta) r(t, \varphi(\theta)(\xi)) d\theta, \ \theta \leq 0, \ \xi \geq 0.$$

Finally let $C \in L(\mathbb{R}, E)$ be defined as

$$Cu(t)(\xi) = d(\xi) u(t), \ t \geq 0, \ \xi \geq 0, \ u \in \mathbb{R}, \ d(\xi) \in E.$$

Then, problem (3.26) takes the abstract neutral functional evolution form (3.23)–(3.24). Furthermore, more appropriate conditions on T, w, P, and r ensure the controllability of mild solutions on $(-\infty, +\infty)$ of the system (3.26) by Theorems 3.20 and 1.30.

3.8   Notes and Remarks

The results of Chap. 3 are taken from [15, 36]. Other results may be found in [108, 141, 145].

Chapter 4
Perturbed Partial Functional Evolution Equations

4.1 Introduction

Perturbed partial functional and neutral functional evolution equations with finite and infinite delay are studied in this chapter on the semi-infinite interval $\mathbb{R}_+$.

4.2 Perturbed Partial Functional Evolution Equations with Finite Delay

4.2.1 Introduction

In this section, we give the existence of mild solutions for the following perturbed partial functional evolution equations with finite delay

$$y'(t) = A(t)y(t) + f(t, y_t) + h(t, y_t), \quad \text{a.e.} \quad t \in J = \mathbb{R}_+ \tag{4.1}$$

$$y(t) = \varphi(t), \quad t \in H, \tag{4.2}$$

where $r > 0$, $f, h : J \times C(H, E) \to E$ and $\varphi \in C(H, E)$ are given functions and $\{A(t)\}_{t \geq 0}$ is a family of linear closed (not necessarily bounded) operators from E into E that generate an evolution system of operators $\{U(t, s)\}_{(t,s) \in J \times J}$ for $0 \leq s \leq t < +\infty$.

Here we are interested to give the existence of mild solutions for the partial functional perturbed evolution equations (4.1)–(4.2). This result is an extension of the problem (2.1) in [33] when the delay is finite.

© Springer International Publishing Switzerland 2015
S. Abbas, M. Benchohra, *Advanced Functional Evolution Equations and Inclusions*, Developments in Mathematics 39,
DOI 10.1007/978-3-319-17768-7_4

4.2.2 Existence of Mild Solutions

Before stating and proving the main result, we give first the definition of mild solution of our perturbed evolution problem (4.1)–(4.2).

Definition 4.1. We say that the continuous function $y(\cdot) : \mathbb{R} \to E$ is a mild solution of (4.1)–(4.2) if $y(t) = \varphi(t)$ for all $t \in H$ and y satisfies the following integral equation

$$y(t) = U(t,0)\,\varphi(0) + \int_0^t U(t,s)\,[f(s,y_s) + h(s,y_s)]\,ds, \qquad \text{for each } t \in \mathbb{R}_+.$$

We introduce the following hypotheses which are assumed hereafter:

(4.1.1) $U(t,s)$ is compact for $t - s > 0$ and there exists a constant $\hat{M} \geq 1$ such that:

$$\|U(t,s)\|_{B(E)} \leq \hat{M} \quad \text{for every } (t,s) \in \Delta.$$

(4.1.2) There exists a function $p \in L^1_{\text{loc}}(J, \mathbb{R}_+)$ and a continuous nondecreasing function $\psi : \mathbb{R}_+ \to (0, \infty)$ such that:

$$|f(t,u)| \leq p(t)\,\psi(\|u\|) \text{ for a.e. } t \in J \text{ and each } u \in C(H,E).$$

(4.1.3) There exists a function $\eta \in L^1(J, \mathbb{R}_+)$ where $\|\eta\|_{L^1} < \dfrac{1}{\hat{M}}$ such that:

$$|h(t,u) - h(t,v)| \leq \eta(t)\|u - v\| \text{ for a.e. } t \in J \text{ and all } u, v \in C(H,E).$$

For every $n \in \mathbb{N}$, we define in $C([-r, +\infty), E)$ the semi-norms by:

$$\|y\|_n := \sup \{ e^{-\tau\, L_n^*(t)} \,|y(t)| : t \in [0,n] \}$$

where $L_n^*(t) = \displaystyle\int_0^t \bar{l}_n(s)\,ds,\ \bar{l}_n(t) = \hat{M}\eta(t)$.

Then $C([-r, +\infty), E)$ is a Fréchet space with the family of semi-norms $\{\|\cdot\|_n\}_{n\in\mathbb{N}}$. In what follows we will choose $\tau > 1$.

Theorem 4.2. *Suppose that hypotheses* (4.1.1)–(4.1.3) *are satisfied and moreover*

$$\int_{c_{15,n}}^{+\infty} \frac{ds}{s + \psi(s)} > \hat{M} \int_0^n \max(p(s), \eta(s))\,ds, \quad \text{for each } n > 0 \qquad (4.3)$$

with

$$c_{15,n} = \hat{M}\|\varphi\| + \hat{M} \int_0^n |h(s,0)|\,ds,$$

then the problem (4.1)–(4.2) *has a mild solution.*

Proof. Transform the problem (4.1)–(4.2) into a fixed point problem. Consider the operator $N_{15} : C([-r, +\infty), E) \to C([-r, +\infty), E)$ defined by:

$$
N_{15}(y)(t) = \begin{cases}
\varphi(t), & \text{if } t \in H; \\
U(t, 0)\,\varphi(0) + \displaystyle\int_0^t U(t, s)\,f(s, y_s)\,ds \\
\quad + \displaystyle\int_0^t U(t, s)\,h(s, y_s)\,ds, & \text{if } t \geq 0.
\end{cases}
$$

Clearly, the fixed points of the operator N_{15} are mild solutions of the problem (4.1)–(4.2).

Define the operators $F, G : C([-r, +\infty), E) \to C([-r, +\infty), E)$ by

$$
(Fy)(t) = \begin{cases}
\varphi(t), & \text{if } t \leq 0; \\
U(t, 0)\,\varphi(0) + \displaystyle\int_0^t U(t, s)\,f(s, y_s)\,ds, & \text{if } t \geq 0.
\end{cases}
$$

and

$$
(Gy)(t) = \int_0^t U(t, s)\,h(s, y_s)\,ds.
$$

Obviously the operator N_{15} has a fixed point is equivalent to $F + G$ has one, so it turns to prove that $F + G$ has a fixed point. The proof will be given in several steps. Let us first show that the operator F is continuous and compact.

Step 1: F is continuous. Let $(y_k)_k$ be a sequence in $C([-r, +\infty), E)$ such that $y_k \to y$ in $C([-r, +\infty), E)$. Then

$$
|F(y_k)(t) - F(y)(t)| \leq \int_0^t \|U(t, s)\|_{B(E)}\,|f(s, y_{k_s}) - f(s, y_s)|\,ds
$$

$$
\leq \hat{M} \int_0^t |f(s, y_{k_s}) - f(s, y_s)|\,ds \to 0 \text{ as } k \to +\infty.
$$

Step 2: F maps bounded sets of $C([-r, +\infty), E)$ into bounded sets. It is enough to show that for any $d > 0$, there exists a positive constant ℓ such that for each $y \in B_d = \{y \in C([-r, +\infty), E) : \|y\|_\infty \leq d\}$ we have $F(y) \in B_\ell$. Let $y \in B_d$. By (4.1.1), (4.1.2) and the nondecreasing character of ψ, we have for each $t \in J$

$$|F(y)(t)| \leq \|U(t,s)\|_{B(E)} |\varphi(0)| + \int_0^t \|U(t,s)\|_{B(E)} |f(s,y_s)| \, ds$$

$$\leq \hat{M}\|\varphi\| + \hat{M} \int_0^t p(s) \, \psi(\|y_s\|) \, ds$$

$$\leq \hat{M}\|\varphi\| + \hat{M} \, \psi(d) \int_0^t p(s) ds.$$

Then we have $[\|F(y)\|_\infty \leq \hat{M}\|\varphi\| + \hat{M} \, \psi(d) \, \|p\|_{L^1} := \ell$. Hence $F(B_d) \subset B_\ell$.

Step 3: F maps bounded sets into equi-continuous sets of $\mathcal{C}([-r, +\infty), E)$. We consider B_d as in Step 2 and we show that $F(B_d)$ is equi-continuous. Let $\tau_1, \tau_2 \in J$ with $\tau_2 > \tau_1$ and $y \in B_d$. Then, by (4.1.1), (4.1.2) and the nondecreasing character of ψ, we get

$$|F(y)(\tau_2) - F(y)(\tau_1)| \leq |U(\tau_2, 0) - U(\tau_1, 0)| \, |\varphi(0)|$$

$$+ \left| \int_0^{\tau_1} [U(\tau_2, s) - U(\tau_1, s)] f(s, y_s) \, ds \right|$$

$$+ \left| \int_{\tau_1}^{\tau_2} U(\tau_2, s) \, |f(s, y_s)| \, ds \right|$$

$$\leq \|U(\tau_2, 0) - U(\tau_1, 0)\|_{B(E)} \, \|\varphi\|$$

$$+ \int_0^{\tau_1} \|U(\tau_2, s) - U(\tau_1, s)\|_{B(E)} \, p(s) \, \psi(\|y_s\|) \, ds$$

$$+ \int_{\tau_1}^{\tau_2} \|U(\tau_2, s)\|_{B(E)} \, p(s) \, \psi(\|y_s\|) ds$$

$$\leq \|U(\tau_2, 0) - U(\tau_1, 0)\|_{B(E)} \, \|\varphi\|$$

$$+ \psi(d) \int_0^{\tau_1} \|U(\tau_2, s) - U(\tau_1, s)\|_{B(E)} \, p(s) \, ds$$

$$+ \hat{M} \, \psi(d) \int_{\tau_1}^{\tau_2} p(s) \, ds.$$

The right-hand of the above inequality tends to zero as $\tau_2 - \tau_1 \to 0$, since $U(t,s)$ is a strongly continuous operator and the compactness of $U(t,s)$ for $t > s$ implies the continuity in the uniform operator topology (see [20, 168]). As a consequence of Steps 1–3 together with the Arzelá–Ascoli theorem it suffices to show that the operator F maps B_d into a precompact set in E.

Let $t \in J$ be fixed and let ϵ be a real number satisfying $0 < \epsilon < t$. For $y \in B_d$ we define

$$F_\epsilon(y)(t) = U(t,0)\,\varphi(0) + \int_0^{t-\epsilon} U(t,s)\,f(s,y_s)\,ds$$

$$= U(t,0)\,\varphi(0) + U(t,t-\epsilon)\int_0^{t-\epsilon} U(t-\epsilon,s)\,f(s,y_s)ds.$$

Since $U(t,s)$ is a compact operator, the set $Z_\epsilon(t) = \{F_\epsilon(y)(t) : \ y \in B_d\}$ is precompact in E for every ϵ, $0 < \epsilon < t$. Moreover by the nondecreasing character of ψ, we get

$$|F(y)(t) - F_\epsilon(y)(t)| \le \int_{t-\epsilon}^t \|U(t,s)\|_{B(E)}|f(s,y_s)|ds$$

$$\le \hat{M}\,\psi(d)\int_{t-\epsilon}^t p(s)ds.$$

Therefore the set $Z(t) = \{F(y)(t) : y \in B_d\}$ is totally bounded. Hence the set $\{F(y)(t) : y \in B_d\}$ is relatively compact E. So we deduce from Steps 1, 2, and 3 that F is a compact operator.

Step 4:   We can show that the operator G is a contraction for all $n \in \mathbb{N}$ as in the proof of Theorem 2.2).

Step 5:   For applying Theorem 1.30, we must check $(S2)$: i.e., it remains to show that the set

$$\mathcal{E} = \left\{ y \in \mathcal{C}([-r,+\infty), E) : y = \lambda F(y) + \lambda G\left(\frac{y}{\lambda}\right) \ \text{for some } 0 < \lambda < 1\right\}$$

is bounded.

Let $y \in \mathcal{E}$. By (4.1.1)–(4.1.3), we have for each $t \in [0,n]$

$$|y(t)| \le \lambda\, U(t,0)\varphi(0) + \lambda \int_0^t \|U(t,s)\|_{B(E)}|f(s,y_s)|ds$$

$$+\lambda \int_0^t \|U(t,s)\|_{B(E)} \left| h\left(s,\frac{y_s}{\lambda}\right) - h(s,0) + h(s,0)\right| ds$$

$$\le \lambda \left\{ \hat{M}\|\varphi\| + \lambda\,\hat{M}\int_0^t p(s)\,\psi\,(\|y_s\|)\,ds \right.$$

$$\left. + \hat{M}\int_0^t \eta(s)\left\|\frac{y_s}{\lambda}\right\| ds + \hat{M}\int_0^t |h(s,0)|ds \right\}.$$

The nondecreasing character of ψ gives with the fact that $0 < \lambda < 1$

$$\frac{|y(t)|}{\lambda} \leq \hat{M}\|\varphi\| + \hat{M} \int_0^n |h(s,0)|ds$$

$$+ \hat{M} \int_0^t \eta(s) \left\|\frac{y_s}{\lambda}\right\| ds + \hat{M} \int_0^t p(s) \, \psi\left(\left\|\frac{y_s}{\lambda}\right\|\right) ds$$

Set $c_{15,n} := \hat{M}\|\varphi\| + \hat{M} \int_0^n |h(s,0)|ds$. Thus

$$\frac{|y(t)|}{\lambda} \leq c_{15,n} + \hat{M} \int_0^t \eta(s) \left\|\frac{y_s}{\lambda}\right\| ds + \hat{M} \int_0^t p(s) \, \psi\left(\left\|\frac{y_s}{\lambda}\right\|\right) ds.$$

Consider the function μ defined by

$$\mu(t) := \sup\left\{\frac{|y(s)|}{\lambda} : 0 \leq s \leq t\right\}, \quad 0 \leq t < +\infty.$$

Let $t^* \in [0,t]$ be such that $\mu(t) = \dfrac{|y(t^*)|}{\lambda}$. By the previous inequality, we have

$$\mu(t) \leq c_{15,n} + \hat{M} \int_0^t \eta(s)\mu(s)ds + \hat{M} \int_0^t p(s)\psi(\mu(s))ds, \quad \text{for } t \in [0,n].$$

Let us take the right-hand side of the above inequality as $v(t)$. Then, we have

$$\mu(t) \leq v(t) \text{ for all } t \in [0,n].$$

From the definition of v, we have

$$v(0) = c_{15,n} \quad \text{and} \quad v'(t) = \hat{M}\eta(t)\mu(t) + \hat{M}p(t)\psi(\mu(t)) \text{ a.e. } t \in [0,n].$$

Using the nondecreasing character of ψ, we get

$$v'(t) \leq \hat{M}p(t)\psi(v(t)) + \hat{M}\eta(t)v(t) \text{ a.e. } t \in [0,n].$$

This implies that for each $t \in [0,n]$ and using (4.3), we get

$$\int_{c_{15,n}}^{v(t)} \frac{ds}{s + \psi(s)} \leq \hat{M} \int_0^t \max(p(s),\eta(s))ds$$

$$\leq \hat{M} \int_0^n \max(p(s),\eta(s))ds$$

$$< \int_{c_{15,n}}^{+\infty} \frac{ds}{s + \psi(s)}.$$

Thus, for every $t \in [0, n]$, there exists a constant Λ_n such that $v(t) \leq \Lambda_n$ and hence $\mu(t) \leq \Lambda_n$. Since $\|y\|_n \leq \mu(t)$, we have $\|y\|_n \leq \Lambda_n$. This shows that the set $\mathcal{E}$ is bounded. Then statement $(S2)$ in Theorem 1.30 does not hold. The nonlinear alternative of Avramescu implies that $(S1)$ holds, we deduce that the operator $F + G$ has a fixed point y^* the fixed point of the operator N_{15}, which is a mild solution of the problem (4.1)–(4.2). □

4.2.3   An Example

As an application of Theorem 4.2, we present the following partial functional differential equation

$$
\begin{cases}
\dfrac{\partial z}{\partial t}(t, x) = a(t, x)\dfrac{\partial^2 z}{\partial x^2}(t, x) \\
\qquad\qquad +Q(t, z(t - r, x)) + P(t, z(t - r, x)) \ t \in [0, +\infty), x \in [0, \pi] \\[2mm]
z(t, 0) = z(t, \pi) = 0 \qquad\qquad\qquad t \in [0, +\infty) \\[2mm]
z(t, x) = \Phi(t, x) \qquad\qquad\qquad\quad t \in H, x \in [0, \pi],
\end{cases}
$$

(4.4)

where $a(t, x) : [0, \infty) \times [0, \pi] \to \mathbb{R}$ is a continuous function and is uniformly Hölder continuous in t, $Q, P : [0, +\infty) \times \mathbb{R} \to \mathbb{R}$ and $\Phi : H \times [0, \pi] \to \mathbb{R}$ are continuous functions.

Consider $E = L^2([0, \pi], \mathbb{R})$ and define $A(t)$ by $A(t)w = a(t, x)w''$ with domain

$$
D(A) = \{w \in E : w, w' \text{ are absolutely continuous, } w'' \in E, \ w(0) = w(\pi) = 0 \}
$$

Then $A(t)$ generates an evolution system $U(t, s)$ satisfying assumption (4.1.1) (see [112, 149]).

For $x \in [0, \pi]$, we set

$$
y(t)(x) = z(t, x), \quad t \in \mathbb{R}_+,
$$

$$
f(t, y_t)(x) = Q(t, z(t - r, x)), \quad t \in \mathbb{R}_+
$$

$$
h(t, y_t)(x) = P(t, z(t - r, x)), \quad t \in \mathbb{R}_+
$$

and

$$
\varphi(t)(x) = \Phi(t, x), \quad -r \leq t \leq 0.
$$

Thus, under the above definitions of f, h, φ, and $A(\cdot)$, the system (4.4) can be represented by the abstract evolution problem (4.1)–(4.2). Furthermore, more appropriate conditions on Q and P ensure the existence of mild solutions for (4.4) by Theorems 4.2 and 1.30.

4.3 Perturbed Neutral Functional Evolution Equations with Finite Delay

4.3.1 Introduction

In this section, we consider the following perturbed neutral functional evolution equations with finite delay

$$\frac{d}{dt}[y(t) - g(t, y_t)] = A(t)y(t) + f(t, y_t) + h(t, y_t), \quad \text{a.e. } t \in J = \mathbb{R}_+ \quad (4.5)$$

$$y(t) = \varphi(t), \quad t \in H, \tag{4.6}$$

where $r > 0$, $A(\cdot), f, h$, and φ are as in problem (4.1)–(4.2) and $g : J \times C(H, E) \to E$ is a given function.

Here we are interested to give the existence of mild solutions for the perturbed neutral functional evolution equations (4.5)–(4.6). This result is an extension of the problem (4.1) for the neutral case.

4.3.2 Existence of Mild Solutions

In this section, we give an existence result for the perturbed neutral functional evolution problem (4.5)–(4.6). Firstly we define the mild solution.

Definition 4.3. We say that the continuous function $y(\cdot) : \mathbb{R} \to E$ is a mild solution of (4.5)–(4.6) if $y(t) = \varphi(t)$ for all $t \in H$ and y satisfies the following integral equation

$$y(t) = U(t, 0)[\varphi(0) - g(0, \varphi)] + g(t, y_t) + \int_0^t U(t, s)A(s)g(s, y_s)ds$$

$$+ \int_0^t U(t, s)[f(s, y_s) + h(s, y_s)]\, ds, \quad \text{for each } t \in \mathbb{R}_+.$$

We consider the hypotheses (4.1.1)–(4.1.3) and in what follows we will need the following additional assumptions:

(G1)   There exists a constant $\overline{M}_0 > 0$ such that:

$$\|A^{-1}(t)\|_{B(E)} \leq \overline{M}_0 \quad for\ all\ t \in J.$$

(G2)   There exists a constant $0 < L < \dfrac{1}{\overline{M}_0}$ such that:

$$|A(t)\,g(t,\varphi)| \leq L\,(\|\varphi\| + 1) \text{ for all } t \in J \text{ and } \varphi \in H.$$

(G3)   There exists a constant $L_* > 0$ such that:

$$|A(s)\,g(s,\varphi) - A(\bar{s})\,g(\bar{s},\overline{\varphi})| \leq L_*\,(|s - \bar{s}| + \|\varphi - \overline{\varphi}\|)$$

for all $s, \bar{s} \in J$ and $\varphi, \overline{\varphi} \in H$.

For every $n \in \mathbb{N}$, we define in $C([-r, +\infty), E)$ the semi-norms by:

$$\|y\|_n := \sup\{\, e^{-\tau\,L_n^*(t)}\,|y(t)| : t \in [0, n]\,\}$$

where $L_n^*(t) = \displaystyle\int_0^t \bar{l}_n(s)\,ds$, $\bar{l}_n(t) = K_n\hat{M}[L_* + \eta(t)]$.

Then $C([-r, +\infty), E)$ is a Fréchet space with the family of semi-norms $\{\|\cdot\|_n\}_{n\in\mathbb{N}}$. Let us fix $\tau > 0$ and assume that $\left[\overline{M}_0 L_* K_n + \dfrac{1}{\tau}\right] < 1$.

Theorem 4.4. *Suppose that hypotheses* (4.1.1)–(4.1.3) *and the assumptions* (G1)– (G3) *are satisfied and moreover*

$$\int_{c_{16,n}}^{+\infty} \frac{ds}{s + \psi(s)} > \frac{\hat{M}}{1 - \overline{M}_0 L} \int_0^n \max(L, \eta(s), p(s))ds, \quad for\ each\ n > 0 \quad (4.7)$$

with

$$c_{16,n} := \frac{1}{1 - \overline{M}_0 L}\left[\overline{M}_0 L(\hat{M} + 1) + \hat{M}Ln + \hat{M}\overline{M}_0 L\|\varphi\| + \hat{M}\int_0^t |h(s,0)|\,ds\right].$$

Then the problem (4.5)–(4.6) *has a mild solution.*

Proof. Transform the problem (4.5)–(4.6) into a fixed point problem. Consider the operator $N_{16} : C([-r, +\infty), E) \to C([-r, +\infty), E)$ defined by:

$$N_{16}(y)(t) = \begin{cases} \varphi(t), & \text{if } t \le 0; \\ \\ U(t,0)\,[\varphi(0) - g(0,\varphi)] + g(t,y_t) \\ + \displaystyle\int_0^t U(t,s)A(s)g(s,y_s)ds \\ + \displaystyle\int_0^t U(t,s)[f(s,y_s) + h(s,y_s)]ds, & \text{if } t \ge 0. \end{cases}$$

Clearly, the fixed points of the operator N_{16} are mild solutions of the problem (4.5)–(4.6).

Define the operators $F, G : \mathcal{C}([-r, +\infty), E) \to \mathcal{C}([-r, +\infty), E)$ by

$$F(y)(t) = \begin{cases} \varphi(t), & \text{if } t \le 0; \\ \\ U(t,0)\,\varphi(0) + \displaystyle\int_0^t U(t,s)\,f(s,y_s)\,ds, & \text{if } t \ge 0. \end{cases}$$

and

$$G(y)(t) = g(t,y_t) - U(t,0)g(0,\varphi) + \int_0^t U(t,s)A(s)g(s,y_s)ds$$

$$+ \int_0^t U(t,s)\,h(s,y_s)\,ds.$$

Obviously the operator N_{16} has a fixed point is equivalent to $F + G$ has one, so it turns to prove that $F + G$ has a fixed point. The proof will be given in several steps.

We can show that the operator F is continuous and compact. We can prove also that the operator G is a contraction for all $n \in \mathbb{N}$ as in the proof of Theorem 2.6.

For applying Theorem 1.30, we must check (S2): i.e., it remains to show that the set

$$\mathcal{E} = \left\{ y \in \mathcal{C}([-r, +\infty), E) : y = \lambda F(y) + \lambda G\left(\frac{y}{\lambda}\right) \text{ for some } 0 < \lambda < 1 \right\}$$

is bounded.

Let $y \in \mathcal{E}$. Then, we have

$$|y(t)| \le \lambda \int_0^t \|U(t,s)\|_{B(E)} |f(s,y_s)|\,ds$$

$$+ \lambda \left\{ \left| g\left(t, \frac{y_t}{\lambda}\right) \right| + \|U(t,0)\|_{B(E)} |g(0,\varphi)| \right.$$

$$+ \int_0^t \|U(t,s)\|_{B(E)} \left| A(s)g\left(s, \frac{y_s}{\lambda}\right) \right| ds$$

$$+ \left. \int_0^t \|U(t,s)\|_{B(E)} \left| h\left(s, \frac{y_s}{\lambda}\right) - h(s,0) + h(s,0) \right| ds \right\}.$$

By the hypotheses (4.1.1)–(4.1.3), $(G1)$, and $(G2)$ we obtain

$$\frac{|y(t)|}{\lambda} \leq \hat{M} \int_0^t f(s, y_s)\, ds + \|A^{-1}(s)\| \left|A(t)g\left(t, \frac{y_t}{\lambda}\right)\right| + \hat{M}\|A^{-1}(s)\| |A(t)g(0, \varphi)|$$

$$+\hat{M} \int_0^t \left|A(s)g\left(s, \frac{y_s}{\lambda}\right)\right|\, ds + \hat{M}\int_0^t \left|h\left(s, \frac{y_s}{\lambda}\right) - h(s, 0)\right|\, ds$$

$$+\hat{M} \int_0^t |h(s, 0)|\, ds$$

$$\leq \hat{M} \int_0^t p(s)\psi(\|y_s\|)\, ds + \overline{M}_0 L \left(\left\|\frac{y_t}{\lambda}\right\| + 1\right) + \hat{M}\overline{M}_0 L(\|\varphi\| + 1)$$

$$+\hat{M} L \int_0^t \left(\left\|\frac{y_s}{\lambda}\right\| + 1\right)\, ds + \hat{M}\int_0^t \eta(s)\left\|\frac{y_s}{\lambda}\right\|\, ds + \hat{M}\int_0^t |h(s, 0)|\, ds$$

$$\leq \hat{M} \int_0^t p(s)\psi(\|y_s\|)\, ds + \overline{M}_0 L \left\|\frac{y_t}{\lambda}\right\| + \overline{M}_0 L(\hat{M} + 1) + \hat{M}Ln + \hat{M}\overline{M}_0 L\|\varphi\|$$

$$+\hat{M} \int_0^t |h(s, 0)|\, ds + \hat{M}\int_0^t L\left\|\frac{y_s}{\lambda}\right\|\, ds + \hat{M}\int_0^t \eta(s)\left\|\frac{y_s}{\lambda}\right\|\, ds.$$

The nondecreasing character of ψ gives with the fact that $0 < \lambda < 1$

$$\frac{|y(t)|}{\lambda} \leq \overline{M}_0 L \left\|\frac{y_t}{\lambda}\right\| + \overline{M}_0 L(\hat{M} + 1) + \hat{M}Ln + \hat{M}\overline{M}_0 L\|\varphi\| + \hat{M}\int_0^t |h(s, 0)|\, ds$$

$$+\hat{M} \int_0^t L\left\|\frac{y_s}{\lambda}\right\|\, ds + \hat{M}\int_0^t \eta(s)\left\|\frac{y_s}{\lambda}\right\|\, ds + \hat{M}\int_0^t p(s)\psi\left(\left\|\frac{y_s}{\lambda}\right\|\right)\, ds.$$

Consider the function μ defined by

$$\mu(t) := \sup\left\{\frac{|y(s)|}{\lambda} : 0 \leq s \leq t\right\}, \quad 0 \leq t < +\infty.$$

Let $t^* \in [0, t]$ be such that $\mu(t) = \dfrac{|y(t^*)|}{\lambda}$, by the previous inequality, we have

$$\left(1 - \overline{M}_0 L\right)\mu(t) \leq \overline{M}_0 L(\hat{M} + 1) + \hat{M}Ln + \hat{M}\overline{M}_0 L\|\varphi\| + \hat{M}\int_0^t |h(s, 0)|\, ds$$

$$+ \hat{M}\int_0^t L\mu(s)\, ds + \hat{M}\int_0^t \eta(s)\mu(s)\, ds$$

$$+ \hat{M}\int_0^t p(s)\psi(\mu(s))\, ds.$$

Set $c_{16,n} := \dfrac{1}{1 - \overline{M}_0 L} \left[\overline{M}_0 L(\hat{M} + 1) + \hat{M} L n + \hat{M} \overline{M}_0 L \|\varphi\| + \hat{M} \int_0^t |h(s,0)| \, ds \right]$.

Thus, for each $t \in [0, n]$ we get

$$\mu(t) \leq c_{16,n} + \frac{\hat{M}}{1 - \overline{M}_0 L} \left[\int_0^t L\mu(s) \, ds + \int_0^t \eta(s)\mu(s) \, ds + \int_0^t p(s)\psi\,(\mu(s)) \, ds \right].$$

Let us take the right-hand side of the above inequality as $v(t)$. Thus, we have

$$\mu(t) \leq v(t) \text{ for all } t \in [0, n].$$

From the definition of v, we have

$$v(0) = c_{16,n} \quad \text{and} \quad v'(t) = \frac{\hat{M}}{1 - \overline{M}_0 L} [L\mu(t) + \eta(t)\mu(t) + p(t)\psi(\mu(t))]$$

$$\text{a.e. } t \in [0, n].$$

Using the nondecreasing character of ψ, we get

$$v'(t) \leq \frac{\hat{M}}{1 - \overline{M}_0 L} [Lv(t) + \eta(t)v(t) + p(t)\psi(v(t))] \quad \text{a.e. } t \in [0, n].$$

This implies that for each $t \in [0, n]$ and using the condition (4.7), we get

$$\int_{c_{16,n}}^{v(t)} \frac{ds,}{s + \psi(s)} \leq \frac{\hat{M}}{1 - \overline{M}_0 L} \int_0^t \max(L, \eta(s), p(s)) ds$$

$$\leq \frac{\hat{M}}{1 - \overline{M}_0 L} \int_0^n \max(L, \eta(s), p(s)) ds$$

$$< \int_{c_{16,n}}^{+\infty} \frac{ds}{s + \psi(s)}.$$

Thus, for every $t \in [0, n]$, there exists a constant Λ_n such that $v(t) \leq \Lambda_n$ and hence $\mu(t) \leq \Lambda_n$. Since $\|y\|_n \leq \mu(t)$, we have $\|y\|_n \leq \Lambda_n$. This shows that the set $\mathcal{E}$ is bounded. Then the statement (S2) in Theorem 1.30 does not hold. A consequence of the nonlinear alternative of Avramescu that (S1) holds, we deduce that the operator $F + G$ has a fixed point y^*. Then $y^*(t) = y^*(t) + x(t), t \in [-r, +\infty)$ is a fixed point of the operator N_{16}, which is the mild solution of the problem (4.5)–(4.6).   □

4.3.3   An Example

Consider the following model

$$
\begin{cases}
\dfrac{\partial}{\partial t}\left[z(t,x) - \displaystyle\int_{-r}^{t}\int_{0}^{\pi} b(s-t,u,x)\,z(s,u)\,du\,ds\right] = a(t,x)\dfrac{\partial^{2} z}{\partial x^{2}}(t,x) \\
\qquad\qquad +Q\left(t, z(t-r,x), \dfrac{\partial z}{\partial x}(t-r,x)\right) \\
\qquad\qquad +P\left(t, z(t-r,x), \dfrac{\partial z}{\partial x}(t-r,x)\right), \quad t\in[0,+\infty), x\in[0,\pi] \\[2mm]
z(t,0) = z(t,\pi) = 0, \qquad\qquad\qquad\qquad\qquad t\in[0,+\infty) \\[2mm]
z(t,x) = \Phi(t,x), \qquad\qquad\qquad\qquad\qquad\quad t\in H, x\in[0,\pi]
\end{cases}
$$
$$(4.8)$$

where $r>0$; $a(t,x)$ is a continuous function and is uniformly Hölder continuous in t, $Q,P:[0,+\infty)\times\mathbb{R}\times\mathbb{R}\to\mathbb{R}$ and $\Phi:H\times[0,\pi]\to\mathbb{R}$ are continuous functions.
Let

$$y(t)(x) = z(t,x),\ t\in[0,\infty),\ x\in[0,\pi],$$

$$g(t,y_t)(x) = \int_{-r}^{t}\int_{0}^{\pi} b(s-t,u,x)z(s,u)\,du\,ds,\ x\in[0,\pi],$$

$$f(t,y_t)(x) = Q\left(t, z(\theta,x), \frac{\partial z}{\partial x}(\theta,x)\right),\ \theta\in H,\ x\in[0,\pi],\ t\ge 0,$$

$$h(t,y_t)(x) = P\left(t, z(\theta,x), \frac{\partial z}{\partial x}(\theta,x)\right),\ \theta\in H,\ x\in[0,\pi],\ t\ge 0$$

and

$$\varphi(\theta)(x) = \Phi(\theta,x),\ \theta\in H,\ x\in[0,\pi].$$

Consider $E = L^{2}([0,\pi],\mathbb{R})$ and define $A(t)$ by $A(t)w = a(t,x)w''$ with domain

$$D(A) = \{w\in E : w, w' \text{ are absolutely continuous}, w''\in E,\ w(0) = w(\pi) = 0\}.$$

Then $A(t)$ generates an evolution system $U(t,s)$ satisfying assumptions (4.1.1) and (G1) (see [112, 149]).
   Here we assume that $\varphi:H\to E$ is Lebesgue measurable and $h(s)|\varphi(s)|^{2}$ is Lebesgue integrable on H where $h:H\to\mathbb{R}$ is a positive integrable function. The norm is defined here by:

$$\|\varphi\| = |\Phi(0)| + \left(\int_{-r}^{0} h(s)|\varphi(s)|^{2}\,ds\right)^{\frac{1}{2}}.$$

The function b is measurable on $[0, \infty) \times [0, \pi] \times [0, \pi]$,

$$b(s, u, 0) = b(s, u, \pi) = 0, \quad (s, u) \in [0, \infty) \times [0, \pi],$$

$$\int_0^\pi \int_{-r}^t \int_0^\pi \frac{b^2(s, u, x)}{h(s)} \, ds \, du \, dx < \infty$$

and $\sup_{t \in [0, \infty)} \mathcal{N}(t) < \infty$, where

$$\mathcal{N}(t) = \int_0^\pi \int_{-r}^t \int_0^\pi \frac{1}{h(s)} \left(a(s, x) \frac{\partial^2}{\partial x^2} b(s, u, x) \right)^2 ds \, du \, dx.$$

Thus, under the above definitions of f, g, h, φ, and $A(\cdot)$, the system (4.8) can be represented by the abstract evolution problem (4.5)–(4.6). Furthermore, more appropriate conditions on Q and P ensure the existence of mild solutions for (4.8) by Theorems 4.4 and 1.30.

4.4 Perturbed Partial Functional Evolution Equations with Infinite Delay

4.4.1 Introduction

The existence of mild solutions is studied here for the following perturbed partial functional evolution equations with infinite delay

$$y'(t) = A(t)y(t) + f(t, y_t) + h(t, y_t), \quad \text{a.e.} \quad t \in J \tag{4.9}$$

$$y_0 = \phi \in \mathcal{B}, \tag{4.10}$$

where $f, h : J \times \mathcal{B} \to E$ and $\phi \in \mathcal{B}$ are given functions and $\{A(t)\}_{0 \leq t < +\infty}$ is a family of linear closed (not necessarily bounded) operators from E into E that generate an evolution system of operators $\{U(t, s)\}_{(t,s) \in J \times J}$ for $0 \leq s \leq t < +\infty$.

4.4.2 Existence of Mild Solutions

Before stating and proving the main result, we give first the definition of mild solution of our perturbed evolution problem (4.9)–(4.10).

Definition 4.5. We say that the continuous function $y(\cdot) : \mathbb{R} \to E$ is a mild solution of (4.9)–(4.10) if $y(t) = \phi(t)$ for all $t \in (-\infty, 0]$ and y satisfies the following integral equation

$$y(t) = U(t, 0)\, \phi(0) + \int_0^t U(t, s)\, [f(s, y_s) + h(s, y_s)]\, ds, \quad \text{for each } t \in \mathbb{R}_+.$$

We introduce the following hypotheses which are assumed hereafter:

(4.1.1) $U(t, s)$ is compact for $t - s > 0$ and there exists a constant $\hat{M} \geq 1$ such that:

$$\|U(t, s)\|_{B(E)} \leq \hat{M} \quad \text{for every } (t, s) \in \Delta.$$

(4.1.2) There exists a function $p \in L^1_{\mathrm{loc}}(J, \mathbb{R}_+)$ and a continuous nondecreasing function $\psi : \mathbb{R}_+ \to (0, \infty)$ and such that:

$$|f(t, u)| \leq p(t)\, \psi(\|u\|_B) \text{ for a.e. } t \in J \text{ and each } u \in \mathcal{B}.$$

(4.1.3) There exists a function $\eta \in L^1(J, \mathbb{R}_+)$ where $\|\eta\|_{L^1} < \dfrac{1}{\hat{M}}$ such that:

$$|h(t, u) - h(t, v)| \leq \eta(t)\|u - v\|_B \text{ for a.e. } t \in J \text{ and all } u, v \in \mathcal{B}.$$

Consider the following space

$$B_{+\infty} = \left\{ y : \mathbb{R} \to E : y|_{[0,T]} \in C([0, T], E), \ y_0 \in \mathcal{B} \right\},$$

where $y|_{[0,T]}$ is the restriction of y to any real compact interval $[0, T]$.

For every $n \in \mathbb{N}$, we define in $B_{+\infty}$ the semi-norms by:

$$\|y\|_n := \sup \left\{ e^{-\tau\, L_n^*(t)}\, |y(t)| : t \in [0, n] \right\}$$

where $L_n^*(t) = \displaystyle\int_0^t \bar{l}_n(s)\, ds$ and $\bar{l}_n(t) = K_n \hat{M} \eta(t)$.

Then $B_{+\infty}$ is a Fréchet space with the family of semi-norms $\| \cdot \|_{n \in \mathbb{N}}$. In what follows let us fix $\tau > 1$.

Theorem 4.6. *Suppose that hypotheses (4.1.1)–(4.1.3) are satisfied and moreover*

$$\int_{c_{17,n}}^{+\infty} \frac{ds}{s + \psi(s)} > K_n \hat{M} \int_0^n \max(p(s), \eta(s))\, ds, \quad \text{for each } n > 0 \qquad (4.11)$$

with

$$c_{17,n} = (K_n \hat{M} H + M_n)\|\phi\|_{\mathcal{B}} + K_n \hat{M} \int_0^n |h(s, 0)|\, ds,$$

then the problem (4.9)–(4.10) has a mild solution.

Proof. Consider the operator $N_{17} : B_{+\infty} \to B_{+\infty}$ defined by:

$$
N_{17}(y)(t) = \begin{cases} \phi(t), & \text{if } t \leq 0; \\[2mm] U(t,0)\,\phi(0) + \displaystyle\int_0^t U(t,s)\,f(s,y_s)\,ds \\[2mm] \quad + \displaystyle\int_0^t U(t,s)\,h(s,y_s)\,ds, & \text{if } t \geq 0. \end{cases}
$$

Clearly, the fixed points of the operator N_{17} are mild solutions of the problem (4.9)–(4.10).

For $\phi \in B$, we will define the function $x(.) : \mathbb{R} \to E$ by

$$
x(t) = \begin{cases} \phi(t), & \text{if } t \in (-\infty, 0]; \\[2mm] U(t,0)\,\phi(0), & \text{if } t \in J. \end{cases}
$$

Then $x_0 = \phi$. For each function $z \in B_{+\infty}$, set $y(t) = z(t) + x(t)$. It is obvious that y satisfies Definition 4.5 if and only if z satisfies $z_0 = 0$ and

$$
z(t) = \int_0^t U(t,s)\,f(s, z_s + x_s)\,ds + \int_0^t U(t,s)\,h(s, z_s + x_s)\,ds, \qquad \text{for } t \in J.
$$

Let $B_{+\infty}^0 = \{z \in B_{+\infty} : z_0 = 0\}$. Define the operators $F, G : B_{+\infty}^0 \to B_{+\infty}^0$ by

$$
F(z)(t) = \int_0^t U(t,s)\,f(s, z_s + x_s)\,ds, \qquad \text{for } t \in J
$$

and

$$
G(z)(t) = \int_0^t U(t,s)\,h(s, z_s + x_s)\,ds, \qquad \text{for } t \in J.
$$

Obviously the operator N_{17} has a fixed point is equivalent to $F + G$ has one, so it turns to prove that $F + G$ has a fixed point. The proof will be given in several steps.

We can show that the operator F is continuous and compact. We can prove also that the operator G is a contraction for all $n \in \mathbb{N}$ as in the proof of Theorem 3.2).

For applying Theorem 1.30, we must check (S2): i.e., it remains to show that the set

$$
\mathcal{E} = \left\{ z \in B_{+\infty}^0 : z = \lambda F(z) + \lambda G\left(\frac{z}{\lambda}\right) \text{ for some } 0 < \lambda < 1 \right\}
$$

is bounded.

Let $z \in \mathcal{E}$. By (4.1.1)–(4.1.3), we have for each $t \in [0, n]$

$$
\begin{aligned}
|z(t)| \leq{}& \lambda \int_0^t \|U(t, s)\|_{B(E)} |f(s, z_s + x_s)| ds \\
&+ \lambda \int_0^t \|U(t, s)\|_{B(E)} \left| h\left(s, \frac{z_s}{\lambda} + x_s \right) - h(s, 0) + h(s, 0) \right| ds \\
\leq{}& \lambda \hat{M} \int_0^t p(s) \psi \left(\|z_s + x_s\|_B \right) ds \\
&+ \lambda \hat{M} \int_0^t \eta(s) \left\| \frac{z_s}{\lambda} + x_s \right\|_B ds + \lambda \hat{M} \int_0^t |h(s, 0)| ds.
\end{aligned}
$$

Using the inequality (3.4) and the nondecreasing character of ψ, we get

$$
\begin{aligned}
\frac{1}{\lambda} |z(t)| \leq{}& \hat{M} \int_0^t p(s) \psi (K_n |z(s)| + \alpha_n) ds \\
&+ \hat{M} \int_0^t \eta(s) \left(\frac{K_n}{\lambda} |z(s)| + \alpha_n \right) ds + \hat{M} \int_0^t |h(s, 0)| ds.
\end{aligned}
$$

The nondecreasing character of ψ gives with the fact that $0 < \lambda < 1$

$$
\begin{aligned}
\frac{K_n}{\lambda} |z(t)| + \alpha_n \leq{}& \alpha_n + K_n \hat{M} \int_0^t |h(s, 0)| ds + K_n \hat{M} \int_0^t p(s) \psi \left(\frac{K_n}{\lambda} |z(s)| + \alpha_n \right) ds \\
&+ K_n \hat{M} \int_0^t \eta(s) \left(\frac{K_n}{\lambda} |z(s)| + \alpha_n \right) ds.
\end{aligned}
$$

Set $c_{17,n} := K_n \hat{M} \int_0^t |h(s, 0)| ds + \alpha_n$, thus

$$
\begin{aligned}
\frac{K_n}{\lambda} |z(t)| + \alpha_n \leq{}& c_{17,n} + K_n \hat{M} \int_0^t p(s) \psi \left(\frac{K_n}{\lambda} |z(s)| + \alpha_n \right) ds \\
&+ K_n \hat{M} \int_0^t \eta(s) \left(\frac{K_n}{\lambda} |z(s)| + \alpha_n \right) ds.
\end{aligned}
$$

We consider the function μ defined by

$$
\mu(t) := \sup \left\{ \frac{K_n}{\lambda} |z(s)| + \alpha_n \ : \ 0 \leq s \leq t \right\}, \quad 0 \leq t < +\infty.
$$

Let $t^* \in [0, t]$ be such that $\mu(t) = \dfrac{K_n}{\lambda} |z(t^*)| + \alpha_n$, by the previous inequality, we have

$$\mu(t) \le c_{17,n} + K_n \hat{M} \int_0^t p(s) \psi(\mu(s)) ds + K_n \hat{M} \int_0^t \eta(s) \mu(s) ds, \quad \text{for} \quad t \in [0, n]$$

Let us take the right-hand side of the above inequality as $v(t)$. Then, we have

$$\mu(t) \le v(t) \text{ for all } t \in [0, n].$$

From the definition of v, we have

$$v(0) = c_{17,n} \quad \text{and} \quad v'(t) = K_n \hat{M} p(t) \psi(\mu(t)) + K_n \hat{M} \eta(t) \mu(t) \quad \text{a.e. } t \in [0, n].$$

Using the nondecreasing character of ψ, we get

$$v'(t) \le K_n \hat{M} p(t) \psi(v(t)) + K_n \hat{M} \eta(t) v(t) \quad \text{a.e. } t \in [0, n].$$

This implies that for each $t \in [0, n]$ and using (4.11), we get

$$\int_{c_{17,n}}^{v(t)} \frac{ds}{s + \psi(s)} \le K_n \hat{M} \int_0^t \max(p(s), \eta(s)) ds$$

$$\le K_n \hat{M} \int_0^n \max(p(s), \eta(s)) ds$$

$$< \int_{c_{17,n}}^{+\infty} \frac{ds}{s + \psi(s)}.$$

Thus, for every $t \in [0, n]$, there exists a constant Λ_n such that $v(t) \le \Lambda_n$ and hence $\mu(t) \le \Lambda_n$. Since $\|z\|_n \le \mu(t)$, we have $\|z\|_n \le \Lambda_n$. This shows that the set $\mathcal{E}$ is bounded. Then statement $(S2)$ in Theorem 1.30 does not hold. The nonlinear alternative of Avramescu implies that $(S1)$ holds, we deduce that the operator $F + G$ has a fixed point z^*. Then $y^*(t) = z^*(t) + x(t)$, $t \in \mathbb{R}$ is a fixed point of the operator N_{17}, which is the mild solution of the problem (4.9)–(4.10). $\quad \square$

4.4.3 An Example

Consider the following model

$$
\begin{cases}
\dfrac{\partial v}{\partial t}(t,\xi) = a(t,\xi)\dfrac{\partial^2 v}{\partial \xi^2}(t,\xi) \\[2mm]
\qquad + \displaystyle\int_{-\infty}^{0} P(\theta)r(t,v(t+\theta,\xi))d\theta \\[2mm]
\qquad + \displaystyle\int_{-\infty}^{0} Q(\theta)s(t,v(t+\theta,\xi))d\theta \qquad t \in \mathbb{R}_+, \quad \xi \in [0,\pi] \\[4mm]
v(t,0) = v(t,\pi) = 0 \qquad\qquad\qquad t \in \mathbb{R}_+ \\[2mm]
v(\theta,\xi) = v_0(\theta,\xi) \qquad\qquad\qquad -\infty < \theta \le 0, \, \xi \in [0,\pi],
\end{cases}
$$
$$(4.12)$$

where $a(t,\xi)$ is a continuous function and is uniformly Hölder continuous in t; $P, Q :$ $(-\infty, 0] \to \mathbb{R}$; $r, s : (-\infty, 0] \times \mathbb{R} \to \mathbb{R}$ and $v_0 : (-\infty, 0] \times [0,\pi] \to \mathbb{R}$ are continuous functions.

Consider $E = L^2([0,\pi], \mathbb{R})$ and define $A(t)$ by $A(t)w = a(t,\xi)w''$ with domain

$$D(A) = \{\, w \in E \; : \; w, w' \text{ are absolutely continuous, } w'' \in E, \, w(0) = w(\pi) = 0 \,\}$$

Then $A(t)$ generates an evolution system $U(t,s)$ satisfying assumption (4.1.1) (see [112, 149]).

For the phase space $\mathcal{B}$, we choose the well-known space $BUC(\mathbb{R}^-, E)$: the space of uniformly bounded continuous functions endowed with the following norm

$$\|\varphi\| = \sup_{\theta \le 0} |\varphi(\theta)| \quad \text{for} \quad \varphi \in \mathcal{B}.$$

If we put for $\varphi \in BUC(\mathbb{R}^-, E)$ and $\xi \in [0,\pi]$

$$y(t)(\xi) = v(t,\xi), \, t \in \mathbb{R}_+, \, \xi \in [0,\pi],$$

$$\phi(\theta)(\xi) = v_0(\theta,\xi), \, -\infty < \theta \le 0, \, \xi \in [0,\pi],$$

$$f(t,\varphi)(\xi) = \int_{-\infty}^{0} P(\theta)r(t,\varphi(\theta)(\xi))d\theta, \, -\infty < \theta \le 0, \, \xi \in [0,\pi]$$

and

$$h(t,\varphi)(\xi) = \int_{-\infty}^{0} Q(\theta)s(t,\varphi(\theta)(\xi))d\theta, \, -\infty < \theta \le 0, \, \xi \in [0,\pi].$$

Then, the problem (4.12) takes the abstract partial perturbed evolution form (4.9)–(4.10). In order to show the existence of mild solutions of problem (4.12), we suppose the following assumptions:

- The function s is Lipschitz continuous with respect to its second argument. Let $lip(s)$ denote the Lipschitz constant of s.
- There exist $p \in L^1(J, \mathbb{R}^+)$ and a nondecreasing continuous function $\psi : [0, \infty) \to [0, \infty)$ such that

$$|r(t, u)| \le p(t)\psi(|u|), \text{ for } \in J, \text{ and } u \in \mathbb{R}.$$

- P and Q are integrable on $(-\infty, 0]$.

By the dominated convergence theorem, one can show that f is a continuous function from B to E. Moreover the mapping h is Lipschitz continuous in its second argument, in fact, we have

$$|h(t, \varphi_1) - h(t, \varphi_2)| \le lip(s) \int_{-\infty}^{0} |Q(\theta)| \, d\theta \, |\varphi_1 - \varphi_2|, \text{ for } \varphi_1, \varphi_2 \in B.$$

On the other hand, we have for $\varphi \in B$ and $\xi \in [0, \pi]$

$$|f(t, \varphi)(\xi)| \le \int_{-\infty}^{0} |p(t)P(\theta)| \, \psi(|(\varphi(\theta))(\xi)|) d\theta.$$

Since the function ψ is nondecreasing, it follows that

$$|f(t, \varphi)| \le p(t) \int_{-\infty}^{0} |P(\theta)| \, d\theta \psi(|\varphi|), \text{ for } \varphi \in B.$$

Proposition 4.7. *Under the above assumptions, if we assume that condition (4.11) in Theorem 4.6 is true, $\varphi \in B$, then the problem (4.12) has a mild solution which is defined in $(-\infty, +\infty)$.*

4.5   Notes and Remarks

The results of Chap. 4 are taken from Adimy et al. [12], Baghli et al. [34], and Balachandran and Anandhi [42]. Other results may be found in [8, 9, 50, 158].

Chapter 5
Partial Functional Evolution Inclusions with Finite Delay

5.1 Introduction

In this chapter, we provide sufficient conditions for the existence of mild solutions on the semi-infinite interval $J = \mathbb{R}_+$ for some classes of first order partial functional and neutral functional differential evolution inclusions with finite delay by using the recent nonlinear alternative of Frigon [114, 115] for contractive multi-valued maps in Fréchet spaces [116], combined with the semigroup theory [16, 20, 168].

5.2 Partial Functional Evolution Inclusions

5.2.1 Introduction

We establish here the existence of mild solutions for the partial functional evolution inclusion of the form

$$y'(t) \in A(t)y(t) + F(t, y_t), \quad \text{a.e.} \quad t \in J = \mathbb{R}_+ \tag{5.1}$$

$$y(t) = \varphi(t), \quad t \in H, \tag{5.2}$$

where $F : J \times C(H, E) \to \mathcal{P}(E)$ is a multi-valued map with nonempty compact values, $\mathcal{P}(E)$ is the family of all subsets of E, $\varphi \in C(H, E)$ is a given function, and $\{A(t)\}_{0 \leq t < +\infty}$ is a family of linear closed (not necessarily bounded) operators from E into E that generate an evolution system of operators $\{U(t, s)\}_{(t,s) \in J \times J}$ for $0 \leq s \leq t < +\infty$.

This result is an extension of the problem (2.1) in [33] for multi-valued case.

© Springer International Publishing Switzerland 2015
S. Abbas, M. Benchohra, *Advanced Functional Evolution Equations
and Inclusions*, Developments in Mathematics 39,
DOI 10.1007/978-3-319-17768-7_5

5.2.2   Existence of Mild Solutions

Let us introduce the definition of the mild solution of our partial functional evolution inclusion system (5.1)–(5.2) before stating and proving our main result.

Definition 5.1. We say that the continuous function $y(\cdot) : [-r, +\infty) \to E$ is a mild solution of the evolution system (5.1)–(5.2) if $y(t) = \varphi(t)$ for all $t \in H$ and the restriction of $y(\cdot)$ to the interval J is continuous and there exists $f(\cdot) \in L^1(J, E)$: $f(t) \in F(t, y_t)$ a.e. in J such that y satisfies the following integral equation:

$$y(t) = U(t, 0)\, \varphi(0) + \int_0^t U(t, s) f(s)\, ds, \quad \text{for each } t \in \mathbb{R}_+.$$

We will introduce the following hypotheses which are assumed afterwards

(5.1.1) There exists a constant $\widehat{M} \geq 1$ such that:

$$\|U(t, s)\|_{B(E)} \leq \widehat{M} \quad \text{for every } (t, s) \in \Delta.$$

(5.1.2) The multi-function $F : J \times C(H, E) \longrightarrow \mathcal{P}(E)$ is L^1_{loc}-Carathéodory with compact and convex values for each $u \in C(H, E)$ and there exist a function $p \in L^1_{\text{loc}}(J, \mathbb{R}_+)$ and a continuous nondecreasing function $\psi : J \to (0, \infty)$ and such that:

$$\|F(t, u)\|_{\mathcal{P}(E)} \leq p(t)\, \psi(\|u\|) \text{ for a.e. } t \in J \text{ and each } u \in C(H, E).$$

(5.1.3) For all $R > 0$, there exists $l_R \in L^1_{\text{loc}}(J, \mathbb{R}_+)$ such that:

$$H_d(F(t, u) - F(t, v)) \leq l_R(t)\, \|u - v\|$$

for each $t \in J$ and for all $u, v \in C(H, E)$ with $\|u\| \leq R$ and $\|v\| \leq R$ and

$$d(0, F(t, 0)) \leq l_R(t) \quad \text{a.e. } t \in J.$$

For every $n \in \mathbb{N}$, we define in $C([-r, +\infty), E)$ the family of semi-norms by

$$\|y\|_n := \sup \{\, e^{-\tau\, L_n^*(t)}\, |y(t)| : t \in [0, n] \,\}$$

where $L_n^*(t) = \int_0^t \bar{l}_n(s)\, ds$, $\bar{l}_n(t) = \widehat{M} l_n(t)$ and l_n is the function from (5.1.3). Then $C([-r, +\infty), E)$ is a Fréchet space with the family of semi-norms $\|\cdot\|_{n \in \mathbb{N}}$. In what follows we will choose $\tau > 1$.

Theorem 5.2. *Suppose that hypotheses* (4.1.1)–(4.1.3) *are satisfied and moreover*

$$\int_{c_{19}}^{+\infty} \frac{ds}{\psi(s)} > \widehat{M} \int_0^n p(s)\, ds, \quad \text{for each } n > 0 \tag{5.3}$$

with $c_{19} = \widehat{M}\,\|\varphi\|$. *Then the evolution inclusion problem* (5.1)–(5.2) *has a mild solution.*

Proof. Transform the problem (5.1)–(5.2) into a fixed point problem. Consider the multi-valued operator $N_{19} : \mathcal{C}([-r, +\infty), E) \to \mathcal{P}(\mathcal{C}([-r, +\infty), E))$ defined by:

$$N_{19}(y) = \left\{ h \in \mathcal{C}([-r, +\infty), E) : h(t) = \begin{cases} \varphi(t), & \text{if } t \in H; \\ U(t,0)\,\varphi(0) \\ \quad + \displaystyle\int_0^t U(t,s) f(s)\, ds, & \text{if } t \geq 0. \end{cases} \right\}$$

where $f \in S_{F,y} = \{v \in L^1(J, E) : v(t) \in F(t, y_t) \text{ for a.e. } t \in J\}$. Clearly, the fixed points of the operator N_{19} are mild solutions of the problem (5.1)–(5.2). We remark also that, for each $y \in \mathcal{C}([-r, +\infty), E)$, the set $S_{F,y}$ is nonempty since, by (5.1.2), F has a measurable selection (see [94], Theorem III.6).

Let y be a possible fixed point of the operator N_{19}. Given $n \in \mathbb{N}$ and $t \leq n$, then y should be solution of the inclusion $y \in \lambda N_{19}(y)$ for some $\lambda \in (0, 1)$ and there exists $f \in S_{F,y} \Leftrightarrow f(t) \in F(t, y_t)$ such that, for each $t \in \mathbb{R}_+$, we have

$$|y(t)| \leq \|U(t,0)\|_{B(E)}\, |\varphi(0)| + \int_0^t \|U(t,s)\|_{B(E)}\, |f(s)|\, ds$$

$$\leq \widehat{M}\,\|\varphi\| + \widehat{M} \int_0^t p(s)\,\psi\,(\|y_s\|)\, ds.$$

Consider the function μ defined by

$$\mu(t) := \sup\{\|y(s)\| \;:\; 0 \leq s \leq t\}, \quad 0 \leq t < +\infty.$$

Let $t^* \in [-r, t]$ be such that $\mu(t) = |y(t^*)|$. If $t^* \in [0, n]$, by the previous inequality, we have

$$\mu(t) \leq \widehat{M}\,\|\varphi\| + \widehat{M} \int_0^t p(s)\,\psi(\mu(s))\, ds, \quad \text{for } t \in [0, n].$$

If $t^* \in H$, then $\mu(t) = \|\varphi\|$ and the previous inequality holds.

Let us take the right-hand side of the above inequality as $v(t)$. Then, we have

$$\mu(t) \leq v(t) \text{ for all } t \in [0, n].$$

From the definition of v, we have

$$c_{19} := v(0) = \widehat{M} \, \|\varphi\| \quad \text{and} \quad v'(t) = \widehat{M} p(t) \, \psi(\mu(t)) \quad \text{a.e. } t \in [0, n].$$

Using the nondecreasing character of ψ, we get

$$v'(t) \leq \widehat{M} \, p(t) \, \psi(v(t)) \quad \text{a.e. } t \in [0, n].$$

This implies that for each $t \in [0, n]$ and using the condition (5.3), we get

$$\int_{c_{19}}^{v(t)} \frac{ds}{\psi(s)} \leq \widehat{M} \int_0^t p(s) \, ds$$

$$\leq \widehat{M} \int_0^n p(s) \, ds$$

$$< \int_{c_{19}}^{+\infty} \frac{ds}{\psi(s)}.$$

Thus, for every $t \in [0, n]$, there exists a constant Λ_n such that $v(t) \leq \Lambda_n$ and hence $\mu(t) \leq \Lambda_n$. Since $\|y_t\| \leq \mu(t)$, we have $\|y\|_n \leq \max\{\|\varphi\|; \Lambda_n\} := \Delta_n$. Set

$$\mathcal{U} = \{ y \in \mathcal{C}([-r, +\infty), E) : \sup\{ |y(t)| : 0 \leq t \leq n \} < \Delta_n + 1 \quad \text{for all } n \in \mathbb{N} \}.$$

Clearly, $\mathcal{U}$ is an open subset of $\mathcal{C}([-r, +\infty), E)$.

We shall show that $N_{19} : \overline{\mathcal{U}} \twoheadrightarrow \mathcal{P}(\mathcal{C}([-r, +\infty), E))$ is a contraction and an admissible operator. First, we prove that N_{19} is a contraction; Let $y, \overline{y} \in \mathcal{C}([-r, +\infty), E)$ and $h \in N_{19}(y)$. Then there exists $f(t) \in F(t, y_t)$ such that for each $t \in [0, n]$

$$h(t) = U(t, 0) \, \varphi(0) + \int_0^t U(t, s) f(s) \, ds.$$

From (5.1.3) it follows that

$$H_d(F(t, y_t), F(t, \overline{y}_t)) \leq l_n(t) \, \|y_t - \overline{y}_t\|.$$

Hence, there is $\rho \in F(t, \overline{y}_t)$ such that

$$|f(t) - \rho| \leq l_n(t) \, \|y_t - \overline{y}_t\|, \quad t \in [0, n].$$

Consider $\mathcal{U}_* : [0, n] \to \mathcal{P}(E)$, given by

$$\mathcal{U}_* = \{ \rho \in E : |f(t) - \rho| \leq l_n(t) \, \|y_t - \overline{y}_t\| \}.$$

Since the multi-valued operator $V(t) = U_*(t) \cap F(t, \overline{y}_t)$ is measurable (in [94], see Proposition III.4), there exists a function $\overline{f}(t)$, which is a measurable selection for V. So, $\overline{f}(t) \in F(t, \overline{y}_t)$ and we obtain for each $t \in [0, n]$

$$|f(t) - \overline{f}(t)| \leq l_n(t) \|y_t - \overline{y}_t\|.$$

Let us define, for each $t \in [0, n]$

$$\overline{h}(t) = U(t, 0) \, \varphi(0) + \int_0^t U(t, s) \overline{f}(s) ds.$$

Then we can show as in previous sections that we have

$$\|h - \overline{h}\|_n \leq \frac{1}{\tau} \|y - \overline{y}\|_n.$$

By an analogous relation, obtained by interchanging the roles of y and $\overline{y}$, it follows that

$$H_d(N_{19}(y), N_{19}(\overline{y})) \leq \frac{1}{\tau} \|y - \overline{y}\|_n.$$

So, for $\tau > 1$, N_{19} is a contraction for all $n \in \mathbb{N}$.

It remains to show that N_{19} is an admissible operator. Let $y \in C([-r, +\infty), E)$. Consider $N_{19} : C([-r, n], E) \to \mathcal{P}(C([-r, n], E))$, given by

$$N_{19}(y) = \left\{ h \in C([-r, +\infty), E) : h(t) = \begin{cases} \varphi(t), & \text{if } t \in H; \\ U(t, 0) \, \varphi(0) \\ + \int_0^t U(t, s) f(s) \, ds, & \text{if } t \in [0, n], \end{cases} \right\}$$

where $f \in S^n_{F,y} = \{v \in L^1([0, n], E) : v(t) \in F(t, y_t) \text{ for a.e. } t \in [0, n]\}$.

From (5.1.1) to (5.1.3) and since F is a multi-valued map with compact values, we can prove that for every $y \in C([-r, n], E)$, $N_{19}(y) \in \mathcal{P}_{cp}(C([-r, n], E))$ and there exists $y_* \in C([-r, n], E)$ such that $y_* \in N_{19}(y_*)$. Let $h \in C([-r, n], E)$, $\overline{y} \in \overline{U}$ and $\epsilon > 0$. Assume that $y_* \in N_{19}(\overline{y})$, then we have

$$\|\overline{y}(t) - y_*(t)\| \leq \|\overline{y}(t) - h(t)\| + \|y_*(t) - h(t)\|$$
$$\leq e^{\tau \, L_n^*(t)} \|\overline{y} - N_{19}(\overline{y})\|_n + \|y_*(t) - h(t)\|.$$

Since h is arbitrary, we may assume that $h \in B(y_*, \epsilon) = \{h \in C([-r, n], E) : \|h - y_*\|_n \leq \epsilon\}$. Therefore,

$$\|\overline{y} - y_*\|_n \leq \|\overline{y} - N_{19}(\overline{y})\|_n + \epsilon.$$

If y is not in $N_{19}(\bar{y})$, then $\|y_* - N_{19}(\bar{y})\| \neq 0$. Since $N_{19}(\bar{y})$ is compact, there exists $x \in N_{19}(\bar{y})$ such that $\|y_* - N_{19}(\bar{y})\| = \|y_* - x\|$. Then we have

$$\|\bar{y}(t) - x(t)\| \leq \|\bar{y}(t) - h(t)\| + \|x(t) - h(t)\|$$
$$\leq e^{\tau \, L_n^*(t)} \|\bar{y} - N_{19}(\bar{y})\|_n + \|x(t) - h(t)\|.$$

Thus,

$$\|\bar{y} - x\|_n \leq \|\bar{y} - N_{19}(\bar{y})\|_n + \epsilon.$$

So, N_{19} is an admissible operator contraction. From the choice of $\mathcal{U}$ there is no $y \in \partial \mathcal{U}$ such that $y = \lambda \, N_{19}(y)$ for some $\lambda \in (0, 1)$. Then the statement $(S2)$ in Theorem 1.31 does not hold. A consequence of the nonlinear alternative of Frigon that $(S1)$ holds, we deduce that the operator N_{19} has a fixed point y^* which is a mild solution of the evolution inclusion problem (5.1)–(5.2).   □

5.2.3 An Example

Consider the following model

$$\begin{cases} \dfrac{\partial v}{\partial t}(t, \xi) \in a(t, \xi) \dfrac{\partial^2 v}{\partial \xi^2}(t, \xi) \\[2mm] \qquad + \displaystyle\int_{-r}^{0} P(\theta) R(t, v(t + \theta, \xi)) d\theta \quad \xi \in [0, \pi] \\[4mm] v(t, 0) = v(t, \pi) = 0 \qquad\qquad\qquad t \in \mathbb{R}_+ \\[2mm] v(\theta, \xi) = v_0(\theta, \xi) \qquad\qquad\qquad -r \leq \theta \leq 0, \xi \in [0, \pi], \end{cases} \quad (5.4)$$

where $r > 0$, $a(t, \xi)$ is a continuous function and is uniformly Hölder continuous in t; $P : H \to \mathbb{R}$ and $v_0 : H \times [0, \pi] \to \mathbb{R}$ are continuous functions and $R : \mathbb{R}_+ \times \mathbb{R} \to \mathcal{P}(\mathbb{R})$ is a multi-valued map with compact convex values.

Consider $E = L^2([0, \pi], \mathbb{R})$ and define $A(t)$ by $A(t)w = a(t, \xi)w''$ with domain

$$D(A) = \{\, w \in E \, : \, w, w' \text{ are absolutely continuous, } w'' \in E, \, w(0) = w(\pi) = 0 \,\}$$

Then $A(t)$ generates an evolution system $U(t, s)$ satisfying assumption (4.1.1), see [112, 149].

For $\xi \in [0, \pi]$, we have

$$y(t)(\xi) = v(t, \xi), \; t \in \mathbb{R}_+,$$

$$\varphi(\theta)(\xi) = v_0(\theta, \xi), \; -r \leq \theta \leq 0,$$

and

$$F(t, \eta)(\xi) = \int_{-r}^{0} P(\theta)R(t, \eta(\theta)(\xi))d\theta, \ -r \le \theta \le 0.$$

Then, the problem (5.4) takes the abstract partial functional evolution inclusion form (5.1)–(5.2). In order to show the existence of mild solutions of problem (5.4), we assume the following assumptions:

–   There exist $p \in L^1(J, \mathbb{R}^+)$ and a nondecreasing continuous function $\psi : \mathbb{R}_+ \to$ $(0, +\infty)$ such that

$$|R(t, \eta)| \le p(t)\psi(|\eta|), \ \text{for} \ \in J, \ \text{and} \ \eta \in \mathbb{R}.$$

– P is integrable on H.

By the dominated convergence theorem, one can show that $f \in S_{F,y}$ is a continuous function from $\mathcal{C}([-r, +\infty), E)$ to E. On the other hand, we have for $\eta \in \mathbb{R}$ and $\xi \in [0, \pi]$

$$|F(t, \eta)(\xi)| \le \int_{-r}^{0} |p(t)P(\theta)| \, \psi(|(\eta(\theta))(\xi)|)d\theta.$$

Since the function ψ is nondecreasing, it follows that

$$\|F(t, \eta)\|_{\mathcal{P}(E)} \le p(t) \int_{-r}^{0} |P(\theta)| \, d\theta \psi(|\eta|), \ \text{for} \ \eta \in \mathbb{R}.$$

Proposition 5.3. *Under the above assumptions, if we assume that condition (5.3) in Theorem 5.2 is true, then the problem (5.4) has a mild solution which is defined in $[-r, +\infty)$.*

5.3   Neutral Functional Evolution Inclusions

5.3.1   Introduction

We investigate in this section the neutral functional evolution inclusion of the form

$$\frac{d}{dt}[y(t) - g(t, y_t)] \in A(t)y(t) + F(t, y_t), \quad \text{a.e.} \ \ t \in J = \mathbb{R}_+ \tag{5.5}$$

$$y(t) = \varphi(t), \quad t \in H, \tag{5.6}$$

where $F : J \times C(H, E) \to \mathcal{P}(E)$ is a multi-valued map with nonempty compact values, $\mathcal{P}(E)$ is the family of all subsets of E, $g : J \times C(H, E) \to E$ and $\varphi \in C(H, E)$ is a given function. This result is an extension of the problem (5.1) for the neutral case and also the multi-valued generalization of the neutral problem (2.8) in [37].

5.3.2　Existence of Mild Solutions

Definition 5.4. We say that the function $y(\cdot) : [-r, +\infty) \to E$ is a mild solution of the neutral functional evolution system (5.5)–(5.6) if $y(t) = \varphi(t)$ for all $t \in H$ and the restriction of $y(\cdot)$ to the interval J is continuous and there exists $f(\cdot) \in L^1(J, E)$: $f(t) \in F(t, y_t)$ a.e. in J such that y satisfies the following integral equation

$$y(t) = U(t, 0)[\varphi(0) - g(0, \varphi)] + g(t, y_t) + \int_0^t U(t, s)A(s)g(s, y_s)ds$$
$$+ \int_0^t U(t, s)f(s)\,ds, \qquad \text{for each } t \in \mathbb{R}_+.$$

We consider the hypotheses (5.1.1)–(5.1.3) and we will need the following assumptions:

(G1)　　There exists a constant $\overline{M}_0 > 0$ such that:

$$\|A^{-1}(t)\|_{B(E)} \leq \overline{M}_0 \quad \text{for all } t \in J.$$

(G2)　　There exists a constant $0 < L < \dfrac{1}{\overline{M}_0}$ such that:

$$|A(t)\, g(t, \varphi)| \leq L\, (\|\varphi\| + 1) \text{ for all } t \in J \text{ and } \varphi \in C(H, E).$$

(G3)　　There exists a constant $L_* > 0$ such that:

$$|A(s)\, g(s, \varphi) - A(\bar{s})\, g(\bar{s}, \overline{\varphi})| \leq L_*\, (|s - \bar{s}| + \|\varphi - \overline{\varphi}\|)$$

for all $s, \bar{s} \in J$ and $\varphi, \overline{\varphi} \in C(H, E)$.

For every $n \in \mathbb{N}$, let us take here $\bar{l}_n(t) = \widehat{M}[L_* + l_n(t)]$ for the family of seminorm $\{\|\cdot\|_n\}_{n \in \mathbb{N}}$ defined in Sect. 5.3. In what follows we fix $\tau > 0$ such that $\left[\overline{M}_0 L_* + \dfrac{1}{\tau}\right] < 1$.

Theorem 5.5. *Suppose that hypotheses* (5.1.1)–(5.1.3) *and the assumptions* (G1)–(G3) *are satisfied and moreover*

$$\int_{c_{20,n}}^{+\infty} \frac{ds}{s + \psi(s)} > \frac{\widehat{M}}{1 - \overline{M}_0 L} \int_0^n \max(L, p(s)) ds, \quad \text{for each } n > 0 \qquad (5.7)$$

with

$$c_{20,n} := \frac{\widehat{M}(1 + \overline{M}_0 L)\|\varphi\| + \overline{M}_0 L(\widehat{M} + 1) + \widehat{M} L n}{1 - \overline{M}_0 L}$$

then the neutral functional evolution problem (5.5)–(5.6) *has a mild solution.*

Proof. Transform the neutral functional evolution problem (5.5)–(5.6) into a fixed point problem. Consider the multi-valued operator N_{20} : $\mathcal{C}([-r, +\infty), E) \to \mathcal{P}(\mathcal{C}([-r, +\infty), E))$ defined by:

$$N_{20}(y) = \left\{ h \in \mathcal{C}([-r, +\infty), E) : h(t) = \begin{cases} \varphi(t), & \text{if } t \le 0; \\ U(t,0)\left[\varphi(0) - g(0,\varphi)\right] + g(t, y_t) \\ + \int_0^t U(t,s)A(s)g(s, y_s) ds \\ + \int_0^t U(t,s)f(s) ds, & \text{if } t \ge 0, \end{cases} \right\}$$

where $f \in S_{F,y} = \{v \in L^1(J, E) : v(t) \in F(t, y_t) \text{ for a.e. } t \in J\}$.

Clearly, the fixed points of the operator N_{20} are mild solutions of the problem (5.5)–(5.6). We remark also that, for each $y \in \mathcal{C}([-r, +\infty), E)$, the set $S_{F,y}$ is nonempty since, by (5.1.2), F has a measurable selection (see [94], Theorem III.6).

Let y be a possible fixed point of the operator N_{20}. Given $n \in \mathbb{N}$ and $t \le n$, then y should be solution of the inclusion $y \in \lambda N_{20}(y)$ for some $\lambda \in (0, 1)$ and there exists $f \in S_{F,y} \Leftrightarrow f(t) \in F(t, y_t)$ such that, for each $t \in \mathbb{R}_+$, we have

$$|y(t)| \le \|U(t,0)\|_{B(E)}|\varphi(0)| + \|U(t,0)\|_{B(E)}\|A^{-1}(0)\|\|A(0) g(0,\varphi)\|$$

$$+ \|A^{-1}(t)\|_{B(E)}\|A(t)g(t, y_t)\| + \int_0^t \|U(t,s)\|_{B(E)}\|A(s)g(s, y_s)\| ds$$

$$+ \int_0^t \|U(t,s)\|_{B(E)}|f(s)| ds$$

$$\le \widehat{M}\|\varphi\| + \widehat{M}\overline{M}_0 L(\|\varphi\| + 1) + \overline{M}_0 L(\|y_t\| + 1) + \widehat{M} \int_0^t L(\|y_s\| + 1) ds$$

$$+ \widehat{M} \int_0^t p(s)\psi(\|y_s\|) ds$$

$$\leq \widehat{M}(1 + \overline{M}_0 L)\|\varphi\| + \overline{M}_0 L(\widehat{M} + 1) + \widehat{M} L n$$

$$+ \overline{M}_0 L \|y_t\| + \widehat{M} \int_0^t L \|y_s\| ds + \widehat{M} \int_0^t p(s)\psi(\|y_s\|) ds.$$

We consider the function μ defined by

$$\mu(t) := \sup\{\, |y(s)| \, : \, 0 \leq s \leq t \}, \quad 0 \leq t < +\infty.$$

Let $t^* \in [-r, t]$ be such that $\mu(t) = |y(t^*)|$. If $t^* \in [0, n]$, by the previous inequality we have for $t \in [0, n]$

$$\mu(t) \leq \widehat{M}(1 + \overline{M}_0 L)\|\varphi\| + \overline{M}_0 L(\widehat{M} + 1) + \widehat{M} L n$$

$$+ \overline{M}_0 L \mu(t) + \widehat{M} \int_0^t L \mu(s) ds + \widehat{M} \int_0^t p(s)\psi(\mu(s)) ds.$$

Then

$$(1 - \overline{M}_0 L)\mu(t) \leq \widehat{M}(1 + \overline{M}_0 L)\|\varphi\| + \overline{M}_0 L(\widehat{M} + 1) + \widehat{M} L n$$

$$+ \widehat{M} \int_0^t L \mu(s) ds + \widehat{M} \int_0^t p(s)\psi(\mu(s)) ds.$$

Set $c_{20,n} := \dfrac{\widehat{M}(1 + \overline{M}_0 L)\|\varphi\| + \overline{M}_0 L(\widehat{M} + 1) + \widehat{M} L n}{1 - \overline{M}_0 L}$, thus

$$\mu(t) \leq c_{20,n} + \frac{\widehat{M}}{1 - \overline{M}_0 L} \int_0^t [L \mu(s) + p(s)\psi(\mu(s))] \, ds.$$

If $t^* \in H$, then $\mu(t) = \|\varphi\|$ and the previous inequality holds.

Let us take the right-hand side of the above inequality as $v(t)$. Then we have

$$\mu(t) \leq v(t) \text{ for all } t \in [0, n].$$

From the definition of v, we have

$$v(0) = c_{20,n} \text{ and } v'(t) = \frac{\widehat{M}}{1 - \overline{M}_0 L} [L \mu(t) + p(t)\psi(\mu(t))] \quad \text{a.e. } t \in [0, n].$$

Using the nondecreasing character of ψ, we get

$$v'(t) \leq \frac{\widehat{M}}{1 - \overline{M}_0 L} [L v(t) + p(t)\psi(v(t))] \quad \text{a.e. } t \in [0, n].$$

This implies that for each $t \in [0, n]$ and using the condition (5.7), we get

$$\int_{c_{20,n}}^{v(t)} \frac{ds}{s + \psi(s)} \leq \frac{\widehat{M}}{1 - \overline{M}_0 L} \int_0^t \max(L, p(s))ds$$

$$\leq \frac{\widehat{M}}{1 - \overline{M}_0 L} \int_0^n \max(L, p(s))ds$$

$$< \int_{c_{20,n}}^{+\infty} \frac{ds}{s + \psi(s)}.$$

Thus, for every $t \in [0, n]$, there exists a constant Λ_n such that $v(t) \leq \Lambda_n$ and hence $\mu(t) \leq \Lambda_n$. Since $\|y_t\| \leq \mu(t)$, we have

$$\|y\|_n \leq \max\{\|\varphi\|, \Lambda_n\} := \Delta_n.$$

We can show that N_{20} is an admissible operator and we shall prove now that $N_{20} : \overline{\mathcal{U}} \to \mathcal{P}(\mathcal{C}([-r, +\infty), E))$ is a contraction.

Let $y, \overline{y} \in \mathcal{C}([-r, +\infty), E)$ and $h \in N_{20}(y)$. Then there exists $f(t) \in F(t, y_t)$ such that for each $t \in [0, n]$

$$h(t) = U(t, 0)[\varphi(0) - g(0, \varphi)] + g(t, y_t) + \int_0^t U(t, s)A(s)g(s, y_s)ds$$

$$+ \int_0^t U(t, s)f(s)ds.$$

From (5.1.3) it follows that

$$H_d(F(t, y_t), F(t, \overline{y}_t)) \leq l_n(t) \|y_t - \overline{y}_t\|.$$

Hence, there is $\rho \in F(t, \overline{y}_t)$ such that

$$|f(t) - \rho| \leq l_n(t) \|y_t - \overline{y}_t\| \quad t \in [0, n].$$

Consider $\mathcal{U}_* : [0, n] \to \mathcal{P}(E)$, given by

$$\mathcal{U}_* = \{\rho \in E : |f(t) - \rho| \leq l_n(t) \|y_t - \overline{y}_t\|\}.$$

Since the multi-valued operator $\mathcal{V}(t) = \mathcal{U}_*(t) \cap F(t, \overline{y}_t)$ is measurable (in [94], see Proposition III.4), there exists a function $\overline{f}(t)$, which is a measurable selection for $\mathcal{V}$. So, $\overline{f}(t) \in F(t, \overline{y}_t)$, and we obtain for each $t \in [0, n]$

$$|f(t) - \overline{f}(t)| \leq l_n(t) \|y_t - \overline{y}_t\|.$$

Let us define, for each $t \in [0, n]$

$$\bar{h}(t) = U(t,0)[\varphi(0) - g(0,\varphi)] + g(t,\bar{y}_t) + \int_0^t U(t,s)A(s)g(s,\bar{y}_s)ds$$

$$+ \int_0^t U(t,s)\bar{f}(s)ds.$$

Then we can show as in previous sections that we have for each $t \in [0, n]$ and $n \in \mathbb{N}$

$$\|h - \bar{h}\|_n \leq \left[\overline{M}_0 L_* + \frac{1}{\tau}\right]\|y - \bar{y}\|_n.$$

By an analogous relation, obtained by interchanging the roles of y and $\bar{y}$, it follows that

$$H_d(N_{20}(y), N_{20}(\bar{y})) \leq \left[\overline{M}_0 L_* + \frac{1}{\tau}\right]\|y - \bar{y}\|_n.$$

So, for $\left[\overline{M}_0 L_* + \frac{1}{\tau}\right] < 1$, the operator N_{20} is a contraction for all $n \in \mathbb{N}$ and an admissible operator. From the choice of $\mathcal{U}$ there is no $y \in \partial\mathcal{U}$ such that $y = \lambda N_{20}(y)$ for some $\lambda \in (0, 1)$. Then the statement $(S2)$ in Theorem 1.31 does not hold. By the nonlinear alternative due to Frigon we get that $(S1)$ holds, we deduce that the operator N_{20} has a fixed point y^* which is a mild solution of the neutral functional evolution inclusion problem (5.5)–(5.6).   □

5.3.3 An Example

Consider the following model

$$\begin{cases} \dfrac{\partial}{\partial t}\left[v(t,\xi) - \displaystyle\int_{-r}^0 T(\theta)u(t,v(t+\theta,\xi))d\theta\right] \\ \qquad \in a(t,\xi)\dfrac{\partial^2 v}{\partial\xi^2}(t,\xi) \\ \qquad + \displaystyle\int_{-r}^0 P(\theta)R(t,v(t+\theta,\xi))d\theta & t \in \mathbb{R}_+, \quad \xi \in [0,\pi] \\ v(t,0) = v(t,\pi) = 0 & t \in \mathbb{R}_+ \\ v(\theta,\xi) = v_0(\theta,\xi) & -r \leq \theta \leq 0, \xi \in [0,\pi], \end{cases}$$

$$(5.8)$$

where $r > 0$, $a(t, \xi)$ is a continuous function and is uniformly Hölder continuous in t; $T, P : H \to \mathbb{R}$; $u : H \times \mathbb{R} \to \mathbb{R}$ and $v_0 : H \times [0, \pi] \to \mathbb{R}$ are continuous functions and $R : \mathbb{R}_+ \times \mathbb{R} \to \mathcal{P}(\mathbb{R})$ is a multi-valued map with compact convex values.

Consider $E = L^2([0, \pi], \mathbb{R})$ and define $A(t)$ by $A(t)w = a(t, \xi)w''$ with domain

$$D(A) = \{\, w \in E \; : \; w, w' \text{ are absolutely continuous, } w'' \in E, \; w(0) = w(\pi) = 0 \,\}$$

Then $A(t)$ generates an evolution system $U(t, s)$ satisfying assumptions (5.1.1) and (G1), see [112, 149].

For $\xi \in [0, \pi]$, we have

$$y(t)(\xi) = v(t, \xi), \; t \in \mathbb{R}_+,$$

$$\varphi(\theta)(\xi) = v_0(\theta, \xi), \; -r \le \theta \le 0,$$

$$g(t, \eta)(\xi) = \int_{-r}^{0} T(\theta)u(t, \eta(\theta)(\xi))d\theta, \; -r \le \theta \le 0,$$

and

$$F(t, \eta)(\xi) = \int_{-r}^{0} P(\theta)R(t, \eta(\theta)(\xi))d\theta, \; -r \le \theta \le 0.$$

Then, the problem (5.8) takes the abstract neutral functional evolution inclusion form (5.5)–(5.6). In order to show the existence of mild solutions of problem (5.8), we suppose the following assumptions:

- u is Lipschitz with respect to its second argument. Let $lip(u)$ denotes the Lipschitz constant of u.
- There exist $p \in L^1(J, \mathbb{R}^+)$ and a nondecreasing continuous function $\psi : \mathbb{R}_+ \to (0, +\infty)$ such that

$$|R(t, \eta)| \le p(t)\psi(|\eta|), \text{ for } \in J, \text{ and } \eta \in \mathbb{R}.$$

- T, P are integrable on H.

By the dominated convergence theorem, one can show that $f \in S_{F,y}$ is a continuous function from $\mathcal{C}(H, E)$ to E. Moreover the mapping g is Lipschitz continuous in its second argument, in fact, we have

$$|g(t, \eta_1) - g(t, \eta_2)| \le \overline{M}_0 L_* lip(u) \int_{-r}^{0} |T(\theta)| \, d\theta \, |\eta_1 - \eta_2|, \text{ for } \eta_1, \eta_2 \in \mathbb{R}.$$

On the other hand, we have for $\eta \in \mathbb{R}$ and $\xi \in [0, \pi]$

$$|F(t, \eta)(\xi)| \le \int_{-r}^{0} |p(t)P(\theta)| \, \psi(|(\eta(\theta))(\xi)|)d\theta.$$

Since the function ψ is nondecreasing, it follows that

$$\|F(t, \eta)\|_{\mathcal{P}(E)} \leq p(t) \int_{-r}^{0} |P(\theta)| \, d\theta \, \psi(|\eta|).$$

Proposition 5.6. *Under the above assumptions, if we assume that condition (5.7) in Theorem 5.5 is true, then the problem (5.8) has a mild solution which is defined in* $[-r, +\infty)$.

5.4 Notes and Remarks

The results of Chap. 5 are taken from Arara et al. [27, 28]. Other results may be found in [1, 3, 51, 53, 74, 75].

Chapter 6
Partial Functional Evolution Inclusions with Infinite Delay

6.1 Introduction

We are interested in this chapter by the study of the existence of mild solutions of two classes of partial functional and neutral functional evolution inclusions with infinite delay on the semi-infinite interval $\mathbb{R}_+$.

It is known that in the modeling of the evolution of some physical, biological, and economic systems using functional and partial functional differential equations, the response of the systems depends not only on the current state of the system but also on the past history of the system. We assume that the histories y_t belongs to some abstract *phase space* $\mathcal{B}$.

Sufficient conditions are provided to get existence results of mild solutions of the partial functional and neutral functional differential evolution problems by applying the recent nonlinear alternative of Frigon [114, 115] for contractive multi-valued maps in Fréchet spaces [116], combined with the semigroup theory [16, 20, 168].

6.2 Partial Functional Evolution Inclusions

6.2.1 Introduction

In this chapter, we consider the partial functional evolution inclusions with infinite delay of the form

$$y'(t) \in A(t)y(t) + F(t, y_t), \quad \text{a.e.} \quad t \in J = \mathbb{R}_+ \tag{6.1}$$

$$y_0 = \phi \in \mathcal{B}, \tag{6.2}$$

© Springer International Publishing Switzerland 2015
S. Abbas, M. Benchohra, *Advanced Functional Evolution Equations and Inclusions*, Developments in Mathematics 39,
DOI 10.1007/978-3-319-17768-7_6

where $F : J \times \mathcal{B} \to \mathcal{P}(E)$ is a multi-valued map with nonempty compact values, $\mathcal{P}(E)$ is the family of all subsets of E, $\phi \in \mathcal{B}$ are given functions, and $\{A(t)\}_{0 \leq t < +\infty}$ is a family of linear closed (not necessarily bounded) operators from E into E that generate an evolution system of operators $\{U(t, s)\}_{(t,s) \in J \times J}$ for $0 \leq s \leq t < +\infty$.

6.2.2   Existence of Mild Solutions

Definition 6.1. We say that the function $y(\cdot) : \mathbb{R} \to E$ is a mild solution of the evolution system (6.1)–(6.2) if $y(t) = \phi(t)$ for all $t \in (-\infty, 0]$ and the restriction of $y(\cdot)$ to the interval J is continuous and there exists $f(\cdot) \in L^1(J, E) : f(t) \in F(t, y_t)$ a.e. in J such that y satisfies the following integral equation:

$$y(t) = U(t, 0) \, \phi(0) + \int_0^t U(t, s) \, f(s) \, ds, \quad \text{for each } t \in \mathbb{R}_+.$$

We will need to introduce the following hypotheses which are assumed hereafter:

(6.1.1) There exists a constant $\widehat{M} \geq 1$ such that:

$$\|U(t, s)\|_{B(E)} \leq \widehat{M} \quad \text{for every } (t, s) \in \Delta.$$

(6.1.2) The multi-function $F : J \times \mathcal{B} \longrightarrow \mathcal{P}(E)$ is L^1_{loc}-Carathéodory with compact and convex values for each $u \in \mathcal{B}$ and there exist a function $p \in L^1_{\text{loc}}(J, \mathbb{R}_+)$ and a continuous nondecreasing function $\psi : J \to (0, \infty)$ and such that:

$$\|F(t, u)\|_{\mathcal{P}(E)} \leq p(t) \, \psi(\|u\|_{\mathcal{B}}) \text{ for a.e. } t \in J \text{ and each } u \in \mathcal{B}.$$

(6.1.3) For all $R > 0$, there exists $l_R \in L^1_{\text{loc}}(J, \mathbb{R}_+)$ such that:

$$H_d(F(t, u) - F(t, v)) \leq l_R(t) \, \|u - v\|_{\mathcal{B}}$$

for each $t \in J$ and for all $u, v \in \mathcal{B}$ with $\|u\|_{\mathcal{B}} \leq R$ and $\|v\|_{\mathcal{B}} \leq R$ and

$$d(0, F(t, 0)) \leq l_R(t) \quad \text{a.e. } t \in J.$$

Consider the following space

$$B_{+\infty} = \left\{ y : \mathbb{R} \to E : y|_{[0,T]} \in C([0, T], E), \ y_0 \in \mathcal{B} \right\},$$

where $y|_{[0,T]}$ is the restriction of y to any real compact interval $[0, T]$.

For every $n \in \mathbb{N}$, we define in $B_{+\infty}$ the family of semi-norms by:

$$\|y\|_n := \sup \{ e^{-\tau \, L_n^*(t)} \, |y(t)| : t \in [0,n] \}$$

where $L_n^*(t) = \displaystyle\int_0^t \bar{l}_n(s) \, ds$, $\bar{l}_n(t) = K_n \widehat{M} l_n(t)$ and l_n is the function from (6.1.3). Then $B_{+\infty}$ is a Fréchet space with the family of semi-norms $\|\cdot\|_{n\in\mathbb{N}}$. In what follows we will choose $\tau > 1$.

Theorem 6.2. *Suppose that hypotheses* (6.1.1)–(6.1.3) *are satisfied and moreover*

$$\int_{c_{21,n}}^{+\infty} \frac{ds}{\psi(s)} > K_n \widehat{M} \int_0^n p(s) \, ds, \quad \text{for each } n > 0 \tag{6.3}$$

with $c_{21,n} = (K_n \widehat{M} H + M_n)\|\phi\|_{\mathcal{B}}$. *Then evolution problem* (6.1)–(6.2) *has a mild solution.*

Proof. Consider the multi-valued operator $N_{21} : B_{+\infty} \to \mathcal{P}(B_{+\infty})$ defined by:

$$N_{21}(y) = \left\{ h \in B_{+\infty} : h(t) = \begin{cases} \phi(t), & \text{if } t \leq 0; \\ U(t,0)\,\phi(0) + \displaystyle\int_0^t U(t,s) f(s) \, ds, & \text{if } t \geq 0. \end{cases} \right\}$$

where $f \in S_{F,y} = \{v \in L^1(J,E) : v(t) \in F(t, y_t) \text{ for a.e. } t \in J\}$.

Clearly, the fixed points of the operator N_{21} are mild solutions of the problem (6.1)–(6.2). We remark also that, for each $y \in B_{+\infty}$, the set $S_{F,y}$ is nonempty since, by (6.1.2), F has a measurable selection (see [94], Theorem III.6).

For $\phi \in \mathcal{B}$, we will define the function $x(.) : \mathbb{R} \to E$ by

$$x(t) = \begin{cases} \phi(t), & \text{if } t \in (-\infty, 0]; \\ U(t,0)\,\phi(0), & \text{if } t \in J. \end{cases}$$

Then $x_0 = \phi$. For each function $z \in B_{+\infty}$, set $y(t) = z(t) + x(t)$. It is obvious that y satisfies Definition 6.1 if and only if z satisfies $z_0 = 0$ and

$$z(t) = \int_0^t U(t,s) f(s) \, ds, \quad \text{for } t \in J.$$

where $f(t) \in F(t, z_t + x_t)$ a.e. $t \in J$.

Let

$$B_{+\infty}^0 = \{z \in B_{+\infty} : z_0 = 0\}.$$

Define in $B_{+\infty}^0$, the multi-valued operator $\mathcal{F} : B_{+\infty}^0 \to \mathcal{P}(B_{+\infty}^0)$ by:

$$\mathcal{F}(z) = \left\{ h \in B_{+\infty}^0 : h(t) = \int_0^t U(t,s) f(s) \, ds, \quad t \in J \right\},$$

where $f \in S_{F,z} = \{ v \in L^1(J, E) : v(t) \in F(t, z_t + x_t) \text{ for a.e. } t \in J \}$.

Obviously the operator inclusion N_{21} has a fixed point is equivalent to the operator inclusion $\mathcal{F}$ has one, so it turns to prove that $\mathcal{F}$ has a fixed point.

Let $z \in B_{+\infty}^0$ be a possible fixed point of the operator $\mathcal{F}$. Given $n \in \mathbb{N}$, then z should be solution of the inclusion $z \in \lambda \, \mathcal{F}(z)$ for some $\lambda \in (0, 1)$ and there exists $f \in S_{F,z} \Leftrightarrow f(t) \in F(t, z_t + x_t)$ such that, for each $t \in [0, n]$, we have

$$|z(t)| \leq \int_0^t \|U(t,s)\|_{B(E)} \, |f(s)| \, ds$$

$$\leq \widehat{M} \int_0^t p(s) \, \psi \, (\|z_s + x_s\|_\mathcal{B}) \, ds.$$

Set $c_{21,n} := (K_n \widehat{M} H + M_n) \|\phi\|_\mathcal{B} = \alpha_n$, then using the inequality (3.4) and the nondecreasing character of ψ, we get

$$|z(t)| \leq \widehat{M} \int_0^t p(s) \, \psi \, (K_n |z(s)| + \alpha_n) \, ds.$$

Then

$$K_n |z(t)| + \alpha_n \leq K_n \widehat{M} \int_0^t p(s) \psi (K_n |z(s)| + \alpha_n) ds + c_{21,n}.$$

We consider the function μ defined by

$$\mu(t) := \sup \{ K_n |z(s)| + \alpha_n : 0 \leq s \leq t \}, \quad 0 \leq t < +\infty.$$

Let $t^* \in [0, t]$ be such that $\mu(t) = K_n |z(t^*)| + \alpha_n$. By the previous inequality, we have

$$\mu(t) \leq K_n \widehat{M} \int_0^t p(s) \, \psi(\mu(s)) \, ds + c_{21,n}, \quad \text{for } t \in [0, n].$$

Let us take the right-hand side of the above inequality as $v(t)$. Then, we have

$$\mu(t) \leq v(t) \text{ for all } t \in [0, n].$$

From the definition of v, we have

$$v(0) = c_{21,n} \quad \text{and} \quad v'(t) = K_n \widehat{M} p(t) \, \psi(\mu(t)) \quad \text{a.e. } t \in [0, n].$$

Using the nondecreasing character of ψ, we get

$$v'(t) \le K_n \widehat{M} \, p(t) \, \psi(v(t)) \quad \text{a.e. } t \in [0, n].$$

This implies that for each $t \in [0, n]$ and using the condition (6.3), we get

$$\int_{c_{21,n}}^{v(t)} \frac{ds}{\psi(s)} \le K_n \widehat{M} \int_0^t p(s) \, ds$$

$$\le K_n \widehat{M} \int_0^n p(s) \, ds$$

$$< \int_{c_{21,n}}^{+\infty} \frac{ds}{\psi(s)}.$$

Thus, for every $t \in [0, n]$, there exists a constant Λ_n such that $v(t) \le \Lambda_n$ and hence $\mu(t) \le \Lambda_n$. Since $\|z\|_n \le \mu(t)$, we have $\|z\|_n \le \Lambda_n$. Set

$$\mathcal{U} = \{ z \in B^0_{+\infty} : \sup\{ |z(t)| : 0 \le t \le n \} < \Lambda_n + 1 \quad \text{for all } n \in \mathbb{N} \}.$$

Clearly, $\mathcal{U}$ is an open subset of $B^0_{+\infty}$.

We shall show that $\mathcal{F} : \overline{\mathcal{U}} \to \mathcal{P}(B^0_{+\infty})$ is a contraction and an admissible operator.

First, we prove that $\mathcal{F}$ is a contraction; Let $z, \bar{z} \in B^0_{+\infty}$ and $h \in \mathcal{F}(z)$. Then there exists $f(t) \in F(t, z_t + x_t)$ such that for each $t \in [0, n]$

$$h(t) = \int_0^t U(t, s) f(s) \, ds.$$

From (5.1.3) it follows that

$$H_d(F(t, z_t + x_t), F(t, \bar{z}_t + x_t)) \le l_n(t) \, \|z_t - \bar{z}_t\|_{\mathcal{B}}.$$

Hence, there is $\rho \in F(t, \bar{z}_t + x_t)$ such that

$$|f(t) - \rho| \le l_n(t) \, \|z_t - \bar{z}_t\|_{\mathcal{B}} \quad t \in [0, n].$$

Consider $\mathcal{U}_* : [0, n] \to \mathcal{P}(E)$, given by

$$\mathcal{U}_* = \{ \rho \in E : |f(t) - \rho| \le l_n(t) \, \|z_t - \bar{z}_t\|_{\mathcal{B}} \}.$$

Since the multi-valued operator $\mathcal{V}(t) = \mathcal{U}_*(t) \cap F(t, \overline{z}_t + x_t)$ is measurable (in [94], see Proposition III.4), there exists a function $\overline{f}(t)$, which is a measurable selection for $\mathcal{V}$. So, $\overline{f}(t) \in F(t, \overline{z}_t + x_t)$ and using (A_1), we obtain for each $t \in [0, n]$

$$
\begin{aligned}
|f(t) - \overline{f}(t)| &\leq l_n(t) \, \|z_t - \overline{z}_t\|_{\mathcal{B}} \\
&\leq l_n(t) \, [K(t) \, |z(t) - \overline{z}(t)| + M(t) \, \|z_0 - \overline{z}_0\|_{\mathcal{B}}] \\
&\leq l_n(t) \, K_n \, |z(t) - \overline{z}(t)|
\end{aligned}
$$

Let us define, for each $t \in [0, n]$

$$
\overline{h}(t) = \int_0^t U(t, s) \overline{f}(s) \, ds.
$$

Then we can show as in previous sections that we have

$$
\|h - \overline{h}\|_n \leq \frac{1}{\tau} \|z - \overline{z}\|_n.
$$

By an analogous relation, obtained by interchanging the roles of z and $\overline{z}$, it follows that

$$
H_d(\mathcal{F}(z), \mathcal{F}(\overline{z})) \leq \frac{1}{\tau} \|z - \overline{z}\|_{\overline{B}}.
$$

So, for $\tau > 1$, $\mathcal{F}$ is a contraction for all $n \in \mathbb{N}$.

It remains to show that $\mathcal{F}$ is an admissible operator. Let $z \in B_{+\infty}^0$. Set, for every $n \in \mathbb{N}$, the space

$$
B_n^0 := \{y : (-\infty, n] \to E : y|_{[0,n]} \in \mathcal{C}([0, n], E), \; y_0 \in \mathcal{B}\},
$$

and let us consider the multi-valued operator $\mathcal{F} : B_n^0 \to \mathcal{P}_{cl}(B_n^0)$ defined by:

$$
\mathcal{F}(z) = \left\{h \in B_n^0 : h(t) = \int_0^t U(t, s) f(s) \, ds, \quad t \in [0, n]\right\}.
$$

where $f \in S_{F,y}^n = \{v \in L^1([0, n], E) : v(t) \in F(t, y_t) \text{ for a.e. } t \in [0, n]\}$.

From (6.1.1) to (6.1.3) and since F is a multi-valued map with compact values, we can prove that for every $z \in B_n^0$, $\mathcal{F}(z) \in \mathcal{P}_{cl}(B_n^0)$ and there exists $z_* \in B_n^0$ such that $z_* \in \mathcal{F}(z_*)$. Let $h \in B_n^0$, $\overline{y} \in \mathcal{U}$ and $\epsilon > 0$. Assume that $z_* \in \mathcal{F}(\overline{z})$, then we have

$$
\begin{aligned}
|\overline{z}(t) - z_*(t)| &\leq |\overline{z}(t) - h(t)| + |z_*(t) - h(t)| \\
&\leq e^{\tau \, L_n^*(t)} \, \|\overline{z} - \mathcal{F}(\overline{z})\|_n + \|z_* - h\|.
\end{aligned}
$$

Since h is arbitrary, we may suppose that $h \in B(z_*, \epsilon) = \{h \in B_n^0 : \|h - z_*\|_n \leq \epsilon\}$. Therefore,

$$\|\overline{z} - z_*\|_n \leq \|\overline{z} - \mathcal{F}(\overline{z})\|_n + \epsilon.$$

If z is not in $\mathcal{F}(\overline{z})$, then $\|z_* - \mathcal{F}(\overline{z})\| \neq 0$. Since $\mathcal{F}(\overline{z})$ is compact, there exists $x \in \mathcal{F}(\overline{z})$ such that $\|z_* - \mathcal{F}(\overline{z})\| = \|z_* - x\|$. Then we have

$$|\overline{z}(t) - z_*(t)| \leq |\overline{z}(t) - h(t)| + |x(t) - h(t)|$$

$$\leq e^{\tau \, L_n^*(t)} \|\overline{z} - \mathcal{F}(\overline{z})\|_n + |x(t) - h(t)|.$$

Thus,

$$\|\overline{z} - x\|_n \leq \|\overline{z} - \mathcal{F}(\overline{z})\|_n + \epsilon.$$

So, $\mathcal{F}$ is an admissible operator contraction. From the choice of $\mathcal{U}$ there is no $z \in \partial \mathcal{U}$ such that $z = \lambda \, \mathcal{F}(z)$ for some $\lambda \in (0, 1)$. Then the statement (S2) in Theorem 1.31 does not hold. A consequence of the nonlinear alternative due to Frigon we get that (S1) holds, we deduce that the operator $\mathcal{F}$ has a fixed point z^*. Then $y^*(t) = z^*(t) + x(t)$, $t \in \mathbb{R}$ is a fixed point of the operator N_{21}, which is a mild solution of the evolution inclusion problem (6.1)–(6.2). $\qquad \square$

6.2.3 An Example

Consider the following model

$$\begin{cases} \dfrac{\partial v}{\partial t}(t, \xi) \in a(t, \xi) \dfrac{\partial^2 v}{\partial \xi^2}(t, \xi) \\[2mm] \qquad + \displaystyle\int_{-\infty}^0 P(\theta) R(t, v(t + \theta, \xi)) d\theta \quad \xi \in [0, \pi] \\[4mm] v(t, 0) = v(t, \pi) = 0 \qquad\qquad\qquad t \in \mathbb{R}_+ \\[2mm] v(\theta, \xi) = v_0(\theta, \xi) \qquad\qquad\qquad -\infty < \theta \leq 0, \xi \in [0, \pi], \end{cases} \qquad (6.4)$$

where $a(t, \xi)$ is a continuous function and is uniformly Hölder continuous in t; $P : (-\infty, 0] \to \mathbb{R}$ and $v_0 : (-\infty, 0] \times [0, \pi] \to \mathbb{R}$ are continuous functions and $R : \mathbb{R}_+ \times \mathbb{R} \to \mathcal{P}(\mathbb{R})$ is a multi-valued map with compact convex values.

Consider $E = L^2([0, \pi], \mathbb{R})$ and define $A(t)$ by $A(t)w = a(t, \xi)w''$ with domain

$$D(A) = \{ w \in E : w, w' \text{ are absolutely continuous, } w'' \in E, w(0) = w(\pi) = 0 \}$$

Then $A(t)$ generates an evolution system $U(t, s)$ satisfying assumption (6.1.1) (see [112, 149]).

For the phase space $\mathcal{B}$, we choose the well-known space $BUC(\mathbb{R}^-, E)$: the space of uniformly bounded continuous functions endowed with the following norm

$$\|\varphi\| = \sup_{\theta \leq 0} |\varphi(\theta)| \quad \text{for} \quad \varphi \in \mathcal{B}.$$

If we put for $\varphi \in BUC(\mathbb{R}^-, E)$ and $\xi \in [0, \pi]$

$$y(t)(\xi) = v(t, \xi), \ t \in \mathbb{R}_+, \ \xi \in [0, \pi],$$

$$\phi(\theta)(\xi) = v_0(\theta, \xi), \ -\infty < \theta \leq 0, \ \xi \in [0, \pi],$$

and

$$F(t, \varphi)(\xi) = \int_{-\infty}^{0} P(\theta) R(t, \varphi(\theta)(\xi)) d\theta, \ -\infty < \theta \leq 0, \ \xi \in [0, \pi].$$

Then, the problem (6.4) takes the abstract partial functional evolution inclusion form (6.1)–(6.2). In order to show the existence of mild solutions of problem (6.4), we suppose the following assumptions:

– There exist $p \in L^1(J, \mathbb{R}^+)$ and a nondecreasing continuous function $\psi : \mathbb{R}_+ \rightarrow (0, +\infty)$ such that

$$|R(t, u)| \leq p(t) \psi(|u|), \ \text{for} \ \in J, \ \text{and} \ u \in \mathbb{R}.$$

– P is integrable on $(-\infty, 0]$.

By the dominated convergence theorem, one can show that $f \in S_{F,y}$ is a continuous function from $\mathcal{B}$ to (E). On the other hand, we have for $\varphi \in \mathcal{B}$ and $\xi \in [0, \pi]$

$$|F(t, \varphi)(\xi)| \leq \int_{-\infty}^{0} |p(t) P(\theta)| \, \psi(|(\varphi(\theta))(\xi)|) d\theta.$$

Since the function ψ is nondecreasing, it follows that

$$\|F(t, \varphi)\|_{\mathcal{P}(E)} \leq p(t) \int_{-\infty}^{0} |P(\theta)| \, d\theta \psi(|\varphi|), \ \text{for} \ \varphi \in \mathcal{B}.$$

Proposition 6.3. *Under the above assumptions, if we assume that condition (6.3) in Theorem 6.2 is true, $\varphi \in \mathcal{B}$, then the problem (6.4) has a mild solution which is defined in $(-\infty, +\infty)$.*

6.3 Neutral Functional Evolution Inclusions

6.3.1 Introduction

A generalization of existence result of mild solutions to the neutral case is developed in Sect. 6.3 where we look for the neutral functional evolution inclusions with infinite delay of the form

$$\frac{d}{dt}[y(t) - g(t, y_t)] \in A(t)y(t) + F(t, y_t), \quad \text{a.e.} \quad t \in J \tag{6.5}$$

$$y_0 = \phi \in \mathcal{B}, \tag{6.6}$$

where $F : J \times \mathcal{B} \to \mathcal{P}(E)$ is a multi-valued map with nonempty compact values, $g : J \times \mathcal{B} \to E$ and $\phi \in \mathcal{B}$ are given functions.

6.3.2 Existence of Mild Solutions

Definition 6.4. We say that the function $y(\cdot) : \mathbb{R} \to E$ is a mild solution of the neutral functional evolution system (6.5)–(6.6) if $y(t) = \phi(t)$ for all $t \in (-\infty, 0]$ and the restriction of $y(\cdot)$ to the interval J is continuous and there exists $f(\cdot) \in L^1(J, E)$: $f(t) \in F(t, y_t)$ a.e. in J such that y satisfies the following integral equation

$$y(t) = U(t, 0)[\phi(0) - g(0, \phi)] + g(t, y_t) + \int_0^t U(t, s)A(s)g(s, y_s)ds$$
$$+ \int_0^t U(t, s)f(s)\, ds, \quad \text{for each } t \in \mathbb{R}_+.$$

We consider the hypotheses (6.1.1)–(6.1.3) and in what follows we will need the following additional assumptions:

(G1) There exists a constant $\overline{M}_0 > 0$ such that:

$$\|A^{-1}(t)\|_{B(E)} \leq \overline{M}_0 \quad \text{for all } t \in J.$$

(G2) There exists a constant $0 < L < \dfrac{1}{\overline{M}_0 K_n}$ such that:

$$|A(t)\, g(t, \phi)| \leq L\, (\|\phi\|_\mathcal{B} + 1) \text{ for all } t \in J \text{ and } \phi \in \mathcal{B}.$$

(G3) There exists a constant $L_* > 0$ such that:

$$|A(s)\, g(s, \phi) - A(\overline{s})\, g(\overline{s}, \overline{\phi})| \leq L_*\, (|s - \overline{s}| + \|\phi - \overline{\phi}\|_\mathcal{B})$$

for all $s, \overline{s} \in J$ and $\phi, \overline{\phi} \in \mathcal{B}$.

Consider the following space

$$B_{+\infty} = \left\{ y : \mathbb{R} \to E : y|_{[0,T]} \in C([0,T], E), \ y_0 \in \mathcal{B} \right\},$$

where $y|_{[0,T]}$ is the restriction of y to any real compact interval $[0,T]$.

For every $n \in \mathbb{N}$, let us take here $\bar{l}_n(t) = K_n \widehat{M}[L_* + l_n(t)]$ for the family of seminorm $\{\| \cdot \|_n\}_{n \in \mathbb{N}}$. In what follows we fix $\tau > 0$ and assume that

$$\left[\overline{M}_0 L_* K_n + \frac{1}{\tau} \right] < 1.$$

Theorem 6.5. *Suppose that hypotheses* (6.1.1)–(6.1.3) *and the assumptions* (G1)–(G3) *are satisfied and moreover*

$$\int_{c_{22,n}}^{+\infty} \frac{ds}{s + \psi(s)} > \frac{K_n \widehat{M}}{1 - \overline{M}_0 L K_n} \int_0^n \max(L, p(s)) ds, \quad \text{for each } n > 0, \quad (6.7)$$

with

$$c_{22,n} = \left[\frac{\overline{M}_0 L K_n \widehat{M}}{1 - \overline{M}_0 L K_n} + (K_n \widehat{M} H + M_n) \left(1 + \frac{\overline{M}_0 L K_n}{1 - \overline{M}_0 L K_n} \right) \right] \|\phi\|_{\mathcal{B}}$$

$$+ \frac{K_n}{1 - \overline{M}_0 L K_n} \left[\overline{M}_0 L(\widehat{M} + 1) + \widehat{M} L n \right],$$

then the neutral functional evolution problem (6.5)–(6.6) *has a mild solution.*

Proof. Consider the multi-valued operator $N_{22} : B_{+\infty} \to \mathcal{P}(B_{+\infty})$ defined by

$$N_{22}(y) = \left\{ h \in B_{+\infty} : h(t) = \begin{cases} \phi(t), & \text{if } t \leq 0; \\[2mm] U(t,0) [\phi(0) - g(0,\phi)] + g(t, y_t) \\[1mm] + \displaystyle\int_0^t U(t,s) A(s) g(s, y_s) ds \\[1mm] + \displaystyle\int_0^t U(t,s) f(s) ds, & \text{if } t \geq 0, \end{cases} \right\}$$

where $f \in S_{F,y} = \{ v \in L^1(J, E) : v(t) \in F(t, y_t) \text{ for a.e. } t \in J \}$.

Clearly, the fixed points of the operator N_{22} are mild solutions of the problem (6.5)–(6.6). We remark also that, for each $y \in B_{+\infty}$, the set $S_{F,y}$ is nonempty since, by (6.1.2), F has a measurable selection (see [94], Theorem III.6).

For $\phi \in \mathcal{B}$, we will define the function $x(.) : \mathbb{R} \to E$ by

$$x(t) = \begin{cases} \phi(t), & \text{if } t \in (-\infty, 0]; \\[2mm] U(t,0) \phi(0), & \text{if } t \in J. \end{cases}$$

Then $x_0 = \phi$. For each function $z \in B_{+\infty}$, set $y(t) = z(t) + x(t)$. It is obvious that y satisfies Definition 6.4 if and only if z satisfies $z_0 = 0$ and

$$z(t) = g(t, z_t + x_t) - U(t, 0)g(0, \phi) + \int_0^t U(t, s)A(s)g(s, z_s + x_s)ds$$

$$+ \int_0^t U(t, s)f(s)ds.$$

where $f(t) \in F(t, z_t + x_t)$ a.e. $t \in J$.

Let

$$B_{+\infty}^0 = \{z \in B_{+\infty} : z_0 = 0\}.$$

Define in $B_{+\infty}^0$, the multi-valued operator $\mathcal{F} : B_{+\infty}^0 \to \mathcal{P}(B_{+\infty}^0)$ by:

$$\mathcal{F}(z) = \left\{ h \in B_{+\infty}^0 : h(t) = g(t, z_t + x_t) - U(t, 0)g(0, \phi) \right.$$

$$\left. + \int_0^t U(t, s)A(s)g(s, z_s + x_s)ds + \int_0^t U(t, s)f(s)ds, \ t \in J \right\}$$

where $f \in S_{F,z} = \{v \in L^1(J, E) : v(t) \in F(t, z_t + x_t) \text{ for a.e. } t \in J\}$.

Obviously the operator inclusion N_{22} has a fixed point is equivalent to the operator inclusion $\mathcal{F}$ has one, so it turns to prove that $\mathcal{F}$ has a fixed point.

Let $z \in B_{+\infty}^0$ be a possible fixed point of the operator $\mathcal{F}$. Given $n \in \mathbb{N}$, then z should be solution of the inclusion $z \in \lambda \mathcal{F}(z)$ for some $\lambda \in (0, 1)$ and there exists $f \in S_{F,z} \Leftrightarrow f(t) \in F(t, z_t + x_t)$ such that, for each $t \in [0, n]$, we have

$$|z(t)| \leq \|A^{-1}(t)\|_{B(E)}\|A(t)g(t, z_t + x_t)\| + \|U(t, 0)\|_{B(E)}\|A^{-1}(0)\|\|A(0) g(0, \phi)\|$$

$$+ \int_0^t \|U(t, s)\|_{B(E)}\|A(s)g(s, z_s + x_s)\|ds + \int_0^t \|U(t, s)\|_{B(E)}|f(s)|ds$$

$$\leq \overline{M}_0 L(\|z_t + x_t\|_B + 1) + \widehat{M}\overline{M}_0 L(\|\phi\|_B + 1)$$

$$+ \widehat{M} \int_0^t L(\|z_s + x_s\|_B + 1)ds + \widehat{M} \int_0^t p(s)\psi(\|z_s + x_s\|_B)ds$$

$$\leq \overline{M}_0 L\|z_t + x_t\|_B + \overline{M}_0 L(\widehat{M} + 1) + \widehat{M}Ln + \widehat{M}\overline{M}_0 L\|\phi\|_B$$

$$+ \widehat{M} \int_0^t L\|z_s + x_s\|_B ds + \widehat{M} \int_0^t p(s)\psi(\|z_s + x_s\|_B)ds.$$

Using the inequality (3.4) and the nondecreasing character of ψ, we obtain

$$|z(t)| \leq \overline{M}_0 L(K_n|z(t)| + \alpha_n) + \overline{M}_0 L(\widehat{M} + 1) + \widehat{M}Ln + \widehat{M}\overline{M}_0 L\|\phi\|_{\mathcal{B}}$$
$$+\widehat{M}\int_0^t L(K_n|z(s)| + \alpha_n)ds + \widehat{M}\int_0^t p(s)\psi(K_n|z(s)| + \alpha_n)ds$$
$$\leq \overline{M}_0 LK_n|z(t)| + \overline{M}_0 L(\widehat{M} + 1) + \widehat{M}Ln + \overline{M}_0 L\alpha_n + \widehat{M}\overline{M}_0 L\|\phi\|_{\mathcal{B}}$$
$$+\widehat{M}\left[\int_0^t L(K_n|z(s)| + \alpha_n)ds + \int_0^t p(s)\psi(K_n|z(s)| + \alpha_n)ds\right].$$

Then

$$(1 - \overline{M}_0 LK_n)|z(t)| \leq (\widehat{M} + 1)\overline{M}_0 L + \widehat{M}Ln + \overline{M}_0 L\alpha_n + \widehat{M}\overline{M}_0 L\|\phi\|_{\mathcal{B}}$$
$$+\widehat{M}\left[\int_0^t L(K_n|z(s)| + \alpha_n)ds + \int_0^t p(s)\psi(K_n|z(s)| + \alpha_n)ds\right].$$

Set

$$c_{22,n} := \alpha_n + \frac{K_n}{1 - \overline{M}_0 LK_n}\left[(\widehat{M} + 1)\overline{M}_0 L + \widehat{M}Ln + \overline{M}_0 L\alpha_n + \widehat{M}\overline{M}_0 L\|\phi\|_{\mathcal{B}}\right].$$

Thus

$$K_n|z(t)| + \alpha_n \leq c_{22,n}$$
$$+\frac{K_n\widehat{M}}{1 - \overline{M}_0 LK_n}\left[\int_0^t L(K_n|z(s)| + \alpha_n)ds\right.$$
$$\left.+\int_0^t p(s)\psi(K_n|z(s)| + \alpha_n)ds\right].$$

We consider the function μ defined by

$$\mu(t) := \sup\{K_n|z(s)| + \alpha_n : 0 \leq s \leq t\}, \quad 0 \leq t < +\infty.$$

Let $t^* \in [0, t]$ be such that $\mu(t) = K_n|z(t^*)| + \alpha_n$. By the previous inequality, we have

$$\mu(t) \leq c_{22,n} + \frac{K_n\widehat{M}}{1 - \overline{M}_0 LK_n}\left[\int_0^t L\mu(s)ds + \int_0^t p(s)\psi(\mu(s))ds\right] \quad \text{for } t \in [0, n].$$

Let us take the right-hand side of the above inequality as $v(t)$. Then, we have

$$\mu(t) \leq v(t) \quad \text{for all } t \in [0, n].$$

From the definition of v, we have $v(0) = c_{22,n}$ and

$$v'(t) = \frac{K_n \widehat{M}}{1 - \overline{M}_0 L K_n} [L\mu(t) + p(t)\psi(\mu(t))] \quad \text{a.e. } t \in [0, n].$$

Using the nondecreasing character of ψ, we get

$$v'(t) \le \frac{K_n \widehat{M}}{1 - \overline{M}_0 L K_n} [Lv(t) + p(t)\psi(v(t))] \quad \text{a.e. } t \in [0, n].$$

This implies that for each $t \in [0, n]$ and using the condition (6.7), we get

$$\int_{c_{22,n}}^{v(t)} \frac{ds}{s + \psi(s)} \le \frac{K_n \widehat{M}}{1 - \overline{M}_0 L K_n} \int_0^t \max(L, p(s)) ds$$

$$\le \frac{K_n \widehat{M}}{1 - \overline{M}_0 L K_n} \int_0^n \max(L, p(s)) ds$$

$$< \int_{c_{22,n}}^{+\infty} \frac{ds}{s + \psi(s)}.$$

Thus, for every $t \in [0, n]$, there exists a constant Λ_n such that $v(t) \le \Lambda_n$ and hence $\mu(t) \le \Lambda_n$. Since $\|z\|_n \le \mu(t)$, we have $\|z\|_n \le \Lambda_n$.

We can show that $\mathcal{F}$ is an admissible operator and we shall prove now that $\tilde{\mathcal{F}}$: $\overline{\mathcal{U}} \to \mathcal{P}(B^0_{+\infty})$ is a contraction.

Let $z, \overline{z} \in B^0_{+\infty}$ and $h \in \mathcal{F}(z)$. Then there exists $f(t) \in F(t, z_t + x_t)$ such that for each $t \in [0, n]$, we have

$$h(t) = g(t, z_t + x_t) - U(t, 0)g(0, \phi) + \int_0^t U(t, s)A(s)g(s, z_s + x_s)ds$$

$$+ \int_0^t U(t, s)f(s)ds.$$

From (5.1.3) it follows that

$$H_d(F(t, z_t + x_t), F(t, \overline{z}_t + x_t)) \le l_n(t) \|z_t - \overline{z}_t\|_{\mathcal{B}}.$$

Hence, there is $\rho \in F(t, \overline{z}_t + x_t)$ such that

$$|f(t) - \rho| \le l_n(t) \|z_t - \overline{z}_t\|_{\mathcal{B}} \quad t \in [0, n].$$

Consider $\mathcal{U}_* : [0, n] \to \mathcal{P}(E)$, given by

$$\mathcal{U}_* = \{\rho \in E : |f(t) - \rho| \le l_n(t) \|z_t - \overline{z}_t\|_{\mathcal{B}}\}.$$

Since the multi-valued operator $V(t) = U_*(t) \cap F(t, \bar{z}_t + x_t)$ is measurable (in [94], see Proposition III.4), there exists a function $\bar{f}(t)$, which is a measurable selection for V. So, $\bar{f}(t) \in F(t, \bar{z}_t + x_t)$ and using (A_1), we obtain for each $t \in [0, n]$

$$
\begin{aligned}
|f(t) - \bar{f}(t)| &\leq l_n(t) \, \|z_t - \bar{z}_t\|_{\mathcal{B}} \\
&\leq l_n(t) \, [K(t) \, |z(t) - \bar{z}(t)| + M(t) \, \|z_0 - \bar{z}_0\|_{\mathcal{B}}] \\
&\leq l_n(t) \, K_n \, |z(t) - \bar{z}(t)|.
\end{aligned}
$$

Let us define, for each $t \in [0, n]$

$$
\bar{h}(t) = g(t, \bar{z}_t + x_t) - U(t, 0)g(0, \phi) + \int_0^t U(t, s)A(s)g(s, \bar{z}_s + x_s)ds
$$

$$
+ \int_0^t U(t, s)\bar{f}(s)ds.
$$

Then we can show as in previous sections that we have for each $t \in [0, n]$ and $n \in \mathbb{N}$

$$
\|h - \bar{h}\|_n \leq \left[\overline{M}_0 L_* K_n + \frac{1}{\tau} \right] \|z - \bar{z}\|_n.
$$

By an analogous relation, obtained by interchanging the roles of z and $\bar{z}$, it follows that

$$
H_d(\mathcal{F}(z), \mathcal{F}(\bar{z})) \leq \left[\overline{M}_0 L_* K_n + \frac{1}{\tau} \right] \|z - \bar{z}\|_{\overline{B}}.
$$

So, for $\left[\overline{M}_0 L_* K_n + \frac{1}{\tau} \right] < 1$, the operator $\mathcal{F}$ is a contraction for all $n \in \mathbb{N}$ and an admissible operator. From the choice of U there is no $z \in \partial U$ such that $z = \lambda \, \mathcal{F}(z)$ for some $\lambda \in (0, 1)$. Then the statement $(S2)$ in Theorem 1.31 does not hold. By the nonlinear alternative due to Frigon we get that $(S1)$ holds, we deduce that the operator $\mathcal{F}$ has a fixed point z^*. Then $y^*(t) = z^*(t) + x(t)$, $t \in \mathbb{R}$ is a fixed point of the operator N_{22}, which is a mild solution of the neutral functional evolution inclusion problem (6.5)–(6.6).   □

6.3.3   An Example

Consider the following model

$$
\begin{cases}
\dfrac{\partial}{\partial t}\left[v(t,\xi) - \displaystyle\int_{-\infty}^{0} T(\theta)u(t, v(t+\theta,\xi))d\theta \right] \\[2mm]
\qquad \in\; a(t,\xi)\,\dfrac{\partial^2 v}{\partial \xi^2}(t,\xi) \\[2mm]
\qquad +\displaystyle\int_{-\infty}^{0} P(\theta)R(t, v(t+\theta,\xi))d\theta \qquad\qquad t\in\mathbb{R}_+, \quad \xi\in[0,\pi] \\[4mm]
v(t,0) = v(t,\pi) = 0 \qquad\qquad\qquad\qquad\qquad t\in\mathbb{R}_+ \\[2mm]
v(\theta,\xi) = v_0(\theta,\xi) \qquad\qquad\qquad\qquad -\infty < \theta \le 0,\, \xi\in[0,\pi],
\end{cases}
$$
$$\tag{6.8}$$

where $a(t,\xi)$ is a continuous function and is uniformly Hölder continuous in t; $T, P :$ $(-\infty, 0] \to \mathbb{R}$; $u : (-\infty, 0]\times\mathbb{R} \to \mathbb{R}$ and $v_0 : (-\infty, 0]\times[0,\pi] \to \mathbb{R}$ are continuous functions and $R : \mathbb{R}_+ \times \mathbb{R} \to \mathcal{P}(\mathbb{R})$ is a multi-valued map with compact convex values.

Consider $E = L^2([0,\pi],\mathbb{R})$ and define $A(t)$ by $A(t)w = a(t,\xi)w''$ with domain

$$D(A) = \{\, w \in E : w, w' \text{ are absolutely continuous, } w'' \in E,\ w(0) = w(\pi) = 0 \,\}$$

Then $A(t)$ generates an evolution system $U(t,s)$ satisfying assumptions (4.1.1) and (G1), see [112, 149].

For the phase space $\mathcal{B}$, we choose the well-known space $BUC(\mathbb{R}^-, E)$: the space of uniformly bounded continuous functions endowed with the following norm

$$\|\varphi\| = \sup_{\theta \le 0} |\varphi(\theta)| \quad \text{for} \quad \varphi \in \mathcal{B}.$$

If we put for $\varphi \in BUC(\mathbb{R}^-, E)$ and $\xi \in [0,\pi]$

$$y(t)(\xi) = v(t,\xi),\ t\in\mathbb{R}_+,\ \xi\in[0,\pi],$$

$$\phi(\theta)(\xi) = v_0(\theta,\xi),\ -\infty < \theta \le 0,\ \xi\in[0,\pi],$$

$$g(t,\varphi)(\xi) = \int_{-\infty}^{0} T(\theta)u(t,\varphi(\theta)(\xi))d\theta,\ -\infty < \theta \le 0,\ \xi\in[0,\pi],$$

and

$$F(t,\varphi)(\xi) = \int_{-\infty}^{0} P(\theta)R(t,\varphi(\theta)(\xi))d\theta,\ -\infty < \theta \le 0,\ \xi\in[0,\pi].$$

Then, the problem (6.8) takes the abstract neutral functional evolution inclusion form (6.5)–(6.6). In order to show the existence of mild solutions of problem (6.8), we suppose the following assumptions:

- u is Lipschitz with respect to its second argument. Let $lip(u)$ denotes the Lipschitz constant of u.
- There exist $p \in L^1(J, \mathbb{R}^+)$ and a nondecreasing continuous function $\psi : \mathbb{R}_+ \to (0, +\infty)$ such that

$$|R(t, x)| \leq p(t)\psi(|x|), \text{ for } \in J, \text{ and } x \in \mathbb{R}.$$

- T, P are integrable on $(-\infty, 0]$.

By the dominated convergence theorem, one can show that $f \in S_{F,y}$ is a continuous function from $\mathcal{B}$ to (E). Moreover the mapping g is Lipschitz continuous in its second argument, in fact, we have

$$|g(t, \varphi_1) - g(t, \varphi_2)| \leq \overline{M}_0 L_* lip(u) \int_{-\infty}^0 |T(\theta)|\, d\theta\, |\varphi_1 - \varphi_2|, \text{ for } \varphi_1, \varphi_2 \in \mathcal{B}.$$

On the other hand, we have for $\varphi \in \mathcal{B}$ and $\xi \in [0, \pi]$

$$|F(t, \varphi)(\xi)| \leq \int_{-\infty}^0 |p(t)P(\theta)|\, \psi(|(\varphi(\theta))(\xi)|)d\theta.$$

Since the function ψ is nondecreasing, it follows that

$$\|F(t, \varphi)\|_{\mathcal{P}(E)} \leq p(t) \int_{-\infty}^0 |P(\theta)|\, d\theta\psi(|\varphi|), \text{ for } \varphi \in \mathcal{B}.$$

Proposition 6.6. *Under the above assumptions, if we assume that condition (6.7) in Theorem 6.5 is true, $\varphi \in \mathcal{B}$, then the problem (6.8) has a mild solution which is defined in $(-\infty, +\infty)$.*

6.4 Notes and Remarks

The results of Chap. 6 are taken from [11, 12, 35, 45]. Other results may be found in [10, 11, 54, 142, 145, 182].

Chapter 7
Densely Defined Functional Differential Inclusions with Finite Delay

7.1 Introduction

In this chapter, we are concerned by the existence of mild and extremal solutions of some first order classes of impulsive semi-linear functional differential inclusions with local and nonlocal conditions when the delay is finite in a separable Banach space $(E, |\cdot|)$.

In the literature devoted to equations with finite delay, the phase space is much of time the space of all continuous functions on H, endowed with the uniform norm topology. We mention, for instance, the books of Ahmed [16], Engel and Nagel [106], Kamenskii et al. [144], Pazy [168], and Wu [184].

7.2 Existence of Mild Solutions with Local Conditions

7.2.1 Introduction

In this section, we consider the following class of semi-linear impulsive differential inclusions:

$$y'(t) - Ay(t) \in F(t, y_t), \quad t \in J := [0, b], t \neq t_k \tag{7.1}$$

$$\Delta y|_{t=t_k} \in I_k(y(t_k^-)), \quad k = 1, \dots, m \tag{7.2}$$

$$y(t) = \phi(t), \quad t \in H, \tag{7.3}$$

where $F : J \times D \to 2^E$ is a closed, bounded, and convex valued multi-valued map,

© Springer International Publishing Switzerland 2015
S. Abbas, M. Benchohra, *Advanced Functional Evolution Equations and Inclusions*, Developments in Mathematics 39,
DOI 10.1007/978-3-319-17768-7_7

$D = \{\psi : H \to E, \psi$ continuous everywhere except for a finite number of points at which $\psi(s^-)$ and $\psi(s^+)$ exist and $\psi(s^-) = \psi(s)\}, \phi \in D, A : D(A) \subset E \to E$ is the infinitesimal generator of a strongly continuous semigroup $T(t)$, $t \geq 0$, E a real separable Banach space endowed with the norm $|.|$, $0 = t_0 < t_1 < \cdots < t_m < t_{m+1} = b$, $I_k \in C(E, E), (k = 1, 2, \ldots, m)$.

7.2.2   Main Result

We assume that F is compact and convex valued multi-valued map. In order to define the mild solution to the problem (7.1)–(7.3), we shall consider the following space

$$PC = \left\{ y : [0, b] \to E : \quad y_k \in C[J_k, E], k = 0, \ldots, m \quad \text{such that} \right.$$
$$\left. y(t_k^-), y(t_k^+) \text{ exist with } y(t_k) = y(t_k^-), k = 1, \ldots, m \right\},$$

which is a Banach space with the norm

$$\|y\|_{PC} := \max\{\|y_k\|_\infty : k = 0, \ldots, m\},$$

where y_k is the restriction of y to $J_k = [t_k, t_{k+1}], k = 0, \ldots, m$.
    Set

$$\Omega = \{y : [-r, b] \to E : y \in D \cap PC\}.$$

Definition 7.1. A function $y \in \Omega$ is said to be a mild solution of system (7.1)–(7.3) if $y(t) = \phi(t)$ for all $t \in H$, the restriction of $y(\cdot)$ to the interval $[0, b]$ is continuous and there exists $v(\cdot) \in L^1(J_k, E)$, $\mathcal{I}_k \in I_k(y(t_k^-))$ such that $v(t) \in F(t, y_t)$ a.e $t \in [0, b]$, and such that y satisfies the integral equation,

$$y(t) = T(t)\phi(0) + \int_0^t T(t - s)v(s)ds + \sum_{0 < t_k < t} T(t - t_k)\mathcal{I}_k, \quad t \in J.$$

We will need the following hypotheses which are assumed hereafter

(7.1.1) $A : D(A) \subset E \to E$ is the infinitesimal generator of a strongly continuous semigroup $\{T(t)\}, t \in J$ which is compact for $t > 0$ in the Banach space E, and there exists a constant $M \geq 1$, such that Let $\|T(t)\|_{B(E)} \leq M; \ t \geq 0$

(7.1.2) There exist constants $c_k \geq 0, k = 1, \ldots, m$ such that

$$H_d(I_k(y) - I_k(x)) \leq c_k|y - x| \quad \text{for each} \quad x, y \in E.$$

(7.1.3) F is L^1-Carathéodory with compact convex values.

(7.1.4) There exist a function $k \in L^1(J, \mathbb{R}_+)$ and a continuous nondecreasing function $\psi : [0, \infty) \to (0, \infty)$ such that

$$\|F(t, x)\| \leq k(t)\psi(\|x\|_{\mathcal{D}}) \text{ for a.e. } t \in J \text{ and each } x \in D,$$

with

$$\int_{c_0}^{\infty} \frac{ds}{\psi(s)} > C_1 \|p\|_{L^1}, \tag{7.4}$$

where

$$C_0 = \frac{M[\|\phi\|_D + \sum_{k=1}^{m} \|I_k(0)\|]}{1 - \sum_{k=1}^{m} c_k} \tag{7.5}$$

$$C_1 = \frac{M}{1 - \sum_{k=1}^{m} c_k}. \tag{7.6}$$

Theorem 7.2. *Assume that (7.1.1)–(7.1.4) hold. If*

$$M \sum_{k=1}^{m} c_k < 1 \tag{7.7}$$

then the IVP (7.1)–(7.3) has at least one mild solution on $[-r, b]$.

Proof. Transform the problem (7.1)–(7.3) into a fixed point problem. Consider the multi-valued operator: $N : \Omega \to \Omega$ defined by

$$N(y) = \left\{ h \in \Omega : h(t) = \begin{cases} \phi(t), & \text{if } t \in H, \\ T(t)\phi(0) + \int_0^t T(t-s)v(s)ds \\ + \sum_{0 < t_k < t} T(t-t_k)\mathcal{I}_k, v \in S_{F,y}, \mathcal{I}_k \in I_k(y(t_k^-)) & \text{if } t \in J. \end{cases} \right\}$$

It is clear that the fixed points of N are mild solutions of the IVP (7.1)–(7.3). Consider these multi-valued operators:

$$\mathcal{A}, \mathcal{B} : \Omega \to \Omega$$

defined by

$$
A(y) := \left\{ h \in \Omega : \ h(t) = \begin{cases} 0, & \text{if } t \in H; \\ \displaystyle\sum_{0 < t_k < t} T(t - t_k)\mathcal{I}_k, \mathcal{I}_k \in I_k(y(t_k^-)) & \text{if } t \in J, \end{cases} \right\}
$$

and

$$
B(y) := \left\{ h \in \Omega : \ h(t) = \begin{cases} \phi(t), & \text{if } t \in H; \\ T(t)\phi(0) \\ \displaystyle + \int_0^t T(t - s)v(s)ds, \ v \in S_{F,y} & \text{if } t \in J. \end{cases} \right\}
$$

The problem of finding mild solutions of (7.1)–(7.3) is then reduced to finding mild solutions of the operator inclusion $y \in A(y) + B(y)$. The proof will be given in several steps.

Step 1: A is a contraction. Let $y_1, y_2 \in \Omega$, then from (7.1.2) we have

$$
H_d(A(y_1), A(y_2)) = H_d\left(\sum_{0 < t_k < t} T(t - t_k)I_k(y_1(t_k^-)), \sum_{0 < t_k < t} T(t - t_k)I_k(y_2(t_k^-)) \right)
$$

$$
\leq M \sum_{k=0}^{k=m} c_k |[y_1(t_k^-) - y_2(t_k^-)|
$$

$$
\leq M \sum_{k=0}^{k=m} c_k \|y_1 - y_2\|_\Omega.
$$

From (7.7) it follows that A is a contraction.

Step 2: B has compact, convex values, and is completely continuous.

   Claim 1: B has compact values.
      The operator B is equivalent to the composition $\mathcal{L} \circ S_F$ of two operators on $L^1(J, E)$, where $\mathcal{L} : L^1(J, E) \to \Omega$ is the continuous operator defined by

$$
\mathcal{L}(v(t)) = T(t)\phi(0) + \int_0^t T(t - s)v(s)ds
$$

   Then, it suffices to show that $\mathcal{L} \circ S_F$ has compact values on Ω.
   Let $y \in \Omega$ arbitrary, v_n a sequence in $S_{F,y}$, then by definition of S_F, $v_n(t)$ belongs to $F(t, y_t), a.e.t \in J$. Since $F(t, y_t)$ is compact, we may pass to a subsequence.
   suppose that $v_n \to v$ in $L^1(J, E)$, where $v(t) \in F(t, y_t), a.e.t \in J$.

From the continuity of $\mathcal{L}$, it follows that $\mathcal{L}v_n(t) \to \mathcal{L}v(t)$ point wise on J and $n \to \infty$.

In order to show that the convergence is uniform, we first show that $\{\mathcal{L}v_n\}$ is an equi-continuous sequence.

Let $\tau_1, \tau_2 \in J$, then we have:

$$
\left| \mathcal{L}(v_n(\tau_1)) - \mathcal{L}(v_n(\tau_2)) \right| = \left| T(\tau_1)\phi(0) - T(\tau_2)\phi(0) + \int_0^{\tau_1} T(\tau_1 - s)v_n(s)ds \right.
$$

$$
\left. - \int_0^{\tau_2} T(\tau_2 - s)v_n(s)ds \right|
$$

$$
\leq \left| (T(\tau_1) - T(\tau_2))\phi(0) \right|
$$

$$
+ \int_0^{\tau_1} \left| (T(\tau_1 - s) - T(\tau_2 - s)) \right| |v_n(s)| ds
$$

$$
+ \int_{\tau_1}^{\tau_2} |T(\tau_2 - s)| |v_n(s)| ds.
$$

As $\tau_1 \to \tau_2$, the right-hand side of the above inequality tends to zero. Since $T(t)$ is a strongly continuous operator and the compactness of $T(t)$; $t > 0$, implies the continuity in uniform topology. Hence $\{\mathcal{L}v_n\}$ is equi-continuous, and an application of Arzéla–Ascoli theorem implies that there exists a subsequence which is uniformly convergent. Then we have $\mathcal{L}v_{n_j} \to \mathcal{L}v \in (\mathcal{L} \circ S_F)(y)$ as $j \mapsto \infty$, and so $(\mathcal{L} \circ S_F)(y)$ is compact. Therefore $\mathcal{B}$ has compact values.

Claim 2: $\mathcal{B}(y)$ is convex for each $y \in \Omega$. Let $h_1, h_2 \in \mathcal{B}(y)$, then there exists $v_1, v_2 \in S_{F,y}$ such that, for each $t \in J$ we have

$$
h_i(t) = \begin{cases} \phi(t), & \text{if } t \in H, \\[2mm] T(t)\phi(0) + \displaystyle\int_0^t T(t - s)v_i(s)ds & \text{if } t \in J, i = 1, 2. \end{cases}
$$

Let $0 \leq \delta \leq 1$. Then, for each $t \in J$, we have

$$
(\delta h_1 + (1 - \delta)h_2)(t) = \begin{cases} \phi(t), & t \in H, \\[2mm] T(t)\phi(0) \\[1mm] \quad + \int_0^t T(t - s)[\delta v_1(s) + (1 - \delta)v_2(s)]ds & t \in J, \end{cases}
$$

Since $F(t, y_t)$ has convex values, one has

$$
\delta h_1 + (1 - \delta)h_2 \in \mathcal{B}(y).
$$

Claim 3: $\mathcal{B}$ maps bounded sets into bounded sets in Ω Let $B = \{y \in \Omega; \|y\|_\infty \le q\}, q \in \mathbb{R}^+$ a bounded set in Ω. For each $h \in \mathcal{B}(y)$, for some $y \in B$, there exists $v \in S_{F,y}$ such that

$$h(t) = T(t)\phi(0) + \int_0^t T(t-s)v(s)ds.$$

Thus

$$|h(t)| \le M|\phi(0)| + M\int_0^t \varphi_q(s)ds$$

$$\le M|\phi(0)| + M\|\varphi_q\|_{L^1},$$

this implies that:

$$\|h\|_\infty \le M|\phi(0)| + M\|\varphi_q\|_{L^1}.$$

Hence $\mathcal{B}(B)$ is bounded.

Claim 4: $\mathcal{B}$ maps bounded sets into equi-continuous sets. Let B is a bounded set as in Claim 3 and $h \in \mathcal{B}(y)$ for some $y \in B$. Then, there exists $v \in S_{F,y}$ such that

$$h(t) = T(t)\phi(0) + \int_0^t T(t-s)v(s)ds, \quad t \in J$$

Let $\tau_1, \tau_2 \in J \setminus \{t_1, t_2, \ldots t_m\}, \tau_1 < \tau_2$. Thus if $\epsilon > 0$, we have

$$|h(\tau_2) - h(\tau_1)| \le |[T(\tau_2) - T(\tau_1)]\phi(0)|$$

$$+ \int_0^{\tau_1 - \epsilon} \|T(\tau_2 - s) - T(\tau_1 - s)\| |v(s)|ds$$

$$+ \int_{\tau_1 - \epsilon}^{\tau_1} \|T(\tau_2 - s) - T(\tau_1 - s)\| |v(s)|ds$$

$$+ \int_{\tau_1}^{\tau_2} \|T(\tau_2 - s)\| |v(s)|ds$$

$$\le |[T(\tau_2) - T(\tau_1)]\phi(0)|$$

$$+ \int_0^{\tau_1 - \epsilon} \|T(\tau_2 - s) - T(\tau_1 - s)\| \varphi_q(s)ds$$

$$+ \int_{\tau_1 - \epsilon}^{\tau_1} \|T(\tau_2 - s) - T(\tau_1 - s)\| \varphi_q(s)ds$$

$$+ M\int_{\tau_1}^{\tau_2} \varphi_q(s)ds.$$

As $\tau_1 \to \tau_2$ and ϵ becomes sufficiently small, the right-hand side of the above inequality tends to zero, since $T(t)$ is a strongly continuous operator and the compactness of $T(t)$ for $t > 0$ implies the continuity in the uniform operator topology.

This proves the equi-continuity for the case where $t \neq t_i, i = 1, \ldots, m + 1$. It remains to examine the equi-continuity at $t = t_i$.

First we prove the equi-continuity at $t = t_i^-$, we have for some $y \in B$, there exists $v \in S_{F,y}$ such that

$$h(t) = T(t)\phi(0) + \int_0^t T(t - s)v(s)ds, \quad t \in J$$

Fix $\delta_1 > 0$ such that $\{t_k, k \neq i\} \cap [t_i - \delta_1, t_i + \delta_1] = \emptyset$. For $0 < \rho < \delta_1$, we have

$$|h(t_i - \rho) - h(t_i)| \leq |[T(t_i - \rho) - T(t_i)]\phi(0)|$$
$$+ \int_0^{t_i - \rho} \|T(t_i - \rho - s) - T(t_i - s)\| |v(s)| ds$$
$$+ \int_{t_i - \rho}^{t_i} M\varphi_q(s) ds$$

Which tends to zero as $\rho \to 0$.

Define

$$\hat{h}_0(t) = h(t), \quad t \in [0, t_1]$$

and

$$\hat{h}_i(t) = \begin{cases} h(t), & \text{if } t \in (t_i, t_{i+1}] \\ h(t_i^+), & \text{if } t = t_i. \end{cases}$$

Next, we prove equi-continuity at $t = t_i^+$. Fix $\delta_2 > 0$ such that $\{t_k, k \neq i\} \cap [t_i - \delta_2, t_i + \delta_2] = \emptyset$. Then

$$\hat{h}(t_i) = T(t_i)\phi(0) + \int_0^{t_i} T(t_i - s)v(s)ds.$$

For $0 < \rho < \delta_2$, we have

$$|\hat{h}(t_i + \rho) - \hat{h}(t_i)| \leq |[T(t_i + \rho) - T(t_i)]\phi(0)|$$
$$+ \int_0^{t_i} \|T(t_i + \rho - s) - T(t_i - s)\| |v(s)| ds$$
$$+ \int_{t_i}^{t_i + \rho} M\varphi_q(s) ds.$$

The right-hand side tends to zero as $\rho \to 0$.

The equi-continuity for the cases $\tau_1 < \tau_2 \leq 0$ and $\tau_1 \leq 0 \leq \tau_2$ follows from the uniform continuity of ϕ on the interval H As a consequence of Claims 1–3 together with Arzelá–Ascoli theorem it suffices to show that $\mathcal{B}$ maps B into a precompact set in E.

Let $0 < t^* < b$ be fixed and let ϵ be a real number satisfying $0 < \epsilon < t^*$. For $y \in B$, we define

$$h_\epsilon(t^*) = T(t^*)\phi(0) + T(\epsilon)\int_0^{t^*-\epsilon} T(t^* - s - \epsilon)v(s)ds,$$

where $v \in S_{F,y}$. Since $T(t^*)$ is a compact operator, the set

$$H^\epsilon(t^*) = \{h_\epsilon(t^*) : \quad h_\epsilon \in \mathcal{B}(y)\}$$

is precompact in E for every ϵ, $0 < \epsilon < t^*$. Moreover, for every $h \in \mathcal{B}(y)$ we have

$$|h(t^*) - h_\epsilon(t^*)| = \left|\int_0^{t^*} T(t^* - s)v(s)ds - T(\epsilon)\int_0^{t^*-\epsilon} T(t^* - s - \epsilon)v(s)ds\right|$$

$$= \left|\int_{t^*-\epsilon}^{t^*} T(t^* - s)v(s)ds\right|$$

$$\leq M\int_{t^*-\epsilon}^{t^*} \varphi_q(s)ds.$$

Therefore, there are precompact sets arbitrarily close to the set $H(t^*) = \{h(t^*) : h \in \mathcal{B}(y)\}$. Hence the set $H(t^*) = \{h(t^*) : h \in \mathcal{B}(B)\}$ is precompact in E. Hence the operator $\mathcal{B}$ is completely continuous.

Claim 5: $\mathcal{B}$ has closed graph. Let $y_n \to y_*$, $h_n \in \mathcal{B}(y_n)$, and $h_n \to h_*$. We shall show that $h_* \in \mathcal{B}(y_*)$. $h_n \in \mathcal{B}(y_n)$ means that there exists $v_n \in S_{F,y_n}$ such that

$$h_n(t) = T(t)\phi(0) + \int_0^t T(t - s)v_n(s)ds, \quad t \in J.$$

We must prove that there exists $v_* \in S_{F,y_*}$ such that

$$h_*(t) = T(t)\phi(0) + \int_0^t T(t - s)v_*(s)ds.$$

Consider the linear and continuous operator $\mathcal{K} : L^1(J, E) \to D$ defined by

$$(\mathcal{K}v)(t) = \int_0^t T(t - s)v(s)ds.$$

We have

$$|(h_n(t)-T(t)\phi(0)) - (h_*(t) - T(t)\phi(0))| = |h_n(t) - h_*(t)|$$
$$\leq \|h_n - h_*\|_\infty \to 0, \quad \text{as } n \mapsto \infty.$$

From Lemma 1.11 it follows that $\mathcal{K} \circ S_F$ is a closed graph operator and from the definition of $\mathcal{K}$ one has

$$h_n(t) - T(t)\phi(0) \in \mathcal{K} \circ S_{F, y_n}.$$

As $y_n \to y_*$ and $h_n \to h_*$, there is a $v_* \in S_{F, y_*}$ such that

$$h_*(t) - T(t)\phi(0) = \int_0^t T(t-s)v_*(s)ds.$$

Hence the multi-valued operator $\mathcal{B}$ is upper semi-continuous.

Step 3: A priori bounds on solutions. Now, it remains to show that the set

$$\mathcal{E} = \{y \in \Omega | \ y \in \lambda \mathcal{A}y + \lambda \mathcal{B}y, \ 0 \leq \lambda \leq 1\}$$

is unbounded.

Let $y \in \mathcal{E}$ be any element. Then there exist $v \in S_{F,y}$ and $\mathcal{I}_k \in I_k(y(t_k^-))$ such that

$$y(t) = \lambda T(t)\phi(0) + \lambda \int_0^t T(t-s)v(s)ds + \lambda \sum_{0 < t_k < t} T(t - t_k)\mathcal{I}_k.$$

Then for each $t \in J$

$$|y(t)| \leq M|\phi(0)| + M \int_0^t |v(s)|ds + M \sum_{k=0}^m |\mathcal{I}_k|$$

$$\leq M\|\phi\|_D + M \int_0^t p(s)\psi(\|y_s\|)ds + M \sum_{k=0}^m |\mathcal{I}_k|$$

$$\leq M\|\phi\|_D + M \int_0^t p(s)\psi(\|y_s\|)ds + M \sum_{k=0}^m c_k|y(t_k^-)| + M \sum_{k=0}^{k=m} |I_k(0)|$$

$$\leq M\|\phi\|_D + M \int_0^t p(s)\psi(\|y_s\|)ds + M \sum_{k=0}^m |I_k(0)|$$

$$+ M \sum_{k=0}^m c_k|y(t_k^-)|.$$

Set

$$C = M\Big[\|\phi\|_D + \sum_{k=0} m|I_k(0)|\Big].$$

Then, we have:

$$|y(t)| \leq C + M \sum_{k=0}^{m} c_k |y(t_k^-)| + M \int_0^t p(s)\psi(\|y_s\|)ds. \tag{7.8}$$

Consider the function $\mu(t)$ defined by

$$\mu(t) = \sup\{|y(s)| : -r \leq s \leq t\}, \quad 0 \leq t \leq b$$

Then, we have, for all $t \in J, \|y_s\| \leq \mu(t)$.
Let $t^* \in J$ such that $\mu(t) = |y(t^*)|$, then by the previous inequality we have, for $t \in J$,

$$\mu(t) \leq C + M \sum_{k=0}^{m} c_k |\mu(t)| + M \int_0^t p(s)\psi(\|\mu_s\|)ds. \tag{7.9}$$

Thus

$$\Big(1 - M \sum_{k=0}^{m} c_k\Big)\mu(t) \leq C + M \int_0^t p(s)\psi(\mu(s))ds. \tag{7.10}$$

It follows that

$$\mu(t) \leq C_0 + C_1 \int_0^t p(s)\psi(\mu(s))ds. \tag{7.11}$$

Then, we have

$$\mu(t) \leq v(t) \quad \text{for all } t \in J,$$

$$v(0) = C_0.$$

Differentiating both sides of the above equality, we obtain

$$v'(t) = C_1 p(t)\psi(\mu(t)), \quad a.e. \ t \in J,$$

and using the nondecreasing character of the function ψ, we obtain

$$v'(t) \leq C_1 p(t)\psi(v(t)), \quad a.e. \ t \in J,$$

that is

$$\frac{v'(t)}{\psi(v(t))} \leq C_1 p(t), \quad a.e. \ t \in J. \tag{7.12}$$

Integrating both sides of the previous inequality from 0 to t we get

$$\int_0^t \frac{v'(s)}{\psi(v(s))} ds \leq C_1 \int_0^t p(s) ds.$$

By a change of variables we get

$$\int_{v(0)}^{v(t)} \frac{du}{\psi(u)} \leq C_1 \|p\|_{L^1} \leq \int_{c_0}^{\infty} \frac{du}{\psi(u)}.$$

Hence there exists a constant K such that

$$\mu(t) \leq v(t) \leq K,$$

for all $t \in J$. Now from the definition of μ it follows that

$$\|y\|_{\Omega} = \sup_{t \in [-r,b]} |y(t)| \leq \mu(b) \leq K$$

for all $y \in \mathcal{E}$. This shows that the set $\mathcal{E}$ is bounded. As a consequence of Theorem 1.32, we deduce that $\mathcal{A} + \mathcal{B}$ has a fixed point y on $[-r, b]$ which is a mild solution of our problem.   □

7.2.3   Existence of Extremal Mild Solutions

In this subsection we prove the existence of maximal and minimal mild solutions of problem (7.1)–(7.3) under suitable monotonicity conditions on the multi-valued functions involved in it. Our proof is based upon the Theorem 1.37 due to Dhage.

Let us introduce the concept of lower and upper mild solutions for problem (7.1)–(7.3).

Definition 7.3. We say that a continuous function $v : [-r, b] \to E$ is a lower mild solution of problem (7.1)–(7.3) if there exist functions $v \in L^1(J, E)$ such that $v(t) \in F(t, y_t)$, a.e. on J, $y(t) = \phi(t), t \in H$, and

$$y(t) \leq T(t)\phi(0) + \int_0^t T(t-s)v(s)ds + \sum_{0 < t_k < t} T(t - t_k)I_k((t_k^-)), \quad t \in J, \ t \neq t_k$$

and $v(t_k^+) - v(t_k^-) \leq I_k(v(t_k)), \ t = t_k, k = 1, \ldots m$. Similarly an upper mild solution w of (7.1)–(7.3) is defined by reversing the order.

Definition 7.4. A solution x_M of IVP (7.1)–(7.3) is said to be maximal if for any other solution x of IVP (7.1)–(7.3) on J, we have that $x(t) \leq x_M(t)$ for each $t \in J$.

Similarly a minimal solution of IVP (7.1)–(7.3) is defined by reversing the order of the inequalities.

We consider the following assumptions in the sequel.

(7.4.1) The multi-valued function $F(t, y)$ and is strictly monotone increasing in y for almost each $t \in J$.

(7.4.2) The IVP (7.1)–(7.3) has a lower mild solution v and an upper mild solution w with $v \leq w$.

(7.4.3) $T(t)$ is preserving the order, that is $T(t)v \geq 0$ whenever $v \geq 0$.

(7.4.4) The multi-valued functions $I_k, k = 1, \ldots m$ are continuous and non-decreasing.

Theorem 7.5. *Assume that assumptions (7.1.11)–(7.1.4) and (7.4.1)–(7.4.4) hold. Then IVP (7.1)–(7.3) has minimal and maximal solutions on $[-r, b]$.*

Proof. It can be shown as in the proof of Theorem 7.2 that $\mathcal{A}$ is completely continuous and $\mathcal{B}$ is a contraction on $[v, w]$. We shall show that $\mathcal{A}$ and $\mathcal{B}$ are isotone increasing on $[v, w]$. Let $y, \bar{y} \in [v, w]$ be such that $y \leq \bar{y}, y \neq \bar{y}$. Then by (7.4.4), we have for each $t \in J$

$$\mathcal{A}(y) = \{h \in \Omega : h(t) = \sum_{0 < t_k < t} T(t - t_k)\mathcal{I}_k, \mathcal{I}_k \in I_k(y(t_k^-))\}$$

$$\leq_{\mathcal{P}} \{h \in \Omega : h(t) = \sum_{0 < t_k < t} T(t - t_k)\mathcal{I}_k, \mathcal{I}_k \in I_k(\bar{y}(t_k^-))\}$$

$$= \mathcal{A}(\bar{y}).$$

Similarly, by (7.4.1), (7.4.3)

$$\mathcal{B}(y) = \{h \in \Omega : h(t) = T(t)\phi(0) + \int_0^t T(t - s)v(s)ds, \ v \in S_{F,y}\}$$

$$\leq_{\mathcal{P}} \{h \in \Omega : h(t) = T(t)\phi(0) + \int_0^t T(t - s)v(s)ds, \ v \in S_{F,\bar{y}}\}$$

$$= \mathcal{B}(\bar{y}).$$

Therefore $\mathcal{A}$ and $\mathcal{B}$ are isotone increasing on $[v, w]$. Finally, let $x \in [v, w]$ be any element. By (7.4.2), (7.4.3) we deduce that

$$v \leq \mathcal{A}(v) + \mathcal{B}(v) \leq \mathcal{A}(x) + \mathcal{B}(x) \leq \mathcal{A}(w) + \mathcal{B}(w) \leq w,$$

which shows that $\mathcal{A}(x) + \mathcal{B}(x) \in [v, w]$ for all $x \in [v, w]$. Thus, $\mathcal{A}$ and $\mathcal{B}$ satisfy all conditions of Theorem 7.5, hence IVP (7.1)–(7.3) has maximal and minimal solutions on J. $\square$

7.3 Existence of Mild Solutions with Nonlocal Conditions

In this section we consider the following class of semi-linear Impulsive differential inclusions:

$$y'(t) - Ay(t) \in F(t, y_t), \quad t \in J := [0, b], t \neq t_k \tag{7.13}$$

$$\Delta y|_{t=t_k} \in I_k(y(t_k^-)), \quad k = 1, \ldots, m \tag{7.14}$$

$$y(t) + h_t(y) = \phi(t), \quad t \in [-r, 0], \tag{7.15}$$

where A F and I_k are as in the previous section, and $h_t : C(H, E) \rightarrow E$ is a given function. The nonlocal Cauchy Problem was introduced by Byszewski in [89], and the importance of nonlocal conditions in different fields has been discussed in [89, 90].

7.3.1 Main Result

Let us start by the definition of the mild solution of the problem (7.13)–(7.15)

Definition 7.6. A function $y \in \Omega$ is said to be a mild solution of problem (7.13)–(7.15) if $y(t) = \phi(t) - h_t(y)$, $t \in [-r, 0]$, and the restriction of $y(\cdot)$ to the interval $[0, b]$ is continuous and there exist $v(\cdot) \in L^1(J_k, E)$ and $\mathcal{I}_k \in I_k(y(t_k^-))$ such that $v(t) \in F(t, y_t)$ a.e $t \in [0, b]$, and y satisfies the integral equation,

$$y(t) = T(t)(\phi(0) - h_0(y)) + \int_0^t T(t-s)v(s)ds + \sum_{0 < t_k < t} T(t - t_k)\mathcal{I}_k$$

Let us introduce the following assumptions.

(7.6.1) The function h is continuous with respect to t, and there exists a constant $\alpha > 0$ such that

$$\|h_t(u)\| \leq \alpha, \quad u \in C(H, E)$$

and for each $k > 0$ the set

$$\{\phi(0) - h_0(y), \ y \in C(H, E), \|y\| \leq k\}$$

is precompact in E.

Theorem 7.7. *Assume that hypotheses (7.1.1)–(7.1.4) and (7.6.1) hold. Then the IVP (7.13)–(7.15) has at least one mild solution on $[-r, b]$.*

Proof. Consider the two multi-valued operators $\mathcal{A}_1 \mathcal{B}_1 \colon \Omega \to \mathcal{P}(\Omega)$

$$\mathcal{B}_1(y) := \left\{ f \in \Omega \colon f(t) = \begin{cases} \phi(t) - h_t(y), & \text{if } t \in H; \\[4pt] T(t)\,(\phi(0) - h_0(y)) \\[2pt] \quad + \displaystyle\int_0^t T(t-s)v(s)ds, \; v \in S_{F,y} \\[4pt] & \text{if } t \in J, \end{cases} \right\}$$

$$\mathcal{A}_1(y) := \left\{ f \in \Omega \colon f(t) = \begin{cases} 0, & \text{if } t \in H; \\[4pt] \displaystyle\sum_{0 < t_k < t} T(t - t_k)\mathcal{I}_k, \mathcal{I}_k \in I_k(y(t_k^-)), & \text{if } t \in J, \end{cases} \right\}$$

Then the problem of finding the solution of problem (8.8)–(8.11) is reduced to finding the solution of the operator inclusion $y \in \mathcal{A}_1(y) + \mathcal{B}_1(y)$. By parallel steps of Theorem 7.2 we can show that the operators $\mathcal{A}_1$ and $\mathcal{B}_1$ satisfy all conditions of Theorem 1.32. $\qquad\qquad\square$

7.4   Application to the Control Theory

This section is devoted to an application of the argument used in previous sections to the controllability of a semi-linear functional differential inclusions. More precisely we will consider the following IVP:

$$y'(t) - Ay(t) \in F(t, y_t) + Bu(t), \quad t \in J := [0, b], t \neq t_k \tag{7.16}$$

$$\Delta y|_{t=t_k} \in I_k(y(t_k^-)), \quad k = 1, \ldots, m \tag{7.17}$$

$$y(t) = \phi(t), \quad t \in H, \tag{7.18}$$

where A and F are as in the previous section, the control function $u(\cdot)$ is given in $L^2(J, U)$, a Banach space of admissible control functions, with U as a Banach space. Finally B is a bounded linear operator from U to E. In the case of single-valued functions I_k, the problem (7.16–7.18) has been recently studied in the monographs by Ahmed [16], and Benchohra et al. [81], and in the papers [18, 19].

7.4.1   Main Result

Before stating and proving our result we give the meaning of mild solution of our problem (7.16)–(7.18).

Definition 7.8. A function $y \in \Omega$ is said to be a mild solution of system (7.16)–(7.18) if $y(t) = \phi(t)$ for all $t \in H$, the restriction of $y(\cdot)$ to the interval $[0, b]$ is continuous and there exists $v(\cdot) \in L^1(J_k, E)$ and $\mathcal{I}_k \in I_k(y(t_k^-))$, such that $v(t) \in F(t, y_t)$ a.e $[0, b]$, and such that y satisfies the integral equation,

$$
y(t) = T(t)\phi(0) + \int_0^t T(t-s)v(s)ds + \int_0^t T(t-s)Bu_y(s)ds
$$
$$
+ \sum_{0 < t_k < t} T(t-t_k)\mathcal{I}_k, \quad t \in J.
$$

Theorem 7.9. *Assume that hypotheses (7.1.1)–(7.1.3) hold. Moreover we suppose that:*

(C1) *the linear operator $W : L^2(J, U) \to E$, defined by*

$$
Wu = \int_0^b T(b-s)Bu(s)ds,
$$

has a bounded inverse operator W^{-1} which takes values in $L^2(J, U) \backslash KerW$, and there exist positive constants $\overline{M}$, $\overline{M}_1$ such that $\|B\| \leq \overline{M}$ and $\|W^{-1}\| \leq \overline{M}_1$.
(C2) *F has closed, bounded and convex values, and there exists a function $l \in L^1(J, \mathbb{R}_+)$ such that*

$$
H_d\big(F(t, y), F(t, x)\big) \leq l(t)\|y - x\|_D, \quad \text{for a.e. } t \in J, x, y \in D
$$

(C3) *There exist a function $k \in L^1(J, \mathbb{R}_+)$ and a continuous nondecreasing function $\psi : [0, \infty) \to (0, \infty)$ such that*

$$
\|F(t, x)\|_{\mathcal{P}} \leq k(t)\psi(\|x\|_D) \quad \text{for a.e. } t \in J \text{ and each } x \in D,
$$

with

$$
\int_{C_0^*}^\infty \frac{ds}{s + \psi(s)} = \infty, \tag{7.19}
$$

where

$$
C_0^* = \frac{C^*}{1 - \sum_{k=0}^m c_k},
$$

with

$$
C^* = M\|\phi\|_D + M\overline{M}\overline{M}_1 b\big[|y_1| + M\|\phi\|_D\big]
$$
$$
+ \big[M^2\overline{M}\overline{M}_1 b + M\big] \sum_{0 < t_k < s} |I_k(0)|.
$$

If

$$MMM_1 b \|\ell\|_{L^1} + M^2 \overline{MM}_1 b \sum_{k=0}^{m} c_k + M \sum_{k=0}^{m} c_k < 1,$$

then the IVP (7.16)–(7.18) is controllable on $[-r, b]$.

Proof. Using hypothesis (C1) for each arbitrary function $y(\cdot)$ define the control

$$u_y(t) = W^{-1} \left[y_1 - T(b)\phi(0) - \int_0^b T(b-s)v(s)ds - \sum_{0<t_k<t} T(b-t_k)\mathcal{I}_k \right](t),$$

where $v \in S_{F,y}$ and $\mathcal{I}_k \in I_k(y(t_k^-))$. We shall show that the operator $N : \Omega \to \mathcal{P}(\Omega)$ defined by

$$N(y) = \left\{ f \in \Omega : f(t) = \begin{cases} \phi(t), & \text{if } t \in H, \\ T(t)\phi(0) + \displaystyle\int_0^t T(t-s)v(s)ds \\ \quad + \displaystyle\int_0^t T(t-s)(Bu_y)(s)ds \\ \quad + \displaystyle\sum_{0<t_k<t} T(t-t_k)\mathcal{I}_k, \mathcal{I}_k \in I_k(y(t_k^-)), v \in S_{F,y} & \text{if } t \in J \end{cases} \right\}$$

has a fixed point. This fixed point is then the mild solution of the IVP (7.16)–(7.18). Consider the multi-valued operators:

$$\mathcal{A}, \mathcal{B} : \Omega \to \mathcal{P}(\Omega)$$

defined by

$$\mathcal{A}(y) := \left\{ f \in \Omega : f(t) = \begin{cases} 0, & \text{if } t \in H; \\ \displaystyle\int_0^t T(t-s)(Bu_y)(s)ds \\ \quad + \displaystyle\sum_{0<t_k<t} T(t-t_k)\mathcal{I}_k, \mathcal{I}_k \in I_k(y(t_k^-)) & \text{if } t \in J, \end{cases} \right\}$$

and

$$\mathcal{B}(y) := \left\{ f \in \Omega : f(t) = \begin{cases} \phi(t), & \text{if } t \in H; \\ T(t)\phi(0) + \displaystyle\int_0^t T(t-s)v(s)ds, v \in S_{F,y} \\ & \text{if } t \in J. \end{cases} \right\}$$

It is clear that

$$N = \mathcal{A} + \mathcal{B}.$$

Similarly, as in Theorem 7.2, we can prove that $\mathcal{A}$ is a contraction operator, and $\mathcal{B}$ is a completely continuous operator with compact convex values. Now, we prove that the set:

$$\mathcal{E} = \{y \in \Omega \mid y \in \lambda\mathcal{A}y + \lambda\mathcal{B}y, \ 0 \le \lambda \le 1\}$$

is unbounded.

Let $y \in \mathcal{E}$ be any element. Then there exist $v \in S_{F,y}$ and $\mathcal{I}_k \in I_k(y(t_k^-))$ such that

$$y(t) = \lambda T(t)\phi(0) + \lambda \int_0^t T(t-s)v(s)ds + \lambda \int_0^t T(t-s)Bu_y(s)ds$$
$$+\lambda \sum_{0<t_k<t} T(t-t_k)\mathcal{I}_k.$$

This implies by (C1)–(C3) that

$$|y(t)| \le M\|\phi\|_D + M \int_0^t k(s)\psi(\|y_s\|)ds + M\overline{MM}_1 b\big[|y_1| + M\|\phi\|_D\big]$$
$$+M^2\overline{MM}_1 b \int_0^t k(s)\psi(\|y_s\|)ds + M^2\overline{MM}_1 \int_0^t \sum_{0<t_k<m} |\mathcal{I}_k|ds$$
$$+M \sum_{k=0}^{k=m} |\mathcal{I}_k|.$$

Thus

$$|y(t)| \le C + M^2\overline{MM}_1 \int_0^t \sum_{0<t_k<s} c_k|y(t_k^-)|ds \qquad (7.20)$$
$$+M \sum_{k=0}^{k=m} c_k|y(t_k^-)| + \big[M^2\overline{MM}_1 b + M\big] \int_0^t k(s)\psi(\|y_s\|)ds.$$

Consider the function $\mu(t)$ defined by

$$\mu(t) = \sup\{|y(s)| : \ -r \le s \le t\}, \ 0 \le t \le b$$

Then, we have, for all $t \in J, \|y_s\| \le \mu(t)$. Let $t^* \in J$ such that $\mu(t) = |y(t^*)|$, then by (7.20) we have, for $t \in J$,

$$\mu(t) \le C^* + M^2 \overline{MM}_1 \int_0^t \sum_{0 < t_k < s} c_k \mu(s) ds \qquad (7.21)$$

$$+ M \sum_{k=0}^{k=m} c_k \mu(t) + \left[M^2 \overline{MM}_1 b + M \right] \int_0^t p(s) \psi(\mu(s)) ds.$$

From (7.21) we obtain

$$\left(1 - M \sum_{k=0}^m c_k \right) \mu(t) \le C^* + M^2 \overline{MM}_1 \int_0^t \sum_{0 < t_k < s} c_k \mu(s) ds \qquad (7.22)$$

$$+ \left[M^2 \overline{MM}_1 b + M \right] \int_0^t p(s) \psi(\mu(s)) ds.$$

Let

$$C_0^* = \frac{C^*}{1 - M \sum_{k=0}^m c_k}, \quad C_1^* = \frac{M^2 \overline{MM}_1}{1 - M \sum_{k=0}^m c_k}, \quad C_2^* = \frac{M^2 \overline{MM}_1 b + M}{1 - M \sum_{k=0}^m c_k} \qquad (7.23)$$

It follows from (7.22) and (7.23) that

$$\mu(t) \le C_0^* + C_1^* \int_0^t \sum_{0 < t_k < s} c_k \mu(s) ds$$

$$+ C_2^* \int_0^t p(s) \psi(\mu(s)) ds$$

$$\le C_0^* + \int_0^t \hat{M}(s) [\mu(s) + p(s) \psi(\mu(s))] ds,$$

where

$$\hat{M}(s) = \max \left(C_1^* \sum_{0 < t_k < s} c_k, C_2^* k(s) \right)(s)$$

Let

$$v(t) = C_0^* + \int_0^t \hat{M}(s) [\mu(s) + \psi(\mu(s))] ds. \qquad (7.24)$$

Then, we have $\mu(t) \leq v(t)$ for all $t \in J$ Differentiating both sides of (7.24), we obtain

$$v'(t) = \hat{M}(t)[\mu(t) + \psi(\mu(t))], \quad a.e. \ t \in J$$

and

$$v(0) = C_0^*.$$

Using the nondecreasing character of the function ψ, we obtain

$$v'(t) \leq \hat{M}(t)[v(t) + \psi(v(t))], \quad a.e. \ t \in J,$$

that is

$$\frac{v'(t)}{v(t) + \psi(v(t))} \leq \hat{M}(t), \quad a.e. \ t \in J. \tag{7.25}$$

Integrating from 0 to t both sides of (7.25) we get

$$\int_0^t \frac{v'(s)}{v(s) + \psi(v(s))} ds \leq \int_0^t \hat{M}(s) ds.$$

By a change of variables we get

$$\int_{v(0)}^{v(t)} \frac{du}{u + \psi(u)} \leq \|\hat{M}\|_{L^1} \leq \infty.$$

From (7.19) there exists a constant K such that

$$\mu(t) \leq v(t) \leq K \quad \text{for all } t \in J.$$

Now from the definition of μ it follows that

$$\|y\|_\Omega = \sup_{t \in [-r,b]} |y(t)| \leq \mu(b) \leq K \quad \text{for all } y \in \mathcal{E}.$$

This shows that the set $\mathcal{E}$ is bounded. As a consequence of Theorem 1.32 $\mathcal{A} + \mathcal{B}$ has a fixed point which is a mild solution of problem (7.16)–(7.18).

Thus, the problem (7.16)–(7.18) is controllable on the interval $[-r, b]$.   □

7.4.2 Example

As an application of our results we consider the following impulsive partial functional differential equation of the form

$$\frac{\partial}{\partial t} z(t, x) = \frac{\partial^2}{\partial x^2} z(t, x)$$

$$+ Q(t, z(t - r, x) + Bu(t), \; x \in [0, \pi], \; t \in [0, b] \setminus \{t_1, t_2, \dots, t_m\}. \tag{7.26}$$

$$z(t_k^+, x) - z(t_k^-, x) \in b_k |z(t_k^-, x)| \overline{B}(0, 1), \; x \in [0, \pi], \; k = 1, \dots, m \tag{7.27}$$

$$z(t, 0) = z(t, \pi) = 0, \; t \in J := [0, b] \tag{7.28}$$

$$z(t, x) = \phi(t, x), \; t \in H, \; x \in [0, \pi], \tag{7.29}$$

where $b_k > 0$, $k = 1, \dots, m$, $\phi \in D = \{\bar{\psi} : H \times [0, \pi] \to \mathbb{R}; \bar{\psi}$ is continuous everywhere except for a countable number of points at which $\bar{\psi}(s^-)$, $\bar{\psi}(s^+)$ exist with $\bar{\psi}(s^-) = \bar{\psi}(s)\}$, $0 = t_0 < t_1 < t_2 < \cdots < t_m < t_{m+1} = b$, $z(t_k^+) = \lim_{(h,x) \to (0^+,x)} z(t_k + h, x)$, $z(t_k^-) = \lim_{(h,x) \to (0^-,x)} z(t_k + h, x)$, where $Q : J \times \mathbb{R} \to \mathcal{P}(\mathbb{R})$, is a multi-valued map with compact values. Here $\overline{B}(0, 1)$ denotes the closure of the unit ball. Let

$$y(t) = z(t, .); \; t \in J,$$

$I_k : \mathbb{R} \to \mathcal{P}\mathbb{R}$ such that

$$I_k(y(t_k^-)) = b_k |z(t_k^-, .)| \overline{B}(0, 1), \; k = 1, \dots, m$$

and

$$F(t, y_t)(x) = Q(t, z(t - r, x)), \; t \in [0, b], \; x \in [0, \pi].$$

Take $E = L^2[0, \pi]$, and define the linear operator $A : D(A) \subset E \to E$ by $Aw = w''$ with domain

$$D(A) = \{w \in E, w, w' \text{ are absolutely continuous, } w'' \in E, w(0) = w(\pi) = 0\}.$$

Then

$$Aw = \sum_{n=1}^{\infty} n^2 (w, w_n) w_n, \; w \in D(A)$$

where $(.,.)$ is the inner product in $L^2[0, \pi]$ and $w_n(s) = \sqrt{\frac{2}{\pi}} \sin ns$. $n = 1, 2$.
is the orthogonal set eigenvectors in A. It is well known (see [168]) that A is the
infinitesimal generator of an analytic semigroup $T(t)$, $t \in (0, b]$ in E given by

$$T(t)w = \sum_{n=1}^{\infty} exp(-n^2 t)(w, w_n)w_n, \quad w \in E.$$

Since the analytic semigroup $T(t)$, $t \in (0, b]$ is compact, there exists a constant
$M \geq 1$ such that

$$\|T(t)\|_{B(E)} \leq M.$$

Assume that $B : U \to Y, U \subset [0, \infty)$ is a bounded linear operator and the operator
W defined by

$$Wu = \int_0^b T(b - s)Bu(s)ds$$

has a bounded invertible operator W^{-1} which takes values in $L^2([0, b], U) \backslash kerW$.
Also assume that there exists an integrable function $\sigma : [0, b] \to \mathbb{R}^+$ such that

$$|Q(t, w(t - r))| \leq \sigma(t)\Omega(|w|)$$

where $\Omega : [0, \infty) \to (0, \infty)$ is continuous and nondecreasing with

$$\int_1^{\infty} \frac{ds}{s + \Omega(s)} = \infty$$

Assume that there exists $\tilde{l} \in L^1([0, b], \mathbb{R}^+)$ such that

$$H_d(Q(t, w(t - r, x)), Q(t, \bar{w}(t - r, x))) \leq \tilde{l}|w - \bar{w}|; \quad t \in [0, b], \ w, \bar{w}\_n\mathbb{R}$$

We can show that problem (7.16)–(7.18) is an abstract formulation of problem (7.26)–(7.29). Since all the conditions of Theorem 7.7 are satisfied, the
problem (7.26)–(7.27) has a solution z on $[-r, b] \times [0, \pi]$.

7.5   Notes and Remarks

The results of Chap. 7 are taken from Abada et al. [4, 6]. Other results may be found
in [53, 54].

Chapter 8
Non-densely Defined Functional Differential Inclusions with Finite Delay

8.1 Introduction

In this chapter, we shall establish sufficient conditions for the existence of integral solutions and extremal integral solutions for some non-densely defined impulsive semi-linear functional differential inclusions in separable Banach spaces with local and nonlocal conditions. We shall rely on a fixed point theorem for the sum of completely continuous and contraction operators. The question of controllability of these inclusions with both multi-valued and single valued jump and the topological structure of the solutions set are considered too.

8.2 Integral Solutions of Non-densely Defined Functional Differential Inclusions with Local Conditions

We will consider the following first order impulsive semi-linear differential inclusions of the form:

$$y'(t) - Ay(t) \in F(t, y_t), \; a.e. \; t \in J = [0, b], \; t \neq t_k, \; k = 1, \dots, m \tag{8.1}$$

$$\Delta y|_{t=t_k} \in I_k(y(t_k^-)), \;\; k = 1, \dots, m \tag{8.2}$$

$$y(t) = \phi(t), \;\; t \in [-r, 0], \tag{8.3}$$

where $F : J \times D \to \mathcal{P}(E)$, D, $I_k : E \to \mathcal{P}(E)$ are as in the previous chapter and $A : D(A) \subset E \to E$ is a non-densely defined closed linear operator on E.

© Springer International Publishing Switzerland 2015
S. Abbas, M. Benchohra, *Advanced Functional Evolution Equations and Inclusions*, Developments in Mathematics 39,
DOI 10.1007/978-3-319-17768-7_8

In order to define a integral solution of problems (8.1)–(8.3) and (8.14)–(8.16), we shall consider the space

$$PC = \left\{ y : [0, b] \to \overline{D(A)} : y_k \in C(J_k, \overline{D(A)}), k = 0, \ldots, m \quad \text{such that} \right.$$
$$\left. y(t_k^-), y(t_k^+) \text{ exist with } y(t_k) = y(t_k^-), k = 1, \ldots, m \right\}$$

which is a Banach space with the norm

$$\|y\|_{PC} = \max\{\|y_k\|_\infty, k = 1, \ldots, m\}$$

where y_k is the restriction of y to $J_k = [t_k, t_{k+1}], k = 0, \ldots, m$.
Set

$$\Omega = \{y : [-r, b] \to \overline{D(A)} : y \in D \cap PC\}.$$

Then Ω is a Banach space with norm

$$\|y\|_\Omega = \max(\|y\|_D, \|y\|_{PC}).$$

8.2.1 Main Results

We assume that the multi-valued F has compact and convex values. Let us first define the concept of integral solution of (8.1)–(8.2).

Definition 8.1. We say that $y : [-r, b] \to E$ is an integral solution of (8.1)–(8.3) if

(i) $y \in \Omega$.

(ii) $\int_0^t y(s)ds \in D(A)$ for $t \in J$,

(iii) $y(t) = \phi(t)$ for all $t \in H$ there exist $v \in L^1(J, E)$ and $\mathcal{I}_k \in I_k \left(y(t_k^-) \right)$ such that $v(t) \in F(t, y_t)$ a.e $t \in J$ and

$$y(t) = S'(t)\phi(0) + \frac{d}{dt}\int_0^t S(t-s)v(s,)ds + \sum_{0<t_k<t} S'(t-t_k)\mathcal{I}_k \quad t \in J. \quad (8.4)$$

We notice also that if y satisfies (8.4), then

$$y(t) = S'(t)\phi(0) + \lim_{\lambda\to\infty}\int_0^t S'(t-s)B_\lambda v(s)ds + \sum_{0<t_k<t} S'(t-t_k)\mathcal{I}_k, \quad t \in J.$$

In what follows we will assume (without lost of generality) that $w > 0$. Let us introduce the following hypotheses:

(8.1.1) A satisfies Hille–Yosida condition;
(8.1.2) There exist constants $c_k > 0$, $k = 1, \ldots, m$ such that for each y, $x \in \overline{D(A)}$

$$H_d(I_k(y), I_k(x)) \leq c_k |y - x|$$

(8.1.3) The multi-valued map F is L^1-Carathéodory, with compact convex values.
(8.1.4) The operator $S'(t)$ is compact in $\overline{D(A)}$ wherever $t > 0$;
(8.1.5) There exist a function $p \in L^1(J, \mathbb{R}_+)$ and a continuous nondecreasing function $\psi : [0, \infty) \to (0, \infty)$ such that

$$\|F(t, x)\| \leq p(t)\psi(\|x\|_D), \quad a.e. \ t \in J, \quad \text{for all } x \in D$$

and

$$C_1 \int_0^b e^{-\omega t} p(t) dt < \int_{C_0}^\infty \frac{du}{\psi(u)}, \tag{8.5}$$

where

$$C_1 = \frac{M e^{\omega b}}{1 - M e^{\omega b} \displaystyle\sum_{k=1}^m e^{-\omega t_k} c_k}, \tag{8.6}$$

$$C_0 = \frac{C}{1 - M e^{\omega b} \displaystyle\sum_{k=1}^m e^{-\omega t_k} c_k}, \tag{8.7}$$

and

$$C = M e^{\omega b} \left(\|\phi\| + \sum_{k=1}^m e^{-\omega t_k} c_k |I_k(0)| \right). \tag{8.8}$$

Theorem 8.2. *Assume that (8.1.1)–(8.1.5) hold and $\phi(0) \in \overline{D(A)}$. If*

$$M e^{\omega b} \sum_{k=1}^m e^{-\omega t_k} c_k < 1, \tag{8.9}$$

then the problem (8.1)–(8.3) has at least one integral solution on $[-r, b]$.

Proof. Consider the multi-valued operator $N : \Omega \to \mathcal{P}(\Omega)$ defined by

$$
N(y) = \left\{ h \in \Omega : h(t) = \begin{cases} \phi(t), & \text{if } t \in H, \\[2mm] \begin{aligned} & S'(t)\phi(0) \\ & + \frac{d}{dt}\int_0^t S(t-s)v(s)ds \\ & + \sum_{0<t_k<t} S'(t-t_k)\mathcal{I}_k, \ \mathcal{I}_k \in I_k(y(t_k^-)); \ v \in S_{F,y} \end{aligned} & \text{if } t \in J. \end{cases} \right\}
$$

Obviously the fixed points of the operator N are integral solutions of the IVP
(8.1)–(8.3). Consider the multi-valued operators $\mathcal{A}, \mathcal{B} : \Omega \to \mathcal{P}(\Omega)$ defined by

$$
\mathcal{A}(y) := \left\{ h \in \Omega : \ h(t) = \begin{cases} 0, & \text{if } t \in H; \\[2mm] \sum_{0<t_k<t} S'(t-t_k)\mathcal{I}_k, \quad \mathcal{I}_k \in I_k(y(t_k^-)), & \text{if } t \in J, \end{cases} \right\}
$$

and

$$
\mathcal{B}(y) := \left\{ h \in \Omega : \ h(t) = \begin{cases} \phi(t), & \text{if } t \in H; \\[2mm] S'(t)\phi(0) + \int_0^t S(t-s)v(s)ds, & v \in S_{F,y} \quad \text{if } t \in J. \end{cases} \right\}
$$

It is clear that

$$
N = \mathcal{A} + \mathcal{B}
$$

The problem of finding integral solutions of (8.1)–(8.3) is reduced to finding integral
solutions of the operator inclusion $y \in \mathcal{A}(y)+\mathcal{B}(y)$. We shall show that the operators
$\mathcal{A}$ and $\mathcal{B}$ satisfy all conditions of the Theorem 1.32. The proof will be given in
several steps.

Step 1: $\mathcal{A}$ is a contraction. Let $y_1, y_2 \in \Omega$, then by (8.1.2) we have

$$
H_d\big(\mathcal{A}(y_1), \mathcal{A}(y_2)\big) = H_d\left(\sum_{0<t_k<t} S'(t-t_k)I_k(y_1(t_k^-)), \sum_{0<t_k<t} S'(t-t_k)I_k(y_2(t_k^-)) \right)
$$

$$
\leq Me^{\omega t} \sum_{k=1}^m e^{-\omega t_k} c_k |y_1(t_k^-) - y_2(t_k^-)|
$$

$$
\leq Me^{\omega b} \sum_{k=1}^m e^{-\omega t_k} c_k \|y_1 - y_2\|_D.
$$

Hence by (8.9), $\mathcal{A}$ is a contraction.

Step 2: $\mathcal{B}$ has compact, convex values, and it is completely continuous. This will be given in several claims.

Claim 1: $\mathcal{B}$ has compact values. The operator $\mathcal{B}$ is equivalent to the composition $\mathcal{L} \circ S_F$ of two operators on $L^1(J, E)$, where $\mathcal{L} : L^1(J, E) \to \Omega$ is the continuous operator defined by

$$\mathcal{L}(v(t)) = S'(t)\phi(0) + \frac{d}{dt}\int_0^t S(t-s)v(s)ds, \ t \in J.$$

Then, it suffices to show that $\mathcal{L} \circ S_F$ has compact values on Ω. Let $y \in \Omega$ arbitrary, v_n a sequence in $S_{F,y}$, that is $v_n(t) \in F(t, y_t)$, a.e. $t \in J$. Since $F(t, y_t)$ is compact, we may pass to a subsequence if necessary to get that $v_n \to v$ weakly in $L^1_w(J, E)$ and $v(t) \in F(t, y_t)$, a.e. $t \in J$. An application of Mazur's Lemma implies that v_n converges strongly to v in $L^1(J, E)$. From the continuity of $\mathcal{L}$, it follows that $\mathcal{L}v_n(t) \to \mathcal{L}v(t)$ pointwise on J as $n \to \infty$. In order to show that the convergence is uniform, we first show that $\{\mathcal{L}v_n\}$ is an equi-continuous sequence. Let $\tau_1, \tau_2 \in J$, then we have:

$$\left|\mathcal{L}(v_n(\tau_1)) - \mathcal{L}(v_n(\tau_2))\right| = \left|S'(\tau_1)\phi(0) - S'(\tau_2)\phi(0)\right.$$

$$+ \frac{d}{dt}\int_0^{\tau_1} S(\tau_1 - s)v_n(s)ds$$

$$\left. - \frac{d}{dt}\int_0^{\tau_2} S(\tau_2 - s)v_n(s)ds\right|$$

$$\leq \left|(S'(\tau_1) - S'(\tau_2))\phi(0)\right|$$

$$+ \left|\lim_{\lambda \to \infty}\int_0^{\tau_1} [S'(\tau_1 - s) - S'(\tau_2 - s)]B_\lambda v_n(s)|ds\right|$$

$$- \left|\lim_{\lambda \to \infty}\int_{\tau_1}^{\tau_2} S'(\tau_2 - s)B_\lambda v_n(s)|ds\right|.$$

As $\tau_1 \to \tau_2$, the right-hand side of the above inequality tends to zero. Since $S'(t)$ is a strongly continuous operator and the compactness of $S'(t)$, $t > 0$, implies the continuity in uniform topology (see [16], Lemma 3.4.1, p. 104, [168]). Hence $\{\mathcal{L}v_n\}$ is equi-continuous, and an application of Arzelá–Ascoli theorem implies that there exists a subsequence which is uniformly convergent. Then we have $\mathcal{L}v_{n_j} \to \mathcal{L}v \in (\mathcal{L} \circ S_F)(y)$ as $j \mapsto \infty$, and so $(\mathcal{L} \circ S_F)(y)$ is compact. Therefore $\mathcal{B}$ is a compact valued multi-valued operator on Ω.

Claim 2: $\mathcal{B}(y)$ is convex for each $y \in \Omega$.
Let $h_1, h_2 \in \mathcal{B}(y)$, then there exist $v_1, v_2 \in S_{F,y}$ such that, for each $t \in J$ we have

$$h_i(t) = \begin{cases} \phi(t), & \text{if } t \in H, \\ S'(t)\phi(0) + \dfrac{d}{dt}\displaystyle\int_0^t S(t-s)v_i(s)ds, & \text{if } t \in J, \quad i = 1,2. \end{cases}$$

Let $0 \le \delta \le 1$. Then, for each $t \in J$, we have

$$(\delta h_1 + (1-\delta)h_2)(t) = \begin{cases} \phi(t), & \text{if } t \in H, \\ S'(t)\phi(0) \\ \quad + \dfrac{d}{dt}\displaystyle\int_0^t S(t-s)[\delta v_1(s) + (1-\delta)v_2(s)]ds & \text{if } t \in J. \end{cases}$$

Since $F(t, y_t)$ has convex values, one has

$$\delta h_1 + (1 - \delta)h_2 \in \mathcal{B}(y).$$

Claim 3: $\mathcal{B}$ maps bounded sets into bounded sets in Ω

Let $B_q = \{y \in \Omega; \|y\|_\Omega \le q\}$, $q > 0$ be a bounded set in Ω. For each $h \in \mathcal{B}(y)$, there exists $v \in S_{F,y}$ such that

$$h(t) = S'(t)\phi(0) + \frac{d}{dt}\int_0^t S(t-s)v(s)ds.$$

Then for each $t \in J$

$$|h(t)| \le Me^{\omega b}|\phi(0)| + Me^{\omega t}\int_0^t e^{-\omega s}\varphi_q(s)ds$$

$$\le Me^{\omega b}|\phi(0)| + Me^{\omega b}\int_0^b e^{-\omega s}\varphi_q(s)ds,$$

this further implies that

$$\|h\|_\infty \le Me^{\omega b}|\phi(0)| + Me^{\omega b}\int_0^b e^{-\omega s}\varphi_q(s)ds.$$

Then, for all $h \in \mathcal{B}(y) \subset \mathcal{B}(B_q) = \bigcup_{y \in B_q} \mathcal{B}(y)$. Hence $\mathcal{B}(B_q)$ is bounded.

Claim 4: $\mathcal{B}$ maps bounded sets into equi-continuous sets.

Let B_q be, as above, a bounded set and $h \in \mathcal{B}(y)$ for some $y \in B_q$. Then, there exists $v \in S_{F,y}$ such that

$$h(t) = S'(t)\phi(0) + \lim_{\lambda \to \infty}\int_0^t S'(t-s)B_\lambda v(s)ds, \quad t \in J.$$

Let $\tau_1, \tau_2 \in J \backslash \{t_1, t_2, \ldots, t_m\}, \tau_1 < \tau_2$. Thus if $\epsilon > 0$, we have

$$|h(\tau_2) - h(\tau_1)| \leq |[S'(\tau_2) - S'(\tau_1)]\phi(0)|$$

$$+ \left| \lim_{\lambda \to \infty} \int_0^{\tau_1 - \epsilon} \|S'(\tau_2 - s) - S'(\tau_1 - s)B_\lambda v(s)ds \right|$$

$$+ \left| \lim_{\lambda \to \infty} \int_{\tau_1 - \epsilon}^{\tau_1} [S'(\tau_2 - s) - S'(\tau_1 - s)]B_\lambda v(s)ds \right|$$

$$+ \left| \lim_{\lambda \to \infty} \int_{\tau_1}^{\tau_2} S'(\tau_2 - s)B_\lambda v(s)ds \right|.$$

As $\tau_1 \to \tau_2$ and ϵ becomes sufficiently small, the right-hand side of the above inequality tends to zero, since $S'(t)$ is a strongly continuous operator and the compactness of $S'(t)$ for $t > 0$ implies the continuity in the uniform operator topology (see [168]).

This proves the equi-continuity for the case where $t \neq t_i, i = 1, \ldots, m+1$. It remains to examine the equi-continuity at $t = t_i$. First we prove the equi-continuity at $t = t_i^-$, we have for some $y \in B_q$, there exists $v \in S_{F,y}$ such that

$$h(t) = S'(t)\phi(0) + \lim_{\lambda \to \infty} \int_0^t S'(t - s)B_\lambda v(s)ds, \quad t \in J.$$

Fix $\delta_1 > 0$ such that $\{t_k, k \neq i\} \cap [t_i - \delta_1, t_i + \delta_1] = \emptyset$. Let $0 < \rho < \delta_1$. First we prove equi-continuity at $t = t_i^-$. Fix $\delta_1 > 0$ such that $\{t_k : k \neq i\} \cap [t_i - \delta_1, t_i + \delta_1] = \emptyset$. For $0 < \rho < \delta_1$ we have

$$|h(t_i - \rho) - h(t_i)| \leq |(S'(t_i - \rho) - S'(t_i))\phi(0)|$$

$$+ \lim_{\lambda \to \infty} \int_0^{t_i - \rho} |(S'(t_i - \rho - s) - S'(t_i - s))B_\lambda v(s)|ds$$

$$+ Me^{\omega b}\psi(q) \int_{t_i - \rho}^{t_i} e^{-\omega s}p(s)\,ds;$$

which tends to zero as $\rho \to 0$. Define

$$\hat{h}_0(t) = h(t), \quad t \in [0, t_1]$$

and

$$\hat{h}_i(t) = \begin{cases} h(t), & \text{if } t \in (t_i, t_{i+1}] \\ h(t_i^+), & \text{if } t = t_i. \end{cases}$$

Next we prove equi-continuity at $t = t_i^+$. Fix $\delta_2 > 0$ such that $\{t_k : k \neq i\} \cap [t_i - \delta_2, t_i + \delta_2] = \emptyset$. For $0 < \rho < \delta_2$ we have

$$|\hat{h}(t_i + \rho) - \hat{h}(t_i)| \leq | \left(S'(t_i + \rho) - S'(t_i) \right) \phi(0)|$$

$$+ \lim_{\lambda \to \infty} \int_0^{t_i} | \left(S'(t_i + \rho - s) - S'(t_i - s) \right) B_\lambda v(s)| ds$$

$$+ M e^{\omega b} \psi(q) \int_{t_i}^{t_i + \rho} e^{-\omega s} p(s) \, ds.$$

The right-hand side tends to zero as $\rho \to 0$. The equi-continuity for the cases $\tau_1 < \tau_2 \leq 0$ and $\tau_1 \leq 0 \leq \tau_2$ follows from the uniform continuity of ϕ on the interval $[-r, 0]$. As consequence of Claims 3 and 4 together with Arzelá–Ascoli theorem it suffices to show that $\mathcal{B}$ maps B_q into a precompact set in E.

Let $0 < t^* < b$ be fixed and let ϵ be a real number satisfying $0 < \epsilon < t^*$. For $y \in B_q$ we define

$$h_\epsilon(t^*) = S'(t)\phi(0) + S'(\epsilon) \lim_{\lambda \to \infty} \int_0^{t-\epsilon} S'(t - s - \epsilon) B_\lambda v(s) ds.$$

where $v \in S_{F,y}$. Since

$$\left| \lim_{\lambda \to \infty} \int_0^{t-\epsilon} S'(t - s - \epsilon) B_\lambda v(s) \, ds \right| \leq M e^{\omega b} \psi(q) \int_0^{t-\epsilon} e^{-\omega s} p(s) ds.$$

and $S'(t)$ is a compact operator for $t > 0$, the set

$$H^\epsilon(t^*) = \{h_\epsilon(t^*) : \quad h_\epsilon \in \mathcal{B}(y)\}$$

is precompact in E for every ϵ, $0 < \epsilon < t^*$. Moreover, for every $h \in \mathcal{B}(y)$ we have

$$|h(t^*) - h_\epsilon(t^*)| \leq M e^{\omega b} \psi(q) \int_{t^* - \epsilon}^{t^*} e^{-\omega s} p(s) ds.$$

Therefore, there are precompact sets arbitrarily close to the set $H^\epsilon(t^*) = \{h(t^*) : \quad h \in \mathcal{B}(y)\}$. Hence the set $H(t^*) = \{h(t^*) : \quad h \in \mathcal{B}(B_q)\}$ is precompact in E. Hence the operator $\mathcal{B} : \Omega \to \mathcal{P}(\Omega)$ is completely continuous.

Claim 5: $\mathcal{B}$ has closed graph. Let $\{y_n\}$ be a sequence such that $y_n \to y_*$ in Ω, $h_n \in \mathcal{B}(y_n)$, and $h_n \to h_*$. We shall show that $h_* \in \mathcal{B}(y_*)$. $h_n \in \mathcal{B}(y_n)$ means that there exists $v_n \in S_{F,y_n}$ such that

$$h_n(t) = S'(t)\phi(0) + \lim_{\lambda \to \infty} \int_0^t S'(t - s) B_\lambda v_n(s) ds, \quad t \in J.$$

We must prove that there exists $v_* \in S_{F,y_*}$ such that

$$h_*(t) = S'(t)\phi(0) + \lim_{\lambda \to \infty} \int_0^t S'(t-s)B_\lambda v_*(s)ds, \quad t \in J.$$

Consider the linear and continuous operator $\mathcal{K} : L^1(J,E) \to C(J,E)$ defined by

$$(\mathcal{K}v)(t) = \lim_{\lambda \to \infty} \int_0^t S'(t-s)B_\lambda v(s)ds, \quad t \in J.$$

Then we have

$$|(h_n(t) - S'(t)\phi(0)) - (h_*(t) - S'(t)\phi(0))| = |h_n(t) - h_*(t)|$$

$$\leq \|h_n - h_*\|_\infty \to 0, \quad \text{as } n \mapsto \infty.$$

From Lemma 1.11 it follows that $\mathcal{K} \circ S_F$ is a closed graph operator and from the definition of $\mathcal{K}$ one has

$$h_n(t) - S'(t)\phi(0) \in \mathcal{K} \circ S_{F,y_n}.$$

As $y_n \to y_*$ and $h_n \to h_*$, there is a $v_* \in S_{F,y_*}$ such that

$$h_*(t) - S'(t)\phi(0) = \lim_{\lambda \to \infty} \int_0^t S'(t-s)B_\lambda v_*(s)ds, \quad t \in J.$$

Hence the multi-valued operator $\mathcal{B}$ is upper semi-continuous.

Step 3: A priori bounds. Now it remains to show that the set

$$\mathcal{E} = \{y \in \Omega : y \in \alpha\mathcal{A}(y) + \alpha\mathcal{B} \quad \text{for some } 0 < \alpha < 1\}$$

is bounded. Let $y \in \mathcal{E}$, then there exist $v \in S_{F,y}$ and $\mathcal{I}_k \in I_k(y(t_k^-))$ such that

$$y(t) = \alpha S'(t)\phi(0) + \alpha \lim_{\lambda \to \infty} \int_0^t S'(t-s)B_\lambda v(s)ds + \alpha \sum_{0 < t_k < t} S'(t-t_k)\mathcal{I}_k$$

for some $0 < \alpha < 1$. Thus, by (8.1.2), (8.1.5) for each $t \in J$, we have

$$|y(t)| \leq Me^{\omega t}|\phi(0)| + Me^{\omega t}\int_0^t e^{-\omega s}p(s)\psi(\|y_s\|)ds$$

$$+ Me^{\omega t}\sum_{k=1}^m e^{-\omega t_k}|\mathcal{I}_k|$$

$$\le Me^{\omega b}\|\phi\| + Me^{\omega b}\int_0^t e^{-\omega s}p(s)\psi(\|y_s\|)ds$$

$$+Me^{\omega b}\sum_{k=1}^{m}e^{-\omega t_k}c_k|y(t_k^-)| + Me^{\omega t}\sum_{k=1}^{m}e^{-\omega t_k}c_k|I_k(0)|$$

$$\le C + Me^{\omega b}\int_0^t e^{-\omega s}p(s)\psi(\|y_s\|)ds$$

$$+Me^{\omega b}\sum_{k=1}^{m}e^{-\omega t_k}c_k|y(t_k^-)|.$$

Now we consider the function μ defined by

$$\mu(t) = \sup\{|y(s)| : \ -r \le s \le t\}, \ \ 0 \le t \le b.$$

Then $\|y_s\| \le \mu(t)$ for all $t \in J$ and there is a point $t^* \in [-r, t]$ such that $\mu(t) = |y(t^*)|$. If $t^* \in [0, b]$, by the previous inequality we have for $t \in [0, b]$ (note $t^* \le t$)

$$\mu(t) \le C + Me^{\omega b}\int_0^t e^{-\omega s}p(s)\psi(\mu(s))ds + Me^{\omega b}\sum_{k=1}^{m}e^{-\omega t_k}c_k\mu(t).$$

Then

$$\left(1 - Me^{\omega b}\sum_{k=1}^{m}e^{-\omega t_k}c_k\right)\mu(t) \le C + Me^{\omega b}\int_0^t e^{-\omega s}p(s)\psi(\mu(s))ds.$$

Thus by (8.6) and (8.7) we have

$$\mu(t) \le C_0 + C_1\int_0^t e^{-\omega s}p(s)\psi(\mu(s))ds. \tag{8.10}$$

Let us take the right-hand side of (8.10) as $v(t)$. Then we have

$$\mu(t) \le v(t) \ \ \text{for all} \ \ t \in J,$$

with

$$v(0) = C_0,$$

and

$$v'(t) = C_1 e^{-\omega t}p(t)\psi(\mu(t)), \ \ a.e. \ \ t \in J.$$

Using the increasing character of ψ we get

$$v'(t) \leq C_1 e^{-\omega t} p(t) \psi(v(t)), \quad a.e. \ t \in J.$$

Integrating from 0 to t we get

$$\int_0^t \frac{(v(s))'}{\psi(v(s))} ds \leq C_1 \int_0^t e^{-\omega t} p(s)) ds.$$

By a change of variable we get

$$\int_{v(0)}^{v(t)} \frac{du}{\psi(u)} \leq C_1 \int_0^t e^{-\omega t} p(s)) ds$$

$$\leq C_1 \int_0^b e^{-\omega t} p(s)) ds.$$

Hence by (8.5) there exist a constant N such that

$$\mu(t) \leq v(t) \leq N \quad \text{for all } t \in J.$$

Now from the definition of μ it follows that

$$\|y\|_{\Omega} \leq \max(\|\phi\|_D, N), \quad \text{for all } y \in \mathcal{E}.$$

This shows that the set $\mathcal{E}$ is bounded. As a consequence of Theorem 1.32 we deduce that $\mathcal{A} + \mathcal{B}$ has a fixed point y defined on the interval $[-r, b]$ which is the integral solution of problem (8.1)–(8.3). $\qquad \square$

We now present another existence result for the problem (8.1)–(8.3) where a Lipschitz condition on the multi-valued F with respect to its second variable is assumed instead of a Wintner growth condition used in Theorem 8.2.

Theorem 8.3. *Assume that (8.1.1)–(8.1.4), $\Phi(0) \in \overline{D(A)}$ hold and the condition*

(8.3.1) There exists a function $l \in L^1(J, \mathbb{R}_+)$ such that:

$$H_d(F(t, u), F(t, \bar{u})) \leq l(t) \|u - \bar{u}\|_D \ a.e. \ t \in J, \ \text{and for all } u, \bar{u} \in D,$$

and

$$H_d(0, F(t, 0)) \leq l(t) \ \text{for a.e. } t \in J,$$

where $\int_0^b e^{-\omega s} l(s) ds < \infty,$

$$C_0^* = \frac{M e^{\omega b} \left(\|\phi\| + \sum_{k=1}^m e^{-\omega t_k} c_k |I_k(0)| + \int_0^b e^{-\omega s} l(s) ds \right)}{1 - M e^{\omega b} \sum_{k=1}^m e^{-\omega t_k} c_k} \tag{8.11}$$

and

$$C_1^* = \frac{Me^{\omega b}}{1 - Me^{\omega b} \sum\limits_{k=1}^{m} e^{-\omega t_k} c_k}. \tag{8.12}$$

If

$$Me^{\omega b} \sum_{k=1}^{m} e^{-\omega t_k} c_k < 1, \tag{8.13}$$

then the problem (8.1)–(8.3) has at least one integral solution on $[-r, b]$.

Proof. Let $\mathcal{A}$ and $\mathcal{B}$ the operators defined in Theorem 8.2. It can be shown, as in the proof of Theorem 8.2 that $\mathcal{B}$ is completely continuous and upper semi-continuous and $\mathcal{A}$ is a contraction. Now we prove that

$$\mathcal{E} = \{y \in \Omega : y \in \alpha \mathcal{A}(y) + \alpha \mathcal{B}(y), \text{ for some } 0 < \alpha < 1\}$$

is bounded.

Let $y \in \mathcal{E}$, then there exist $v \in S_{F,y}$ and $\mathcal{I}_k \in I_k(y(t_k^-))$ such that for each $t \in J$

$$y(t) = \alpha S'(t)\phi(0) + \alpha \frac{d}{dt} \int_0^t S(t-s)v(s)ds + \alpha \sum_{0 < t_k < t} S'(t-t_k) \mathcal{I}_k,$$

for some $0 < \alpha < 1$. Thus, by (8.1.2), (8.3.1), for each $t \in J$, we have

$$|y(t)| \le Me^{\omega t}|\phi(0)| + Me^{\omega t} \int_0^t e^{-\omega s}|v(s)|ds + Me^{\omega t} \sum_{k=1}^{m} e^{-\omega t_k}|\mathcal{I}_k|$$

$$\le Me^{\omega t}|\phi(0)| + Me^{\omega t} \int_0^t e^{-\omega s}l(s)\|y_s\|ds$$

$$+ Me^{\omega t} \int_0^t e^{-\omega s}l(s)ds + Me^{\omega t} \sum_{k=1}^{m} e^{-\omega t_k}c_k|y(t_k^-)|$$

$$+ Me^{\omega t} \sum_{k=1}^{m} e^{-\omega t_k}c_k|I_k(0)|$$

$$\le Me^{\omega b}\left(\|\phi\| + \int_0^t e^{-\omega s}l(s)ds + \sum_{k=1}^{m} e^{-\omega t_k}c_k|I_k(0)|\right)$$

$$+ Me^{\omega b} \int_0^t e^{-\omega s}l(s)\|y_s\|ds + Me^{\omega b} \sum_{k=1}^{m} e^{-\omega t_k}c_k y(t_k^-).$$

Now we consider the function μ defined by

$$\mu(t) = \sup\{|y(s)| : -r \le s \le t\}, \ 0 \le t \le b.$$

Then $\|y_s\| \le \mu(t)$ for all $t \in J$ and there is a point $t^* \in [-r, t]$ such that $\mu(t) = |y(t^*)|$. If $t^* \in [0, b]$, by the previous inequality we have for $t \in [0, b]$ (note $t^* \le t$)

$$\mu(t) \le Me^{\omega b}\left(\|\phi\| + \int_0^t e^{-\omega s}l(s)ds + \sum_{k=1}^m e^{-\omega t_k}c_k|I_k(0)|\right)$$

$$+Me^{\omega b}\int_0^t e^{-\omega s}l(s)\mu(s)ds + Me^{\omega b}\sum_{k=1}^m e^{-\omega t_k}c_k\mu(t).$$

Then

$$\mu(t) \le C_0^* + C_1^*\int_0^t e^{-\omega s}l(s)\mu(s)ds.$$

By Gronwall inequality ([131]) we get for each $t \in J$

$$\mu(t) \le C_0^*\exp\left(C_1^*\int_0^t e^{-\omega s}l(s)ds\right).$$

Hence

$$\|\mu\|_\infty \le C_0^*\exp\left(C_1^*\int_0^b e^{-\omega s}l(s)ds\right) := M^*.$$

Thus

$$\|y\|_\Omega \le \max(\|\phi\|_D, M^*).$$

This shows that the set $\mathcal{E}$ is bounded. As a consequence of Theorem 1.32 we deduce that $\mathcal{A} + \mathcal{B}$ has a fixed point which is a integral solution of problem (8.1)–(8.3).

The following result concerns the compactness property of the solutions set of problem (8.1)–(8.3).   □

Theorem 8.4. *Under assumptions (8.1.1)–(8.1.4), and*

(8.4.1) There exists $p \in C(J, \mathbb{R}_+)$ such that

$$\|F(t, u)\| \le p(t) \ \text{for each } t \in J, \ \text{and each } u \in D.$$

the solution set of (8.1)–(8.3) in not empty and compact in Ω.

Proof. Let

$$S = \{y \in \Omega : y \text{ is solution of } (8.1)\text{–}(8.3)\}.$$

From Theorem 8.2, $S \neq \emptyset$. Now, we prove that S is compact. Let $(y_n)_{n \in \mathbb{N}} \in S$, then there exist $v_n \in S_{F,y_n}$ and $\mathcal{I}_k^n \in I_k(y_n(t_k^-))$ such that

$$y_n(t) = S'(t)\phi(0) + \frac{d}{dt}\int_0^t S(t-s)v_n(s)ds + \sum_{0<t_k<t} S'(t-t_k)\mathcal{I}_k^n.$$

From (8.1.2), (8.4.1), we can prove that there exists an $M_1 > 0$ such that

$$\|y_n\|_\infty \leq M_1, \quad \text{for every } n \geq 1.$$

As in Claim 4 in Theorem 8.2, we can easily show using (7.1.2), (8.4.1) that the set $\{y_n : n \geq 1\}$ is equi-continuous in Ω, hence by Arzelá–Ascoli Theorem we can conclude that, there exists a subsequence (denoted again by $\{y_n\}$) of $\{y_n\}$ such that y_n converges to y in Ω. We shall show that there exist $v(.) \in F(.,y.)$ and $\mathcal{I}_k \in I_k(y(t_k^-))$ such that

$$y(t) = S'(t)\phi(0) + \frac{d}{dt}\int_0^t S(t-s)v(s)ds + \sum_{0<t_k<t} S'(t-t_k)\mathcal{I}_k.$$

Since $F(t,.)$ is upper semi-continuous, then for every $\varepsilon > 0$, there exists $n_0(\epsilon) \geq 0$ such that for every $n \geq n_0$, we have

$$v_n(t) \in F(t, y_{n_t}) \subset F(t, y_t) + \varepsilon B(0,1), \quad \text{a.e. } t \in J.$$

Since $F(.,.)$ has compact values, there exists subsequence $v_{n_m}(.)$ such that

$$v_{n_m}(.) \to v(.) \text{ as } m \to \infty$$

and

$$v(t) \in F(t, y_t), \quad \text{a.e. } t \in J.$$

It is clear that

$$|v_{n_m}(t)| \leq p(t), \quad \text{a.e. } t \in J.$$

By Lebesgue's dominated convergence theorem, we conclude that $v \in L^1(J, E)$ which implies that $v \in S_{F,y}$. Also, since I_k has closed graph we get $\mathcal{I}_k \in I_k(y(t_k^-))$. Thus

$$y(t) = S'(t)\phi(0) + \frac{d}{dt}\int_0^t S(t-s)v(s)ds + \sum_{0<t_k<t} S'(t-t_k)\mathcal{I}_k.$$

Then $S \in \mathcal{P}_{cp}(\Omega)$. $\square$

8.3 Extremal Integral Solutions with Local Conditions

In this section we shall prove the existence of maximal and minimal integral solutions of problem (8.1)–(8.3) under suitable monotonicity conditions on the functions involved in it. Let us give the definition of the extremal integral solutions of the problem (8.1)–(8.3)

Definition 8.5. We say that a continuous function $u : [-r, b] \to E$ is a lower integral solution of problem (8.1)–(8.3) if there exist $v \in L^1(J, E)$ and $\mathcal{I}_k \in I_k(u(t_k^-))$ such that $v(t) \in F(t, u_t)$ a.e. on J, $y(t) = \phi(t)$, $t \in H$, and

$$u(t) \leq S'(t)\phi(0) + \frac{d}{dt}\int_0^t S(t-s)v(s)ds + \sum_{0<t_k<t} S'(t-t_k)\mathcal{I}_k, \ t \in J, \ t \neq t_k$$

and $u(t_k^+) - u(t_k^-) \leq \mathcal{I}_k$, $k = 1, \ldots, m$. Similarly an upper integral solution w of problem (8.1)–(8.3) is defined by reversing the order.

Definition 8.6. A solution x_M of problem (8.1)–(8.3) is said to be maximal if for any other solution x of problem (8.1)–(8.3) on J, we have that $x(t) \leq x_M(t)$ for each $t \in J$. Similarly a minimal solution of problem (8.1)–(8.3) is defined by reversing the order of the inequalities.

Definition 8.7. A multi-valued function $F(t, x)$ is called strictly monotone increasing in x almost everywhere for $t \in J$, if $F(t, x) \leq F(t, y)$ a.e. $t \in J$ for all $x, y \in D$ with $x < y$. Similarly $F(t, x)$ is called strictly monotone decreasing in x almost everywhere for $t \in J$, if $F(t, x) \geq F(t, y)$ a.e. $t \in J$ for all $x, y \in D$ with $x < y$.

Let us the following assumptions.

(8.7.1) The multi-valued function $F(t, y)$ is strictly monotone increasing in y for almost each $t \in J$,
(8.7.2) $S'(t)$ is preserving the order, that is $S'(t)v \geq 0$ whenever $v \geq 0$.
(8.7.3) The multi-valued functions I_k, $k = 1, \ldots, m$ are strictly monotone increasing.
(8.7.4) The problem (8.1)–(8.3) has a lower integral solution v and an upper integral solution w with $v \leq w$.

Theorem 8.8. *Assume that assumptions (8.1.1)–(8.1.4) and (8.7.1)–(8.7.4) hold. Then problem (8.1)–(8.3) has a minimal and a maximal integral solutions on $[-r, b]$.*

Proof. It can be shown, as in the proof of Theorem 8.3, that $\mathcal{B}$ is completely continuous and upper semi-continuous and $\mathcal{A}$ is a contraction on $[v, w]$. We shall show that $\mathcal{A}$ and $\mathcal{B}$ are isotone increasing on $[v, w]$. Let $y, \bar{y} \in [v, w]$ be such that $y \leq \bar{y}$, $y \neq \bar{y}$. Then by (8.7.2), (8.7.3), we have for each $t \in J$

$$A(y) = \left\{ h \in \Omega : h(t) = \sum_{0 < t_k < t} S'(t - t_k) \mathcal{I}_k, \ \mathcal{I}_k \in I_k(y(t_k^-)), \right\}$$

$$\leq \left\{ h \in \Omega : h(t) = \sum_{0 < t_k < t} S'(t - t_k) \mathcal{I}_k, \ \mathcal{I}_k \in I_k(\bar{y}(t_k^-)), \right\}$$

$$= A(\bar{y}).$$

Similarly, by (8.7.1) and (8.7.2) and

$$B(y) := \left\{ h \in \Omega : \ h(t) = S'(t)\phi(0) + \frac{d}{dt} \int_0^t S(t - s)v(s)ds, \quad v \in S_{F,y} \right\}$$

$$\leq \left\{ h \in \Omega : \ h(t) = S'(t)\phi(0) + \frac{d}{dt} \int_0^t S(t - s)v(s)ds, \quad v \in S_{F,\bar{y}} \right\}$$

$$= B(\bar{y}).$$

Therefore A and B are isotone increasing on $[v, w]$. Finally, let $x \in [v, w]$ be any element. By (8.7.4) we deduce that

$$v \leq A(v) + B(v) \leq A(x) + B(x) \leq A(w) + B(w) \leq w,$$

which shows that $A(x) + B(x) \in [v, w]$ for all $x \in [v, w]$. Thus, A and B satisfy all conditions of Theorem 1.32, hence problem (8.1)–(8.3) has a maximal and a minimal integral solutions on $[-r, b]$. This completes the proof. □

8.4   Integral Solutions with Nonlocal Conditions

In this section we prove existence results for problem of the form

$$y'(t) - Ay(t) \in F(t, y_t), \quad a.e. \ t \in J = [0, b], \ t \neq t_k, \ k = 1, \ldots, m \qquad (8.14)$$

$$\Delta y|_{t=t_k} \in I_k(y(t_k^-)), \quad k = 1, \ldots, m \qquad (8.15)$$

$$y(t) + h_t(y) = \phi(t), \quad t \in [-r, 0], \qquad (8.16)$$

where $h_t : \Omega \to \overline{D(A)}$ is a given function, A, F, and I_k are as above.

8.4.1 Main Result

Definition 8.9. A function $y \in \Omega$ is said to be an integral solution of problem (8.14)–(8.16) if $y(t) = \phi(t) - h_t(y)$, $t \in [-r, 0]$, and there exist $v(.) \in L^1(J, E)$ and $\mathcal{I}_k \in I_k(y(t_k^-))$ such that $v(t) \in F(t, y_t)$ a.e. $t \in J$, and y satisfies the integral equation,

$$y(t) = S'(t)\left(\phi(0) - h_0(y)\right) + \frac{d}{dt}\int_0^t S(t-s)v(s)ds + \sum_{0 < t_k < t} S'(t - t_k)\mathcal{I}_k.$$

Theorem 8.10. *Assume that hypotheses (8.1.1)–(8.1.4) hold and moreover*

(A1)   The function h is continuous with respect to t, and there exists a constant $\alpha > 0$ such that

$$|h_t(u)| \leq \alpha, \quad u \in \Omega$$

and for each $k > 0$ the set

$$\{\phi(0) - h_0(y), \ y \in \Omega, \ \|y\|_\Omega \leq k\}$$

is precompact in E,

(A2)   There exist a function $p \in L^1(J, \mathbb{R}_+)$ and a continuous nondecreasing function $\psi : [0, \infty) \to (0, \infty)$ such that

$$\|F(t, x)\| \leq p(t)\psi(\|x\|_D), \quad a.e. \ t \in J, \quad \text{for all } x \in D$$

with

$$\int_{\tilde{C}_0}^\infty \frac{du}{\psi(u)} > \tilde{C}_1 \int_0^b e^{-\omega s}p(s)ds, \tag{8.17}$$

where

$$\tilde{C}_0 = \frac{Me^{\omega b}\left[\|\phi\|_D + \alpha + \sum_{k=1}^m e^{-\omega t_k} c_k |I_k(0)|\right]}{1 - Me^{\omega b}\sum_{k=1}^m e^{-\omega t_k} c_k}, \tag{8.18}$$

and

$$\tilde{C}_1 = \frac{Me^{\omega b}}{1 - Me^{\omega b}\sum_{k=1}^m e^{-\omega t_k} c_k}. \tag{8.19}$$

Moreover, we suppose that

$$Me^{\omega b} \sum_{k=1}^{m} e^{-\omega t_k} c_k < 1. \tag{8.20}$$

Then the problem (8.14)–(8.16) has at least one integral solution on $[-r, b]$.

Proof. Consider the multi-valued operators $\mathcal{A}_1, \mathcal{B}_1 : \Omega \to \mathcal{P}(\Omega)$:

$$\mathcal{B}_1(y) := \left\{ f \in \Omega : f(t) = \begin{cases} \phi(t) - h_t(y), & \text{if } t \in H; \\ S'(t)(\phi(0) - h_0(y)) \\ + \dfrac{d}{dt} \displaystyle\int_0^t S(t-s)v(s)ds, & v \in S_{F,y} \quad \text{if } t \in J, \end{cases} \right\}$$

and

$$\mathcal{A}_1(y) := \left\{ f \in \Omega : f(t) = \begin{cases} 0, & \text{if } t \in H; \\ \displaystyle\sum_{0<t_k<t} S'(t-t_k)\mathcal{I}_k, \ \mathcal{I}_k \in I_k(y(t_k^-)) & \text{if } t \in J. \end{cases} \right\}$$

Then the problem of finding the solution of problem (8.14)–(8.16) is reduced to finding the solution of the operator inclusion $y \in \mathcal{A}_1(y) + \mathcal{B}_1(y)$. As in the previous section, it can be shown that the operators $\mathcal{A}_1$ and $\mathcal{B}_1$ satisfy all conditions of Theorem 1.32. $\square$

8.5 Application to the Control Theory

In this section we treat the controllability of impulsive functional differential inclusions using the argument of the previous sections. More precisely we will consider the following problem:

$$y'(t) - Ay(t) \in F(t, y_t) + Bu(t), \ a.e. \ t \in J = [0, b], \ t \neq t_k, \ k = 1, \dots, m \tag{8.21}$$

$$\Delta y|_{t=t_k} \in I_k(y(t_k^-)), \ k = 1, \dots, m \tag{8.22}$$

$$y(t) = \phi(t), \ t \in [-r, 0], \tag{8.23}$$

where A, F, and I_k are as above, the control function $u(\cdot)$ is given in $L^2(J, U)$ a Banach space of admissible control functions with U as a Banach. Finally B is a bounded linear operator from U to $\overline{D(A)}$.

Definition 8.11. A function $y \in \Omega$ is said to be an integral solution of problem (8.21)–(8.23) if $y(t) = \phi(t)$, $t \in [-r, 0]$, and there exist $v(.) \in L^1(J, E)$ and $\mathcal{I}_k \in I_k(y(t_k^-))$ such that $v(t) \in F(t, y_t)$ a.e. $t \in J$, and y satisfies the impulsive integral equation,

$$y(t) = S'(t)\phi(0) + \frac{d}{dt} \int_0^t S(t - s)v(s)ds$$

$$+ \frac{d}{dt} \int_0^t S(t - s)Bu(s)ds + \sum_{0 < t_k < t} S'(t - t_k)\mathcal{I}_k.$$

Definition 8.12. The system (8.21)–(8.23) is said to be controllable on the interval $[-r, b]$ if for every initial function $\phi \in D$ and every $y_1 \in \overline{D(A)}$, there exists a control $u \in L^2(J, U)$, such that the integral solution $y(t)$ of system (8.21)–(8.23) satisfies $y(b) = y_1$.

8.5.1 Main Result

Let us the following assumptions

(B1) The linear operator $W : L^2(J, U) \to \overline{D(A)}$, defined by

$$Wu = \frac{d}{dt} \int_0^b S(b - s)Bu(s)ds,$$

has a bounded inverse operator W^{-1} which takes values in $L^2(J, U) \backslash KerW$, and there exist positive constants $\overline{M}, \overline{M_1}$, such that $\|B\| \leq \overline{M}$ and $\|W^{-1}\| \leq \overline{M_1}$.

(B2) F has compact and convex values, and there exists a function $l \in L^1(J, \mathbb{R}_+)$ such that

$$H_d(F(t, x), F(t, y)) \leq l(t)\|x - y\|_D \text{ for a.e. } t \in J, \text{ and for all } x, y \in D,$$

with

$$H_d(0, F(t, 0)) \leq l(t), \ a.e. \ t \in J.$$

(B3) There exist a function $p \in L^1(J, \mathbb{R}_+)$ and a continuous nondecreasing function $\psi : [0, \infty) \to (0, \infty)$ such that

$$\|F(t, x)\| \leq p(t)\psi(\|x\|_D), \quad a.e. \ t \in J, \quad \text{for all } x \in D$$

with $\int_0^b e^{-\omega s}p(s)ds < \infty$,

$$\int_{C_0^*}^{\infty} \frac{ds}{s + \psi(s)} = \infty \tag{8.24}$$

where

$$C_0^* = \frac{C^*}{1 - Me^{\omega b}\sum\limits_{k=1}^{m}e^{-\omega t_k}c_k},$$ (8.25)

$$C^* = Me^{\omega b}(1 + M\overline{MM_1}e^{\omega b}b)\|\phi\|_D + Me^{\omega b}\overline{MM_1}b|y_1|$$
$$+ (Me^{\omega b} + M\overline{MM_1}e^{2\omega b}b)\sum\limits_{k=1}^{m}e^{-\omega t_k}c_k|I_k(0)|.$$

Theorem 8.13. *Assume that hypotheses (8.1.1)–(8.1.4) hold. Moreover we suppose that*

$$M^2 e^{2\omega b}\overline{MM_1}b\int_0^b e^{-\omega s}l(s)ds + Me^{\omega b}(1 + Me^{\omega b}\overline{MM_1}b)\sum\limits_{k=1}^{m}e^{-\omega t_k}c_k < 1.$$ (8.26)

Then the problem (8.21)–(8.23) is controllable on $[-r, b]$.

Remark 8.14. The construction of operator W^{-1} and its properties are discussed in [170].

Proof. Using hypothesis (B1) for an arbitrary function $y(.)$ we define the control

$$u_y(t) = W^{-1}\left[y_1 - S'(b)\phi(0) - \lim_{\lambda \to \infty}\int_0^b S'(b - s)B_\lambda v(s)ds \right.$$
$$\left. - \sum\limits_{k=1}^{m}S'(b - t_k)\mathcal{I}_k\right](t),$$

where $v \in S_{F,y}$ and $\mathcal{I}_k \in I_k(y(t_k^-))$. Consider the multi-valued operators defined from Ω to $\mathcal{P}(\Omega)$ by:

$$A(y) := \left\{f \in \Omega : f(t) = \begin{cases} 0, & \text{if } t \in H; \\ \dfrac{d}{dt}\displaystyle\int_0^t S(t - s)(Bu_y)(s)ds \\ \quad + \displaystyle\sum_{0 < t_k < t}S'(t - t_k)\mathcal{I}_k, & \text{if } t \in J, \end{cases}\right\}$$

and

$$B(y) := \left\{f \in \Omega : f(t) = \begin{cases} \phi(t), & \text{if } t \in H; \\ S'(t)\phi(0) + \dfrac{d}{dt}\displaystyle\int_0^t S(t - s)v(s)ds & \text{if } t \in J. \end{cases}\right\}$$

As in Theorem 8.2, we can prove that the operators $\mathcal{A}$ is a contraction operator, and $\mathcal{B}$ is completely continuous and upper semi-continuous with compact convex values.

Now, we prove that the set

$$\mathcal{E} = \{y \in \Omega \mid y \in \alpha\mathcal{A}y + \alpha\mathcal{B}y, \ 0 \le \alpha \le 1\}$$

is bounded.

Let $y \in \mathcal{E}$. be any element, then there exists $v \in S_{F,y}$ and such that

$$y(t) = \alpha S'(t)\phi(0) + \alpha\frac{d}{dt}\int_0^t S(t-s)v(s)ds$$

$$+\alpha\frac{d}{dt}\int_0^t S(t-s)Bu_y(s)ds + \alpha \sum_{0<t_k<t} S'(t-t_k)\mathcal{I}_k.$$

This implies by (B1)–(B3) that, for each $t \in J$, we have

$$|y(t)| \le Me^{\omega b}\Bigg[(1 + M\overline{MM}_1 be^{\omega b})\|\phi\|_D$$

$$+\overline{MM}_1 b|y_1| + (Me^{\omega b} + M^2\overline{MM}_1 be^{\omega b})\sum_{k=1}^{m} e^{-\omega t_k}c_k|I_k(0)|\Bigg]$$

$$+[Me^{\omega b} + M^2\overline{MM}_1 be^{2\omega b}]\int_0^t e^{-\omega s}p(s)\psi(\|y_s\|)ds$$

$$+M^2\overline{MM}_1 be^{2\omega b}\int_0^t e^{-\omega s}\sum_{0<t_k<s} e^{-\omega t_k}c_k|y(t_k^-))|ds$$

$$+Me^{\omega b}\sum_{k=1}^{m} e^{-\omega t_k}c_k|y(t_k^-)|.$$

Consider the function μ defined by

$$\mu(t) = \sup\{|y(s)| : \ -r \le s \le t\}, \ 0 \le t \le b.$$

Then $\|y_s\| \le \mu(t)$ for all $t \in J$ and there is a point $t^* \in [-r, t]$ such that $\mu(t) = |y(t^*)|$. If $t^* \in [0, b]$, by the previous inequality we have for $t \in [0, b]$ (note $t^* \le t$)

$$\mu(t) \le C^* + [Me^{\omega b} + M^2\overline{MM}_1 be^{2\omega b}]\int_0^t e^{-\omega s}p(s)\psi(\mu(s))ds$$

$$+Me^{\omega b}\sum_{k=1}^{m} e^{-\omega t_k}c_k\mu(t)$$

$$+M^2\overline{MM}_1 be^{2\omega b}\int_0^t e^{-\omega s}\sum_{0<t_k<s} e^{-\omega t_k}c_k\mu(s)ds.$$

Then

$$\left[1 - Me^{\omega b}\sum_{k=1}^{m}e^{-\omega t_k}c_k\right]\mu(t) \leq C^* + M^2\overline{MM}_1 be^{2\omega b}\int_0^t e^{-\omega s}\sum_{0<t_k<s}e^{-\omega t_k}c_k\mu(s)ds$$

$$+\left[Me^{\omega b} + M^2\overline{MM}_1 be^{2\omega b}\right]\int_0^t e^{-\omega s}p(s)\psi(\mu(s))ds.$$

Thus we have

$$\mu(t) \leq C_0^* + C_1^*\int_0^t e^{-\omega s}\sum_{0<t_k<s}e^{-\omega t_k}c_k\mu(s)ds$$

$$+C_2^*\int_0^t e^{-\omega s}p(s)\psi(\mu(s))ds$$

$$\leq C_0^* + \int_0^t \hat{M}(s)[\mu(s) + \psi(\mu(s))]ds,$$

where

$$\hat{M}(s) = \max(C_1^* e^{-\omega s}\sum_{0<t_k<s}e^{-\omega t_k}c_k, C_2^* e^{-\omega s}p(s)).$$

Set

$$v(t) = C_0^* + \int_0^t \hat{M}(s)[\mu(s) + \psi(\mu(s))]ds. \tag{8.27}$$

Then we have

$$\mu(t) \leq v(t) \quad \text{for all } t \in J.$$

Differentiating the both sides of (8.27) we get

$$v'(t) = \hat{M}(t)[\mu(t) + \psi(\mu(t))], \quad a.e. \ t \in J,$$

and

$$v(0) = C_0^*.$$

Using the nondecreasing character of ψ we obtain

$$v'(t) \leq \hat{M}(t)[v(t) + \psi(v(t))], \quad a.e. \ t \in J,$$

that is

$$\frac{v'(t)}{v(t) + \psi(v(t))} \leq \hat{M}(t).$$

Integrating from 0 to t both sides of this inequality, we get

$$\int_0^t \frac{v'(s)}{v(s) + \psi(v(s))} ds \le \int_0^t \hat{M}(s) ds.$$

By a change of variables we get

$$\int_{v(0)}^{v(t)} \frac{du}{u + \psi(u)} \le \|\hat{M}\|_{L^1} < \infty.$$

Consequently, by (8.24), there exists a constant d such that $\mu(t) \le v(t) \le d$, $t \in J$ and hence from the definition of μ it follows that

$$\|y\|_\Omega \le \max(\|\phi\|_D, d).$$

This shows that the set $\mathcal{E}$ is bounded. As a consequence of Theorem 1.32 we deduce that $\mathcal{A} + \mathcal{B}$ has a fixed point which is a integral solution of problem (8.21)–(8.23). Thus the system (8.21)–(8.23) is controllable on $[-r, b]$. □

8.5.2 An Example

As an application of our results we consider the following impulsive partial functional differential equation of the form

$$\frac{\partial}{\partial t} z(t, x) \in \frac{\partial^2}{\partial x^2} z(t, x)$$

$$+ [Q_1(t, z(t - r, x)), Q_2(t, z(t - r, x))], \; x \in [0, \pi], \; t \in [0, b] \backslash \{t_1, t_2, \ldots, t_m\}. \tag{8.28}$$

$$z(t_k^+, x) - z(t_k^-, x) \in b_k |z(t_k^-, x)| \bar{B}(0, 1), \; x \in [0, \pi], \; k = 1, \ldots, m \tag{8.29}$$

$$z(t, 0) = z(t, \pi) = 0, \; t \in J := [0, b] \tag{8.30}$$

$$z(t, x) = \phi(t, x), \; t \in H, \; x \in [0, \pi], \tag{8.31}$$

where $b_k > 0$, $k = 1, \ldots, m$, $\phi \in \mathcal{D} = \{\bar{\psi} : H \times [0, \pi] \to \mathbb{R}; \bar{\psi}$ is continuous everywhere except for a countable number of points at which $\bar{\psi}(s^-), \bar{\psi}(s^+)$ exist with $\bar{\psi}(s^-) = \bar{\psi}(s)\}$, $0 = t_0 < t_1 < t_2 < \cdots < t_m < t_{m+1} = b$, $z(t_k^+) = \lim_{(h,x) \to (0^+, x)} z(t_k + h, x), z(t_k^-) = \lim_{(h,x) \to (0^-, x)} z(t_k + h, x)$, where $Q_1, Q_2 : J \times \mathbb{R} \to \mathbb{R}$, are given functions, and $\bar{B}(0, 1)$ the closed unit ball. We assume that for each $t \in J$, $Q_1(t, \cdot)$ is lower semi-continuous (i.e, the set $\{y \in \mathbb{R} : Q_1(t, y) > \mu\}$ is open for each $\mu \in \mathbb{R}$), and assume that for each $t \in J$, $Q_2(t, \cdot)$ is upper semi-continuous (i.e., the set $\{y \in \mathbb{R} : Q_2(t, y) < \mu\}$ is open for each $\mu \in \mathbb{R}$).

Let

$$y(t)(x) = z(t, x), \ t \in J, \ x \in [0, \pi],$$

$$I_k(y(t_k^-))(x) = b_k z(t_k^-, x), \ x \in [0, \pi], \ k = 1, \ldots, m$$

$$F(t, \phi)(x) = [Q_1(t, \phi(\theta, x)), Q_2(t, \phi(\theta, x))], \ \theta \in H, \ x \in [0, \pi],$$

and

$$\phi(\theta)(x) = \phi(\theta, x), \ \theta \in H, \ x \in [0, \pi].$$

It is clear that F is compact and convex valued, and it is upper semi-continuous (see [101]). Assume that there are $p \in C(J, \mathbb{R}^+)$ and $\psi : [0, \infty) \to (0, \infty)$ continuous and nondecreasing such that

$$\max(|Q_1(t, y)|, |Q_2(t, y)|) \le p(t)\psi(|y|), \quad t \in J, \ \text{and} \ y \in \mathbb{R},$$

and

$$\int_1^\infty \frac{ds}{\psi(s)} = +\infty.$$

Consider $E = C([0, \pi])$, the Banach space of continuous function on $[0, \pi]$ with values in $\mathbb{R}$. Define the linear operator A on E by

$$Az = \frac{\partial^2}{\partial x^2} z,$$

on

$$D(A) = \{z \in C([0, \pi]) : z(0) = z(\pi) = 0, \ \frac{\partial^2}{\partial x^2} z \in C([0, \pi])\}.$$

Now, we have

$$\overline{D(A)} = C_0([0, \pi]) = \{v \in C([0, \pi]) : v(0) = v(\pi) = 0\} \ne C([0, \pi]).$$

It is well known from [100] that A is sectorial, $(0, +\infty) \subseteq \rho(A)$ and for $\lambda > 0$

$$\|R(\lambda, A)\|_{B(E)} \le \frac{1}{\lambda}.$$

It follows that A generates an integrated semigroup $(S(t))_{t \ge 0}$ and that $\|S'(t)\|_{B(E)} \le e^{-\mu t}$ for $t \in J$ for some constant $\mu > 0$ and A satisfied the Hille–Yosida condition. Assume that there exist functions $\widetilde{l_1}, \widetilde{l_2} \in L^1(J, \mathbb{R}^+)$ such that

$$|Q_1(t, w) - Q_1(t, \overline{w})| \le \widetilde{l_1}(t)|w - \overline{w}|, \ t \in J, \ w, \overline{w} \in \mathbb{R},$$

and

$$|Q_2(t, w) - Q_2(t, \overline{w})| \leq \widetilde{l_2}(t)|w - \overline{w}|, \ t \in J, \ w, \overline{w} \in \mathbb{R}.$$

We can show that problem (8.21)–(8.23) is an abstract formulation of problem (8.28)–(8.31). Since all the conditions of Theorem 8.2 are satisfied, the problem (8.28)–(8.31) has a solution z on $[-r, b] \times [0, \pi]$.

8.6 Notes and Remarks

The results of Chap. 8 are taken from Abada et al. [2, 4]. Other results may be found in [51, 54, 74, 109, 110].

Chapter 9
Impulsive Semi-linear Functional Differential Equations

9.1 Introduction

In this chapter, we shall prove the existence of mild solutions of first order impulsive functional equations in a separable Banach space. Our approach will be based for the existence of mild solutions, on a fixed point theorem of Burton and Kirk [88] for the sum of a contraction map and a completely continuous map.

9.2 Semi-linear Differential Evolution Equations with Impulses and Delay

9.2.1 Introduction

In this section, we shall establish sufficient conditions for the existence of mild and extremal mild solutions of first order impulsive functional equations in a separable Banach space $(E. |.|)$ of the form:

$$y'(t) - Ay(t) = f(t, y_t), \ a.e. \ t \in J = [0, b] \, , \ t \neq t_k, \ k = 1, \ldots, m \tag{9.1}$$

$$\Delta y|_{t=t_k} = I_k(y(t_k^-)), \ k = 1, \ldots, m \tag{9.2}$$

$$y(t) = \phi(t), \ t \in [-r, 0] \, , \tag{9.3}$$

where $f : J \times D \to E$ is a given function, $D = \{\psi : [-r, 0] \to E, \psi$ is continuous everywhere except for a finite number of points s at which $\psi(s^-), \psi(s^+)$ exist and $\psi(s^-) = \psi(s)\}, \phi \in D, 0 < r < \infty, 0 = t_0 < t_1 < \cdots < t_m < t_{m+1} = b,$

© Springer International Publishing Switzerland 2015
S. Abbas, M. Benchohra, *Advanced Functional Evolution Equations and Inclusions*, Developments in Mathematics 39,
DOI 10.1007/978-3-319-17768-7_9

$I_k \in C(E, E)$, $k = 1, 2, \ldots, m$, $A : D(A) \subset E \to E$ is the infinitesimal generator of a C_0-semigroup $T(t)$, $t \geq 0$, and E a real separable Banach space with norm $|.|$. In the case where the impulses are absent (i.e., $I_k = 0, k = 1, \ldots, m$) and F is a single or multi-valued map and A is a densely defined linear operator generating a C_0-semigroup of bounded linear operators the problem (9.1)–(9.3) has been investigated on compact intervals in, for instance, the monographs by Ahmed [16], Hu and Papageorgiou [143], Kamenskii et al. [144], and Wu [184], and the papers of Benchohra and Ntouyas [58, 60, 63].

Next, we study the impulsive functional differential equations with nonlocal initial conditions of the form

$$y'(t) - Ay(t) = f(t, y_t), \quad a.e.\ t \in J = [0, b], t \neq t_k, k = 1, \ldots, m \tag{9.4}$$

$$\Delta y|_{t=t_k} = I_k(y(t_k^-)), \quad k = 1, \ldots, m \tag{9.5}$$

$$y(t) + h_t(y) = \phi(t), \quad t \in [-r, 0], \tag{9.6}$$

where $h_t : PC([-r, b], E) \to E$ is a given function. The nonlocal condition can be applied in physics with better effect than the classical initial condition $y(0) = y_0$. For example, $h_t(y)$ may be given by

$$h_t(y) = \sum_{i=1}^{p} c_i y(t_i + t), \quad t \in [-r, 0]$$

where $c_i, i = 1, \ldots, p$, are given constants and $0 < t_1 < \cdots < t_p \leq b$.

9.2.2   Existence of Mild Solutions

Definition 9.1. A function $y \in PC([-r, b], E)$ is said to be a mild solution of problem (11.15)–(11.17) if $y(t) = \phi(t), t \in [-r, 0]$, and y is a solution of impulsive integral equation

$$y(t) = T(t)\phi(0) + \int_0^t T(t - s)f(t, y_s)ds + \sum_{0 < t_k < t} T(t - t_k)I_k(y(t_k^-)), \quad t \in J.$$

Let us introduce the following hypotheses:

(9.1.1) $A : D(A) \subset E \to E$ is the infinitesimal generator of a C_0-semigroup $\{T(t)\}$, $t \in J$ which is compact for $t > 0$ in the Banach space E. Let $M = \sup\{\|T(t)\|_{B(E)} : t \in J\}$;

(9.1.2) There exist constants $d_k > 0$, $k = 1, \ldots, m$ with $M \sum_{k=1}^{m} d_k < 1$ such that for each $y, x \in E$

$$|I_k(y) - I_k(x)| \leq d_k |y - x|$$

(9.1.3) The function $f : J \times D \to E$ is Carathéodory;
(9.1.4) There exist a function $p \in L^1(J, \mathbb{R}_+)$ and a continuous nondecreasing function $\psi : [0, \infty) \to (0, \infty)$ such that

$$|f(t, x)| \leq p(t)\psi(\|x\|_D), \quad a.e.\ t \in J, \quad \text{for all } x \in D,$$

with

$$\int_{D_o}^{\infty} \frac{ds}{\psi(s)} > D_1 \|p\|_{L^1},$$

where

$$D_0 = \frac{M(\|\phi\| + \sum_{k=1}^{m} |I_k(0)|)}{1 - M \sum_{k=1}^{m} d_k}, \qquad D_1 = \frac{M}{1 - M \sum_{k=1}^{m} d_k}.$$

Theorem 9.2. *Assume that (9.1.1)–(9.1.4) hold. Then the problem (9.1)–(9.3) has at least one mild solution on* $[-r, b]$.

Proof. Consider the two operators:

$$\mathcal{A}, \mathcal{B} : PC([-r, b], E) \to PC([-r, b], E).$$

defined by

$$\mathcal{A}(y)(t) := \begin{cases} 0, & \text{if } t \in H; \\ \sum_{0 < t_k < t} T(t - t_k) I_k(y(t_k^-)), & \text{if } t \in J, \end{cases}$$

and

$$\mathcal{B}(y)(t) := \begin{cases} \phi(t), & \text{if } t \in H; \\ T(t)\phi(0) + \int_0^t T(t - s)f(s, y_s)\,ds, & \text{if } t \in J. \end{cases}$$

Then, the problem of finding the solution of problem (9.1)–(9.3) is reduced to finding the solution of the operator equation $\mathcal{A}(y)(t) + \mathcal{B}(y)(t) = y(t), t \in [-r, b]$. We shall show that the operators $\mathcal{A}$ and $\mathcal{B}$ satisfy all the conditions of Theorem 1.32 For better readability, we break the proof into a sequence of steps.

Step 1: $\mathcal{B}$ is continuous.

Let $\{y_n\}$ be a sequence such that $y_n \to y$ in $PC([-r, b], E)$. Then for $t \in J$

$$|\mathcal{B}(y_n)(t) - \mathcal{B}(y)(t)| = \left| \int_0^t T(t-s)[f(s, y_{n_s}) - f(s, y_s)]ds \right|$$

$$\leq M \int_0^b |f(s, y_{n_s}) - f(s, y_s)| \, ds.$$

Since $f(s, \cdot)$ is continuous for a.e. $s \in J$, we have by the Lebesgue dominated convergence theorem

$$|\mathcal{B}(y_n)(t) - \mathcal{B}(y)(t)| \to 0 \text{ as } n \to \infty.$$

Thus $\mathcal{B}$ is continuous.

Step 2: $\mathcal{B}$ maps bounded sets into bounded sets in $PC([-r, b], E)$.

It is enough to show that for any $q > 0$ there exists a positive constant l such that for each $y \in B_q = \{y \in PC([-r, b], E) : \|y\| \leq q\}$ we have $\|\mathcal{B}(y)\| \leq l$. So choose $y \in B_q$, then we have for each $t \in J$,

$$|\mathcal{B}(y)(t)| = \left| T(t)\phi(0) + \int_0^t T(t-s)f(s, y_s)ds \right|$$

$$\leq M|\phi(0)| + M\psi(q) \int_0^b p(s) \, ds.$$

Then we have

$$\|\mathcal{B}(y)\| \leq M\|\phi\| + M\psi(q)\|p\|_{L^1} := l.$$

Step 3: $\mathcal{B}$ maps bounded sets into equi-continuous sets of $PC([-r, b], E)$.

We consider B_q as in step 2 and let $\tau_1, \tau_2 \in J \backslash \{t_1, \ldots, t_m\}$, $\tau_1 < \tau_2$. Thus if $\epsilon > 0$ and $\epsilon \leq \tau_1 < \tau_2$ we have

$$|\mathcal{B}(y)(\tau_2) - \mathcal{B}(y)(\tau_1)| \leq |T(\tau_2)\phi(0) - T(\tau_1)\phi(0)|$$

$$+ \psi(q) \int_0^{\tau_1 - \epsilon} \|T(\tau_2 - s) - T(\tau_1 - s)\|_{B(E)} p(s) ds$$

$$+ \psi(q) \int_{\tau_1 - \epsilon}^{\tau_1} \|T(\tau_2 - s) - T(\tau_1 - s)\|_{B(E)} p(s) ds$$

$$+ \psi(q) \int_{\tau_1}^{\tau_2} \|T(\tau_2 - s)\|_{B(E)} p(s) ds.$$

As $\tau_1 \to \tau_2$ and ϵ become sufficiently small, the right-hand side of the above inequality tends to zero, since $T(t)$ is a strongly continuous operator and the compactness of $T(t)$ for $t > 0$ implies the continuity in the uniform operator topology [16]. This proves the equi-continuity for the case where $t \neq t_i, k = 1, 2, \ldots, m + 1$. It remains to examine the equi-continuity at $t = t_i$.

First we prove equi-continuity at $t = t_i^-$. Fix $\delta_1 > 0$ such that $\{t_k : k \neq i\} \cap [t_i - \delta_1, t_i + \delta_1] = \emptyset$. For $0 < h < \delta_1$ we have

$$|\mathcal{B}(y)(t_i - h) - \mathcal{B}(y)(t_i)| \leq |(T(t_i - h) - T(t_i))\,\phi(0)|$$
$$+ \int_0^{t_i - h} |(T(t_i - h - s) - T(t_i - s))f(s, y_s)|ds$$
$$+ \psi(q)M \int_{t_i - h}^{t_i} p(s)ds;$$

which tends to zero as $h \to 0$. Define

$$\hat{\mathcal{B}}_0(y)(t) = \mathcal{B}(y)(t),\ t \in [0, t_1],$$

and

$$\hat{\mathcal{B}}_i(y)(t) = \begin{cases} \mathcal{B}(y)(t), & \text{if } t \in (t_i, t_{i+1}] \\ \mathcal{B}(y)(t_i^+), & \text{if } t = t_i. \end{cases}$$

Next we prove equi-continuity at $t = t_i^+$. Fix $\delta_2 > 0$ such that $\{t_k : k \neq i\} \cap [t_i - \delta_2, t_i + \delta_2] = \emptyset$. For $0 < h < \delta_2$ we have

$$|\hat{\mathcal{B}}(y)(t_i + h) - \hat{\mathcal{B}}(y)(t_i)| \leq |(T(t_i + h) - T(t_i))\,\phi(0)|$$
$$+ \int_0^{t_i} |(T(t_i + h - s) - T(t_i - s))f(s, y_s)|ds$$
$$+ \psi(q)M \int_{t_i}^{t_i + h} p(s)ds.$$

The right-hand side tends to zero as $h \to 0$. The equi-continuity for the cases $\tau_1 < \tau_2 \leq 0$ and $\tau_1 \leq 0 \leq \tau_2$ follows from the uniform continuity of ϕ on the interval H.

As a consequence of Steps 1–3 together with Arzelá–Ascoli theorem, it suffices to show that $\mathcal{B}$ maps B into a precompact set in E.

Let $0 < t < b$ be fixed and let ϵ be a real number satisfying $0 < \epsilon < t$. For $y \in B_q$ we define

$$\mathcal{B}_\epsilon(y)(t) = T(t)\phi(0) + T(\epsilon) \int_0^{t-\epsilon} T(t - s - \epsilon)f(s, y_s)ds.$$

Since $T(t)$ is a compact operator, the set

$$Y_\epsilon(t) = \{\mathcal{B}_\epsilon(y)(t) : \ y \in B_q\}$$

is precompact in E for every ϵ, $0 < \epsilon < t$. Moreover, for every $y \in B_q$ we have

$$|\mathcal{B}(y)(t) - \mathcal{B}_\epsilon(y)(t)| \leq \psi(q) \int_{t-\epsilon}^t \|T(t-s)\|_{B(E)} p(s)ds$$

$$\leq \psi(q) M \int_{t-\epsilon}^t p(s)ds.$$

Therefore, there are precompact sets arbitrarily close to the set $Y_\epsilon(t) = \{\mathcal{B}_\epsilon(y)(t) : y \in B_q\}$. Hence the set $Y(t) = \{\mathcal{B}(y)(t) : \ y \in B_q\}$ is precompact in E. Hence the operator $\mathcal{B} : PC([-r, b], E) \to PC([-r, b], E)$ is completely continuous.

Step 4: $\mathcal{A}$ is a contraction

Let $x, y \in PC([-r, b], E)$. Then for $t \in J$

$$|\mathcal{A}(y)(t) - \mathcal{A}(x)(t)| = \left| \sum_{0 < t_k < t} T(t - t_k) \left(I_k \left(y \left(t_k^- \right) \right) - I_k \left(x \left(t_k^- \right) \right) \right) \right|$$

$$\leq M \sum_{0 < t_k < t} \left| I_k \left(y \left(t_k^- \right) \right) - I_k \left(x \left(t_k^- \right) \right) \right|$$

$$\leq M \sum_{k=1}^m d_k \left| y \left(t_k^- \right) - x \left(t_k^- \right) \right|$$

$$\leq M \sum_{k=1}^m d_k \|y - x\|.$$

Then

$$\|\mathcal{A}(y) - \mathcal{A}(x)\| \leq M \sum_{k=1}^m d_k \|y - x\|,$$

which is a contraction, since $M \sum_{k=1}^m d_k < 1$.

Step 5: A priori bounds. Now it remains to show that the set

$$\mathcal{E} = \left\{ y \in PC([-r, b], E) : y = \lambda \mathcal{B}(y) + \lambda \mathcal{A}\left(\frac{y}{\lambda}\right) \ \text{for some} \ 0 < \lambda < 1 \right\}$$

is bounded. Let $y \in \mathcal{E}$, then $y = \lambda \mathcal{B}(y) + \lambda \mathcal{A}\left(\frac{y}{\lambda}\right)$ for some $0 < \lambda < 1$. Thus, for each $t \in J$,

$$y(t) = \lambda T(t)\phi(0) + \lambda \int_0^t T(t-s)f(s, y_s)ds + \lambda \sum_{0<t_k<t} T(t-t_k) I_k \left(\frac{y}{\lambda}(t_k^-)\right).$$

This implies by (9.1.2) and (9.1.4) that, for each $t \in J$, we have

$$|y(t)| \leq \lambda M |\phi(0)| + \lambda M \int_0^t p(s)\psi(\|y_s\|)ds + \lambda M \sum_{k=1}^m \left| I_k\left(\frac{y}{\lambda}(t_k^-)\right)\right|$$

$$\leq \lambda M \|\phi\| + \lambda M \int_0^t p(s)\psi(\|y_s\|)ds$$

$$+ \lambda M \sum_{k=1}^m \left| I_k\left(\frac{y}{\lambda}(t_k^-)\right) - I_k(0)\right| + \lambda M \sum_{k=1}^m |I_k(0)|$$

$$\leq \lambda M \left(\|\phi\| + \sum_{k=1}^m |I_k(0)|\right) + \lambda M \int_0^t p(s)\psi(\|y_s\|)ds$$

$$+ \lambda M \sum_{k=1}^m d_k \left|\frac{y}{\lambda}(t_k^-)\right|$$

$$\leq M \left(\|\phi\| + \sum_{k=1}^m |I_k(0)|\right) + M \left[\int_0^t p(s)\psi(\|y_s\|)ds + \sum_{k=1}^m d_k |y(t_k^-)|\right].$$

Now we consider the function μ defined by

$$\mu(t) = \sup\{|y(s)| : -r \leq s \leq t\}, \quad 0 \leq t \leq b.$$

Then $\|y_s\| \leq \mu(t)$ for all $t \in J$ and there is a point $t^* \in [-r, t]$ such that $\mu(t) = |y(t^*)|$. If $t^* \in [0, b]$, by the previous inequality we have for $t \in [0, b]$ (note $t^* \leq t$)

$$\mu(t) \leq M \left(\|\phi\| + \sum_{k=1}^m |I_k(0)|\right) + M \int_0^t p(s)\psi(\mu(s))ds + M \sum_{k=1}^m d_k \mu(t).$$

Then

$$\left(1 - M \sum_{k=1}^m d_k\right) \mu(t) \leq M \left(\|\phi\| + \sum_{k=1}^m |I_k(0)|\right) + M \int_0^t p(s)\psi(\mu(s))ds.$$

Thus we have

$$\mu(t) \le D_0 + D_1 \int_0^t p(s)\psi(\mu(s))ds.$$

Let us take the right-hand side of the above inequality as $v(t)$. Then we have

$$\mu(t) \le v(t) \quad \text{for all } t \in J,$$

$$v(0) = D_0,$$

and

$$v'(t) = D_1 p(t)\psi(\mu(t)), \quad a.e. \ t \in J.$$

Using the nondecreasing character of ψ we get

$$v'(t) \le D_1 p(t)\psi(v(t)), \quad a.e. \ t \in J.$$

That is

$$\frac{v'(t)}{\psi(v(t))} \le D_1 p(t), \quad a.e. \ t \in J.$$

Integrating from 0 to t we get

$$\int_0^t \frac{v'(s)}{\psi(v(s))}ds \le D_1 \int_0^t p(s)\,ds.$$

By a change of variable we get

$$\int_{v(0)}^{v(t)} \frac{du}{\psi(u)} \le D_1 \int_0^b p(s)\,ds = D_1 \|p\|_{L^1} < \int_{D_0}^{\infty} \frac{du}{\psi(u)}.$$

Hence there exists a constant N such that

$$\mu(t) \le v(t) \le N \quad \text{for all } t \in J.$$

Now from the definition of μ it follows that

$$\|y\| = \sup_{t \in [-r,b]} |y(t)| \le \mu(b) \le N, \quad \text{for all } y \in \mathcal{E}.$$

This shows that the set $\mathcal{E}$ is bounded. As a consequence of Theorem 1.32 we deduce that $\mathcal{A} + \mathcal{B}$ has a fixed point which is a mild solution of problem (11.15)–(11.17).

$\square$

9.2.3 Existence of Extremal Mild Solutions

In this section we shall prove the existence of maximal and minimal solutions of problem (9.1)–(9.3) under suitable monotonicity conditions on the functions involved in it.

We need the following definitions in the sequel.

Definition 9.3. We say that a function $v \in PC([-r, b], E)$ is a lower mild solution of problem (9.1)–(9.3) if $v(t) = \phi(t)$, $t \in H$, and

$$v(t) \le T(t)\phi(0) + \int_0^t T(t - s)f(s, v_s)\, ds + \sum_{0 < t_k < t} T(t - t_k)I_k(v(t_k^-)), \ t \in J, \ t \ne t_k$$

and $v(t_k^+) - v(t_k^-) \le I_k(v(t_k))$, $t = t_k$, $k = 1, \ldots, m$. Similarly an upper mild solution w of problem (9.1)–(9.3) is defined by reversing the order.

Definition 9.4. A solution x_M of problem (9.1)–(9.3) is said to be maximal if for any other solution x of problem (9.1)–(9.3) on J, we have that $x(t) \le x_M(t)$ for each $t \in J$.

Similarly a minimal solution of problem (9.1)–(9.3) is defined by reversing the order of the inequalities.

Definition 9.5. A function $f(t, x)$ is called strictly monotone increasing in x almost everywhere for $t \in J$, if $(t, x) \le f(t, y)$ a.e. $t \in J$ for all $x, y \in D$ with $x < y$. Similarly $f(t, x)$ is called strictly monotone decreasing in x almost everywhere for $t \in J$, if $f(t, x) \ge f(t, y)$ a.e. $t \in J$ for all $x, y \in D$ with $x < y$.

We consider the following assumptions in the sequel.

(9.10.1) The function $f(t, y)$ is strictly monotone increasing in y for almost each $t \in J$.
(9.10.2) $T(t)$ is preserving the order, that is $T(t)v \ge 0$ whenever $v \ge 0$.
(9.10.3) The function I_k, $k = 1, \ldots, m$ are continuous and nondecreasing.
(9.10.4) The problem (9.1)–(9.3) has a lower mild solution v and an upper mild solution w with $v \le w$.

Theorem 9.6. *Assume that assumptions (9.1.1)–(9.1.4) and (9.10.1)–(9.10.4) hold. Then problem (9.1)–(9.3) has minimal and maximal solutions on $[-r, b]$.*

Proof. It can be shown, as in the proof of Theorem 9.2, that $\mathcal{B}$ is completely continuous and $\mathcal{A}$ is a contraction on $[v, w]$. We shall show that $\mathcal{A}$ and $\mathcal{B}$ are isotone increasing on $[v, w]$. Let $y, \bar{y} \in [a, b]$ be such that $y \le \bar{y}$, $y \ne \bar{y}$. Then by (9.10.1), (9.10.2), we have for each $t \in J$

$$\mathcal{B}(y)(t) = T(t)\phi(0) + \int_0^t T(t - s)f(s, y_s)\, ds$$

$$\le T(t)\phi(0) + \int_0^t T(t - s)f(s, \bar{y}_s)\, ds$$

$$= \mathcal{B}(\bar{y})(t).$$

and by (9.10.3), we have for each $t \in J$

$$
\mathcal{A}(y)(t) = \sum_{0 < t_k < t} T(t - t_k) I_k \left(y \left(t_k^- \right) \right)
$$

$$
\leq \sum_{0 < t_k < t} T(t - t_k) I_k \left(\bar{y} \left(t_k^- \right) \right)
$$

$$
= \mathcal{A}(\bar{y})(t).
$$

Therefore $\mathcal{A}$ and $\mathcal{B}$ are isotone increasing on $[v, w]$. Finally, let $x \in [v, w]$ be any element. By (9.10.4) we deduce that

$$
v \leq \mathcal{A}(v) + \mathcal{B}(v) \leq \mathcal{A}(x) + \mathcal{B}(x) \leq \mathcal{A}(w) + \mathcal{B}(w) \leq w,
$$

which shows that $\mathcal{A}(x) + \mathcal{B}(x) \in [v, w]$ for all $x \in [v, w]$. Thus, $\mathcal{A}$ and $\mathcal{B}$ satisfy all conditions of Theorem 1.32, hence problem (9.1)–(9.3) has maximal and minimal solutions on $[-r, b]$.   □

9.2.4 Impulsive Differential Equations with Nonlocal Conditions

In this section we shall prove the existence results for problem (9.4)–(9.6). Nonlocal conditions were initiated by Byszewski [89] when he proved the existence and uniqueness of mild and classical solutions of nonlocal Cauchy problems.

Definition 9.7. A function $y \in PC([-r, b], E)$ is said to be a mild solution of problem (9.4)–(9.6) if $y(t) = \phi(t) - h_t(y)$, $t \in [-r, 0]$, and

$$
y(t) = T(t)(\phi(0) - h_0(y)) + \int_0^t T(t - s) f(s, y_s)\, ds
$$

$$
+ \sum_{0 < t_k < t} T(t - t_k) I_k \left(y \left(t_k^- \right) \right), \quad t \in J.
$$

Theorem 9.8. *Assume that hypotheses (9.1.1)–(9.1.3) hold and moreover*

(A1) The function h is continuous with respect to t, and there exists a constant $\alpha > 0$ such that

$$
|h_t(u)| \leq \alpha, \quad u \in PC([-r, b], E)
$$

and for each $k > 0$ the set

$$
\{\phi(0) - h_0(y),\ y \in PC([-r, b], E),\ \|y\| \leq k\}
$$

is precompact in E

(A2) *There exists a function* $p \in L^1(J, \mathbb{R}_+)$ *and a continuous nondecreasing function* $\psi : [0, \infty) \to (0, \infty)$ *such that*

$$|f(t, x)| \leq p(t)\psi(\|x\|_D), \quad a.e. \ t \in J, \quad \text{for all } x \in D$$

with

$$\int_{\tilde{D}_0}^{\infty} \frac{ds}{\psi(s)} > D_1 \|p\|_{L^1},$$

and

$$\tilde{D}_0 = \frac{M[\|\phi\|_D + \alpha + \sum_{k=1}^{m} |I_k(0)|]}{1 - M \sum_{k=1}^{m} d_k}.$$

Then the problem 9.4)–(9.6) has at least one mild solution on $[-r, b]$.

Proof. Consider the two operators: $\mathcal{B}_1 : PC([-r, b], E) \to PC([-r, b], E)$ defined by

$$\mathcal{B}_1(y)(t) = \begin{cases} \phi(t) - h_t(y), & \text{if } t \in H; \\ T(t)(\phi(0) - h_0(y)) + \displaystyle\int_0^t T(t-s)f(s, y_s)\,ds, & \text{if } t \in J, \end{cases}$$

and

$$\mathcal{A}_1(y)(t) = \begin{cases} 0, & \text{if } t \in H; \\ \displaystyle\sum_{0 < t_k < t} T(t - t_k)I_k(y(t_k^-)), & \text{if } t \in J. \end{cases}$$

Then the problem of finding the solution of problem (9.4)–(9.6) is reduced to finding the solution of the operator equation $\mathcal{A}_1(y)(t) + \mathcal{B}_2(y)(t) = y(t), t \in [-r, b]$. As in Sect. 9.3, we can show that the operators $\mathcal{A}_1$ and $\mathcal{B}_1$ satisfy all conditions of Theorem 1.32. □

9.2.5 An Example

As an application of our results we consider the following impulsive partial functional differential equation of the form

$$\frac{\partial}{\partial t}z(t,x) = \frac{\partial^2}{\partial x^2}z(t,x) \tag{9.7}$$

$$+Q(t,z(t-r,x)), \ x \in [0,\pi], \ t \in [0,b]\backslash\{t_1,t_2,\dots,t_m\}.$$

$$z(t_k^+,x) - z(t_k^-,x) = b_k z(t_k^-,x), \ x \in [0,\pi], \ k = 1,\dots,m \tag{9.8}$$

$$z(t,0) = z(t,\pi) = 0, \ t \in [0,b] \tag{9.9}$$

$$z(t,x) = \phi(t,x), \ t \in H, \ x \in [0,\pi], \tag{9.10}$$

where $b_k > 0$, $k = 1,\dots,m$, $\phi \in \mathcal{D} = \{\psi : H \times [0,\pi] \to \mathbb{R}; \psi$ is continuous everywhere except for a countable number of points at which $\psi(s^-), \psi(s^+)$ exist with $\psi(s^-) = \psi(s)\}$, $0 = t_0 < t_1 < t_2 < \cdots < t_m < t_{m+1} = b$, $z(t_k^+) = \lim\limits_{(h,x)\to(0^+,x)} z(t_k + h, x), z(t_k^-) = \lim\limits_{(h,x)\to(0^-,x)} z(t_k + h, x)$ and $Q : [0,b] \times \mathbb{R} \to \mathbb{R}$ is a given function.

Let

$$y(t)(x) = z(t,x), \ t \in J, \ x \in [0,\pi],$$

$$I_k(y(t_k^-))(x) = b_k z(t_k^-,x), \ x \in [0,\pi], \ k = 1,\dots,m$$

$$F(t,\phi)(x) = Q(t,\phi(\theta,x)), \ \theta \in H, \ x \in [0,\pi],$$

$$\phi(\theta)(x) = \phi(\theta,x), \ \theta \in H, \ x \in [0,\pi].$$

Take $E = L^2[0,\pi]$ and define $A : D(A) \subset E \to E$ by $Aw = w''$ with domain $D(A) = \{w \in E, w, w'$ are absolutely continuous, $w'' \in E, w(0) = w(\pi) = 0\}$. Then

$$Aw = \sum_{n=1}^{\infty} n^2(w,w_n)w_n, \ w \in D(A)$$

where $(\ , \ )$ is the inner product in L^2 and $w_n(s) = \sqrt{\frac{2}{\pi}}\sin ns$, $n = 1,2,\dots$ is the orthogonal set of eigenvectors in A. It is well known (see [168]) that A is the infinitesimal generator of an analytic semigroup $T(t)$, $t \in [0,b]$ in E and is given by

$$T(t)w = \sum_{n=1}^{\infty} \exp(-n^2 t)(w,w_n)w_n, \ w \in E.$$

Since the analytic semigroup $T(t)$ is compact, there exists a constant $M \geq 1$ such that

$$\|T(t)\|_{B(E)} \leq M.$$

Also assume that there exists an integrable function $\sigma : [0,b] \to \mathbb{R}^+$ such that

$$|Q(t,w(t-r,x))| \leq \sigma(t)\Omega(|w|)$$

where $\Omega : [0, \infty) \to (0, \infty)$ is continuous and nondecreasing with

$$\int_1^\infty \frac{ds}{s + \Omega(s)} = +\infty.$$

Assume that there exists a function $\tilde{l} \in L^1([0, b], \mathbb{R}^+)$ such that

$$|Q(t, w) - Q(t, \overline{w})| \leq \tilde{l}(t)|w - \overline{w}|, \ t \in [0, b], \ w, \overline{w} \in \mathbb{R}.$$

We can show that problem (11.15)–(11.17) is an abstract formulation of problem (9.7)–(9.10). Since all the conditions of Theorem 9.2 are satisfied, the problem (9.7)–(9.10) has a solution z on $[-r, b] \times [0, \pi]$.

9.3 Impulsive Semi-linear Functional Differential Equations with Non-densely Defined Operators

9.3.1 Introduction

In this section, we shall be concerned with the existence of integral solutions and extremal integral solutions defined on a compact real interval for first order impulsive semi-linear functional equations in a separable Banach space. We will consider the following first order impulsive semi-linear differential equations of the form:

$$y'(t) - Ay(t) = f(t, y_t), \ a.e. \ t \in J = [0, b], \ t \neq t_k, \ k = 1, \ldots, m \quad (9.11)$$

$$\Delta y|_{t=t_k} = I_k(y(t_k^-)), \ \ k = 1, \ldots, m \quad (9.12)$$

$$y(t) = \phi(t), \ \ t \in [-r, 0], \quad (9.13)$$

where $f : J \times D \to E$ is a given function, $D = \{\psi : [-r, 0] \to E, \psi$ is continuous everywhere except for a finite number of points s at which $\psi(s^-), \psi(s^+)$ exist and $\psi(s^-) = \psi(s)\}, \phi \in D, (0 < r < \infty), 0 = t_0 < t_1 < \cdots < t_m < t_{m+1} = b$, $I_k : E \to E \ (k = 1, 2, \ldots, m), A : D(A) \subset E \to E$ is a non-densely defined closed linear operator on E, and E a real separable Banach space with norm $|.|$.

We shall prove the existence of extremal integral solutions of the problem (11.15)–(11.17), and our approach here is based on the concept of upper and lower solutions combined with a fixed point theorem on ordered Banach spaces established recently by Dhage [102]. Next, we study the impulsive functional differential equations with nonlocal initial conditions of the form

$$y'(t) - Ay(t) = f(t, y_t), \ \ a.e. \ t \in J = [0, b], \ t \neq t_k, \ k = 1, \ldots, m \quad (9.14)$$

$$\Delta y|_{t=t_k} = I_k(y(t_k^-)), \ \ k = 1, \ldots, m \quad (9.15)$$

$$y(t) + h_t(y) = \phi(t), \ \ t \in [-r, 0], \quad (9.16)$$

where $h_t : PC([-r, b], \overline{D(A)}) \to \overline{D(A)}$ is a given function. The nonlocal condition can be applied in physics with better effect than the classical initial condition $y(0) = y_0$. For example, $h_t(y)$ may be given by

$$h_t(y) = \sum_{i=1}^{p} c_i y(t_i + t), \quad t \in [-r, 0] \tag{9.17}$$

where $c_i, i = 1, \ldots, p$, are given constants and $0 < t_1 < \cdots < t_p \leq p$.

9.3.2 Examples of Operators with Non-dense Domain

In this section we shall present examples of linear operators with non-dense domain satisfying the Hille–Yosida estimate. More details can be found in the paper by Da Prato and Sinestrari [100].

Example 9.9. Let $E = C([0, 1], \mathbb{R})$ and the operator $A : D(A) \to E$ defined by $Ay = y'$, where

$$D(A) = \{y \in C^1((0, 1), \mathbb{R}) : y(0) = 0\}.$$

Then

$$\overline{D(A)} = \{y \in C((0, 1), \mathbb{R}) : y(0) = 0\} \neq E.$$

Example 9.10. Let $E = C([0, 1], \mathbb{R})$ and the operator $A : D(A) \to E$ defined by $Ay = y''$, where

$$D(A) = \{y \in C^2((0, 1), \mathbb{R}) : y(0) = y(1) = 0\}.$$

Then

$$\overline{D(A)} = \{y \in C((0, 1), \mathbb{R}) : y(0) = y(1) = 0\} \neq E.$$

Example 9.11. Let us set for some $\alpha \in (0, 1)$

$$E = C_0^\alpha([0, 1], \mathbb{R}) = \{y : [0, 1] \to \mathbb{R} : y(0) = 0 \text{ and } \sup_{0 \leq t < s \leq 1} \frac{|y(t) - y(s)|}{|t - s|^\alpha} < \infty\}$$

and the operator $A : D(A) \to E$ defined by $Ay = -y'$, where

$$D(A) = \{y \in C^{1+\alpha}((0, 1), \mathbb{R}) : y(0) = y'(0) = 0\}.$$

Then

$$\overline{D(A)} = h_0^\alpha(0, 1), \mathbb{R}) = \{y : [0, 1] \to \mathbb{R} : \lim_{\delta \to 0} \sup_{0 < |t-s| \leq \delta} \frac{|y(t) - y(s)|}{|t - s|^\alpha} = 0\} \neq E.$$

Here

$$C^{1+\alpha}([0, 1], \mathbb{R}) = \{y : [0, 1] \rightarrow \mathbb{R} : y' \in C^{\alpha}([0, 1], \mathbb{R})\}.$$

The elements of $h^{\alpha}((0, 1), \mathbb{R})$ are called little Holder functions and it can be proved that the closure of $C^1((0, 1), \mathbb{R})$ in $C^{\alpha}((0, 1), \mathbb{R})$ is $h^{\alpha}((0, 1), \mathbb{R})$ (see [175], Theorem 5.3).

Example 9.12. Let $\Omega \subset \mathbb{R}^n$ be a bounded open set with regular boundary Γ and define $E = C(\overline{\Omega}, \mathbb{R})$ and the operator $A : D(A) \rightarrow E$ defined by $Ay = \Delta y$, where

$$D(A) = \{y \in C(\overline{\Omega}, \mathbb{R}) : y = 0 \text{ on } \Gamma; \; \Delta y \in C(\overline{\Omega}, \mathbb{R})\}.$$

Here Δ is the Laplacian in the sense of distributions on Ω. In this case we have

$$\overline{D(A)} = \{y \in C(\overline{\Omega}, \mathbb{R}) : y = 0 \text{ on } \Gamma\} \neq E.$$

9.3.3 Existence of Integral Solutions

Definition 9.13. We say that $y : [-r, T] \rightarrow E$ is an integral solution of (9.11)–(9.13) if

(i) $y(t) = \phi(0) + A \int_0^t y(s)ds + \int_0^t f(s, y_s)ds + \sum_{0 < t_k < t} I_k \left(y\left(t_k^-\right)\right), \quad t \in J.$

(ii) $\int_0^t y(s)ds \in D(A)$ for $t \in J$, and $y(t) = \phi(t), \; t \in H.$

From the definition it follows that $y(t) \in \overline{D(A)}$, for each $t \geq 0$, in particular $\phi(0) \in \overline{D(A)}$. Moreover, y satisfies the following variation of constants formula:

$$y(t) = S'(t)\phi(0) + \frac{d}{dt} \int_0^t S(t-s)f(s, y_s)ds + \sum_{0 < t_k < t} S'(t-t_k) I_k \left(y\left(t_k^-\right)\right) \quad t \geq 0.$$

$$(9.18)$$

We notice also that if y satisfies (9.18), then

$$y(t) = S'(t)\phi(0) + \lim_{\lambda \to \infty} \int_0^t S'(t-s)B_\lambda f(s, y_s)ds$$

$$+ \sum_{0 < t_k < t} S'(t-t_k) I_k \left(y\left(t_k^-\right)\right), \quad t \geq 0.$$

Let us introduce the following hypotheses:

(9.17.1) A satisfies Hille–Yosida condition;

(9.17.2) There exist constants $d_k > 0$, $k = 1, \ldots, m$ such that for each y, $x \in \overline{D(A)}$

$$|I_k(y) - I_k(x)| \le d_k |y - x|$$

(9.17.3) The function $f : J \times D \to E$ is Carathéodory;
(9.17.4) The operator $S'(t)$ is compact in $\overline{D(A)}$ wherever $t > 0$;
(9.17.5) There exist a function $p \in L^1(J, \mathbb{R}_+)$ and a continuous nondecreasing function $\psi : [0, \infty) \to (0, \infty)$ such that

$$|f(t, x)| \le p(t)\psi(\|x\|_D), \quad a.e. \ t \in J, \quad \text{for all } x \in D$$

with $\displaystyle\int_0^b e^{-\omega s}p(s)ds < \infty$,

$$\int_{c_0}^{\infty} \frac{du}{\psi(u)} > c_1 \int_0^b e^{-\omega s}p(s)\,ds. \tag{9.19}$$

where

$$c_0 = \frac{e^{\omega b}M\left(\|\phi\| + \sum_{k=1}^{m} |I_k(0)|\right)}{1 - Me^{\omega b}\sum_{k=1}^{m} d_k}, \tag{9.20}$$

and

$$c_1 = \frac{Me^{\omega b}}{1 - Me^{\omega b}\sum_{k=1}^{m} d_k}. \tag{9.21}$$

Theorem 9.14. *Assume that (9.17.1)–(9.17.5) hold. If*

$$Me^{\omega b}\sum_{k=1}^{m} d_k < 1, \tag{9.22}$$

then the problem (9.11)–(9.13) has at least one integral solution on $[-r, b]$.

Proof. Consider the two operators:

$$\mathcal{A}, \mathcal{B} : PC\left([-r, b], \overline{D(A)}\right) \to PC\left([-r, b], \overline{D(A)}\right)$$

defined by

$$\mathcal{A}(y)(t) := \begin{cases} 0, & \text{if } t \in H; \\ \displaystyle\sum_{0 < t_k < t} S'(t - t_k) I_k\left(y\left(t_k^-\right)\right), & \text{if } t \in J, \end{cases}$$

and

$$B(y)(t) := \begin{cases} \phi(t), & \text{if } t \in H; \\ S'(t)\phi(0) \\ \quad + \dfrac{d}{dt} \displaystyle\int_0^t S(t-s)f(s,y_s)\,ds, & \text{if } t \in J. \end{cases}$$

The problem of finding the solution of problem (9.11)–(9.13) is reduced to finding the solution of the operator equation $A(y)(t) + B(y)(t) = y(t), t \in [-r, b]$. We shall show that the operators A and B satisfy all the conditions of Theorem 1.32. For better readability, we break the proof into a sequence of steps.

Step 1: B is continuous. Let $\{y_n\}$ be a sequence such that $y_n \to y$ in $PC([-r, b], \overline{D(A)})$. Then for $\omega > 0$ (if $\omega < 0$ one has $e^{\omega t} < 1$)

$$|B(y_n)(t) - B(y)(t)| = \left| \frac{d}{dt} \int_0^t S(t-s)[f(s, y_{n_s}) - f(s, y_s)]ds \right|$$

$$\leq Me^{\omega b} \int_0^b e^{-\omega s} |f(s, y_{n_s}) - f(s, y_s)|\, ds.$$

Since $f(s, \cdot)$ is continuous for a.e. $s \in J$, we have by the Lebesgue dominated convergence theorem

$$|B(y_n)(t) - B(y)(t)| \to 0 \ as \ n \to \infty.$$

Thus B is continuous.

Step 2: B maps bounded sets into bounded sets in $PC([-r, b], \overline{D(A)})$. It is enough to show that for any $q > 0$ there exists a positive constant l such that for each $y \in B_q = \{y \in PC([-r, b], \overline{D(A)}) : \|y\| \leq q\}$ we have $\|B(y)\| \leq l$. So choose $y \in B_q$, then we have for each $t \in J$

$$|B(y)(t)| = \left| S'(t)\phi(0) + \frac{d}{dt} \int_0^t S(t-s)f(s, y_s)ds \right|$$

$$\leq Me^{\omega b}|\phi(0)| + Me^{\omega b}\psi(q) \int_0^b e^{-\omega s}p(s)\, ds.$$

Then we have

$$|B(y)(t)| \leq Me^{\omega b}\|\phi\| + Me^{\omega b}\psi(q) \int_0^b e^{-\omega s}p(s)\, ds := l.$$

Step 3: B maps bounded sets into equi-continuous sets of $PC([-r, b], \overline{D(A)})$.

We consider B_q as in Step 2 and let $\tau_1, \tau_2 \in J \setminus \{t_1, \ldots, t_m\}$, $\tau_1 < \tau_2$. Thus if $\epsilon > 0$ and $\epsilon \le \tau_1 < \tau_2$ we have

$$|B(y)(\tau_2) - B(y)(\tau_1)| \le |S'(\tau_2)\phi(0) - S'(\tau_1)\phi(0)|$$

$$+ \left| \lim_{\lambda \to \infty} \int_0^{\tau_1 - \epsilon} [S'(\tau_2 - s) - S'(\tau_1 - s)]B_\lambda f(s, y_s)ds \right|$$

$$+ \left| \lim_{\lambda \to \infty} \int_{\tau_1 - \epsilon}^{\tau_1} [S'(\tau_2 - s) - S'(\tau_1 - s)]B_\lambda f(s, y_s)\, ds \right|$$

$$+ \left| \lim_{\lambda \to \infty} \int_{\tau_1}^{\tau_2} S'(\tau_2 - s)B_\lambda f(s, y_s)\, ds \right|.$$

As $\tau_1 \to \tau_2$ and ϵ become sufficiently small, the right-hand side of the above inequality tends to zero, since $S'(t)$ is a strongly continuous operator and the compactness of $S'(t)$ for $t > 0$ implies the continuity in the uniform operator topology. This proves the equi-continuity for the case where $t \ne t_i, k = 1, 2, \ldots, m + 1$. It remains to examine the equi-continuity at $t = t_i$.

First we prove equi-continuity at $t = t_i^-$. Fix $\delta_1 > 0$ such that

$$\{t_k : k \ne i\} \cap [t_i - \delta_1, t_i + \delta_1] = \emptyset.$$

For $0 < h < \delta_1$ we have

$$|B(y)(t_i - h) - B(y)(t_i)| \le | \left(S'(t_i - h) - S'(t_i) \right) \phi(0)|$$

$$+ \lim_{\lambda \to \infty} \int_0^{t_i - h} \| \left(S'(t_i - h - s) - S'(t_i - s) \right) B_\lambda f(s, y_s) \| ds$$

$$+ Me^{\omega b} \psi(q) \int_{t_i - h}^{t_i} e^{-\omega s} p(s)\, ds;$$

which tends to zero as $h \to 0$. Define

$$\hat{B}_0(y)(t) = B(y)(t), \ t \in [0, t_1]$$

and

$$\hat{B}_i(y)(t) = \begin{cases} B(y)(t), & \text{if } t \in (t_i, t_{i+1}] \\ B(y)(t_i^+), & \text{if } t = t_i. \end{cases}$$

Next we prove equi-continuity at $t = t_i^+$. Fix $\delta_2 > 0$ such that $\{t_k : k \ne i\} \cap [t_i - \delta_2, t_i + \delta_2] = \emptyset$. For $0 < h < \delta_2$ we have

$$|\hat{B}(y)(t_i + h) - \hat{B}(y)(t_i)| \le | \left(S'(t_i + h) - S'(t_i) \right) \phi(0)|$$

$$+ \lim_{\lambda \to \infty} \int_0^{t_i} \| \left(S'(t_i + h - s) - S'(t_i - s) \right) B_\lambda f(s, y_s) \| ds$$

$$+ Me^{\omega b} \psi(q) \int_{t_i}^{t_i + h} e^{-\omega s} p(s)\, ds.$$

The right-hand side tends to zero as $h \to 0$. The equi-continuity for the cases $\tau_1 < \tau_2 \leq 0$ and $\tau_1 \leq 0 \leq \tau_2$ follows from the uniform continuity of ϕ on the interval $[-r, 0]$. As a consequence of steps 1–3 together with Arzelá–Ascoli theorem it suffices to show that $\mathcal{B}$ maps B_q into a precompact set in E.

Let $0 < t < b$ be fixed and let ϵ be a real number satisfying $0 < \epsilon < t$. For $y \in B_q$ we define

$$\mathcal{B}_\epsilon(y)(t) = S'(t)\phi(0) + S'(\epsilon) \lim_{\lambda \to \infty} \int_0^{t-\epsilon} S'(t - s - \epsilon) B_\lambda f(s, y_s) ds.$$

Note

$$\left\{ \lim_{\lambda \to \infty} \int_0^{t-\epsilon} S'(t - s - \epsilon) B_\lambda f(s, y_s) \, ds : y \in B_q \right\}$$

is a bounded set since

$$\left| \lim_{\lambda \to \infty} \int_0^{t-\epsilon} S'(t - s - \epsilon) B_\lambda f(s, y_s) \, ds \right| \leq M e^{\omega b} \psi(q) \int_0^{t-\epsilon} e^{-\omega s} p(s) ds.$$

Since $S'(t)$ is a compact operator, the set

$$Y_\epsilon(t) = \{ \mathcal{B}_\epsilon(y)(t) : \ y \in B_q \}$$

is precompact in E for every ϵ, $0 < \epsilon < t$. Moreover, for every $y \in B_q$ we have

$$|\mathcal{B}(y)(t) - \mathcal{B}_\epsilon(y)(t)| \leq M e^{\omega b} \psi(q) \int_t^{t-\epsilon} e^{-\omega s} p(s) ds.$$

Therefore, there are precompact sets arbitrarily close to the set $Y_\epsilon(t) = \{\mathcal{B}_\epsilon(y)(t) : y \in B_q\}$. Hence the set $Y(t) = \{\mathcal{B}(y)(t) : \ y \in B_q\}$ is precompact in E. Hence the operator $\mathcal{B} : PC\left([-r, b], \overline{D(A)}\right) \to PC\left([-r, b], \overline{D(A)}\right)$ is completely continuous.

Step 4: $\mathcal{A}$ is a contraction. Let $x, y \in PC([-r, b], \overline{D(A)})$. Then for $t \in J$

$$|\mathcal{A}(y)(t) - \mathcal{A}(x)(t)| = \left| \sum_{0 < t_k < t} S'(t - t_k) \left(I_k \left(y\left(t_k^-\right)\right) - I_k \left(x\left(t_k^-\right)\right)\right)\right|$$

$$\leq M e^{\omega b} \sum_{0 < t_k < t} \left| I_k \left(y\left(t_k^-\right)\right) - I_k \left(x\left(t_k^-\right)\right)\right|$$

$$\leq M e^{\omega b} \sum_{k=1}^m d_k \left| y\left(t_k^-\right) - x\left(t_k^-\right)\right|$$

$$\leq M e^{\omega b} \sum_{k=1}^m d_k \, \|y - x\|.$$

Then

$$\|\mathcal{A}(y) - \mathcal{A}(x)\| \leq Me^{\omega b} \sum_{k=1}^{m} d_k \|y - x\|,$$

which is a contraction from (9.22).

Step 5: A priori bounds. Now it remains to show that the set

$$\mathcal{E} = \left\{ y \in PC([-r, b], \overline{D(A)}) : y = \lambda B(y) + \lambda \mathcal{A}\left(\frac{y}{\lambda}\right) \text{ for some } 0 < \lambda < 1 \right\}$$

is bounded.

Let $y \in \mathcal{E}$. Then $y = \lambda B(y) + \lambda \mathcal{A}\left(\frac{y}{\lambda}\right)$ for some $0 < \lambda < 1$. Thus, for each $t \in J$,

$$y(t) = \lambda S'(t)\phi(0) + \lambda \frac{d}{dt} \int_0^t S(t - s)f(s, y_s)ds + \lambda \sum_{0 < t_k < t} S'(t - t_k) I_k\left(\frac{y}{\lambda}(t_k^-)\right).$$

This implies by (9.17.2), (9.17.5) that, for each $t \in J$, we have

$$|y(t)| \leq \lambda Me^{\omega t}|\phi(0)| + \lambda Me^{\omega t} \int_0^t e^{-\omega s}p(s)\psi(\|y_s\|)ds$$

$$+ \lambda Me^{\omega t} \sum_{k=1}^{m} \left| I_k\left(\frac{y}{\lambda}(t_k^-)\right)\right|$$

$$\leq \lambda Me^{\omega t}\|\phi\| + \lambda Me^{\omega t} \int_0^t e^{-\omega s}p(s)\psi(\|y_s\|)ds$$

$$+ \lambda Me^{\omega t} \sum_{k=1}^{m} \left| I_k\left(\frac{y}{\lambda}(t_k^-)\right) - I_k(0)\right|$$

$$+ \lambda Me^{\omega t} \sum_{k=1}^{m} |I_k(0)|$$

$$\leq \lambda Me^{\omega t} \left(\|\phi\| + \sum_{k=1}^{m} |I_k(0)|\right)$$

$$+ \lambda Me^{\omega t} \int_0^t e^{-\omega s}p(s)\psi(\|y_s\|)ds$$

$$+ \lambda Me^{\omega t} \sum_{k=1}^{m} d_k \left|\frac{y}{\lambda}(t_k^-)\right|$$

$$\leq ce^{\omega t} + Me^{\omega t}\left[\int_0^t e^{-\omega s}p(s)\psi(\|y_s\|)ds + \sum_{k=1}^{m} d_k |y(t_k^-)|\right],$$

where

$$c = M \left(\|\phi\| + \sum_{k=1}^{m} |I_k(0)| \right).$$ (9.23)

Now we consider the function μ defined by

$$\mu(t) = \sup\{|y(s)| : -r \leq s \leq t\}, \quad 0 \leq t \leq b.$$

Then $\|y_s\| \leq \mu(t)$ for all $t \in J$ and there is a point $t^* \in [-r, t]$ such that $\mu(t) = |y(t^*)|$. If $t^* \in [0, b]$, by the previous inequality and (9.23) we have for $t \in [0, b]$ (note $t^* \leq t$).

$$\mu(t) \leq ce^{\omega b} + Me^{\omega b} \int_0^t e^{-\omega s} p(s) \psi(\mu(s)) ds + Me^{\omega b} \sum_{k=1}^{m} d_k \mu(t).$$

Then

$$\left(1 - Me^{\omega b} \sum_{k=1}^{m} d_k \right) \mu(t) \leq ce^{\omega b} + Me^{\omega b} \int_0^t e^{-\omega s} p(s) \psi(\mu(s)) ds.$$

Thus from (9.20) and (9.21) we have

$$\mu(t) \leq c_0 + c_1 \int_0^t e^{-\omega s} p(s) \psi(\mu(s)) ds.$$ (9.24)

Let us take the right-hand side of (9.24) as $v(t)$. Then we have

$$\mu(t) \leq v(t) \quad \text{for all } t \in J,$$

$$v(0) = c_0,$$

and

$$v'(t) = c_1 e^{-\omega t} p(t) \psi(\mu(t)), \quad \text{a.e. } t \in J.$$

Using the nondecreasing character of ψ we get

$$v'(t) \leq c_1 e^{-\omega t} p(t) \psi(v(t)), \quad \text{a.e. } t \in J.$$

That is

$$\frac{v'(t)}{\psi(v(t))} \leq c_1 e^{-\omega t} p(t), \quad \text{a.e. } t \in J.$$

Integrating from 0 to t we get

$$\int_0^t \frac{v'(s)}{\psi(v(s))} ds \leq c_1 \int_0^t e^{-\omega s} p(s) ds.$$

By a change of variable and (9.19) we get

$$\int_{v(0)}^{v(t)} \frac{du}{\psi(u)} \le c_1 \int_0^b e^{-\omega s} p(s)\, ds < \int_{c_0}^{\infty} \frac{du}{\psi(u)}.$$

Hence there exists a constant N such that

$$\mu(t) \le v(t) \le N \quad \text{for all } t \in J.$$

Now from the definition of μ it follows that

$$\|y\| = \sup_{t \in [-r,b]} |y(t)| \le \max(\|\phi\|, N) \text{ for all } y \in \mathcal{E}.$$

This shows that the set $\mathcal{E}$ is bounded. As a consequence of Theorem 1.32 we deduce that $\mathcal{A} + \mathcal{B}$ has a fixed point which is a integral solution of problem (9.11)–(9.13). $\quad\Box$

Now we give a result where f Lipschitz with respect to y.

Theorem 9.15. *Assume that (9.17.1)–(9.17.4) hold and the condition*

(9.19.1) There exists a function $l \in L^1(J, \mathbb{R}_+)$ such that:

$$|f(t, x) - f(t, y)| \le l(t)\|x - y\|_D \text{ a.e. } t \in J, \text{ and for all } x, y \in D,$$

with $\displaystyle\int_0^b e^{-\omega s} l(s)\, ds < \infty,$

$$c_0^* = \frac{Me^{\omega b}\left(\|\phi\| + \displaystyle\sum_{k=1}^m |I_k(0)| + \int_0^b e^{-\omega s}|f(s,0)|\, ds\right)}{1 - Me^{\omega b}\displaystyle\sum_{k=1}^m d_k} \tag{9.25}$$

and

$$c_1^* = \frac{Me^{\omega b}}{1 - Me^{\omega b}\displaystyle\sum_{k=1}^m d_k}. \tag{9.26}$$

If

$$Me^{\omega b}\sum_{k=1}^m d_k < 1,$$

then the problem (9.11)–(9.13) has at least one integral solution on $[-r, b]$.

Proof. Let $\mathcal{A}$ and $\mathcal{B}$ the operator defined in Theorem 9.14. It can be shown that $\mathcal{B}$ is completely continuous and $\mathcal{A}$ is a contraction. Now we prove that

$$\mathcal{E} = \left\{ y \in PC([-r, b], \overline{D(A)}) : y = \lambda \mathcal{B}(y) + \lambda \mathcal{A}\left(\frac{y}{\lambda}\right) \text{ for some } 0 < \lambda < 1 \right\}$$

is bounded.

Let $y \in \mathcal{E}$. Then $y = \lambda \mathcal{B}(y) + \lambda \mathcal{A}\left(\frac{y}{\lambda}\right)$ for some $0 < \lambda < 1$. Thus, for each $t \in J$,

$$y(t) = \lambda S'(t)\phi(0) + \lambda \frac{d}{dt} \int_0^t S(t - s) f(s, y_s) ds + \lambda \sum_{0 < t_k < t} S'(t - t_k) I_k\left(\frac{y}{\lambda}(t_k^-)\right).$$

This implies by (9.17.2) and (9.19.1) that, for each $t \in J$, we have

$$|y(t)| \le \lambda M e^{\omega t} |\phi(0)| + \lambda M e^{\omega t} \int_0^t e^{-\omega s} |f(s, y_s) - f(s, 0)| ds$$

$$+ \lambda M e^{\omega t} \int_0^t e^{-\omega s} |f(s, 0)| ds + \lambda M e^{\omega t} \sum_{k=1}^m \left| I_k\left(\frac{y}{\lambda}(t_k^-)\right) \right|$$

$$\le \lambda M e^{\omega t} \|\phi\| + \lambda M e^{\omega t} \int_0^t e^{-\omega s} l(s) \|y_s\| ds + \lambda M e^{\omega t} \int_0^t e^{-\omega s} |f(s, 0)| ds$$

$$+ \lambda M e^{\omega t} \sum_{k=1}^m \left| I_k\left(\frac{y}{\lambda}(t_k^-)\right) - I_k(0) \right| + \lambda M e^{\omega t} \sum_{k=1}^m |I_k(0)|$$

$$\le M e^{\omega t} \left(\|\phi\| + \int_0^t e^{-\omega s} |f(s, 0)| ds + \sum_{k=1}^m |I_k(0)| \right)$$

$$+ M e^{\omega t} \int_0^t e^{-\omega s} l(s) \|y_s\| ds + M e^{\omega t} \sum_{k=1}^m d_k y(t_k^-).$$

Now we consider the function μ defined by

$$\mu(t) = \sup\{|y(s)| : -r \le s \le t\}, \quad 0 \le t \le b.$$

Then $\|y_s\| \leq \mu(t)$ for all $t \in J$ and there is a point $t^* \in [-r, t]$ such that $\mu(t) = |y(t^*)|$. If $t^* \in [0, b]$, by the previous inequality we have for $t \in [0, b]$ (note $t^* \leq t$)

$$\mu(t) \leq M e^{\omega t} \left(\|\phi\| + \int_0^t e^{-\omega s} |f(s, 0)| ds + \sum_{k=1}^m |I_k(0)| \right)$$

$$+ M e^{\omega t} \int_0^t e^{-\omega s} l(s) \mu(s) ds + M e^{\omega t} \sum_{k=1}^m d_k \mu(t).$$

Then

$$\left(1 - M e^{\omega b} \sum_{k=1}^m d_k \right) \mu(t) \leq M e^{\omega t} \left(\|\phi\| + \int_0^t e^{-\omega s} |f(s, 0)| ds \right.$$

$$\left. + \sum_{k=1}^m |I_k(0)| \right)$$

$$+ M e^{\omega b} \int_0^t e^{-\omega s} l(s) \mu(s) ds.$$

Thus by (9.25) and (9.26) we have

$$\mu(t) \leq c_0^* + c_1^* \int_0^t e^{-\omega s} l(s) \mu(s) ds.$$

By Gronwall inequality [131] we get for each $t \in J$

$$\mu(t) \leq c_0^* \exp \left(c_1^* \int_0^t e^{-\omega s} l(s) ds \right).$$

Thus

$$\|y\| \leq c_0^* \exp \left(c_1^* \int_0^b e^{-\omega s} l(s) ds \right) := M^*.$$

This shows that the set $\mathcal{E}$ is bounded. As a consequence of Theorem 1.32 we deduce that $\mathcal{A} + \mathcal{B}$ has a fixed point which is a integral solution of problem (9.11)–(9.13).

$\square$

9.3.4 Existence of Extremal Integral Solutions

In this section we shall prove the existence of maximal and minimal integral solutions of problem (9.11)–(9.13) under suitable monotonicity conditions on the functions involved in it. We need the following definitions in the sequel.

Definition 9.16. We say that a continuous function $v : [-r, b] \to E$ is a lower integral solution of problem (9.11)–(9.13) if $v(t) = \phi(t), t \in H$, and

$$v(t) \leq S'(t)\phi(0) + A \int_0^t v(s)ds + \int_0^t f(s, v_s)ds$$
$$+ \sum_{0 < t_k < t} S'(t - t_k) I_k \left(v\left(t_k^-\right) \right), \; t \in J, \; t \neq t_k$$

and $v(t_k^+) - v(t_k^-) \leq I_k(v(t_k))$, $t = t_k$, $k = 1, \ldots, m$. Similarly an upper integral solution w of problem (11.15)–(11.17) is defined by reversing the order.

Definition 9.17. A solution x_M of problem (9.11)–(9.13) is said to be maximal if for any other solution x of problem (9.11)–(9.13) on J, we have that $x(t) \leq x_M(t)$ for each $t \in J$.

Similarly a minimal solution of problem (9.11)–(9.13) is defined by reversing the order of the inequalities.

Definition 9.18. A function $f(t, x)$ is called strictly monotone increasing in x almost everywhere for $t \in J$, if $f(t, x) \leq f(t, y)$ a.e. $t \in J$ for all $x, y \in D$ with $x < y$. Similarly $f(t, x)$ is called strictly monotone decreasing in x almost everywhere for $t \in J$, if $f(t, x) \geq f(t, y)$ a.e. $t \in J$ for all $x, y \in D$ with $x < y$.

We consider the following assumptions in the sequel.

(9.22.1) The function $f(t, y)$ is strictly monotone increasing in y for almost each $t \in J$.
(9.22.2) $S'(t)$ is preserving the order, that is $S'(t)v \geq 0$ whenever $v \geq 0$.
(9.22.3) The functions I_k, $k = 1, \ldots, m$ are continuous and nondecreasing.
(9.22.4) The problem (11.15)–(11.17) has a lower integral solution v and an upper integral solution w with $v \leq w$.

Theorem 9.19. *Assume that assumptions (9.17.1)–(9.17.5) and (9.12.1)–(9.12.4) hold. Then problem (9.11)–(9.13) has a minimal and a maximal integral solutions on $[-r, b]$.*

Proof. It can be shown, as in the proof of Theorem 9.14, that $\mathcal{B}$ is completely continuous and $\mathcal{A}$ is a contraction on $[v, w]$. We shall show that $\mathcal{A}$ and $\mathcal{B}$ are isotone increasing on $[v, w]$. Let $y, \bar{y} \in [a, b]$ be such that $y \leq \bar{y}, y \neq \bar{y}$. Then by (9.22.1), (9.22.2), we have for each $t \in J$

$$\mathcal{B}(y)(t) = S'(t)\phi(0) + \frac{d}{dt}\int_0^t S(t-s)f(s,y_s)\,ds$$

$$\leq S'(t)\phi(0) + \frac{d}{dt}\int_0^t S(t-s)f(s,\bar{y}_s)\,ds$$

$$= \mathcal{B}(\bar{y})(t).$$

and by (9.22.3), we have for each $t \in J$

$$\mathcal{A}(y)(t) = \sum_{0<t_k<t} S'(t-t_k) I_k\left(y\left(t_k^-\right)\right)$$

$$\leq \sum_{0<t_k<t} S'(t-t_k) I_k\left(\bar{y}\left(t_k^-\right)\right)$$

$$= \mathcal{A}(\bar{y})(t).$$

Therefore $\mathcal{A}$ and $\mathcal{B}$ are isotone increasing on $[v, w]$. Finally, let $x \in [v, w]$ be any element. By (9.22.4) we deduce that

$$v \leq \mathcal{A}(v) + \mathcal{B}(v) \leq \mathcal{A}(x) + \mathcal{B}(x) \leq \mathcal{A}(w) + \mathcal{B}(w) \leq w,$$

which shows that $\mathcal{A}(x) + \mathcal{B}(x) \in [v, w]$ for all $x \in [v, w]$. Thus, problem (9.11)–(9.13) has a maximal and a minimal integral solutions on $[-r, b]$.   □

9.3.5 Impulsive Differential Equations with Nonlocal Conditions

In this section we shall prove existence results for problem (9.14)–(9.16).

Definition 9.20. A function $y \in PC\left([-r, b], \overline{D(A)}\right)$ is said to be a integral solution of problem (9.14)–(9.16) if $y(t) = \phi(t) - h_t(y)$, $t \in H$, and

$$y(t) = S'(t)(\phi(0) - h_0(y)) + \int_0^t T(t-s)f(s,y_s)\,ds$$

$$+ \sum_{0<t_k<t} T(t-t_k) I_k\left(y\left(t_k^-\right)\right), \quad t \in J.$$

Theorem 9.21. *Assume that hypotheses (9.17.1)–(9.17.4) and the following hypotheses hold:*

(A1) The function h is continuous with respect to t, and there exists a constant $\alpha > 0$ such that

$$|h_t(u)| \leq \alpha, \quad u \in PC([-r, b], \overline{D(A)})$$

and for each k > 0 the set

$$\{\phi(0) - h_0(y), \; y \in PC([-r, b], \overline{D(A)}), \; \|y\| \leq k\}$$

is precompact in E

(A2) *There exist a function* $p \in L^1(J, \mathbb{R}_+)$ *and a continuous nondecreasing function* $\psi : [0, \infty) \rightarrow (0, \infty)$ *such that*

$$|f(t, x)| \leq p(t)\psi(\|x\|_D), \quad a.e. \; t \in J, \quad \text{for all } x \in D$$

with

$$\int_{\tilde{c}_0}^{\infty} \frac{du}{\psi(u)} > \tilde{c}_1 \int_0^b e^{-\omega s} p(s) ds,$$

where

$$\tilde{c}_0 = \frac{Me^{\omega b}[\|\phi\|_D + \alpha + \sum_{k=1}^m |I_k(0)|]}{1 - Me^{\omega b} \sum_{k=1}^m d_k},$$

and

$$\tilde{c}_1 = \frac{Me^{\omega b}}{1 - Me^{\omega b} \sum_{k=1}^m d_k}.$$

Moreover, we suppose that

$$Me^{\omega b} \sum_{k=1}^m d_k < 1,$$

then the problem (9.14)–(9.16) has at least one integral solution on $[-r, b]$.

Proof. Consider the two operators: $\mathcal{B}_1 : PC\left([-r, b], \overline{D(A)}\right) \rightarrow PC\left([-r, b], \overline{D(A)}\right)$

$$\mathcal{B}_1(y)(t) := \begin{cases} \phi(t) - h_t(y), & \text{if } t \in H; \\ S'(t)(\phi(0) - h_0(y)) + \dfrac{d}{dt} \int_0^t S(t-s)f(s, y_s)\, ds & \text{if } t \in J, \end{cases}$$

$$A_1(y)(t) = \begin{cases} 0, & \text{if } t \in H; \\ \displaystyle\sum_{0 < t_k < t} S'(t - t_k)I_k(y(t_k^-)), & \text{if } t \in J. \end{cases}$$

Then the problem of finding the solution of problem (9.14)–(9.16) is reduced to finding the solution of the operator equation $A_1(y)(t) + B_2(y)(t) = y(t)$, $t \in [-r, b]$. As in Sect. 9.3, we can show that the operators A_1 and B_1 satisfy all conditions of Theorem 1.32. □

9.3.6 Applications to Control Theory

This section is devoted to an application of the argument used in the previous sections to the controllability of impulsive functional differential equations. More precisely we will consider the following problem:

$$y'(t) - Ay(t) = f(t, y_t) + Bu(t), \text{ a.e. } t \in J = [0, b], \ t \neq t_k, \ k = 1, \ldots, m \quad (9.27)$$

$$\Delta y|_{t=t_k} = I_k(y(t_k^-)), \ k = 1, \ldots, m \quad (9.28)$$

$$y(t) = \phi(t), \ t \in [-r, 0], \quad (9.29)$$

where A, f, and I_k are as in Sect. 9.3, the control function $u(\cdot)$ is given in $L^2(J, U)$ a Banach space of admissible control functions with U as a Banach. Finally B is a bounded linear operator from U to $\overline{D(A)}$.

Definition 9.22. A function $y \in PC\left([-r, b], \overline{D(A)}\right)$ is said to be a integral solution of problem (9.27)–(9.29) if $y(t) = \phi(t)$, $t \in [-r, 0]$, and y is a solution of impulsive integral equation

$$y(t) = S'(t)\phi(0) + \frac{d}{dt}\int_0^t S(t-s)f(s, y_s) \, ds + \frac{d}{dt}\int_0^t S(t-s)Bu(s) \, ds$$
$$+ \sum_{0 < t_k < t} S'(t - t_k) I_k\left(y\left(t_k^-\right)\right), \ t \in J.$$

Definition 9.23. The system (9.27)–(9.29) is said to be controllable on the interval $[-r, b]$ if for every initial function $\phi \in D$ and every $y_1 \in \overline{D(A)}$, there exists a control $u \in L^2(J, U)$, such that the mild solution $y(t)$ of system (9.27)–(9.29) satisfies $y(b) = y_1$.

Our main result in this section is the following.

Theorem 9.24. *Assume that hypotheses (9.17.1)–(9.17.4) hold. Moreover we suppose that*

(B1) The linear operator $W : L^2(J, U) \rightarrow \overline{D(A)}$ defined by

$$Wu = \int_0^b T(b-s)Bu(s)\,ds,$$

has a bounded inverse operator W^{-1} which takes values in $L^2(J, U) \setminus KerW$, and there exist positive constants $\overline{M}, \overline{M_1}$, such that $\|B\| \leq \overline{M}$ and $\|W^{-1}\| \leq \overline{M_1}$.

(B2) There exists a function $l \in L^1(J, \mathbb{R}_+)$ such that

$$|f(t,x) - f(t,y)| \leq l(t)\|x - y\|_D \text{ for a.e. } t \in J, \text{ and for all } x,y \in D,$$

with

$$M^2 e^{2\omega b}\overline{M}\overline{M_1}b \int_0^b e^{-\omega s}l(s)ds + Me^{\omega b}(1 + Me^{\omega b}\overline{M}\overline{M_1}b) \sum_{k=0}^m d_k < 1$$

(B3) There exist a function $p \in L^1(J, \mathbb{R}_+)$ and a continuous nondecreasing function $\psi : [0, \infty) \rightarrow (0, \infty)$ such that

$$|f(t,x)| \leq p(t)\psi(\|x\|_D), \quad a.e. \ t \in J, \quad \text{for all } x \in D$$

with $\int_0^b e^{-\omega s}p(s)ds < \infty$,

$$\int_{c_3}^{\infty} \frac{ds}{s + \psi(s)} > \|\hat{m}\|_{L^1} \tag{9.30}$$

where

$$c_3 = \frac{c_2}{1 - Me^{\omega b}(1 + M\overline{M}\overline{M_1}be^{\omega b})\sum_{k=1}^m d_k}, \tag{9.31}$$

$$c_2 = M(1 + M\overline{M}\overline{M_1}e^{\omega b}b)\|\phi\|$$

$$+ M\overline{M}\overline{M_1}b|y_1| + M(1 + M\overline{M}\overline{M_1}e^{\omega b}b)\sum_{k=1}^m |I_k(0)|, \tag{9.32}$$

$$\hat{m}(s) = \max\{\omega, c_4 p(s)\}, \tag{9.33}$$

and

$$c_4 = \frac{M + M^2 \overline{MM_1} e^{\omega b} b}{1 - M e^{\omega b} (1 + M \overline{MM_1} b e^{\omega b}) \sum_{k=1}^{m} d_k}. \tag{9.34}$$

Then the problem (9.27)–(9.29) is controllable on $[-r, b]$.

Remark 9.25. The construction of operator W^{-1} and its properties are discussed in [93].

Proof. Using hypothesis (B1) for an arbitrary function $y(.)$ we define the control

$$u_y(t) = W^{-1} \left[y_1 - S'(b) \phi(0) - \lim_{\lambda \to \infty} \int_0^b S'(b-s) B_\lambda f(s, y_s) ds \right.$$

$$\left. - \sum_{0 < t_k < t} S'(b - t_k) I_k \left(y \left(t_k^- \right) \right) \right] (t).$$

Consider the two operators:

$$\overline{A}, \overline{B} : PC\left([-r, b], \overline{D(A)} \right) \to PC\left([-r, b], \overline{D(A)} \right)$$

defined by

$$\overline{A}(y)(t) := \begin{cases} 0, & \text{if } t \in H; \\ \dfrac{d}{dt} \displaystyle\int_0^t S(t-s) B u(s) ds \\ \quad + \displaystyle\sum_{0 < t_k < t} S'(t - t_k) I_k \left(y \left(t_k^- \right) \right), & \text{if } t \in J, \end{cases}$$

and

$$\overline{B}(y)(t) := \begin{cases} \phi(t), & \text{if } t \in H; \\ S'(t) \phi(0) \\ \quad + \dfrac{d}{dt} \displaystyle\int_0^t S(t-s) f(s, y_s) ds, & \text{if } t \in J. \end{cases}$$

We can prove that $\overline{A}$ is a contraction operator and $\overline{B}$ is completely continuous. Now, we prove that

$$\mathcal{E} = \left\{ y \in PC([-r, b], \overline{D(A)}) : y = \lambda \overline{B}(y) + \lambda \overline{A} \left(\frac{y}{\lambda} \right) \text{ for some } 0 < \lambda < 1 \right\}$$

is bounded.

Let $y \in \mathcal{E}$. Then $y = \lambda \overline{\mathcal{B}}(y) + \lambda \overline{\mathcal{A}} \left(\frac{y}{\lambda} \right)$ for some $0 < \lambda < 1$. Thus, for each $t \in J$,

$$y(t) = \lambda S'(t)\phi(0) + \lambda \frac{d}{dt} \int_0^t S(t-s)f(s, y_s)ds$$

$$+ \lambda \frac{d}{dt} \int_0^t S(t-s)Bu_y(s)\,ds$$

$$+ \lambda \sum_{0 < t_k < t} S'(t-t_k) I_k \left(\frac{y}{\lambda}(t_k^-) \right).$$

This implies by (B1)–(B3) that, for each $t \in J$, we have

$$|y(t)| \leq \lambda M e^{\omega t}|\phi(0)| + \lambda M e^{\omega t} \int_0^t e^{-\omega s}p(s)\psi(\|y_s\|)ds$$

$$+ \lambda M e^{\omega t} \int_0^t e^{-\omega s} |Bu_y(s)|\,ds + \lambda M e^{\omega t} \sum_{k=0}^m \left| I_k \left(\frac{y}{\lambda}(t_k^-) \right) \right|$$

$$\leq \lambda M e^{\omega t} \left[(1 + M\overline{M}M_1 b e^{\omega b})\|\phi\| + \overline{M}M_1 b|y_1| \right]$$

$$+ \lambda M e^{\omega t} \int_0^t e^{-\omega s}p(s)\psi(\|y_s\|)ds$$

$$+ \lambda M^2 \overline{M_1 M} b e^{\omega t} e^{\omega b} \int_0^t e^{-\omega s}p(s)\psi(\|y_s\|)ds$$

$$+ \lambda M^2 \overline{M_1 M} b e^{\omega t} e^{\omega b} \sum_{k=1}^m |I_k(\frac{y}{\lambda}(t_k^-))|$$

$$+ \lambda M e^{\omega t} \sum_{k=1}^m \left| I_k \left(\frac{y}{\lambda}(t_k^-) \right) \right|$$

$$\leq \lambda M e^{\omega t}[(1 + M\overline{M}M_1 b e^{\omega b})\|\phi\|$$

$$+ \overline{M}M_1 b|y_1| + M e^{\omega t}(1 + M\overline{M}M_1 b e^{\omega b}) \sum_{k=1}^m |I_k(0)|]$$

$$+\lambda M e^{\omega t} \int_0^t e^{-\omega s} p(s)\psi(\|y_s\|)ds$$

$$+\lambda M^2 e^{\omega t}\overline{MM_1}be^{\omega b}\int_0^t e^{-\omega s} p(s)\psi(\|y_s\|)ds$$

$$+\lambda M e^{\omega t}(1+M\overline{MM_1}be^{\omega b})\sum_{k=1}^m |y(t_k^-))|.$$

Set

$$\alpha = M(1+M\overline{MM_1}e^{\omega b}b)\|\phi\| + M\overline{MM_1}b|y_1|$$

$$+M(1+M\overline{MM_1}e^{\omega b}b)\sum_{k=1}^m |I_k(0)|.$$

Consider the function μ defined by

$$\mu(t) = \sup\{|y(s)| : -r \le s \le t\}, \quad 0 \le t \le b.$$

Then $\|y_s\| \le \mu(t)$ for all $t \in J$ and there is a point $t^* \in [-r, t]$ such that $\mu(t) = |y(t^*)|$. If $t^* \in [0, b]$, by the previous inequality we have for $t \in [0, b]$ (note $t^* \le t$)

$$\mu(t) \le \alpha e^{\omega t} + M e^{\omega t}\int_0^t e^{-\omega s} p(s)\psi(\mu(s))ds$$

$$+M^2 e^{\omega t}\overline{MM_1}be^{\omega b}\int_0^t e^{-\omega s} p(s)\psi(\mu(s))ds$$

$$+\lambda M e^{\omega t}(1+M\overline{MM_1}be^{\omega b})\sum_{k=1}^m d_k\mu(t).$$

Then

$$[1 - M e^{\omega b}(1+M\overline{MM_1}be^{\omega b})\sum_{k=1}^m d_k]\mu(t) \le \alpha e^{\omega t}$$

$$+M e^{\omega t}(1+M\overline{MM_1}be^{\omega b})\int_0^t p(s)\psi(\mu(s))ds.$$

Thus by (9.31), (9.32), (9.34) we have

$$e^{-\omega t}\mu(t) \le c_3 + c_4 \int_0^t p(s)\psi(\mu(s))ds. \tag{9.35}$$

Let us take the right-hand side of (9.35) as $v(t)$. Then we have

$$\mu(t) \le e^{\omega t}v(t) \quad \text{for all } t \in J,$$

$$v(0) = c_3,$$

and

$$v'(t) = c_4 p(t)\psi(\mu(t)), \quad a.e. \ t \in J.$$

Using the nondecreasing character of ψ we get

$$v'(t) \le c_4 p(t)\psi(e^{\omega t}v(t)), \quad a.e. \ t \in J.$$

Then by (9.33) for a.e. $t \in J$ we have

$$(e^{\omega t}v(t))' = \omega e^{\omega t}v(t) + v'(t)e^{\omega t}$$
$$\le \omega e^{\omega t}v(t) + c_4 p(t)e^{\omega t}\psi(e^{\omega t}v(t))$$
$$\le \hat{m}(t)[e^{\omega t}v(t) + \psi(e^{\omega t}v(t))].$$

Thus (9.30) gives

$$\int_{v(0)}^{e^{\omega t}v(t)} \frac{du}{u + \psi(u)} \le \int_0^b \hat{m}(s)ds = \|\hat{m}\|_{L^1} < \int_{c_3}^{\infty} \frac{du}{u + \psi(u)}.$$

Consequently, by (B3), there exists a constant d such that $e^{\omega t}v(t) \le d$, $t \in J$ and hence $\|y\| \le d$. This shows that the set $\mathcal{E}$ is bounded. As a consequence of Theorem 9.14 we deduce that $\overline{A} + \overline{B}$ has a fixed point which is a integral solution of problem (9.27)–(9.29). Thus the system (9.27)–(9.29) is controllable on $[-r, b]$. $\quad\square$

9.3.7 An Example

As an application of our results we consider the following impulsive partial functional differential equation of the form

$$\frac{\partial}{\partial t}z(t,x) = \frac{\partial^2}{\partial x^2}z(t,x)$$

$$+Q(t,z(t-r,x))+Bu(t), \ x \in [0,\pi], t \in [0,b]\setminus\{t_1,t_2,\ldots,t_m\}. \quad (9.36)$$

$$z(t_k^+,x) - z(t_k^-,x) = b_k z(t_k^-,x), \ x \in [0,\pi], \ k = 1,\ldots,m \quad (9.37)$$

$$z(t,0) = z(t,\pi) = 0, \ t \in [0,b] \quad (9.38)$$

$$z(t,x) = \phi(t,x), \ t \in H, \ x \in [0,\pi], \quad (9.39)$$

where $b_k > 0$, $k = 1,\ldots,m$, $\phi \in \mathcal{D} = \{\psi : H \times [0,\pi] \to \mathbb{R}; \psi$ is continuous everywhere except for a countable number of points at which $\psi(s^-), \psi(s^+)$ exist with $\psi(s^-) = \psi(s)\}$, $0 = t_0 < t_1 < t_2 < \cdots < t_m < t_{m+1} = b$, $z(t_k^+) = \lim\limits_{(h,x)\to(0^+,x)} z(t_k + h, x), z(t_k^-) = \lim\limits_{(h,x)\to(0^-,x)} z(t_k + h, x)$ and $Q : [0,b] \times \mathbb{R} \to \mathbb{R}$, is a given function.

Let

$$y(t)(x) = z(t,x), \ t \in [0,b], \ x \in [0,\pi],$$

$$I_k(y(t_k^-))(x) = b_k z(t_k^-,x), \ x \in [0,\pi], \ k = 1,\ldots, m$$

$$F(t,\phi)(x) = Q(t,\phi(\theta,x)), \ \theta \in H, \ x \in [0,\pi],$$

and

$$\phi(\theta)(x) = \phi(\theta,x), \ \theta \in H, \ x \in [0,\pi].$$

Consider $E = C(\overline{\Omega})$, the Banach space of continuous function on $\overline{\Omega}$ with values in $\mathbb{R}$. Define the linear operator A on E by

$$Az = \frac{\partial^2}{\partial x^2}z, \quad \text{in} \quad D(A) = \{z \in C(\overline{\Omega}) : z = 0 \text{ on } \partial\Omega, \frac{\partial^2}{\partial x^2}z \in C(\overline{\Omega}\}$$

Now, we have

$$\overline{D(A)} = C_0(\overline{\Omega}) = \{v \in C(\overline{\Omega}) : v = 0 \text{ on } \partial\Omega\} \neq C(\overline{\Omega}).$$

It is well known from [100] that A is sectorial, $(0, +\infty) \subseteq \rho(A)$ and for $\lambda > 0$

$$\|R(\lambda, A)\|_{B(E)} \leq \frac{1}{\lambda}.$$

It follows that A generates an integrated semigroup $(S(t))_{t\geq 0}$ and that $\|S'(t)\|_{B(E)} \leq e^{-\mu t}$ for $t \geq 0$ for some constant $\mu > 0$ and A satisfied the Hille–Yosida condition.

Assume that the operator $B : U \to Y, U \subset [0,\infty)$, is a bounded linear operator and the operator

$$Wu = \int_0^b T(b-s)Bu(s)ds$$

has a bounded inverse operator W^{-1} which takes values in $L^2([0,b], U)\backslash kerW$. Also assume that there exists an integrable function $\sigma : [0,b] \rightarrow \mathbb{R}^+$ such that

$$|Q(t, w(t-r,x))| \le \sigma(t)\Omega(|w|)$$

where $\Omega : [0,\infty) \rightarrow (0,\infty)$ is continuous and nondecreasing with

$$\int_1^\infty \frac{ds}{s + \Omega(s)} = +\infty.$$

Assume that there exists a function $\tilde{l} \in L^1([0,b], \mathbb{R}^+)$ such that

$$|Q(t, w(t-r,x)) - Q(t, \overline{w}(t-r,x))| \le \tilde{l}(t)|w - \overline{w}|, \ t \in [0,b], \ w, \overline{w} \in \mathbb{R}.$$

We can show that problem (8.21)–(8.23) is an abstract formulation of problem (9.36)–(9.39). Since all the conditions of Theorem 9.24 are satisfied, the problem (9.36)–(9.39) has a solution z on $[-r,b] \times [0,\pi]$.

9.4 Impulsive Semi-linear Neutral Functional Differential Equations with Infinite Delay

9.4.1 Introduction

In this section we shall be concerned with the existence of mild solutions as well as integral solutions defined on a compact real interval for first order impulsive semi-linear functional equations in a separable Banach space. More precisely we consider the initial value problem

$$\frac{d}{dt}[y(t) - g(t, y_t)] = A[y(t) - g(t, y_t)] \tag{9.40}$$

$$+f(t, y_t), \ a.e. \ t \in J = [0,b], \ t \ne t_k, \ k = 1,\ldots,m$$

$$\Delta y|_{t=t_k} = I_k(y(t_k^-)), \ k = 1,\ldots,m \tag{9.41}$$

$$y(t) = \phi(t), \ t \in (-\infty, 0], \tag{9.42}$$

where $f, g : J \times D \rightarrow E$ is a given function, $D = \{\psi : [-\infty, 0] \rightarrow E, \psi$ is continuous everywhere except for a finite number of points s at which $\psi(s^-), \psi(s^+)$ exist and $\psi(s^-) = \psi(s)\}, \phi \in D, (0 < r < \infty), 0 = t_0 < t_1 < \cdots < t_m < t_{m+1} = b,$ $I_k \in C(E, E) \ (k = 1, 2,\ldots,m), A$ is a closed linear operator on E, and E a real separable Banach space with norm $|.|$.

9.4.2   Existence of Mild Solutions

This section is devoted to the case when the operator A generates a (C_0)-semigroup. Before starting and proving our main result, we will give the definition of the mild solution.

Definition 9.26. We say that a function $y : (-\infty, b] \to E$ is a mild solution of problem (9.40)–(9.42) if $y_0 = \phi$ and the restriction of $y(\cdot)$ to the interval $[0, b]$ is continuous; and

$$
\begin{aligned}
y(t) = {} & g(t, y_t) + T(t)[\phi(0) - g(0, \phi)] + \int_0^t T(t-s)f(s, y_s)ds \\
& + \sum_{0 < t_k < t} T(t - t_k)I_k\left(y\left(t_k^-\right)\right) - \sum_{0 < t_k < t} T(t - t_k)\Delta g(t_k, y_{t_k}), \quad t \in J.
\end{aligned}
\tag{9.43}
$$

Let us introduce the following hypotheses:

(9.30.1) A is the infinitesimal generator of a (C_0)–semigroup $\{T(t)\}_{t \in J}$, which is compact for $t > 0$ in the Banach space E. Let $M = \sup\{\|T(t)\|_{B(E)} : t \in J\}$.

(9.30.2) There exist constants $\alpha_1, \alpha_2 \geq 0$ and $l_g \geq 0$ such that:

(i) $|g(t, u) - g(t, \bar{u})| \leq l_g \|u - \bar{u}\|_D$, $t \in J$, and $u, \bar{u} \in D$, and

(ii) $|g(t, u)| \leq \alpha_1 \|u\|_D + \alpha_2$, $t \in J$, for a.e. $t \in J$, and each $u \in D$.

(9.30.3) $f : J \times D \to E$ is Carathéodory function;

(9.30.4) There exist constants $d_k > 0$, $k = 1, \ldots, m$ such that for each $y, x \in E$

$$
|I_k(y) - I_k(x)| \leq d_k |y - x|
$$

(9.30.5) There exists a function $p \in L^1(J, \mathbb{R}_+)$ and a continuous nondecreasing function $\psi : [0, \infty) \to [0, \infty)$ such that

$$
|f(t, u)| \leq p(t)\psi(\|u\|_D), \quad a.e.\ t \in J, \quad \text{for all } u \in D.
$$

with

$$
\int_{C_1^*}^{\infty} \frac{ds}{\psi(s)} > C_2^* \|p\|_{L^1},
$$

where

$$
C_1^* = K_b \frac{C_1}{C} + (MK_b + M_b)\|\phi\|, \qquad C_2^* = \frac{MK_b}{C}
\tag{9.44}
$$

with

$$
C = 1 - \alpha_1 K_b - M \sum_{k=1}^{m} d_k - 2Mml_g K_b,
$$

and

$$C_1 = \alpha_1(MK_b + M_b)\|\phi\| + (1 + M)\alpha_2 + M\sum_{k=1}^{m} d_k\|\phi\|$$

$$+ M\sum_{k=1}^{m}|I_k(0)| + 2Mml_g(MK_b + M_b)\|\phi\| + M\sum_{k=1}^{m}|\Delta g(t_k, 0)|$$

Theorem 9.27. *Assume that (9.30.1)–(9.30.5) hold. Suppose that*

$$K_b \max(l_g, \alpha_1) + M\sum_{k=1}^{m} d_k + 2mMl_g K_b < 1 \tag{9.45}$$

Then the problem (9.40)–(9.42) has at least one mild solution.

Proof. Consider the operator $N : D_b \to D_b$ defined by:

$$(Ny)(t) = \begin{cases} \phi(t), & t \in (-\infty, 0], \\ T(t)[\phi(0) - g(0, \phi)] + g(t, y_t) \\ \quad + \int_0^t T(t-s)f(s, y_s)ds + \sum_{0<t_k<t} T(t-t_k)I_k\left(y\left(t_k^-\right)\right) \\ \quad - \sum_{0<t_k<t} T(t-t_k)\Delta g(t_k, y_{t_k}), & t \in J. \end{cases}$$

For $\phi \in D$, we define the function:

$$\tilde{\phi}(t) = \begin{cases} \phi(t); & t \in (-\infty, 0], \\ T(t)\phi(0); & t \in J, \end{cases}$$

Then $\tilde{\phi} \in D_b$. Set

$$y(t) = x(t) + \tilde{\phi}(t).$$

It is clear to see that y satisfies (9.43) if and only if x satisfies $x_0 = 0$, and

$$x(t) = g(t, x_t + \tilde{\phi}_t) - T(t)g(0, \phi)$$

$$+ \int_0^t T(t-s)f(s, x_s + \tilde{\phi}_s)ds + \sum_{0<t_k<t} T(t-t_k)I_k\left(x\left(t_k^-\right) + \tilde{\phi}\left(t_k^-\right)\right)$$

$$- \sum_{0<t_k<t} T(t-t_k)\Delta g(t_k, x_{t_k} + \tilde{\phi}_{t_k}), \quad t \in J.$$

Let

$$D_b^0 = \{x \in D_b : x_0 = 0 \in D\}.$$

For any $x \in D_b^0$, we have

$$\|x\|_b = \|x_0\|_D + \sup\{|x(s)| : 0 \le s \le b\} = \sup\{|x(s)| : 0 \le s \le b\}.$$

Thus $(D_b^0, \|\cdot\|_b)$ is a Banach space. Define the two operators $\mathcal{A}, \mathcal{B} : D_b^0 \to D_b^0$ by:

$$\mathcal{B}(x)(t) = \int_0^t T(t-s)f(s, x_s + \tilde{\phi}_s)ds, \ t \in J$$

and

$$\mathcal{A}(x)(t) = g(t, x_t + \tilde{\phi}_t) - T(t)g(0, \phi) + \sum_{0 < t_k < t} T(t - t_k)I_k\left(x\left(t_k^-\right) + \tilde{\phi}\left(t_k^-\right)\right)$$

$$- \sum_{0 < t_k < t} T(t - t_k)\Delta g(t_k, x_{t_k} + \tilde{\phi}_{t_k}), \ t \in J.$$

Obviously the operator N has a fixed point is equivalent to $\mathcal{A} + \mathcal{B}$ has one, so it turns to prove that $\mathcal{A} + \mathcal{B}$ has a fixed point. We shall show that the operators $\mathcal{A}$ and $\mathcal{B}$ satisfy all the conditions of Theorem 1.32. The proof will be given in several steps.

Step 1: $\mathcal{B}$ is continuous. Let $\{x_n\}$ be a sequence such that $x_n \to x$ in D_b^0. Then

$$|\mathcal{B}(x_n)(t) - \mathcal{B}(x)(t)| = \left|\int_0^t T(t-s)[f(s, x_{n_s} + \tilde{\phi}_s) - f(s, x_s + \tilde{\phi}_s)]ds\right|$$

$$\le M \int_0^b \left|f(s, x_{n_s} + \tilde{\phi}_s) - f(s, x_s + \tilde{\phi}_s)\right| ds.$$

Since $f(s, \cdot)$ is continuous for a.e. $s \in J$, we have by the Lebesgue dominated convergence theorem

$$|\mathcal{B}(x_n)(t) - \mathcal{B}(x)(t)| \to 0 \ as \ n \to \infty.$$

Thus $\mathcal{B}$ is continuous.

Step 2: $\mathcal{B}$ maps bounded sets into bounded sets in D_b^0. It is enough to show that for any $q > 0$ there exists a positive constant l such that for each $x \in B_q = \{x \in D_b^0 : \|x\| \le q\}$ we have $\|\mathcal{B}(x)\| \le l$. Let $x \in B_q$, then

$$\|x_s + \tilde{\phi}_s\|_{\mathcal{D}} \le \|x_s\|_{\mathcal{D}} + \|\tilde{\phi}_s\|_{\mathcal{D}}$$
$$\le K_b q + K_b M |\phi(0)| + M_b \|\phi\|_{\mathcal{D}}$$
$$= q_*.$$

Then we have for each $t \in J$

$$|\mathcal{B}(x)(t)| = |\int_0^t T(t-s)f(s, x_s + \tilde{\phi}_s)ds|$$

$$\le M \int_0^t p(s)\psi(\|x_s + \tilde{\phi}_s\|)ds$$

$$\le M \int_0^t p(s)\psi(q_*)ds$$

$$\le M\psi(q_*) \int_0^t p(s)ds.$$

Taking the supremum over t we obtain

$$\|\mathcal{B}(x)\|_b \le M\psi(q_*)\|p\|_{L^1} := l.$$

Step 3: $\mathcal{B}$ maps bounded sets into equi-continuous sets in $\mathcal{D}_b^0$.

We consider B_q as in step 2 and let $\tau_1, \tau_2 \in J \setminus \{t_1, \ldots, t_m\}$, $\tau_1 < \tau_2$. Thus if $\epsilon > 0$ and $\epsilon \le \tau_1 < \tau_2$ we have We consider B_q as in step 2 and let $\tau_1, \tau_2 \in J \setminus \{t_1, \ldots, t_m\}$, $\tau_1 < \tau_2$. Thus if $\epsilon > 0$ and $\epsilon \le \tau_1 < \tau_2$ we have

$$|\mathcal{B}(x)(\tau_2) - \mathcal{B}(x)(\tau_1)| \le \psi(q_*) \int_0^{\tau_1 - \epsilon} \|T(\tau_2 - s) - T(\tau_1 - s)\|_{B(E)} p(s)ds$$

$$+\psi(q_*) \int_{\tau_1 - \epsilon}^{\tau_1} \|T(\tau_2 - s) - T(\tau_1 - s)\|_{B(E)} p(s)ds$$

$$+\psi(q_*) \int_{\tau_1}^{\tau_2} \|T(\tau_2 - s)\|_{B(E)} p(s)ds.$$

As $\tau_1 \to \tau_2$ and ϵ become sufficiently small, the right-hand side of the above inequality tends to zero, since $T(t)$ is a strongly continuous operator and the compactness of $T(t)$ for $t > 0$ implies the continuity in the uniform operator topology. This proves the equi-continuity for the case where $t \ne t_i, k = 1, 2, \ldots, m+1$. It remains to examine the equi-continuity at $t = t_i$.

First we prove equi-continuity at $t = t_i^-$. Fix $\delta_1 > 0$ such that $\{t_k : k \ne i\} \cap [t_i - \delta_1, t_i + \delta_1] = \emptyset$. For $0 < h < \delta_1$ we have

$$|\mathcal{B}(x)(t_i - h) - \mathcal{B}(x)(t_i)| \leq \int_0^{t_i-h} |(T(t_i - h - s) - T(t_i - s))f(s, x_s + \tilde{\phi}_s)| ds$$

$$+ \psi(q_*)M \int_{t_i-h}^{t_i} p(s) ds;$$

which tends to zero as $h \to 0$. Define

$$\hat{\mathcal{B}}_0(x)(t) = \mathcal{B}(x)(t), \ t \in [0, t_1]$$

and

$$\hat{\mathcal{B}}_i(x)(t) = \begin{cases} \mathcal{B}(x)(t), & \text{if } t \in (t_i, t_{i+1}] \\ \mathcal{B}(x)(t_i^+), & \text{if } t = t_i. \end{cases}$$

Next we prove equi-continuity at $t = t_i^+$. Fix $\delta_2 > 0$ such that $\{t_k : k \neq i\} \cap [t_i - \delta_2, t_i + \delta_2] = \emptyset$. For $0 < h < \delta_2$ we have

$$|\hat{\mathcal{B}}(x)(t_i + h) - \hat{\mathcal{B}}(x)(t_i)| \leq \int_0^{t_i} |(T(t_i + h - s) - T(t_i - s))f(s, x_s + \tilde{\phi}_s)| ds$$

$$+ \psi(q_*)M \int_{t_i}^{t_i+h} p(s) ds.$$

The right-hand side tends to zero as $h \to 0$. The equi-continuity for the cases $\tau_1 < \tau_2 \leq 0$ and $\tau_1 \leq 0 \leq \tau_2$ follows from the uniform continuity of ϕ on the interval $[-r, 0]$.

As a consequence of steps 1–3 together with Arzelá–Ascoli theorem it suffices to show that $\mathcal{B}$ maps B into a precompact set in E.

Let $0 < t < b$ be fixed and let ϵ be a real number satisfying $0 < \epsilon < t$. For $y \in B_q$ we define

$$\mathcal{B}_\epsilon(x)(t) = T(\epsilon) \int_0^{t-\epsilon} T(t - s - \epsilon)f(s, x_s + \tilde{\phi}_s) ds.$$

Since $T(t)$ is a compact operator, the set

$$X_\epsilon(t) = \{\mathcal{B}_\epsilon(x)(t) : \ x \in B_q\}$$

is precompact in E for every ϵ, $0 < \epsilon < t$. Moreover, for every $y \in B_q$ we have

$$|\mathcal{B}(x)(t) - \mathcal{B}_\epsilon(x)(t)| \leq \psi(q_*) \int_{t-\epsilon}^t \|T(t - s)\|_{B(E)} p(s) ds$$

$$\leq \psi(q_*)M \int_{t-\epsilon}^t p(s) ds.$$

Therefore, there are precompact sets arbitrarily close to the set $X_\epsilon(t) = \{\mathcal{B}_\epsilon(x)(t) : x \in B_q\}$. Hence the set $X(t) = \{\mathcal{B}(x)(t) : x \in B_q\}$ is precompact in E. Hence the operator $\mathcal{B}$ is completely continuous.

Step 4: $\mathcal{A}$ is a contraction

Let $x_1, x_2 \in D_b^0$. Then for $t \in J$

$$|\mathcal{A}(x_1)(t) - \mathcal{A}(x_2)(t)| \leq |g(t, x_{1t} + \tilde{\phi}_t) - g(t, x_{2t} + \tilde{\phi}_t)|$$

$$+ M \sum_{0 < t_k < t} |I_k(x_1(t_k^-)) - I_k(x_2(t_k^-))|$$

$$+ M \sum_{0 < t_k < t} |\Delta g(t_k, x_{1t_k} + \tilde{\phi}_{t_k}) - \Delta g(t_k, x_{2t_k} + \tilde{\phi}_{t_k})|$$

$$\leq l_g \|x_{1t} - x_{2t}\| + M \sum_{k=1}^{m} d_k |x_1(t_k^-) - x_2(t_k^-)|$$

$$+ 2M l_g \sum_{k=1}^{m} |x_{1t_k} - x_{2t_k}|$$

$$\leq \left(M \sum_{k=1}^{m} d_k + l_g K_b + 2m M l_g K_b \right) \|x_1 - x_2\|.$$

Then

$$\|\mathcal{A}(x_1) - \mathcal{A}(x_2)\| \leq (M \sum_{k=1}^{m} d_k + l_g K_b + 2m M l_g K_b) \|x_1 - x_2\|,$$

which is a contraction, since

$$K_b l_g + M \sum_{k=1}^{m} d_k + 2m M l_g K_b \leq K_b \max(l_g, \alpha_1) + M \sum_{k=1}^{m} d_k + 2m M l_g K_b < 1.$$

Step 5: A priori bounds.

Now it remains to show that the set $\mathcal{E}$ is bounded.
Let $x \in \mathcal{E}$, then $x = \lambda \mathcal{B}(x) + \lambda \mathcal{A}\left(\dfrac{x}{\lambda}\right)$ for some $0 < \lambda < 1$. Thus, for each $t \in J$,

$$x(t) = \lambda \int_0^t T(t-s)f(s, x_s + \tilde{\phi}_s)ds - \lambda T(t)g(0, \phi) + \lambda g(t, \frac{x_t}{\lambda} + \tilde{\phi}_t)$$

$$+ \lambda \sum_{0 < t_k < t} T(t - t_k) I_k\left(\frac{x}{\lambda}(t_k^-) + \tilde{\phi}(t_k)\right)$$

$$- \lambda \sum_{0 < t_k < t} T(t - t_k) \Delta g(t_k, \frac{x_{t_k}}{\lambda} + \tilde{\phi}_{t_k}).$$

This implies by (9.30.2) and (9.30.4) that, for each $t \in J$, we have

$$|x(t)| \leq \lambda M \int_0^t p(s)\psi(\|x_s + \tilde{\phi}_s\|)ds$$

$$+ \lambda |g(t, \frac{x_t}{\lambda} + \tilde{\phi}_t)| + \lambda M \sum_{k=1}^m \left| I_k\left(\frac{x}{\lambda}(t_k^-) + \tilde{\phi}(t_k^-)\right)\right|$$

$$+ \lambda M \sum_{k=1}^m |\Delta g(t_k, \frac{x_{t_k}}{\lambda} + \tilde{\phi}_{t_k})| + \lambda M g(t_k, \frac{x_{t_k}}{\lambda} + \tilde{\phi}_{t_k})| + \lambda M |g(0, \phi)|$$

$$\leq \lambda M \int_0^t p(s)\psi(K_b|x_s| + (MK_b + M_b)\|\phi\|)ds + \lambda \alpha_1\|\frac{x_t}{\lambda} + \tilde{\phi}_t\| + \lambda \alpha_2$$

$$+ \lambda M \sum_{k=1}^m |I_k\left(\frac{x}{\lambda}(t_k^-) + \tilde{\phi}(t_k^-)\right) - I_k(0)| + \lambda M \sum_{k=1}^m |I_k(0)|$$

$$+ 2\lambda M \sum_{k=1}^m l_g|\frac{x_{t_k}}{\lambda} + \tilde{\phi}_{t_k}| + \lambda M \sum_{k=1}^m |\Delta g(t_k, 0)| + \lambda M \alpha_2$$

$$\leq \lambda M \int_0^t p(s)\psi(K_b|x(s)| + (MK_b + M_b)\|\phi\|)ds$$

$$+ \lambda \alpha_1\left(K_b\left|\frac{x(t)}{\lambda}\right| + (MK_b + M_b)\|\phi\|\right)$$

$$+ \lambda \alpha_2 + \lambda M \sum_{k=1}^m d_k\left(\left|\frac{x(t)}{\lambda}\right| + \|\phi\|\right) + \lambda M \sum_{k=1}^m |I_k(0)|$$

$$+ 2\lambda M \sum_{k=1}^m l_g\left(K_b\left|\frac{x(t)}{\lambda}\right| + (MK_b + M_b)\|\phi\|\right)$$

$$+ \lambda M \sum_{k=1}^m |\Delta g(t_k, 0)| + \lambda M \alpha_2$$

$$\leq M \int_0^t p(s)\psi(K_b|x(s)| + (MK_b + M_b)\|\phi\|)ds$$

$$+ \alpha_1\left(K_b|x(t)| + (MK_b + M_b)\|\phi\|\right)$$

$$+\alpha_2 + M \sum_{k=1}^{m} d_k |x(t)| + M \sum_{k=1}^{m} d_k \|\phi\| + M \sum_{k=1}^{m} |I_k(0)| + M\alpha_2$$

$$+2Mml_gK_b|x(t)| + 2Mml_g(MK_b + M_b)\|\phi\| + M \sum_{k=1}^{m} |\Delta g(t_k, 0)|.$$

Then

$$\left(1 - \alpha_1 K_b - M \sum_{k=1}^{m} d_k - 2Mml_gK_b\right) |x(t)|$$

$$\leq M \int_0^t p(s)\psi(K_b|x(s)| + (MK_b + M_b)\|\phi\|)ds$$

$$+\alpha_1 (MK_b + M_b) \|\phi\| + (1 + M)\alpha_2$$

$$+M \sum_{k=1}^{m} d_k \|\phi\| + M \sum_{k=1}^{m} |I_k(0)|$$

$$+2Mml_g (MK_b + M_B) \|\phi\| + M \sum_{k=1}^{m} |\Delta g(t_k, 0)|.$$

Thus

$$K_b|x(t)| + (MK_b + M_b) \|\phi\| \leq C_1^* + C_2^* \int_0^t p(s)\psi(K_b|x(s)| + (MK_b + M_b)\|\phi\|)ds$$

We consider the function μ defined by

$$\mu(t) = \sup\{K_b|x(s)| + (MK_b + M_b)\|\phi\| : 0 \leq s \leq t\}, \ 0 \leq t \leq b.$$

Let $t^* \in [0, t]$ be such that $\mu(t) = K_b|x(t^*)| + (MK_b + M_b)\|\phi\|$, by the previous inequality we have for $t \in [0, b]$

$$\mu(t) \leq C_1^* + C_2^* \int_0^t p(s)\psi(\mu(t))ds. \tag{9.46}$$

Let us take the right-hand side of (9.46) as $v(t)$. Then we have

$$v(0) = C_1^*, \qquad \mu(t) \leq v(t) \ \text{for all} \ t \in J,$$

and

$$v'(t) = C_2^* p(t) \psi(\mu(t)), \quad a.e. \ t \in J.$$

Using the nondecreasing character of ψ we get

$$v'(t) \leq p(t) \psi(v(t)), \quad a.e. \ t \in J.$$

Thus

$$\int_{C_1^*}^{v(t)} \frac{ds}{\psi(s)} \leq \int_0^b p(s) ds = \|p\|_{L^1} < \int_{C_1^*}^{\infty} \frac{ds}{\psi(s)}.$$

Consequently, by assumption (9.30.5), there exists a constant N such that $v(t) \leq N, \ t \in J$ and hence there exists a constant Λ such that $\|z\|_b \leq \Lambda$. This shows that the set $\mathcal{E}$ is bounded. As a consequence of Theorem 1.32, we deduce that $F + G$ has a fixed point z^*. Then $z^*(t) = x^*(t) + \tilde{\phi}(t), \ t \in (-\infty, b]$ is a fixed point of the operator N, which gives rise to a mild solution of the problem (9.40)–(9.42).   □

9.4.3 Existence of Integral Solutions

In the previous section we considered the same problem, when the operator was non-densely defined. However, as indicated in [100], we sometimes need to deal with non-densely defined operators. For example, when we look at a one-dimensional heat equation with Dirichlet conditions on $[0, 1]$ and consider $A = \dfrac{\partial^2}{\partial x^2}$ in $C([0, 1], \mathbb{R})$ in order to measure the solutions in the sup-norm, then the domain,

$$D(A) = \{\phi \in C^2([0, 1], \mathbb{R}) : \phi(0) = \phi(1) = 0\},$$

is not dense in $C([0, 1], \mathbb{R})$ with the sup-norm. Before starting and proving this one, we give the definition of its integral solution.

Definition 9.28. We say that $y : (-\infty, b] \to E$ is an integral solution of (9.40)–(9.42) if

(i) $\displaystyle\int_0^t [y(s) - g(s, y_s)] ds \in D(A)$ for $t \in J$,

(ii) $y(t) = \phi(t), \ t \in (-\infty, 0]$.

(iii) $y(t) = \phi(0) - g(0, \phi) + g(t, y_t) + A \displaystyle\int_0^t [y(s) - g(s, y_s)] ds + \int_0^t f(s, y_s) ds +$
$\displaystyle\sum_{0 < t_k < t} I_k(y(t_k^-)) - \sum_{0 < t_k < t} \Delta g(t_k, y_{t_k}), \quad t \in J.$

From the definition it follows that $y(t) - g(t, y_t) \in \overline{D(A)}, \forall\, t \geq 0$, in particular $\phi(0) - g(0, \phi) \in \overline{D(A)}$. Moreover, y satisfies the following variation of constants formula:

$$
\begin{aligned}
y(t) = {} & g(t, y_t) + S'(t)(\phi(0) - g(0, \phi)) + \frac{d}{dt} \int_0^t S(t - s) f(s, y_s) ds \\
& + \sum_{0 < t_k < t} S'(t - t_k) I_k(y(t_k^-)) - \sum_{0 < t_k < t} S'(t - t_k) \Delta g(t_k, y_{t_k}), \quad t \geq 0.
\end{aligned}
\tag{9.47}
$$

Let $B_\lambda = \lambda R(\lambda, A) := \lambda(\lambda I - A)^{-1}$. Then [146] for all $x \in \overline{D(A)}, B_\lambda x \to x$ as $\lambda \to \infty$. Also from the Hille–Yosida condition (with $n = 1$) it easy to see that $\lim\limits_{\lambda \to \infty} |B_\lambda x| \leq M|x|$, since

$$
|B_\lambda| = |\lambda(\lambda I - A)^{-1}| \leq \frac{M\lambda}{\lambda - \omega}.
$$

Thus $\lim\limits_{\lambda \to \infty} |B_\lambda| \leq M$. Also if y is given by (9.47), then

$$
\begin{aligned}
y(t) = {} & g(t, y_t) + S'(t)(\phi(0) - g(0, \phi)) + \lim_{\lambda \to \infty} \int_0^t S'(t - s) B_\lambda f(s, y_s) ds \\
& + \sum_{0 < t_k < t} S'(t - t_k) I_k(y(t_k^-)) - \sum_{0 < t_k < t} S'(t - t_k) \Delta g(t_k, y_{t_k}), \quad t \in J.
\end{aligned}
\tag{9.48}
$$

The key tool in our approach is the following form of the fixed point theorem of Dhage [102].

Let D_{b_*} the set of all functions that belong in D_b and have values in $\overline{D(A)}$. Let us introduce the following hypotheses:

(C1) A satisfies Hille–Yosida condition;
(C2) The operator $S'(t)$ is compact in $\overline{D(A)}$ whenever $t > 0$;
(C3) There exists a function $p \in L^1(J, \mathbb{R}_+)$ and a continuous nondecreasing function $\psi : [0, \infty) \to [0, \infty)$ such that

$$
|f(t, u)| \leq p(t) \psi(\|u\|_{\mathcal{D}}), \quad \text{for a.e. } t \in J, \text{ and each } u \in \mathcal{D}.
$$

with

$$
\int_{C_1^*}^{\infty} \frac{du}{\psi(u)} > C_2^* \int_0^b e^{-\omega t} p(t) dt
$$

where

$$
C_1^* = K_b \frac{C_1}{C} + (MK_b + M_b)\|\phi\|, \qquad C_2^* = \frac{Me^{\omega b} K_b}{C}
\tag{9.49}
$$

with

$$C = 1 - \alpha_1 K_b - Me^{\omega b} \sum_{k=1}^{m} e^{-\omega t_k} d_k - 2Me^{\omega b} l_g K_b \sum_{k=1}^{m} e^{-\omega t_k} \qquad (9.50)$$

and

$$C_1 = \alpha_1 (MK_b + M_b)\|\phi\| + \alpha_2 + Me^{\omega b} \sum_{k=1}^{m} e^{-\omega t_k} d_k \|\phi\|$$

$$+ Me^{\omega b} \sum_{k=1}^{m} e^{-\omega t_k} |I_k(0)| + Me^{\omega b} \alpha_2$$

$$+ 2Me^{\omega b} l_g (MK_b + M_b) \sum_{k=1}^{m} e^{-\omega t_k} \|\phi\| + Me^{\omega b} \sum_{k=1}^{m} e^{-\omega t_k} |\Delta g(t_k, 0)|.$$

Theorem 9.29. *Assume that (9.30.2)–(9.30.4) and (C1)–(C3) hold. If*

$$K_b \max (l_g, \alpha_1) + Me^{\omega b} \sum_{k=1}^{m} e^{-\omega t_k} d_k + 2Ml_g K_b e^{\omega b} \sum_{k=1}^{m} e^{-\omega t_k} < 1, \qquad (9.51)$$

then the problem (9.40)–(9.42) has at least one integral solution on $(-\infty, b]$.

Proof. Transform the problem (9.40)–(9.42) into a fixed point problem. Consider the operator $N : D_{b*} \to D_{b*}$ defined by:

$$(Ny)(t) = \begin{cases} \phi(t), & t \in (-\infty, 0], \\[2mm] S'(t)[\phi(0) - g(0, \phi)] + g(t, y_t) \\[2mm] \quad + \dfrac{d}{dt} \displaystyle\int_0^t S(t-s)f(s, y_s)ds + \sum_{0 < t_k < t} S'(t - t_k) I_k \left(y \left(t_k^- \right) \right) \\[4mm] \quad - \displaystyle\sum_{0 < t_k < t} S'(t - t_k) \Delta g(t_k, y_{t_k}), & t \in J. \end{cases}$$

For $\phi \in \mathcal{D}$, we define the function:

$$\tilde{\phi}(t) = \begin{cases} \phi(t), & t \in (-\infty, 0], \\[2mm] S'(t)\phi(0), & t \in J, \end{cases}$$

Then $\tilde{\phi} \in D_b$. Set

$$y(t) = x(t) + \tilde{\phi}(t).$$

It is clear that x satisfies $x_0 = 0$ and

$$
\begin{aligned}
x(t) = {} & g(t, x_t + \tilde{\phi}_t) - S'(t)g(0, \phi) \\
& + \frac{d}{dt} \int_0^t S(t - s)f(s, x_s + \tilde{\phi}_s)ds + \sum_{0 < t_k < t} S'(t - t_k)I_k \left(x\left(t_k^-\right) + \tilde{\phi}\left(t_k^-\right)\right) \\
& - \sum_{0 < t_k < t} S'(t - t_k)\Delta g(t_k, x_{t_k} + \tilde{\phi}_{t_k}), \quad t \in J.
\end{aligned}
$$

Let $D_{b_*}^0 = \{x \in D_{b_*} : x_0 = 0\}$. For any $x \in D_{b_*}^0$ we have

$$
\|x\|_b = \|x_0\|_{\mathcal{D}} + \sup\{|z(s)| : 0 \le s \le b\} = \sup\{|x(s)| : 0 \le s \le b\}.
$$

Thus $(D_{b_*}^0, \|\cdot\|_b)$ is a Banach space.

Define the two operators $\mathcal{A}, \mathcal{B} : D_{b_*}^0 \to D_{b_*}^0$ by:

$$
\mathcal{B}(x)(t) = \frac{d}{dt} \int_0^t S(t - s)f(s, x_s + \tilde{\phi}_s)ds, \quad t \in J
$$

and

$$
\begin{aligned}
\mathcal{A}(z)(t) = {} & g(t, x_t + \tilde{\phi}_t) - S'(t)g(0, \phi) + \sum_{0 < t_k < t} S'(t - t_k)I_k \left(x\left(t_k^-\right) + \tilde{\phi}\left(t_k^-\right)\right) \\
& - \sum_{0 < t_k < t} S'(t - t_k)\Delta g(t_k, x_{t_k} + \tilde{\phi}_{t_k}), \quad t \in J.
\end{aligned}
$$

Obviously the operator N has a fixed point is equivalent to $\mathcal{A} + \mathcal{B}$ has one, so it turns to prove that $\mathcal{A} + \mathcal{B}$ has a fixed point. We shall show that the operators $\mathcal{A}$ and $\mathcal{B}$ satisfy all the conditions of Theorem 1.32. The proof will be given in several steps.

Step 1: $\mathcal{B}$ is continuous.

Let $\{x_n\}$ be a sequence such that $x_n \to x$ in $D_{b_*}^0$. Then

$$
\begin{aligned}
|\mathcal{B}(x_n)(t) - \mathcal{B}(x)(t)| &= \left| \frac{d}{dt} \int_0^t S(t - s)[f(s, x_{n_s} + \tilde{\phi}_s) - f(s, x_s + \tilde{\phi}_s)]ds \right| \\
&\le Me^{\omega b} \int_0^b e^{\omega s} \left| f(s, x_{n_s} + \tilde{\phi}_s) - f(s, x_s + \tilde{\phi}_s) \right| ds.
\end{aligned}
$$

Since $f(s, \cdot)$ is continuous for a.e. $s \in J$, we have by the Lebesgue dominated convergence theorem

$$
|\mathcal{B}(x_n)(t) - \mathcal{B}(x)(t)| \to 0 \text{ as } n \to \infty.
$$

Thus $\mathcal{B}$ is continuous.

Step 2: $\mathcal{B}$ maps bounded sets into bounded sets in D_{b*}^0.

It is enough to show that for any $q > 0$ there exists a positive constant l such that for each $x \in B_q = \{y \in D_{b*}^0 : \|x\| \leq q\}$ we have $\|\mathcal{B}(x)\| \leq l$. Let $x \in B_q$, then

$$
\begin{aligned}
\|x_s + \tilde{\phi}_s\|_B &\leq \|x_s\|_D + \|\tilde{\phi}_s\|_D \\
&\leq K_b q + K_b M |\phi(0)| + M_b \|\phi\|_D \\
&= q_*.
\end{aligned}
$$

Then we have for each $t \in J$

$$
\begin{aligned}
|\mathcal{B}(x)(t)| &= |\frac{d}{dt} \int_0^t S(t-s) f(s, x_s + \tilde{\phi}_s) ds| \\
&\leq M e^{\omega b} \int_0^b e^{\omega s} p(s) \psi(\|x_s + \tilde{\phi}_s\|) ds \\
&\leq M e^{\omega b} \psi(q_*) \int_0^b e^{\omega s} p(s) ds := l.
\end{aligned}
$$

Step 3: $\mathcal{B}$ maps bounded sets into equi-continuous sets in D_{b*}^0.

We consider B_q as in step 2 and let $\tau_1, \tau_2 \in J \setminus \{t_1, \ldots, t_m\}$, $\tau_1 < \tau_2$. Thus if $\epsilon > 0$ and $\epsilon \leq \tau_1 < \tau_2$ we have   We consider B_q as in step 2 and let $\tau_1, \tau_2 \in J \setminus \{t_1, \ldots, t_m\}$, $\tau_1 < \tau_2$. Thus if $\epsilon > 0$ and $\epsilon \leq \tau_1 < \tau_2$ we have

$$
\begin{aligned}
|\mathcal{B}(x)(\tau_2) - \mathcal{B}(x)(\tau_1)| &\leq \left| \lim_{\lambda \to \infty} \int_0^{\tau_1 - \epsilon} [S'(\tau_2 - s) - S'(\tau_1 - s)] B_\lambda f(s, x_s + \tilde{\phi}_s) ds \right| \\
&+ \left| \lim_{\lambda \to \infty} \int_{\tau_1 - \epsilon}^{\tau_1} [S'(\tau_2 - s) - S'(\tau_1 - s)] B_\lambda f(s, x_s + \tilde{\phi}_s) ds \right| \\
&+ \left| \lim_{\lambda \to \infty} \int_{\tau_1}^{\tau_2} S'(\tau_2 - s) B_\lambda f(s, x_s + \tilde{\phi}_s) ds \right| \\
&\leq \psi(q_*) \int_0^{\tau_1 - \epsilon} \|S'(\tau_2 - s) - S'(\tau_1 - s)\|_{B(E)} p(s) ds \\
&+ \psi(q_*) \int_{\tau_1 - \epsilon}^{\tau_1} \|S'(\tau_2 - s) - S'(\tau_1 - s)\|_{B(E)} p(s) ds \\
&+ \psi(q_*) \int_{\tau_1}^{\tau_2} \|S'(\tau_2 - s)\|_{B(E)} p(s) ds.
\end{aligned}
$$

As $\tau_1 \to \tau_2$ and ϵ become sufficiently small, the right-hand side of the above inequality tends to zero, since $S'(t)$ is a strongly continuous operator and the compactness of $S'(t)$ for $t > 0$ implies the continuity in the uniform operator topology. This proves the equi-continuity for the case where $t \neq t_i, k = 1, 2, \ldots, m + 1$. It remains to examine the equi-continuity at $t = t_i$.

First we prove equi-continuity at $t = t_i^-$. Fix $\delta_1 > 0$ such that $\{t_k : k \neq i\} \cap [t_i - \delta_1, t_i + \delta_1] = \emptyset$. For $0 < h < \delta_1$ we have

$$|\mathcal{B}(x)(t_i - h) - \mathcal{B}(x)(t_i)| \leq \lim_{\lambda \to \infty} \int_0^{t_i - h} |\left(S'(t_i - h - s) - S'(t_i - s)\right) B_\lambda f(s, x_s + \tilde{\phi}_s)| ds$$

$$+ M e^{\omega b} \psi(q_*) \int_{t_i - h}^{t_i} e^{\omega s} p(s) ds;$$

which tends to zero as $h \to 0$. Define

$$\hat{\mathcal{B}}_0(x)(t) = \mathcal{B}(x)(t), \, t \in [0, t_1]$$

and

$$\hat{\mathcal{B}}_i(x)(t) = \begin{cases} \mathcal{B}(x)(t), & \text{if } t \in (t_i, t_{i+1}] \\ \mathcal{B}(x)(t_i^+), & \text{if } t = t_i. \end{cases}$$

Next we prove equi-continuity at $t = t_i^+$. Fix $\delta_2 > 0$ such that $\{t_k : k \neq i\} \cap [t_i - \delta_2, t_i + \delta_2] = \emptyset$. For $0 < h < \delta_2$ we have

$$|\hat{\mathcal{B}}(x)(t_i + h) - \hat{\mathcal{B}}(x)(t_i)| \leq \lim_{\lambda \to \infty} \int_0^{t_i} \|\left(S'(t_i + h - s) - S'(t_i - s)\right) B_\lambda f(s, x_s + \tilde{\phi}_s)\| ds$$

$$+ M e^{\omega b} \psi(q) \int_{t_i}^{t_i + h} e^{-\omega s} p(s) ds.$$

The right-hand side tends to zero as $h \to 0$. The equi-continuity for the cases $\tau_1 < \tau_2 \leq 0$ and $\tau_1 \leq 0 \leq \tau_2$ follows from the uniform continuity of ϕ on the interval $[-r, 0]$. As a consequence of steps 1–3 together with Arzelá–Ascoli theorem it suffices to show that $\mathcal{B}$ maps B into a precompact set in E.

Let $0 < t < b$ be fixed and let ϵ be a real number satisfying $0 < \epsilon < t$. For $y \in B_q$ we define

$$\mathcal{B}_\epsilon(x)(t) = S'(\epsilon) \lim_{\lambda \to \infty} \int_0^{t - \epsilon} S'(t - s - \epsilon) B_\lambda f(s, x_s + \tilde{\phi}_s) ds.$$

Since $S'(t)$ is a compact operator, the set

$$X_\epsilon(t) = \{\mathcal{B}_\epsilon(x)(t) : \, x \in B_q\}$$

is precompact in E for every ϵ, $0 < \epsilon < t$. Moreover, for every $y \in B_q$ we have

$$|\mathcal{B}(x)(t) - \mathcal{B}_\epsilon(x)(t)| \leq Me^{\omega b}\psi(q_*)\int_{t-\epsilon}^t e^{-\omega s}p(s)ds.$$

Therefore, there are precompact sets arbitrarily close to the set $X_\epsilon(t) = \{\mathcal{B}_\epsilon(x)(t) : x \in B_q\}$. Hence the set $X(t) = \{\mathcal{B}(x)(t) : x \in B_q\}$ is precompact in E. Hence the operator $\mathcal{B}$ is completely continuous.

Step 4: $\mathcal{A}$ is a contraction. Let $x_1, x_2 \in D_{b*}^0$. Then for $t \in J$

$$|\mathcal{A}(x_1)(t) - \mathcal{A}(x_2)(t)| \leq |g(t, x_{1t} + \tilde{\phi}_t) - g(t, x_{2t} + \tilde{\phi}_t)|$$

$$+ Me^{\omega b}\sum_{0<t_k<t} e^{-\omega t_k}\left|I_k\left(x_1\left(t_k^-\right)\right) - I_k\left(x_2\left(t_k^-\right)\right)\right|$$

$$+ Me^{\omega b}\sum_{0<t_k<t} e^{-\omega t_k}|\Delta g(t_k, x_{1t_k} + \tilde{\phi}_{t_k})$$

$$- \Delta g(t_k, x_{2t_k} + \tilde{\phi}_{t_k})|$$

$$\leq l_g\|x_{1t} - x_{2t}\| + Me^{\omega b}\sum_{k=1}^m e^{-\omega t_k}d_k\left|x_1\left(t_k^-\right) - x_2\left(t_k^-\right)\right|$$

$$+ 2Me^{\omega b}l_g\sum_{k=1}^m e^{-\omega t_k}|x_{1t_k} - x_{2t_k}|$$

$$\leq \left(Me^{\omega b}\sum_{k=1}^m e^{-\omega t_k}(d_k + 2l_gK_b) + l_gK_b\right)\|x_1 - x_2\|_B.$$

Then

$$\|\mathcal{A}(x_1) - \mathcal{A}(x_2)\| \leq \left(M\sum_{k=1}^m d_k + l_gK_b + 2mMl_gK_b\right)\|x_1 - x_2\|,$$

which is a contraction, since

$$K_bl_g + M\sum_{k=1}^m d_k + 2mMl_gK_b \leq K_b\max(l_g, \alpha_1) + M\sum_{k=1}^m d_k + 2mMl_gK_b < 1.$$

Step 5: A priori bounds.

Now it remains to show that the set

$$\mathcal{E} = \left\{x \in D_b^0 : x = \lambda\mathcal{B}(x) + \lambda\mathcal{A}\left(\frac{x}{\lambda}\right) \text{ for some } 0 < \lambda < 1\right\}$$

is bounded.

Let $x \in \mathcal{E}$, then $x = \lambda \mathcal{B}(x) + \lambda \mathcal{A}\left(\dfrac{x}{\lambda}\right)$ for some $0 < \lambda < 1$. Thus, for each $t \in J$,

$$x(t) = \lambda \int_0^t T(t-s)f(s, x_s + \tilde{\phi}_s)ds - \lambda T(t)g(0, \phi) + \lambda g(t, \frac{x_t}{\lambda} + \tilde{\phi}_t)$$

$$+ \lambda \sum_{0 < t_k < t} T(t - t_k) I_k \left(\frac{x}{\lambda}(t_k^-) + \tilde{\phi}(t_k)\right)$$

$$- \lambda \sum_{0 < t_k < t} T(t - t_k) \Delta g(t_k, \frac{x_{t_k}}{\lambda} + \tilde{\phi}_{t_k}).$$

This implies by (9.30.2) and (9.30.4) that, for each $t \in J$, we have

$$|x(t)| \le \lambda M \int_0^t p(s) \psi(\|x_s + \tilde{\phi}_s\|)ds$$

$$+ \lambda |g(t, \frac{x_t}{\lambda} + \tilde{\phi}_t)| + \lambda M \sum_{k=1}^m \left| I_k \left(\frac{x}{\lambda}(t_k^-) + \tilde{\phi}(t_k^-)\right)\right|$$

$$+ \lambda M \sum_{k=1}^m |\Delta g(t_k, \frac{x_{t_k}}{\lambda} + \tilde{\phi}_{t_k})| + \lambda M g(t_k, \frac{x_{t_k}}{\lambda} + \tilde{\phi}_{t_k})| + \lambda M |g(0, \phi)|$$

$$\le \lambda M \int_0^t p(s) \psi(K_b |x_s| + (M K_b + M_b)\|\phi\|)ds$$

$$+ \lambda \alpha_1 \|\frac{x_t}{\lambda} + \tilde{\phi}_t\| + \lambda \alpha_2$$

$$+ \lambda M \sum_{k=1}^m |I_k \left(\frac{x}{\lambda}(t_k^-) + \tilde{\phi}(t_k^-)\right) - I_k(0)| + \lambda M \sum_{k=1}^m |I_k(0)|$$

$$+ 2\lambda M \sum_{k=1}^m l_g |\frac{x_{t_k}}{\lambda} + \tilde{\phi}_{t_k}| + \lambda M \sum_{k=1}^m |\Delta g(t_k, 0)| + \lambda M \alpha_2$$

$$\le \lambda M \int_0^t p(s) \psi(K_b |x(s)| + (M K_b + M_b)\|\phi\|)ds$$

$$+ \lambda \alpha_1 \left(K_b \left|\frac{x(t)}{\lambda}\right| + (M K_b + M_b)\|\phi\|\right)$$

$$+ \lambda \alpha_2 + \lambda M \sum_{k=1}^m d_k \left(\left|\frac{x(t)}{\lambda}\right| + \|\phi\|\right) + \lambda M \sum_{k=1}^m |I_k(0)|$$

$$+2\lambda M \sum_{k=1}^{m} l_g \left(K_b \left| \frac{x(t)}{\lambda} \right| + (MK_b + M_b)\|\phi\| \right)$$

$$+\lambda M \sum_{k=1}^{m} |\Delta g(t_k, 0)| + \lambda M \alpha_2$$

$$\leq M \int_0^t p(s)\psi(K_b|x(s)| + (MK_b + M_b)\|\phi\|)ds$$

$$+\alpha_1 (K_b |x(t)| + (MK_b + M_b)\|\phi\|)$$

$$+\alpha_2 + M \sum_{k=1}^{m} d_k |x(t)| + M \sum_{k=1}^{m} d_k \|\phi\| + M \sum_{k=1}^{m} |I_k(0)| + M\alpha_2$$

$$+2Mml_g K_b|x(t)| + 2Mml_g (MK_b + M_b)\|\phi\| + M \sum_{k=1}^{m} |\Delta g(t_k, 0)|$$

Then

$$\left(1 - \alpha_1 K_b - M \sum_{k=1}^{m} d_k - 2Mml_g K_b \right) |x(t)|$$

$$\leq M \int_0^t p(s)\psi(K_b|x(s)| + (MK_b + M_b)\|\phi\|)ds$$

$$+\alpha_1 (MK_b + M_b) \|\phi\| + (1 + M)\alpha_2$$

$$+M \sum_{k=1}^{m} d_k\|\phi\| + M \sum_{k=1}^{m} |I_k(0)|$$

$$+2Mml_g (MK_b + M_B) \|\phi\| + M \sum_{k=1}^{m} |\Delta g(t_k, 0)|.$$

Thus by (9.49) we have

$$K_b|x(t)| + (MK_b + M_b) \|\phi\| \leq C_1^*$$

$$+C_2^* \int_0^t p(s)\psi(K_b|x(s)| + (MK_b + M_b)\|\phi\|)ds$$

We consider the function μ defined by

$$\mu(t) = \sup\{K_b|x(s)| + (MK_b + M_b)\|\phi\| : 0 \leq s \leq t\}, \quad 0 \leq t \leq b.$$

Let $t^* \in [0, t]$ be such that $\mu(t) = K_b|x(t^*)| + (MK_b + M_b)\|\phi\|$, by the previous inequality we have for $t \in [0, b]$

$$\mu(t) \leq C_1^* + C_2^* \int_0^t p(s)\psi(\mu(t))ds \qquad (9.52)$$

Let us take the right-hand side as $v(t)$. Then we have

$$v(0) = C_1^*, \qquad \mu(t) \leq v(t) \quad \text{for all } t \in J,$$

and

$$v'(t) = C_2^* p(t)\psi(\mu(t)), \quad a.e. \ t \in J.$$

Using the nondecreasing character of ψ we get

$$v'(t) \leq p(t)\psi(v(t)), \quad a.e. \ t \in J,$$

Thus

$$\int_{C_1^*}^{v(t)} \frac{ds}{\psi(s)} \leq \int_0^b p(s)ds = \|p\|_{L^1} < \int_{C_1^*}^{\infty} \frac{ds}{\psi(s)}.$$

Consequently, by assumption (9.30.5), there exists a constant N such that $v(t) \leq N$, $t \in J$ and hence there exists a constant Λ such that $\|z\|_b \leq \Lambda$. This shows that the set $\mathcal{E}$ is bounded. As a consequence of Theorem 1.32, we deduce that $F + G$ has a fixed point z^*. Then $z^*(t) = x^*(t) + \tilde{\phi}(t)$, $t \in (-\infty, b]$ is a fixed point of the operator N, which gives rise to a mild solution of the problem (9.40)–(9.42). □

9.4.4 An Example

In this section we apply some of the results established in this section. We begin by mentioning an example of phase space.

The Phase Space

Let $h(.) : (-\infty, -r] \to \mathbb{R}$ be a positive Lebesgue integrable function and $\mathcal{D} := PC_r \times L^2(h, E), r \geq 0$, be the space formed of all classes of functions $\varphi : (-\infty, 0] \to E$ such that $\varphi|_H \in PC(H, E)$, $\varphi(.)$ is Lebesgue-measurable on $(-\infty, -r]$ and $h|\varphi|^p$ is Lebesgue integrable on $(-\infty, -r]$. The semi-norm in $\|.\|_\mathcal{D}$ is defined by

$$\|\varphi\|_{\mathcal{D}} = \sup_{\theta \in H} \|\varphi(\theta)\| + \left(\int_{-\infty}^{-r} h(\theta)\|\varphi(\theta)\|^p d\theta \right)^{1/p} \tag{9.53}$$

Assume that $h(.)$ satisfies conditions (g-6) and (g-7) in the terminology of [142]. Proceeding as in the proof of ([142], Theorem 1.3.8) it follows that $\mathcal{D}$ is a phase space which verifies the axioms (A1)–(A2) and (A3). Moreover, when $r = 0$ this space coincides with $C^0 \times L^2(h, E)$ and the parameters $H = 1; M(t) = \gamma(-t)^{1/2}$ and $K(t) = 1 + \left(\int_{-t}^0 h(\xi)d\xi \right)^{1/2}$, for $t \geq 0$ (see [142]). Let $E = L^2([0, \pi])$ and let A be the operator given by $Af = f''$ with domain

$$D(A) := \{ f \in L^2([0, \pi]) : f'' \in L^2([0, \pi]), f(0) = f(\pi) = 0 \} \tag{9.54}$$

It is well known that A is the infinitesimal generator of a C_0-semigroup on E, which will be denoted by $(T(t))_{t \geq 0}$. Moreover, A has discrete spectrum, the eigenvalues are $-n^2, n \in \mathbb{N}$, with corresponding normalized eigenvectors $z_n(\xi) := (\frac{2}{\pi})^{1/2}\sin(n\xi)$ and the following properties hold:

(a) $\{z_n : n \in \mathbb{N}\}$ is an orthonormal basis of E.
(b) For $f \in E, T(t)f = \sum_{n=1}^{\infty} e^{-n^2 t}\langle f, z_n \rangle$ and $Af = \sum_{n=1}^{\infty} -n^2\langle f, z_n \rangle z_n$ when $f \in D(A)$.

A First Order Neutral Equation

We study the first order neutral differential equation with unbounded delay

$$\frac{d}{dt}\left[u(t, \xi) + \int_{-\infty}^t \int_0^\pi b(t - s, \eta, \xi)u(s, \eta)d\eta ds \right]$$

$$= \frac{\partial^2}{\partial \xi^2}\left[u(t, \xi) + \int_{-\infty}^t \int_0^\pi b(t - s, \eta, \xi)u(s, \eta)d\eta ds \right]$$

$$+ \int_{-\infty}^t F(t, t - s, \xi, u(s, \eta))ds, \quad t \in [0, a], \xi \in [0, \pi] \tag{9.55}$$

$$u(t, 0) = u(t, \pi) = 0, \quad t \in [0, a], \tag{9.56}$$

$$u(\tau, \xi) = \varphi(\tau, \xi), \quad \tau \leq 0, 0 \leq \xi \leq \pi, \tag{9.57}$$

$$\Delta u(t_i)(\xi) = \int_{-\infty}^{t_i} a_i(t_i - s)u(s, \xi)ds, \tag{9.58}$$

where $\varphi \in C_0 \times L^2(h, E), 0 < t_1 < \cdots < t_n < a$ and

(a) The function $b(s, \eta, \xi)$, $\frac{\partial b(s, \eta, \xi)}{\partial \xi}$ are measurable, $b(s, \eta, \pi) = b(s, \eta, 0) = 0$ and

$$l_g := \max \left\{ \left(\int_0^\pi \int_{-\infty}^0 \int_0^\pi \frac{1}{h(s)} \left(\frac{\partial^i b(s, \eta, \xi)}{\partial \xi^i} \right) d\eta ds d\xi \right)^{1/2} : i = 0, 1 \right\} < \infty$$
(9.59)

(b) The function $F : \mathbb{R}^4 \rightarrow \mathbb{R}$ is continuous and there are continuous functions $v : \mathbb{R}^3 \rightarrow \mathbb{R}$ and $\mu : \mathbb{R}^2 \rightarrow \mathbb{R}$ such that

$$|F(t, s, \xi, x)| \leq v(t, s, \xi) + \mu(t, s)|x|, \quad (t, s, \xi, x) \in \mathbb{R}^3;$$
(9.60)

(c) The functions $a_k \in C([0, \infty), \mathbb{R})$ and $d_k := (\int_{-\infty}^0 \frac{(a_k)^2}{h(s)} ds)^{1/2} < \infty$ for all $i = 1, \ldots, m$.

Assuming that conditions $(a) - -(c)$ are verified, our problem can be modeled as the abstract impulsive problem (9.40)–(9.42) by defining

$$g(t, \psi)(\xi) := \int_{-\infty}^0 \int_0^\pi b(s, \eta, \xi) \psi(s, v) dv ds,$$
(9.61)

$$f(t, \psi)(\xi) := \int_{-\infty}^0 F(t, s, \xi, \psi(s, \xi)) ds,$$
(9.62)

$$I_k(\psi)(\xi) := \int_{-\infty}^0 a_k(s) \psi(s, \xi) ds.$$
(9.63)

Moreover, $f(t, .), I_k, i = 1, \ldots, m$, are bounded linear operators, and $\|g(t, \psi)\| \leq \alpha_2 + \alpha_1 \|\psi\|_B$, where

$$\alpha_1 := \left(\int_0^\pi \int_{-\infty}^0 \int_0^\pi \frac{1}{h(s)} |b(s, \eta, \xi)|^2 d\eta ds d\xi \right)^{1/2}, \quad \alpha_2 := 0.$$

Hence, the problem has a mild solution in $(-\infty, b]$.

9.5 Non-densely Defined Impulsive Semi-linear Functional Differential Equations with State-Dependent Delay

9.5.1 Introduction

In this section, we shall be concerned with existence of integral solutions defined on a compact real interval for first order impulsive semi-linear functional equations with state-dependent delay in a separable Banach space of the form:

$$y'(t) = Ay(t) + f(t, y_{\rho(t, y_t)}), \quad t \in I = [0, b], \tag{9.64}$$

$$y(t) = \phi \quad t \in (-\infty, 0], \tag{9.65}$$

$$\Delta \qquad y(t_i) = I_k(y_{t_k}), \quad k = 1, 2, \ldots, m, \tag{9.66}$$

where $f : J \times \mathcal{D} \to E$ is a given function, $\mathcal{D} = \{\psi : (-\infty, 0] \to E, \psi$ is continuous everywhere except for a finite number of points s at which $\psi(s^-)$, $\psi(s^+)$ exist and $\psi(s^-) = \psi(s)\}$, $\phi \in D$, $(0 < r < \infty)$, $0 = t_0 < t_1 < \cdots < t_m < t_{m+1} = b$, $I_k : \mathcal{D} \to E$ $(k = 1, 2, \ldots, m)$, $\rho : I \times \mathcal{D} \to (-\infty, b]$, $A : D(A) \subset E \to E$ is a non-densely defined closed linear operator on E, and E a real separable Banach space with norm $|.|$.

9.5.2 Existence of Integral Solutions

Definition 9.30. We say that $y : (-\infty, T] \to E$ is an integral solution of (9.64)–(9.66) if

(i) $y(t) = \phi(0) + A \int_0^t y(s)ds + \int_0^t f(s, y_{\rho(s, y_s)})ds + \sum_{0 < t_k < t} I_k(y_{t_k}), \quad t \in J.$

(ii) $\int_0^t y(s)ds \in D(A)$ for $t \in J$, and $y(t) = \phi(t), \ t \in (-\infty, 0].$

From the definition it follows that $y(t) \in \overline{D(A)}$, for each $t \geq 0$, in particular $\phi(0) \in \overline{D(A)}$. Moreover, y satisfies the following variation of constants formula:

$$y(t) = S'(t)\phi(0) + \frac{d}{dt}\int_0^t S(t-s)f(s, y_{\rho(s, y_s)})ds + \sum_{0 < t_k < t} S'(t-t_k)I_k(y_{t_k}) \quad t \geq 0. \tag{9.67}$$

We notice also that if y satisfies (9.67), then

$$y(t) = S'(t)\phi(0) + \lim_{\lambda \to \infty}\int_0^t S'(t-s)B_\lambda f(s, y_{\rho(s, y_s)})ds$$

$$+ \sum_{0 < t_k < t} S'(t-t_k)I_k(y_{t_k}), \quad t \geq 0.$$

Our main result in this section is based upon the fixed point theorem due to Burton and Kirk [88]. We always assume that $\rho : I \times \mathcal{D} \to (-\infty, b]$ is continuous. Additionally, we introduce the following hypotheses:

(Hφ) The function $t \to \varphi_t$ is continuous from $\mathcal{R}(\rho^-) = \{\rho(s, \varphi) : (s, \varphi) \in J \times \mathcal{D}, \rho(s, \varphi) \leq 0\}$ into $\mathcal{D}$ and there exists a continuous and bounded function $L^\phi : \mathcal{R}(\rho^-) \to (0, \infty)$ such that $\|\phi_t\|_\mathcal{D} \leq L^\phi(t)\|\phi\|_\mathcal{D}$ for every $t \in \mathcal{R}(\rho^-)$.

(9.34.1) A satisfies Hille–Yosida condition;

(9.34.2) There exist constants $d_k > 0$, $k = 1, \ldots, m$ such that for each y, $x \in \mathcal{D}$

$$\|I_k(y) - I_k(x)\| \leq d_k \|y - x\|_{\mathcal{D}}$$

(9.34.3) The function $f : J \times \mathcal{D} \to E$ is Carathéodory;

(9.34.4) The operator $S'(t)$ is compact in $\overline{D(A)}$ wherever $t > 0$;

(9.34.5) There exist a function $p \in L^1(J, \mathbb{R}_+)$ and a continuous nondecreasing function $\psi : [0, \infty) \to (0, \infty)$ such that

$$|f(t, x)| \leq p(t)\psi(\|x\|_{\mathcal{D}}), \quad a.e. \ t \in J, \quad \text{for all } x \in \mathcal{D}$$

with $\int_0^b e^{-\omega s} p(s)\,ds < \infty$,

$$\int_{c_1}^{\infty} \frac{du}{\psi(u)} > c_2 \int_0^b e^{-\omega s} p(s)\,ds. \tag{9.68}$$

where

$$c_1 = \frac{ce^{\omega b} K_b}{1 - Me^{\omega b} K_b \sum\limits_{k=1}^{m} d_k} + \left(M_b + L^\phi + MK_b\right) \|\phi\|, \tag{9.69}$$

and

$$c = \sum_{k=1}^{m} \left[|I_k(0)| + d_k \left(M_b + L^\phi + MK_b\right) \|\phi\|_{\mathcal{D}}\right]. \tag{9.70}$$

$$c_2 = \frac{MK_b e^{\omega b}}{1 - Me^{\omega b} K_b \sum\limits_{k=1}^{m} d_k}. \tag{9.71}$$

The next result is a consequence of the phase space axioms.

Lemma 9.31 ([139], Lemma 2.1). *If $y : (-\infty, b] \to E$ is a function such that $y_0 = \phi$ and $y|_J \in PC(J : D(A))$, then*

$$\|y_s\|_{\mathcal{D}} \leq (M_a + L^\phi)\|\phi\|_{\mathcal{D}} + K_a \sup\{\|y(\theta)\|; \ \theta \in [0, \max\{0, s\}]\}, \quad s \in \mathcal{R}(\rho^-) \cup J,$$

where $L^\phi = \sup_{t \in \mathcal{R}(\rho^-)} L^\phi(t)$, $M_a = \sup_{t \in J} M(t)$ and $K_a = \sup_{t \in J} K(t)$.

Theorem 9.32. *Assume that (Hφ) and (9.34.1)–(9.34.5) hold. If*

$$Me^{\omega b}K_b \sum_{k=1}^{m} d_k < 1, \tag{9.72}$$

then the problem (9.64)–(9.66) has at least one integral solution on $(-\infty, b]$.

Proof. Consider the operator $N : PC\left((-\infty, b], \overline{D(A)}\right) \to PC\left((-\infty, b], \overline{D(A)}\right)$ defined by:

$$(Ny)(t) = \begin{cases} \phi(t), & t \in (-\infty, 0], \\[2mm] S'(t)\phi(0) + \dfrac{d}{dt}\displaystyle\int_0^t S(t-s)f\left(s, y_{\rho(s,y_s)}\right) ds \\[2mm] \quad + \displaystyle\sum_{0<t_k<t} S'(t-t_k)I_k(y_{t_k}), & t \in J. \end{cases}$$

Let $\tilde{\phi}(.) : (-\infty, b] \to E$ be the function defined by

$$\tilde{\phi}(t) = \begin{cases} \phi(t), & t \in (-\infty, 0], \\[2mm] S'(t)\phi(0), & t \in J. \end{cases}$$

Then $\tilde{\phi}_0 = \phi$. For each $x \in B_b$ with $x(0) = 0$, we denote by $\bar{x}$ the function defined by

$$\bar{x}(t) = \begin{cases} 0, & t \in (-\infty, 0], \\[2mm] x(t), & t \in J, \end{cases}$$

We can decompose it as $y(t) = \tilde{\phi}(t) + x(t)$, $0 \le t \le b$, which implies $y_t = x_t + \tilde{\phi}_t$, for every $0 \le t \le b$ and the function $x(.)$ satisfies

$$x(t) = \frac{d}{dt}\int_0^t S(t-s)f\left(s, x_{\rho(s,x_s+\tilde{\phi}_s)} + \tilde{\phi}_{\rho(s,x_s+\tilde{\phi}_s)}\right) ds$$

$$+ \sum_{0<t_k<t} S'(t-t_k)I_k\left(x_{t_k} + \tilde{\phi}_{t_k}\right) \quad t \in J.$$

Let

$$B_b^0 = \{x \in B_b : x_0 = 0 \in \mathcal{D}\}.$$

For any $x \in B_b^0$ we have

$$\|x\|_b = \|x_0\|_{\mathcal{D}} + \sup\{|x(s)| : 0 \le s \le b\} = \sup\{|x(s)| : 0 \le s \le b\}.$$

Thus $(\mathcal{B}_b^0, \|\cdot\|_b)$ is a Banach space. We define the two operators $\mathcal{A}, \mathcal{B} : \mathcal{B}_b^0 \to \mathcal{B}_b^0$ by:

$$\mathcal{B}(x)(t) = \frac{d}{dt} \int_0^t S(t-s) f\left(s, x_{\rho(s,x_s+\tilde{\phi}_s)} + \tilde{\phi}_{\rho(s,x_s+\tilde{\phi}_s)}\right) ds, \quad t \in J$$

and

$$\mathcal{A}(x)(t) = \sum_{0<t_k<t} S'(t-t_k) I_k \left(x_{t_k} + \tilde{\phi}_{t_k}\right), \quad t \in J.$$

Obviously the operator N has a fixed point is equivalent to $\mathcal{A} + \mathcal{B}$ has one, so it turns to prove that $\mathcal{A} + \mathcal{B}$ has a fixed point. We shall show that the operators $\mathcal{A}$ and $\mathcal{B}$ satisfies all the conditions of Theorem 1.32. For better readability, we break the proof into a sequence of steps.

Step 1: $\mathcal{B}$ is continuous.
Let $\{x_n\}$ be a sequence such that $x_n \to x$ in $\mathcal{B}_b^0$. Then for $\omega > 0$ (if $\omega < 0$ one has $e^{\omega t} < 1$).
At first, we study the convergence of the sequences $(x_{\rho(s,x_s^n)}^n)_{n\in\mathbb{N}}$, $s \in J$. If $s \in J$ is such that $\rho(s,x_s) > 0$ for every $n > N$. In the case, for $n > N$ we see that

$$\|x_{\rho(s,x_s^n)}^n - x_{\rho(s,x_s)}\|_{\mathcal{D}} \leq \|x_{\rho(s,x_s^n)}^n - x_{\rho(s,x_s^n)}\|_{\mathcal{D}} + \|x_{\rho(s,x_s^n)} - x_{\rho(s,x_s)}\|_{\mathcal{D}}$$

$$\leq K_b \|x_n - x\|_{\mathcal{D}} + \|x_{\rho(s,x_s^n)} - x_{\rho(s,x_s)}\|_{\mathcal{D}}.$$

Which prove that $x_{\rho(s,x_s^n)}^n \to x_{\rho(s,x_s)}$ in $\mathcal{D}$ as $n \to \infty$ for every $s \in J$ such that $\rho(s,x_s) > 0$. Similarly, if $\rho(s,x_s) < 0$ and $n \in \mathbb{N}$ is such that $\rho(s,x_s^n) < 0$ for every $n > N$, we get

$$\|x_{\rho(s,x_s^n)}^n - x_{\rho(s,x_s)}\|_{\mathcal{D}} = \|\phi_{\rho(s,x_s^n)} - \phi_{\rho(s,x_s)}\|_{\mathcal{D}} = 0$$

Which also shows that $x_{\rho(s,x_s^n)}^n \to x_{\rho(s,x_s)}$ in $\mathcal{D}$ as $n \to \infty$ for every $s \in J$ such that $\rho(s,x_s) < 0$. Combining the previous arguments, we can prove that $x_{\rho(s,x_s^n)}^n \to \phi$ for every $s \in J$ such that $\rho(s,x_s) = 0$. Finalely,

$$|\mathcal{B}(x_n)(t) - \mathcal{B}(x)(t)| = \left| \frac{d}{dt} \int_0^t S(t-s) \left[f\left(s, x_{\rho(s,x_s^n+\tilde{\phi}_s)}^n + \tilde{\phi}_{\rho(s,x_s^n+\tilde{\phi}_s)}\right) \right. \right.$$

$$\left. \left. -f\left(s, x_{\rho(s,x_s+\tilde{\phi}_s)} + \tilde{\phi}_{\rho(s,x_s+\tilde{\phi}_s)}\right) \right] ds \right|$$

$$\leq M e^{\omega b} \int_0^t e^{-\omega s} \left| f\left(s, x_{\rho(s,x_s^n+\tilde{\phi}_s)}^n + \tilde{\phi}_{\rho(s,x_s^n+\tilde{\phi}_s)}\right) \right.$$

$$\left. -f\left(s, x_{\rho(s,x_s+\tilde{\phi}_s)} + \tilde{\phi}_{\rho(s,x_s+\tilde{\phi}_s)}\right) \right| ds$$

$$\leq M e^{\omega b} \int_0^t e^{-\omega s} \left| f\left(s, x^n_{\rho(s,x_s^n+\tilde{\phi}_s)} + \tilde{\phi}_{\rho(s,x_s^n+\tilde{\phi}_s)} \right) \right.$$

$$\left. - f\left(s, x_{\rho(s,x_s^n+\tilde{\phi}_s)} + \tilde{\phi}_{\rho(s,x_s^n+\tilde{\phi}_s)} \right) \right| ds$$

$$+ M e^{\omega b} \int_0^t e^{-\omega s} \left| f\left(s, x_{\rho(s,x_s^n+\tilde{\phi}_s)} + \tilde{\phi}_{\rho(s,x_s^n+\tilde{\phi}_s)} \right) \right.$$

$$\left. - f\left(s, x_{\rho(s,x_s+\tilde{\phi}_s)} + \tilde{\phi}_{\rho(s,x_s+\tilde{\phi}_s)} \right) \right| ds$$

We infer that $f(s, x^n_{\rho(s,x_s^n)}) \to f(s, x_{\rho(s,x_s)})$ as $n \to \infty$, for every $s \in J$. Now, a standard application of the Lebesgue dominated convergence theorem proves that

$$\|\mathcal{B}(x_n)(t) - \mathcal{B}(x)(t)\|_b \to 0 \text{ as } n \to \infty.$$

Thus $\mathcal{B}$ is continuous.

Step 2: $\mathcal{B}$ maps bounded sets into bounded sets in $\mathcal{B}_b^0$.

It is enough to show that for any $q > 0$ there exists a positive constant l such that for each $x \in B_q = \{x \in \mathcal{B}_b^0 : \|x\|_b \leq q\}$ we have $\|\mathcal{B}(y)\|_b \leq l$. So choose $x \in B_q$, then

$$\|x_{\rho(t,x_t+\tilde{\phi}_t)} + \tilde{\phi}_{\rho(t,x_t+\tilde{\phi}_t)}\|_{\mathcal{D}} \leq K_b q + (M_b + L^\phi)\|\phi\|_{\mathcal{D}} + K_b M |\phi(0)| = q_*$$

Then we have for each $t \in J$

$$|\mathcal{B}(x)(t)| = \left| \frac{d}{dt} \int_0^t S(t-s) f\left(s, x_{\rho(s,x_s+\tilde{\phi}_s)} + \tilde{\phi}_{\rho(s,x_s+\tilde{\phi}_s)} \right) ds \right|$$

$$\leq M e^{\omega b} \int_0^b e^{-\omega s} p(s) \psi\left(\|x_{\rho(t,x_t+\tilde{\phi}_t)} + \tilde{\phi}_{\rho(t,x_t+\tilde{\phi}_t)}\|_{\mathcal{D}} \right).$$

Then we have

$$\|\mathcal{B}(x)(t)\|_b \leq M e^{\omega b} \psi(q_*) \int_0^b e^{-\omega s} p(s) ds := l.$$

Step 3: $\mathcal{B}$ maps bounded sets into equi-continuous sets of $\mathcal{B}_b^0$.

We consider B_q as in Step 2 and let $\tau_1, \tau_2 \in J \setminus \{t_1, \ldots, t_m\}$, $\tau_1 < \tau_2$. Thus if $\epsilon > 0$ and $\epsilon \leq \tau_1 < \tau_2$ we have

$$|\mathcal{B}(x)(\tau_2) - \mathcal{B}(x)(\tau_1)|$$

$$\leq \left| \lim_{\lambda \to \infty} \int_0^{\tau_1 - \epsilon} [S'(\tau_2 - s) - S'(\tau_1 - s)] B_\lambda f\left(s, x_{\rho(s,x_s+\tilde{\phi}_s)} + \tilde{\phi}_{\rho(s,x_s+\tilde{\phi}_s)} \right) ds \right|$$

$$+ \left| \lim_{\lambda \to \infty} \int_{\tau_1-\epsilon}^{\tau_1} [S'(\tau_2 - s) - S'(\tau_1 - s)] B_\lambda f\left(s, x_{\rho(s,x_s+\tilde{\phi}_s)} + \tilde{\phi}_{\rho(s,x_s+\tilde{\phi}_s)}\right) ds \right|$$

$$+ \left| \lim_{\lambda \to \infty} \int_{\tau_1}^{\tau_2} S'(\tau_2 - s) B_\lambda f\left(s, x_{\rho(s,x_s+\tilde{\phi}_s)} + \tilde{\phi}_{\rho(s,x_s+\tilde{\phi}_s)}\right) ds \right|$$

$$\leq \psi(q_*) \int_0^{\tau_1-\epsilon} |S'(\tau_2 - s) - S'(\tau_1 - s)| p(s) ds$$

$$+ \psi(q_*) \int_{\tau_1-\epsilon}^{\tau_1} |S'(\tau_2 - s) - S'(\tau_1 - s)| p(s) ds$$

$$+ M e^{\omega b} \psi(q_*) \int_{\tau_1}^{\tau_2} e^{-\omega s} p(s) ds.$$

As $\tau_1 \to \tau_2$ and ϵ become sufficiently small, the right-hand side of the above inequality tends to zero, since $S'(t)$ is a strongly continuous operator and the compactness of $S'(t)$ for $t > 0$ implies the continuity in the uniform operator topology. This proves the equi-continuity for the case where $t \neq t_i, k = 1, 2, \ldots, m + 1$. It remains to examine the equi-continuity at $t = t_i$.

First we prove equi-continuity at $t = t_i^-$. Fix $\delta_1 > 0$ such that

$$\{t_k : k \neq i\} \cap [t_i - \delta_1, t_i + \delta_1] = \emptyset.$$

For $0 < h < \delta_1$ we have

$$|\mathcal{B}(x)(t_i - h) - \mathcal{B}(x)(t_i)|$$

$$\leq \lim_{\lambda \to \infty} \int_0^{t_i-h} \| \left(S'(t_i - h - s) - S'(t_i - s)\right)$$

$$B_\lambda f\left(s, x_{\rho(s,x_s+\tilde{\phi}_s)} + \tilde{\phi}_{\rho(s,x_s+\tilde{\phi}_s)}\right) \| ds$$

$$+ M e^{\omega b} \psi(q_*) \int_{t_i-h}^{t_i} e^{-\omega s} p(s) \, ds;$$

which tends to zero as $h \to 0$. Define

$$\hat{\mathcal{B}}_0(x)(t) = \mathcal{B}(x)(t), \, t \in [0, t_1]$$

and

$$\hat{\mathcal{B}}_i(x)(t) = \begin{cases} \mathcal{B}(x)(t), & \text{if } t \in (t_i, t_{i+1}] \\ \mathcal{B}(x)(t_i^+), & \text{if } t = t_i. \end{cases}$$

Next we prove equi-continuity at $t = t_i^+$. Fix $\delta_2 > 0$ such that $\{t_k : k \neq i\} \cap [t_i - \delta_2, t_i + \delta_2] = \emptyset$. For $0 < h < \delta_2$ we have

$$|\hat{B}(x)(t_i + h) - \hat{B}(x)(t_i)|$$

$$\leq \lim_{\lambda \to \infty} \int_0^{t_i} \| \left(S'(t_i + h - s) - S'(t_i - s)\right)$$

$$B_\lambda f \left(s, x_{\rho(s,x_s+\tilde{\phi}_s)} + \tilde{\phi}_{\rho(s,x_s+\tilde{\phi}_s)}\right) \|ds$$

$$+ Me^{\omega b} \psi(q_*) \int_{t_i}^{t_i+h} e^{-\omega s} p(s)\, ds;$$

The right-hand side tends to zero as $h \to 0$. The equi-continuity for the cases $\tau_1 < \tau_2 \leq 0$ and $\tau_1 \leq 0 \leq \tau_2$ follows from the uniform continuity of ϕ on the interval $[-r, 0]$. As a consequence of steps 1–3 together with Arzelá–Ascoli theorem it suffices to show that $\mathcal{B}$ maps B_q into a precompact set in E. Let $0 < t < b$ be fixed and let ϵ be a real number satisfying $0 < \epsilon < t$. For $x \in B_q$ we define

$$\mathcal{B}_\epsilon(x)(t) = S'(\epsilon) \lim_{\lambda \to \infty} \int_0^{t-\epsilon} S'(t - s - \epsilon) B_\lambda f(s, x_{\rho(s,x_s+\tilde{\phi}_s)} + \tilde{\phi}_{\rho(s,x_s+\tilde{\phi}_s)}) ds.$$

Note

$$\left\{ \lim_{\lambda \to \infty} \int_0^{t-\epsilon} S'(t - s - \epsilon) B_\lambda f(s, x_{\rho(s,x_s+\tilde{\phi}_s)} + \tilde{\phi}_{\rho(s,x_s+\tilde{\phi}_s)}) ds : y \in B_q \right\}$$

is a bounded set since

$$\left| \lim_{\lambda \to \infty} \int_0^{t-\epsilon} S'(t - s - \epsilon) B_\lambda f(s, x_{\rho(s,x_s+\tilde{\phi}_s)} + \tilde{\phi}_{\rho(s,x_s+\tilde{\phi}_s)}) ds \right|$$

$$\leq Me^{\omega b} \psi(q_*) \int_0^{t-\epsilon} e^{-\omega s} p(s) ds.$$

Since $S'(t)$ is a compact operator, the set

$$X_\epsilon(t) = \{\mathcal{B}_\epsilon(x)(t) : \ x \in B_q\}$$

is precompact in E for every ϵ, $0 < \epsilon < t$. Moreover, for every $y \in B_q$ we have

$$|\mathcal{B}(x)(t) - \mathcal{B}_\epsilon(x)(t)| \leq Me^{\omega b} \psi(q_*) \int_t^{t-\epsilon} e^{-\omega s} p(s) ds.$$

Therefore, there are precompact sets arbitrarily close to the set $X_\epsilon(t) = \{\mathcal{B}_\epsilon(x)(t) : x \in B_q\}$. Hence the set $X(t) = \{\mathcal{B}(x)(t) : x \in B_q\}$ is precompact in E. Hence the operator $\mathcal{B} : \mathcal{B}_b^0 \to \mathcal{B}_b^0$ is completely continuous.

Step 4: $\mathcal{A}$ is a contraction

Let $x_1, x_2 \in \mathcal{B}_b^0$. Then for $t \in J$

$$|\mathcal{A}(x_1)(t) - \mathcal{A}(x_2)(t)| = \left| \sum_{0 < t_k < t} S'(t - t_k) \left(I_k(x_{t_k}^1 + \tilde{\phi}_{t_k}) - I_k(x_{t_k}^2 + \tilde{\phi}_{t_k}) \right) \right|$$

$$\leq M e^{\omega b} \sum_{0 < t_k < t} \left| I_k(x_{t_k}^1) - I_k(x_{t_k}^2) \right|$$

$$\leq M e^{\omega b} \sum_{k=1}^{m} d_k \| x_{t_k}^1 - x_{t_k}^2 \|_{\mathcal{D}}$$

$$\leq M e^{\omega b} K_b \sum_{k=1}^{m} d_k \| x_1 - x_2 \|_{\mathcal{D}}.$$

Then

$$\| \mathcal{A}(x_1) - \mathcal{A}(x_2) \|_b \leq M e^{\omega b} K_b \sum_{k=1}^{m} d_k \| x_1 - x_2 \|_{\mathcal{D}},$$

which is a contraction from (9.72).

Step 5: A priori bounds.

Now it remains to show that the set

$$\mathcal{E} = \left\{ x \in \mathcal{B}_b^0 : x = \lambda \mathcal{B}(x) + \lambda \mathcal{A}\left(\frac{x}{\lambda}\right) \text{ for some } 0 < \lambda < 1 \right\}$$

is bounded.

Let $x \in \mathcal{E}$. Then $x = \lambda \mathcal{B}(x) + \lambda \mathcal{A}\left(\frac{x}{\lambda}\right)$ for some $0 < \lambda < 1$. Thus, for each $t \in J$,

$$x(t) = \lambda \frac{d}{dt} \int_0^t S(t - s) f(s, x_{\rho(s, x_s + \tilde{\phi}_s)} + \tilde{\phi}_{\rho(s, x_s + \tilde{\phi}_s)}) ds$$

$$+ \lambda \sum_{0 < t_k < t} S'(t - t_k) I_k\left(\frac{x_{t_k}}{\lambda} + \tilde{\phi}_{t_k}\right).$$

This implies by (9.34.2), (9.34.5) that, for each $t \in J$, we have

$$|x(t)| \leq \lambda M e^{\omega t} \int_0^t e^{-\omega s} p(s) \psi (\|x_{\rho(s,x_s + \tilde{\phi}_s)} + \tilde{\phi}_{\rho(s,x_s + \tilde{\phi}_s)}\|_{\mathcal{D}}) ds$$

$$+ \lambda M e^{\omega t} \sum_{k=1}^m \left| I_k \left(\frac{x_{t_k}}{\lambda} + \tilde{\phi}_{t_k} \right) \right|$$

$$\leq \lambda M e^{\omega t} \int_0^t e^{-\omega s} p(s) \psi \left(K_b |x(s)| + (M_b + L^\phi + MK_b) \|\phi\|_{\mathcal{D}} \right) ds$$

$$+ \lambda M e^{\omega t} \sum_{k=1}^m \left| I_k \left(\frac{x_{t_k}}{\lambda} + \tilde{\phi}_{t_k} \right) - I_k(0) \right|$$

$$+ \lambda M e^{\omega t} \sum_{k=1}^m |I_k(0)|$$

$$\leq \lambda M e^{\omega t} \int_0^t e^{-\omega s} p(s) \psi \left(K_b |x(s)| + (M_b + L^\phi + MK_b) \|\phi\|_{\mathcal{D}} \right) ds$$

$$+ \lambda M e^{\omega t} \sum_{k=1}^m |I_k(0)| + \lambda M e^{\omega t} \sum_{k=1}^m d_k \left(K_b |x(s)| + (M_b + L^\phi + MK_b) \|\phi\|_{\mathcal{D}} \right)$$

$$\leq c e^{\omega t} + M e^{\omega t} \left[\int_0^t e^{-\omega s} p(s) \psi \left(K_b |x(s)| + (M_b + L^\phi + MK_b) \|\phi\|_{\mathcal{D}} \right) ds \right.$$

$$\left. + K_b \sum_{k=1}^m d_k |x(t)| \right],$$

where

$$c = M \left(\sum_{k=1}^m \left[|I_k(0)| + d_k(M_b + L^\phi + MK_b) \|\phi\|_{\mathcal{D}} \right] \right). \tag{9.73}$$

Therefore,

$$\left(1 - M e^{\omega b} K_b \sum_{k=1}^m d_k \right) |x(t)| \leq c e^{\omega t} + M e^{\omega t} \int_0^t e^{-\omega s} p(s) \psi \left(K_b |x(s)| \right.$$

$$\left. + (M_b + L^\phi + MK_b) \|\phi\|_{\mathcal{D}} \right) ds.$$

Thus from (9.69) and (9.71) we have

$$(M_b + L^\phi + MK_b)\|\phi\|_{\mathcal{D}} + K_b|x(s)| \le c_1 + c_2 \int_0^t e^{-\omega s} p(s)\psi\,(K_b|x(t)|$$

$$+ (M_b + L^\phi + MK_b)\|\phi\|_{\mathcal{D}})\,ds.$$

We consider the function μ defined by

$$\mu(t) = \sup\{K_b|x(s)| + (M_b + L^\phi + MK_b)\|\phi\|_{\mathcal{D}} : 0 \le s \le t\},\ 0 \le t \le b.$$

$t^* \in [0, t]$ be such that $\mu(t) = K_b|x(t^*)| + (M_b + L^\phi + MK_b)\|\phi\|_{\mathcal{D}}$, by the previous inequality we have for $t \in [0, b]$

$$\mu(t) \le c_1 + c_2 \int_0^t e^{-\omega s} p(s)\psi(\mu(s))ds. \tag{9.74}$$

Let us take the right-hand side of (9.74) as $v(t)$. Then we have

$$\mu(t) \le v(t)\ \text{ for all } t \in J,$$
$$v(0) = c_1,$$

and

$$v'(t) = c_2 e^{-\omega t} p(t)\psi(\mu(t)),\ \text{ a.e. } t \in J.$$

Using the nondecreasing character of ψ we get

$$v'(t) \le c_2 e^{-\omega t} p(t)\psi(v(t)),\ \text{ a.e. } t \in J.$$

That is

$$\frac{v'(t)}{\psi(v(t))} \le c_2 e^{-\omega t} p(t),\ \text{ a.e. } t \in J.$$

Integrating from 0 to t we get

$$\int_0^t \frac{v'(s)}{\psi(v(s))} ds \le c_2 \int_0^t e^{-\omega s} p(s) ds.$$

By a change of variable and (9.68) we get

$$\int_{v(0)}^{v(t)} \frac{du}{\psi(u)} \le c_2 \int_0^b e^{-\omega s} p(s)\,ds < \int_{c_1}^\infty \frac{du}{\psi(u)}.$$

Hence there exists a constant N such that

$$\mu(t) \le v(t) \le N \quad \text{for all } t \in J.$$

Now from the definition of μ it follows that

$$\|x\|_b \le N^* \text{ for all } x \in \mathcal{E}.$$

This shows that the set $\mathcal{E}$ is bounded. As a consequence of Theorem 1.32 we deduce that $\mathcal{A} + \mathcal{B}$ has a fixed point which is a integral solution of problem (9.64)–(9.66).

$\square$

Phase Spaces

Let $g : (-\infty, 0] \to [1, \infty)$ be a continuous, nondecreasing function with $g(0) = 1$, which satisfies the conditions (g-1), (g-2) of [142]. This means that the function

$$G(t) = \sup_{-\infty < \theta \le -t} \frac{g(t + \theta)}{g(\theta)}$$

is locally bounded for $t \ge 0$ and that $\lim_{\theta \to -\infty} g(\theta) = \infty$.

We said that $\phi : [-\infty, 0] \to E$ is normalized piecewise continuous, if ϕ is left continuous and the restriction of ϕ to any interval H is piecewise continuous.

Next we modify slightly the definition of the spaces C_g, C_g^0 of [142]. We denote by $\mathcal{PC}_g(E)$ the space formed by the normalized piecewise continuous functions ϕ such that $\dfrac{\phi}{g}$ is bounded on $(-\infty, 0]$ and by $\mathcal{PC}_g^0$ the subspace of $\mathcal{PC}_g(E)$ formed by the functions ϕ such that

$$\lim_{\theta \to -\infty} \frac{\phi(\theta)}{g(\theta)} = 0.$$

It is easy to see that $\mathcal{D} = \mathcal{PC}_g(E)$ and $\mathcal{D} = \mathcal{PC}_g^0(E)$ endowed with the norm

$$\|\phi\|_{\mathcal{D}} = \sup_{\theta \in (-\infty, 0]} \frac{\|\phi(\theta)\|}{g(\theta)}$$

are phase spaces. Moreover, in these cases $K(s) = 1$ for $s \ge 0$.

Let $1 \le p < \infty$, $0 \le r < \infty$, and $g(\cdot)$ is a Borel nonnegative measurable function on $(-\infty, r)$ which satisfies the conditions (g-5)–(g-6) in the terminology of [142]. This means that $g(\cdot)$ is locally integrable on $(-\infty, -r)$ and there exists a nonnegative and locally bounded function G on $(-\infty, 0]$ such that $g(\xi + \theta) \le G(\xi) g(\theta)$ for all $\xi \le 0$ and $\theta \in (-\infty, -r) \backslash N_\xi$, where $N_\xi \subset (-\infty, -r)$ is a set with

Lebesgue measure 0. Let $\mathcal{D} := \mathcal{P}C_r \times L^p(g, E)$, $r \geq 0$, $p > 1$, be the space formed of all classes of functions $\phi : (-\infty, 0] \to E$ such that $\phi|_H \in PC(H, E)$, $\phi(\cdot)$ is Lebesgue measurable on $(-\infty, -r]$ and $g|\phi|^p$ is Lebesgue integrable on $(-\infty, -r]$. The seminorm in $\|\cdot\|_{\mathcal{D}}$ is defined by

$$\|\phi\|_{\mathcal{D}} := \sup_{\theta \in H} \|\phi(\theta)\| + \left(\int_{-\infty}^{-r} g(\theta) \|\phi(\theta)\|^p d\theta \right)^{\frac{1}{p}}.$$

Proceeding as in the proof of ([142], Theorem 1.3.8), it follows that $\mathcal{D}$ is a phase space which satisfies Axioms (A) and (B). Moreover, for $r = 0$ and $p = 2$ this space coincides (see [142]) with $C_0 \times L^2(g, E)$, $H = 1$, $M(t) = G(-t)^{\frac{1}{2}}$ and

$$K(t) = 1 + \left(\int_{-t}^{0} g(s) ds \right)^{\frac{1}{2}}, \text{ for } t \geq 0.$$

A First Order Partial Functional Differential Equations

To apply our abstract results, we consider the partial functional differential equations with state dependent delay of the form

$$\frac{\partial}{\partial t} v(t, \xi) = -\frac{\partial}{\partial \xi} v(t, \xi) + m(t) a(v(t - \sigma(v(t, 0)), \xi)), \, \xi \in [0, \pi], \, t \in [0, b],$$

$$\tag{9.75}$$

$$v(t, 0) = v(t, \pi) = 0, \, t \in [0, b], \tag{9.76}$$

$$v(\theta, \xi) = v_0(\theta, \xi), \, \xi \in [0, \pi], \, \theta \in (-\infty, 0], \tag{9.77}$$

$$\Delta v(t_i)(\xi) = \int_{-\infty}^{t_i} \gamma_i(t_i - s) v(s, \xi) ds \tag{9.78}$$

where $v_0 : (-\infty, 0] \times [0, \pi] \to \mathbb{R}$ is an appropriate function, $\gamma_i \in C[0, \infty), \mathbb{R})$, $0 < t_1 < t_2 < \cdots < t_n < b$. The functions $m : [0, b] \to \mathbb{R}$, $a : \mathbb{R} \times J \to \mathbb{R}$, $\sigma : \mathbb{R} \to \mathbb{R}^+$ are continuous and we assume the existence of positive constants b_1, b_2 such that $|b(t)| \leq b_1 |t| + b_2$ for every $t \in \mathbb{R}$.

Let A be the operator defined on $E = C([0, \pi], \mathbb{R})$ by

$$D(A) = \{g \in C^1([0, \pi], \mathbb{R}) : g(0) = 0\}; \, Ag = g'.$$

Then

$$\overline{D(A)} = C_0([0, \pi], \mathbb{R}) = \{g \in C([0, \pi], \mathbb{R}) : g(0) = 0\}.$$

It is well known from [100] that A is sectorial, $(0, +\infty) \subseteq \rho(A)$ and for $\lambda > 0$

$$\|R(\lambda, A)\|_{B(E)} \leq \frac{1}{\lambda}.$$

It follows that A generates an integrated semigroup $(S(t))_{t \geq 0}$ and that $\|S'(t)\|_{B(E)} \leq e^{-\mu t}$ for $t \geq 0$ for some constant $\mu > 0$ and A satisfied the Hille–Yosida condition.

Set $\gamma > 0$. For the phase space, we choose $\mathcal{D}$ to be defined by

$$\mathcal{D} = C_\gamma = \{\phi \in C((-\infty, 0], E) : \lim_{\theta \to -\infty} e^{\gamma \theta} \phi(\theta) \text{ exists in } E\}$$

with norm

$$\|\phi\|_\gamma = \sup_{\theta \in (-\infty, 0]} e^{\gamma \theta} |\phi(\theta)|, \ \phi \in C_\gamma.$$

By making the following change of variables

$$y(t)(\xi) = v(t, \xi), \ t \geq 0, \ \xi \in (0, \pi],$$
$$\phi(\theta)(\xi) = v_0(\theta, \xi), \ \theta \leq 0, \ \xi \in [0, 1],$$
$$F(t, \varphi)(\xi) = m(t)b(\varphi(0, \xi)), \ \xi \in [0, \pi], \ \phi \in C_\gamma$$
$$\rho(t, \varphi) = t - \sigma(\varphi(0, 0))$$
$$I_k(y_{t_k}) = \int_{-\infty}^0 \gamma_k(s)v(s, \xi)ds,$$

the problem (9.75)–(9.78) takes the abstract form (9.64)–(9.66). Moreover, a simple estimate shows that

$$\|f(t, \varphi)\| \leq m(t)[b_1 \|\varphi\|_\mathcal{D} + b_2 \pi^{1/2}] \text{ for all } (t, \varphi) \in I \times \mathcal{D}$$

with

$$\int_1^\infty \frac{ds}{\psi(s)} = \int_1^\infty \frac{ds}{b_1 s + b_2 \pi^{1/2}} = +\infty.$$

and

$$d_k = \left(\int_{-\infty}^0 \frac{(\gamma_k(s))^2}{e^{\theta s}} ds\right)^{1/2} < \infty$$

Theorem 9.33. *Let $\varphi \in \mathcal{D}$ be such that H_φ is valid and $t \to \varphi_t$ is continuous on $\mathcal{R}(\rho^-)$, then there exists a integral solution of (9.75)–(9.78) whenever*

$$\left(1 + \left(\int_{-\infty}^{0} e^{\theta s} ds\right)^{1/2}\right) \sum_{k=1}^{m} d_k < 1.$$

9.6   Notes and Remarks

The results of Chap. 9 are taken from Abada et al. [1, 3, 4]. Other results may be found in [124, 151, 152, 159].

Chapter 10
Impulsive Functional Differential Inclusions with Unbounded Delay

10.1 Introduction

In this chapter, we shall establish sufficient conditions for the existence of mild, extremal mild, integral, and extremal integral solutions for some impulsive semi-linear neutral functional differential inclusions in separable Banach spaces. We shall rely on a fixed point theorem for the sum of completely continuous and contraction operators.

10.2 Densely Defined Impulsive Functional Differential Inclusions

10.2.1 Introduction

We shall be concerned with existence of mild solutions, integral, and extremal integral solutions defined on a compact real interval for first order impulsive semi-linear neutral functional inclusions in a separable Banach space. We will consider the following first order impulsive semi-linear neutral functional differential inclusions of the form:

$$\frac{d}{dt}[y(t) - g(t, y_t)] - A[y(t) - g(t, y_t)] \in F(t, y_t),$$

$$a.e.\ t \in J = [0, b],\ t \neq t_k,\ k = 1, \ldots, m \tag{10.1}$$

© Springer International Publishing Switzerland 2015
S. Abbas, M. Benchohra, *Advanced Functional Evolution Equations and Inclusions*, Developments in Mathematics 39,
DOI 10.1007/978-3-319-17768-7_10

$$\Delta y|_{t=t_k} \in I_k(y(t_k^-)), \quad k = 1, \ldots, m \tag{10.2}$$

$$y(t) = \phi(t), \quad t \in (-\infty, 0], \tag{10.3}$$

where $F : J \times D \to 2^E$ is a closed, bounded, and convex valued multi-valued map, $g : J \times D \to E$ is a given function, $\phi \in D$ where D is the phase space that will be specified later $I_k \in C(E, E), (k = 1, 2, \ldots, m), A : D(A) \subset E \to E$ is a densely defined closed linear operator on E, and E a real separable Banach space with norm $|.|$. Consider the space

$$PC = \Big\{ y : (-\infty, b] \to E, \ y(t_k^-), y(t_k^+), \text{ exist with } y(t_k) = y(t_k^-),$$

$$y(t) = \phi(t), t \le 0, \ y_k \in C(J_k, E) \Big\},$$

where y_k is the restriction of y to $J_k = (t_k, t_{k+1}], \ k = 0, \ldots, m$. Let $\| \cdot \|_{PC}$ be the norm in PC defined by

$$\|y\|_{PC} = \sup\{|y(s)| : 0 \le s \le b\}, \ y \in PC.$$

We will assume that D satisfies the following axioms:

(A) If $y : (-\infty, b] \to E, b > 0$ and $y(t_k^-)$, $y(t_k^+)$, exist with $y(t_k) = y(t_k^-)$, $k = 1, \ldots, m$ and $y_0 \in D$, then for every t in $[0, b) \backslash \{t_1, \ldots, t_m\}$ the following conditions hold:

  (i) y_t is in D; and y_t is continuous on $[0, b] \backslash \{t_1, \ldots, t_m\}$
  (ii) $\|y_t\|_D \le K(t) \sup\{|y(s)| : 0 \le s \le t\} + M(t)\|y_0\|_D$,
  (iii) $|y(t)| \le H\|y_t\|_D$

  where $H \ge 0$ is a constant, $K : [0, \infty) \to [0, \infty)$ is continuous, $M : [0, \infty) \to [0, \infty)$ is locally bounded and H, K, M are independent of $y(\cdot)$.

(A-1) For the function $y(\cdot)$ in (A), y_t is a D-valued continuous function on $[0, b) \backslash \{t_1, \ldots, t_m\}$.

(A-2) The space D is complete.

Set

$$D_b = \{y : (-\infty, b] \to E| \ y \in PC \cap D\},$$

and let $\| \cdot \|_b$ be the seminorm in D_b defined by

$$\|y\|_b := \|y_0\|_D + \sup\{|y(t)| : 0 \le s \le b\}, \ y \in D_b.$$

Denote

$$K_b = \sup\{K(t) : \ t \in J\} \text{ and } M_b = \sup\{M(t) : t \in J\}.$$

10.2.2 Mild Solutions

In order to define the mild solutions of the problem (10.1)–(10.3) we assume that F is compact and convex valued multi-valued map.

Now, we can define a meaning of the mild solution of problem (10.1)–(10.3).

Definition 10.1. A function $y \in D_b$ is said to be a mild solution of system (10.1)–(10.3) if $y(t) = \phi(t)$ for all $t \in (-\infty, 0]$, the restriction of $y(\cdot)$ to the interval $[0, b]$ is continuous, and there exist $v(\cdot) \in L^1(J_k, E)$ and $\mathcal{I}_k \in I_k(y(t_k^-))$, such that $v(t) \in F(t, y_t)$ a.e $t \in [0, b]$, and y satisfies the integral equation,

$$y(t) = T(t)\,(\phi(0) - g(0, \phi(0))) + g(t, y_t) + \int_0^t T(t-s)v(s)ds$$
$$+ \sum_{0 < t_k < t} T(t - t_k)\mathcal{I}_k, \quad t \in J.$$

We introduce the following hypotheses:

(10.1.1) $A : D(A) \subset E \to E$ is the infinitesimal generator of a strongly continuous semigroup $\{T(t)\}, t \in J$ which is compact for $t > 0$ in the Banach space E, and there exist constant M, such that:

$$\|T(t)\|_{B(E)} \le M; \quad t \in J$$

(10.1.2) There exist constants $c_k \ge 0, k = 1, \ldots, m$ such that

$$H_d(I_k(y), I_k(x))| \le c_k|y - x| \quad \text{for each} \quad x, y \in E.$$

(10.1.3) F is L^1-Carathéodory with compact convex values.
(10.1.4) There exist a function $p \in L^1(J, \mathbb{R}_+)$ and a continuous nondecreasing function $\psi : [0, \infty) \to (0, \infty)$ such that

$$\|F(t, x)\| = \sup\{|v|/v \in F(t, x)\} \le p(t)\psi(\|x\|_D) \quad \text{for a.e. } t \in J \text{ and each } x \in D,$$

with

$$\int_{C_0}^{\infty} \frac{ds}{\psi(s)} > C_1\|p\|_{L^1},$$

where

$$C_0 = \left[M\alpha_1 + M^2 \sum_{k=0}^{m} c_k \right] \|\phi\|_D$$

$$+ M \sum_{k=0}^{m} |I_k(0)| + \alpha_2(1 + M)$$

$$C_1 = \frac{C_0}{1 - M \displaystyle\sum_{k=0}^{m} c_k - K_b \alpha_1}$$

$$+ \frac{(MK_b + M_b)M \displaystyle\sum_{k=0}^{m} c_k}{(1 - M \displaystyle\sum_{k=0}^{m} c_k - K_b \alpha_1)} ()\|\phi\|_D$$

$$C_2 = \frac{M}{1 - M \displaystyle\sum_{k=0}^{m} c_k - K_b \alpha_1}$$

(10.1.5) The function $g(t, .)$ is continuous on J and there exists a constant $l_g > 0$ such that

$$|g(t, u) - g(t, v)| \le l_g \|u - v\| \quad \text{for each} \quad u, v \in D.$$

(10.1.6) There exist constants α_1 and α_2 such that

$$\|g(t, u)\| \le \alpha_1 \|u\|_D + \alpha_2 \quad \text{for each} \quad (t, u) \in [0, b] \times D.$$

Theorem 10.2. *Assume that (10.1.1)–(10.1.6) hold. If $l_g + M \displaystyle\sum_{1}^{m} c_k < 1$, and*

$$\alpha_1 K_b + \sum_{1}^{m} c_k < 1, \text{ then the IVP (10.1)–(10.3) has at least one mild solution on}$$

$(-\infty, b]$.

Proof. Consider the multi-valued operator:
$N : D \to \mathcal{P}(D)$ defined by

$$N(y) = \left\{ h \in D : h(t) = \begin{cases} \phi(t), & \text{if } t \le 0, \\ T(t)(\phi(0) - g(0, \phi(0))) + g(t, y_t) + \displaystyle\int_{0}^{t} T(t - s)v(s)ds \\ \quad + \displaystyle\sum_{0 < t_k < t} T(t - t_k)\mathcal{I}_k, v \in S_{F,y}, \mathcal{I}_k \in I_k(y(t_k^{-})) & \text{if } t \in J, \end{cases} \right\}$$

Has a fixed point. This fixed point is then the mild solution of the IVP (10.1)–(10.3).
For $\phi \in D$ define the function $x(\cdot) : (-\infty, b] \to E$ such that:

$$x(t) = \begin{cases} \phi(t), & \text{if } t \le 0 \\ T(t)\phi(0), & \text{if } t \in J \end{cases}$$

Then $x(\cdot)$ is an element of D_b, and $x_0 = \phi(0)$.
Set

$$y(t) = z(t) + x(t).$$

Obviously if y satisfies the integral equation

$$y(t) = T(t)\left(\phi(0) - g(0, \phi(0))\right) + g(t, y_t) + \int_0^t T(t-s)v(s)ds$$
$$+ \sum_{0<t_k<t} T(t-t_k)\mathcal{I}_k, \quad t \in J,$$

then z satisfies $z_0 = 0$ and

$$z(t) = g(t, z_t + x_t) - T(t)g(0, \phi(0))$$
$$+ \int_0^t T(t-s)v(s)ds + \sum_{0<t_k<t} T(t-t_k)\mathcal{I}_k, \quad t \in J.$$

where $v(t) \in F(t, z_t + x_t)$ a.e. $t \in [0, b]$ and $\mathcal{I}_k \in I_k(z(t_k^-) + x(t_k^-))$.
Let

$$D_b^0 = \{z \in D_b : z_0 = 0\}.$$

For any $z \in D_b^0$, we have

$$\|z\|_b = \|z_0\|_D + \sup\{|z(s)| : 0 \le s \le b\} = \sup\{|z(s)| : 0 \le s \le b\}.$$

Thus $(D_b^0, \|\cdot\|_b)$ is a Banach space.
Let the operator $P : D_b^0 \to \mathcal{P}(D_b^0)$ defined by

$$P(z) = \left\{ h \in D_b^0 \middle| h(t) = \begin{cases} 0 & \text{if } t \in (-\infty, 0]; \\ g(t, z_t + x_t) - T(t)g(0, \phi(0)) \\ + \int_0^t T(t-s)v(s)ds + \sum_{0<t_k<t} T(t-t_k)\mathcal{I}_k, & \text{if } t \in J. \end{cases} \right\}$$

The operator N has a fixed point is equivalent to P has one, so it turns to prove that P has a fixed point. Consider these multi-valued operators:

$$\mathcal{A}, \mathcal{B} : D_b^0 \to \mathcal{P}(D_b^0)$$

defined by

$$
\mathcal{A}(z) := \left\{ h \in D_b^0 : \ h(t) = \left\{ \begin{array}{ll} 0, & \text{if } t \le 0; \\[2mm] g(t, z_t + x_t) - T(t)g(0, \phi(0)) & \\ \ \ + \displaystyle\sum_{0 < t_k < t} T(t - t_k)\mathcal{I}_k, \mathcal{I}_k \in I_k(z(t_k^-) + x(t_k^-)) & \text{if } t \in J, \end{array} \right. \right\}
$$

and

$$
\mathcal{B}(z) := \left\{ h \in D_b^0 : \ h(t) = \left\{ \begin{array}{ll} 0, & \text{if } t \le 0; \\[4mm] \displaystyle\int_0^t T(t - s)v(s)ds & \\[2mm] & \text{if } t \in J, \end{array} \right. \right\}
$$

where

$$
v \in S_{F,z} = \{v \in L^1([0, b], E) : v(t) \in F(t, z_t + x_t) \text{ for a.e. } t \in [0, b]\}.
$$

It is clear that

$$
P = \mathcal{A} + \mathcal{B}
$$

Then the problem of finding mild solutions of (10.1)–(10.3) is then reduced to finding mild solutions of the operator inclusion $z \in \mathcal{A}(z) + \mathcal{B}(z)$. We shall show that the operators $\mathcal{A}$ and $\mathcal{B}$ satisfy all conditions of the Theorem 1.32. The proof will be given in several steps.

Step 1: $\mathcal{A}$ is a contraction
Let $z_1, z_2 \in D_b^0$, then from (10.1.1)

$$
H_d\big(\mathcal{A}(z_1), \mathcal{A}(z_2)\big) \le \|g(t, z_{1_t} + x_t) - g(t, z_{2_t} + x_t)\|
$$

$$
+ H_d\Bigg(\sum_{0 < t_k < t} T(t - t_k)I_k(z_1(t_k^-) + x(t_k^-)),
$$

$$
\sum_{0 < t_k < t} T(t - t_k)I_k(z_2(t_k^-) + x(t_k^-)) \Bigg)
$$

$$
\le l_g\|z_1 - z_2\| + M \sum_1^m |z_1(t_k^-) - z_2(t_k^-)|
$$

$$
\le \Big(l_g + M \sum_1^m c_k\Big)\|z_1 - z_2\|,
$$

which is a contraction since

$$l_g + M \sum_1^m c_k < 1.$$

Step 2: $\mathcal{B}$ has compact, convex values, and it is completely continuous. This will be given in several claims.

Claim 1: $\mathcal{B}$ has compact values. The operator $\mathcal{B}$ is equivalent to the composition $\mathcal{L} \circ S_F$ of two operators on $L^1(J, E)$ $\mathcal{L} : L^1(J, E) \to D_b^0$ is the continuous operator defined by

$$\mathcal{L}(v(t)) = \int_0^t T(t - s)v(s)ds, \qquad t \in J.$$

Then, it suffices to show that $\mathcal{L} \circ S_F$ has compact values on D_b^0.
Let $z \in D_b^0$ arbitrary, v_n a sequence in $S_{F,z}$, then by definition of S_F, $v_n(t)$ belongs to $F(t, z_t), a.e.t \in J$. Since $F(t, z_t)$ is compact, we may pass to a subsequence. Suppose that $v_n \to v$ in $L^1(J, E)$, where $v(t) \in F(t, z_t), a.e.t \in J$.
From the continuity of $\mathcal{L}$, it follows that $\mathcal{L}v_n(t) \to \mathcal{L}v(t)$ pointwise on J as $n \to \infty$.
In order to show that the convergence is uniform, we first show that $\{\mathcal{L}v_n\}$ is an equi-continuous sequence.
Let $\tau_1, \tau_2 \in J$, then we have:

$$\left| \mathcal{L}(v_n(\tau_1)) - \mathcal{L}(v_n(\tau_2)) \right| = \Big| \int_0^{\tau_1} T(\tau_1 - s)v_n(s)ds$$

$$- \int_0^{\tau_2} T(\tau_2 - s)v_n(s)ds \Big|$$

$$\leq \int_0^{\tau_1} \left| (T(\tau_1 - s) - T(\tau_2 - s)) \right| |v_n(s)|ds$$

$$+ \int_{\tau_1}^{\tau_2} |T(\tau_2 - s)||v_n(s)|ds$$

As $\tau_1 \to \tau_2$, the right-hand side of the above inequality tends to zero. Since $T(t)$ is a strongly continuous operator and the compactness of $T(t), t > 0$, implies the continuity in uniform topology. Hence $\{\mathcal{L}v_n\}$ is equi-continuous, and an application of Arzéla-Ascoli theorem implies that there exists a subsequence which is uniformly convergent. Then we have $\mathcal{L}v_{n_j} \to \mathcal{L}v \in (\mathcal{L} \circ S_F)(z)$ as $j \mapsto \infty$, and so $(\mathcal{L} \circ S_F)(z)$ is compact . Therefore $\mathcal{B}$ is a compact valued multi-valued operator on D_b^0.

Claim 2: $\mathcal{B}(z)$ is convex for each $z \in D_b^0$. Let $h_1, h_2 \in \mathcal{B}(z)$, then there exists $v_1, v_2 \in S_{F,z}$ such that, for each $t \in J$ we have

$$h_i(t) = \begin{cases} 0, & \text{if } t \in (-\infty, 0], \\ \displaystyle\int_0^t T(t-s)v_i(s)ds & \\ & \text{if } t \in J, i = 1, 2. \end{cases}$$

Let $0 \le \delta \le 1$. Then, for each $t \in J$, we have

$$(\delta h_1 + (1-\delta)h_2)(t) = \begin{cases} 0, & \text{if } t \in (-\infty, 0], \\ \displaystyle\int_0^t T(t-s)[\delta v_1(s) + (1-\delta)v_2(s)]ds & \\ & \text{if } t \in J, \end{cases}$$

Since $F(t, z_t)$ has convex values, one has

$$\delta h_1 + (1-\delta)h_2 \in \mathcal{B}(z).$$

Claim 3: $\mathcal{B}$ maps bounded sets into bounded sets in D_b^0

Let $B = \{z \in D_b^0; \|z\|_\infty \le q\}, q \in \mathbb{R}^+$ a bounded set in D_b^0. We know that for each $h \in \mathcal{B}(z)$, for some $z \in B$, there exists $v \in S_{F,z}$ such that

$$h(t) = \int_0^t T(t-s)v(s)ds.$$

$v \in S_{F,z} = \{v \in L^1([0, b], E) : v(t) \in F(t, z_t + x_t)$

From (10.1.4) we have

$$\|z_s + x_s\|_D \le \|z_s\|_D + \|x_s\|_D$$
$$\le K_b q + K_b M|\phi(0)| + M_b\|\phi\|_D$$
$$= q_*.$$

Then

$$|h(t)| \le M\psi(q_*)\int_0^t p(s)ds$$
$$\le M\psi(q_*)\|p\|_{L^1} = l,$$

This further implies that:

$$\|h\|_\infty \le l$$

Then, for all $h \in \mathcal{B}(z) \subset \mathcal{B}(B) = \bigcup_{z \in B} \mathcal{B}(z)$. Hence $\mathcal{B}(B)$ is bounded.

Claim 4: $\mathcal{B}$ maps bounded sets into equi-continuous sets.

Let B be, as above, a bounded set and $h \in \mathcal{B}(z)$ for some $z \in B$. Then, there exists $v \in S_{F,z}$ such that

$$h(t) = \int_0^t T(t-s)v(s)ds, \quad t \in J$$

Let $\tau_1, \tau_2 \in J\backslash\{t_1, t_2, \ldots, t_m\}, \tau_1 < \tau_2$. Thus if $\epsilon > 0$, we have

$$|h(\tau_2) - h(\tau_1)| \leq \int_0^{\tau_1-\epsilon} \|T(\tau_2 - s) - T(\tau_1 - s)\||v(s)|ds$$

$$+ \int_{\tau_1-\epsilon}^{\tau_1} \|T(\tau_2 - s) - T(\tau_1 - s)\||v(s)|ds$$

$$+ \int_{\tau_1}^{\tau_2} \|T(\tau_2 - s)\||v(s)|ds$$

$$\leq \psi(q_*)\int_0^{\tau_1-\epsilon} \|T(\tau_2 - s) - T(\tau_1 - s)\|_{B(E)}p(s)ds$$

$$+\psi(q_*)\int_{\tau_1-\epsilon}^{\tau_1} \|T(\tau_2 - s) - T(\tau_1 - s)\|_{B(E)}p(s)ds$$

$$+\psi(q_*)\int_{\tau_1}^{\tau_2} \|T(\tau_2 - s)\|_{B(E)}p(s)ds.$$

As $\tau_1 \to \tau_2$ and ϵ becomes sufficiently small, the right-hand side of the above inequality tends to zero, since $T(t)$ is a strongly continuous operator and the compactness of $T(t)$ for $t > 0$ implies the continuity in the uniform operator topology.

This proves the equi-continuity for the case where $t \neq t_i, i = 1, \ldots, m + 1$. It remains to examine the equi-continuity at $t = t_i$.

First we prove the equi-continuity at $t = t_i^-$, we have for some $z \in B$, there exists $v \in S_{F,z}$ such that

$$h(t) = \int_0^t T(t-s)v(s)ds, \quad t \in J$$

Fix $\delta_1 > 0$ such that $\{t_k, k \neq i\} \cap [t_i - \delta_1, t_i + \delta_1] = \emptyset$. For $0 < \rho < \delta_1$, we have

$$|h(t_i - \rho) - h(t_i)| \leq \int_0^{t_i-\rho} \|T(t_i - \rho - s) - T(t_i - s)\||v(s)|ds$$

$$+\psi(q_*)M\int_{t_i-\rho}^{t_i} p(s)ds,$$

which tends to zero as $\rho \to 0$.

Define

$$\hat{h}_0(t) = h(t), \quad t \in [0, t_1]$$

and

$$\hat{h}_i(t) = \left\{ \begin{array}{ll} h(t), & \text{if } t \in (t_i, t_{i+1}] \\ h(t_i^+), & \text{if } \quad t = t_i \end{array} \right\}$$

Next, we prove equi-continuity at $t = t_i^+$. Fix $\delta_2 > 0$ such that $\{t_k, k \neq i\} \cap [t_i - \delta_2, t_i + \delta_2] = \emptyset$. Then

$$\hat{h}(t_i) = \int_0^{t_i} T(t_i - s)v(s)ds,$$

For $0 < \rho < \delta_2$, we have

$$|\hat{h}(t_i + \rho) - \hat{h}(t_i)| \leq \int_0^{t_i} \|T(t_i + \rho - s) - T(t_i - s)\| |v(s)| ds$$

$$+ \psi(q_*)M \int_{t_i}^{t_i + \rho} p(s)ds$$

The right-hand side tends to zero as $\rho \to 0$.

The equi-continuity for the cases $\tau_1 < \tau_2 \leq 0$ and $\tau_1 \leq 0 \leq \tau_2$ follows from the uniform continuity of ϕ on the interval $(-\infty, 0]$ As a consequence of Claims 1–3 together with Arzelá–Ascoli theorem it suffices to show that B maps B into a precompact set in E.

Let $0 < t < b$ be fixed and let ϵ be a real number satisfying $0 < \epsilon < t$. For $z \in B$, we define

$$h_\epsilon(t) = T(\epsilon) \int_0^{t-\epsilon} T(t - s - \epsilon)v(s)ds,$$

where $v \in S_{F,z}$. Since $T(t)$ is a compact operator, the set

$$H_\epsilon(t) = \{h_\epsilon(t) : h_\epsilon \in B(z)\}$$

is precompact in E for every ϵ, $0 < \epsilon < t$. Moreover, for every $h \in B(z)$ we have

$$|h(t) - h_\epsilon(t)| = \left| \int_0^t T(t - s)v(s)ds - T(\epsilon) \int_0^{t-\epsilon} T(t - s - \epsilon)v(s)ds \right|$$

$$= \left| \int_{t-\epsilon}^t T(t - s)v(s)ds \right|$$

$$\leq M\psi(q_*) \int_{t-\epsilon}^t p(s)ds$$

Therefore, there are precompact sets arbitrarily close to the set $H(t) = \{h(t) : h \in \mathcal{B}(z)\}$. Hence the set $H(t) = \{h(t) : h \in \mathcal{B}(B)\}$ is precompact in E. Hence the operator $\mathcal{B}$ is totally bounded.

Claim 5: $\mathcal{B}$ has closed graph.

Let $z_n \to z_*$, $h_n \in \mathcal{B}(z_n)$, and $h_n \to h_*$. We shall show that $h_* \in \mathcal{B}(z_*)$. $h_n \in \mathcal{B}(z_n)$ means that there exists $v_n \in S_{F,z_n}$ such that

$$h_n(t) = \int_0^t T(t-s)v_n(s)ds, \quad t \in J.$$

We must prove that there exists $v_* \in S_{F,z_*}$ such that

$$h_*(t) = \int_0^t T(t-s)v_*(s)ds.$$

Consider the linear and continuous operator $\mathcal{K} : L^1(J,E) \to D_b^0$ defined by

$$(\mathcal{K}v)(t) = \int_0^t T(t-s)v(s)ds.$$

We have

$$|(h_n(t) - (h_*(t)) \le \|h_n - h_*\|_\infty \to 0, \quad \text{as } n \mapsto \infty.$$

From Lemma 1.11 it follows that $\mathcal{K} \circ S_F$ is a closed graph operator and from the definition of $\mathcal{K}$ one has

$$h_n(t) \in \mathcal{K} \circ S_{F,z_n}.$$

As $z_n \to z_*$ and $h_n \to h_*$, there is a $v_* \in S_{F,z_*}$ such that

$$h_*(t) = \int_0^t T(t-s)v_*(s)ds.$$

Hence the multi-valued operator $\mathcal{B}$ is upper semi-continuous.

Step 3: A priori bounds on solutions. Now, it remains to show that the set

$$\mathcal{E} = \{z \in D_b^0| \ z \in \lambda\mathcal{A}z + \lambda\mathcal{B}z, \ 0 \le \lambda \le 1\}$$

is unbounded.

Let $z \in \mathcal{E}$ be any element. Then there exist $v \in S_{F,z}$ and $\mathcal{I}_k \in I_k(z(t_k^-))$ such that

$$z(t) = \lambda g(t, z_t + x_t) - \lambda T(t)g(0, \phi(0))$$

$$+ \lambda \int_0^t T(t-s)v(s)ds + \lambda \sum_{0 < t_k < t} T(t - t_k)\mathcal{I}_k.$$

Then

$$|z(t)| \leq \alpha_1 \|z_t + x_t\| + \alpha_2 + M(\alpha_1 \|\phi\|_D + \alpha_2)$$

$$+M \int_0^t p(s)\psi(\|z_s + x_s\|)ds + M \sum_{k=0}^{m} c_k |z(t_k^-) + x(t_k^-)| + M \sum_{k=0}^{k=m} |I_k(0)|$$

$$\leq \alpha_1 \big(K_b |z(t)| + (MK_b + M_b)\|\phi\|_D\big) + \alpha_2$$

$$+M(\alpha_1\|\phi\|_D + \alpha_2) + M \int_0^t p(s)\psi\big(K_b|z(s)| + (MK_b + M_b)\|\phi\|_D\big))ds$$

$$+M \sum_{k=0}^{m} c_k |z(t)| + M \sum_{k=0}^{m} c_k |x(t)|$$

$$+M \sum_{k=0}^{k=m} |I_k(0)|$$

$$\leq \alpha_1 \big(K_b |z(t)| + (MK_b + M_b)\|\phi\|_D\big) + \alpha_2 + M(\alpha_1 \|\phi\|_D + \alpha_2)$$

$$+M \int_0^t p(s)\psi(K_b|z(s)| + (MK_b + M_b)\|\phi\|_D))ds + M \sum_{k=0}^{m} c_k |z(t)|$$

$$+M^2 \sum_{k=0}^{m} c_k \|\phi\|_D + M \sum_{k=0}^{k=m} |I_k(0)|.$$

Then, we have:

$$|z(t)| \leq C_0 + \alpha_1\big((K_b|z(t)| + (MK_b + M_b)\|\phi\|_D)\big)$$

$$+M \int_0^t p(s)\psi(K_b|z(s)| + (MK_b + M_b)\|\phi\|_D))ds + M \sum_{k=0}^{m} c_k |z(t)|$$

Since

$$M \sum_{k=0}^{m} c_k < 1,$$

then

$$K_b|z(t)| + (MK_b + M_b)\|\phi\|_D \leq C_1 + C_2 \int_0^t p(s)\psi(K_b|z(s)|$$

$$+(MK_b + M_b)\|\phi\|_D)ds$$

Consider the function $\mu(t)$ defined by

$$\mu(t) = \sup\{K_b|z(s)| + (MK_b + M_b)\|\phi\|_D : \ 0 \le s \le t\}, \ 0 \le t \le b$$

Then, we have, for all $t \in J, \|K_b|z(t)| + (MK_b + M_b)\|\phi\|_D\| \le \mu(t)$.

Let $t^* \in J$ such that $\mu(t) = K_b|z(t^*)| + (MK_b + M_b)\|\phi\|_D$, then by the previous inequality we have, for $t \in J$,

$$\mu(t) \le C_1 + C_2 \int_0^t p(s)\psi(\|\mu_s\|)ds.$$

Let us note the right-hand side of the above inequality by $v(t)$, i.e.,

$$v(t) = C_1 + C_2 \int_0^t p(s)\psi(\mu(s))ds.$$

Then, we have

$$\mu(t) \le v(t) \quad \text{for all } t \in J$$

$$v(0) = C_1$$

Differentiating both sides of the above equality, we obtain

$$v'(t) = C_2 p(t)\psi(\mu(t)), \quad a.e. \ t \in J$$

and using the nondecreasing character of the function ψ, we obtain

$$v'(t) \le C_2 p(t)\psi(v(t)), \quad a.e. \ t \in J,$$

that is

$$\frac{v'(t)}{\psi(v(t))} \le C_2 p(t), \quad a.e. \ t \in J.$$

Integrating from 0 to t we get

$$\int_0^t \frac{v'(s)}{\psi(v(s))}ds \le C_2 \int_0^t p(s)ds.$$

By a change of variables we get

$$\int_{v(0)}^{v(t)} \frac{du}{\psi(u)} \le C_2\|p\|_{L^1} \le \int_{c_1}^{\infty} \frac{du}{\psi(u)}.$$

Hence there exists a constant K such that

$$\mu(t) \leq v(t) \leq K \quad \text{for all } t \in J.$$

Now from the definition of μ it follows that

$$\|K_b|z(t)| + (MK_b + M_b)\|\phi\|_D\| \leq \mu(b) \leq K \quad \text{for all } z \in \mathcal{E},$$

which means that $\mathcal{E}$ is bounded. As a consequence of Theorem 1.32, $\mathcal{A}(z) + \mathcal{B}(z)$ has a fixed point z^* on the interval $(-\infty, b]$, so $y^* = z^* + x$ is a fixed point of the operator N which is the mild solution of problem (10.1)–(10.3). □

10.2.3　Extremal Mild Solutions

In this section we shall prove the existence of maximal and minimal solutions of problem (10.1)–(10.3) under suitable monotonicity conditions on the multi-valued functions involved in it. We need the following definitions in the sequel.

Definition 10.3. We say that a continuous function $\tilde{v} \in D_b$ is a lower mild solution of problem (10.1)–(10.3) if $\tilde{v}(t) = \phi(t)$, $t \in (-\infty, 0]$, and there exist $v(\cdot) \in L^1(J_k, E)$ and $\mathcal{I}_k \in I_k(\tilde{v}(t_k^-))$, such that $v(t) \in F(t, \tilde{v}_t)$ a.e $t \in [0, b]$, and $\tilde{v}$ satisfies,

$$\tilde{v}(t) \leq T(t)(\phi(0) - g(0, \phi(0))) + g(t, \tilde{v}_t) + \int_0^t T(t-s)v(s)ds$$
$$+ \sum_{0 < t_k < t} T(t - t_k)\mathcal{I}_k, \quad t \in J, \ t \neq t_k.$$

and $\tilde{v}(t_k^+) - \tilde{v}(t_k^-) \leq \mathcal{I}_k$ where $\mathcal{I}_k \in I_k(\tilde{v}(t_k))$, $t = t_k$, $k = 1, \ldots, m$ Similarly an upper mild solution $\tilde{w}$ of IVP (10.1)–(10.3) is defined by reversing the order.

Definition 10.4. A solution x_M of IVP (10.1)–(10.3) is said to be maximal if for any other solution x of IVP (10.1)–(10.3) on J, we have that $x(t) \leq x_M(t)$ for each $t \in J$.

Similarly a minimal solution of IVP (10.1)–(10.3) is defined by reversing the order of the inequalities.

We consider the following assumptions in the sequel.

(10.4.1)　The multi-valued function $F(t, y)$ is strictly monotone increasing in y for almost each $t \in J$.

(10.4.2)　The IVP (10.1)–(10.3) has a lower mild solution $\tilde{v}$ and an upper mild solution $\tilde{w}$ with $\tilde{v} \leq \tilde{w}$.

(10.4.3)　$T(t)$ is preserving the order, that is $T(t)v \geq 0$ whenever $v \geq 0$.

(10.4.4)　The functions $I_k, k = 1, \ldots, m$ are continuous and nondecreasing.

Theorem 10.5. *Assume that assumptions (10.1.1)–(10.1.6) and (10.4.1)–(10.4.4) hold. Then IVP (10.1)–(10.3) has minimal and maximal solutions on D_b.*

Proof. We can write $\tilde{v}$ and $\tilde{w}$ as

$$\tilde{v}(t) = v^*(t) + x(t)$$
$$\tilde{w}(t) = w^*(t) + x(t)$$

where $v^* \in D_b^0$ and $w^* \in D_b^0$ and $x(t)$ is defined in the above section.
   Then $\tilde{v}$ is lower solution to IVP (10.1)–(10.3) if v^* satisfies

$$v^*(t) \leq -T(t)g(0, \phi(0)) + g(t, v_t^* + x(t)) + \int_0^t T(t-s)v(s)ds$$
$$+ \sum_{0<t_k<t} T(t-t_k)\mathcal{I}_k, \quad t \in J, \ t \neq t_k.$$

and $v^*(t_k^+) - v^*(t_k^- \leq \mathcal{I}_k$ such that $\mathcal{I}_k I_k(v^*(t_k))$, $t = t_k$, $k = 1, \ldots, m$ respectively $(\tilde{w})$ is upper solution to IVP (10.1)–(10.3) if w^* satisfies the reversed inequality. It can be shown, as in the proof of Theorem 10.2, that $\mathcal{A}$ is completely continuous and $\mathcal{B}$ is a contraction on $[v^*, w^*]$. We shall show that $\mathcal{A}$ and $\mathcal{B}$ are isotone increasing on $[v^*, w^*]$. Let $z, \bar{z} \in [v^*, w^*]$ be such that $z \leq \bar{z}$, $z \neq \bar{z}$. Then by (10.4.4), we have for each $t \in J$

$$\mathcal{A}(z) = \left\{ h \in D_b^0 : h(t) = -T(t)g(0, \phi(0)) + g(t, z_t + x_t) \right.$$

$$+ \sum_{0<t_k<t} T(t-t_k)\mathcal{I}_k, \in \mathcal{I}_k z(t_k^-) \left. \right\}$$

$$\leq \left\{ h \in D_b^0 : h(t) = -T(t)g(0, \phi(0)) + g(t, \bar{z}_t + x_t) \right.$$

$$+ \sum_{0<t_k<t} T(t-t_k)\mathcal{I}_k, \mathcal{I}_k \in I_k(\bar{z}(t_k^-)) \left. \right\}$$

$$= \mathcal{A}(\bar{z}).$$

Similarly, by (10.4.1), (10.4.3)

$$\mathcal{B}(z) = \left\{ h \in D_b^0 : h(t) = \int_0^t T(t-s)v(s)ds, \ v \in S_{F,z} \right\}$$

$$\leq \left\{ h \in D_b^0 : h(t) = \int_0^t T(t-s)v(s)ds, \ f \in S_{F,\bar{z}} \right\}$$

$$= \mathcal{B}(\bar{z}).$$

Therefore $\mathcal{A}$ and $\mathcal{B}$ are isotone increasing on $[v^*, w^*]$. Finally, let $y \in [v^*, w^*]$ be any element. By (10.4.2), (10.4.3) we deduce that

$$v^* \le \mathcal{A}(v^*) + \mathcal{B}(v^*) \le \mathcal{A}(y) + \mathcal{B}(y) \le \mathcal{A}(w^*) + \mathcal{B}(w^*) \le w^*,$$

which shows that $\mathcal{A}(y) + \mathcal{B}(y) \in [v^*, w^*]$ for all $y \in [v^*, w^*]$. Thus, $\mathcal{A}$ and $\mathcal{B}$ satisfy all the conditions of Theorem 1.37, hence IVP (10.1)–(10.3) has maximal and minimal solutions on J. $\qquad\square$

10.2.4 Example

As an application of our results we consider the following impulsive partial functional differential equation of the form

$$\frac{\partial}{\partial t}\left[v(t,\xi) - \int_{-\infty}^{0} K_1(\theta)g_1(t+\theta,\xi)d\theta\right]$$

$$= \frac{\partial^2}{\partial \xi^2}\left[v(t,\xi) - \int_{-\infty}^{0} K_1(\theta)g_1(t+\theta,\xi)d\theta\right]$$

$$+ \int_{-\infty}^{0} K_2(\theta)\left[Q_1(t, v(t+\theta,\xi), Q_2(t, v(t+\theta,\xi)d\theta]\right]$$

for $\xi \in [0,\pi]$, $t \in [0,b]\backslash\{t_1, t_2, \ldots, t_m\}$.   (10.4)

$$v(t_k^+, \xi) - z(t_k^-, \xi) \in b_k|z(t_k^-, \xi)|\bar{B}(0,1), \; \xi \in [0,\pi], \; k = 1, \ldots, m   (10.5)$$

$$v(t,0) - \int_{-\infty}^{0} K_1(\theta)g_1(t+\theta, 0)d\theta = 0, \; t \in J := [0,b]   (10.6)$$

$$v(t,\pi) - \int_{-\infty}^{0} K_1(\theta)g_1(t+\theta, \pi)d\theta = \phi(t,x), \; t \in J := [0,b], \; \xi \in [0,\pi],   (10.7)$$

$$v(\theta, \xi,) = v_0(\theta, \xi) \; fot \; -\infty < \theta \le 0 \; and \; \xi \in [0,\pi],   (10.8)$$

where $b_k > 0$, $k = 1, \ldots, m$, $K_1 : (-\infty, 0] \to \mathbb{R}, K_2 : (-\infty, 0] \to \mathbb{R}$ and $g_1 : J \times \mathbb{R} \to \mathbb{R}$ and $v_0 : (-\infty, 0]X[0,\pi] \to \mathbb{R}$ are continuous functions, $0 = t_0 < t_1 < t_2 < \cdots < t_m < t_{m+1} = b$, $v(t_k^+) = \lim\limits_{(h,x) \to (0^+,x)} v(t_k + h, x), v(t_k^-) = \lim\limits_{(h,x) \to (0^-,x)} v(t_k + h, x)$, where $Q_1, Q_2 : J \times \mathbb{R} \to \mathbb{R}$, are given functions, and $\bar{B}(0,1)$ the closed unit ball. We assume that for each $t \in J$, $Q_1(t, \cdot)$ is lower semi-continuous (i.e., the set $\{y \in \mathbb{R} : Q_1(t,y) > \mu\}$ is open for each $\mu \in \mathbb{R}$), and assume that for each $t \in J$, $Q_2(t, \cdot)$ is upper semi-continuous (i.e., the set $\{y \in \mathbb{R} : Q_2(t,y) < \mu\}$ is open for each $\mu \in \mathbb{R}$).

Let

$$y(t)(\xi) = v(t,\xi), \ t \in J, \ \xi \in [0,\pi],$$
$$I_k(y(t_k^-))(\xi) = b_k v(t_k^-,\xi), \ \xi \in [0,\pi], \ k = 1,\ldots,m$$
$$F(t,\phi)(x) = \int_{-\infty}^0 K_2(\theta)\big[Q_1(t, v(t+\theta,\xi), Q_2(t, v(t+\theta,\xi)d\theta\big],$$
$$\theta \in (-\infty, 0], \ \xi \in [0,\pi],$$
$$h((t,\phi) = \int_{-\infty}^0 K_1(\theta)g_1(t+\theta,\pi)d\theta,$$

and

$$\phi(\theta)(\xi) = \phi(\theta,\xi), \ \theta \in (-\infty,0], \ \xi \in [0,\pi].$$

$E = L^2[0,\pi]$ and define $A : D(A) \subset E \to E$ by $Aw = w''$ with domain

$$D(A) = \{w \in E, w, w' \text{ are absolutely continuous, } w'' \in E, w(0) = w(\pi) = 0\}.$$

Then

$$Aw = \sum_{n=1}^{\infty} n^2(w, w_n)w_n, \ w \in D(A)$$

where $(\cdot,\cdot)$ is the inner product in L^2 and $w_n(s) = \sqrt{\frac{2}{\pi}}\sin ns, \ n = 1,2,\ldots$ is the orthogonal set of eigenvectors in A. It is well known (see [168]) that A is the infinitesimal generator of an analytic semigroup $T(t), \ t \in (0,b]$ in E and is given by

$$T(t)w = \sum_{n=1}^{\infty} exp(-n^2 t)(w, w_n)w_n, \ w \in E.$$

Since the analytic semigroup $T(t), \ t \in (0,b]$ is compact, there exists a constant $M \geq 1$ such that

$$\|T(t)\|_{B(E)} \leq M.$$

It is clear that F is compact and convex valued, and it is upper semi-continuous (see [101]). Assume that there are $p \in C(J, \mathbb{R}^+)$ and $\psi : [0,\infty) \to (0,\infty)$ continuous and nondecreasing such that

$$\max(|Q_1(t,y)|, |Q_2(t,y)|) \leq p(t)\psi(|y|), \quad t \in J, \text{ and } y \in \mathbb{R}.$$

Assume that there exist functions $\tilde{l}_1$, $\tilde{l}_2 \in L^1(J, \mathbb{R}^+)$ such that

$$|Q_1(t, w) - Q_1(t, \overline{w})| \le \tilde{l}_1(t)|w - \overline{w}|, \ t \in J, \ w, \overline{w} \in \mathbb{R},$$

and

$$|Q_2(t, w) - Q_2(t, \overline{w})| \le \tilde{l}_2(t)|w - \overline{w}|, \ t \in J, \ w, \overline{w} \in \mathbb{R}.$$

We can show that problem (10.1)–(10.3) is an abstract formulation of problem (10.4)–(10.8). Since all the conditions of Theorem 10.2 are satisfied, the problem (10.4)–(10.8) has a solution z on $(-\infty, b] \times [0, \pi]$.

10.3   Non-densely Defined Impulsive Neutral Functional Differential Inclusions

In this section, we use the extrapolation method combined with a fixed point theorem for the sum of completely continuous and contraction operators, to establish sufficient conditions for the existence of mild solutions and extremal mild solutions for some classes of non-densely defined impulsive semi-linear neutral functional differential inclusions in separable Banach spaces with infinite delay. More precisely, we will consider the following first order impulsive semi-linear neutral functional differential inclusions of the form:

$$\frac{d}{dt}[y(t) - g(t, y_t)] - A[y(t) - g(t, y_t)] \in F(t, y_t),$$

$$a.e. \ t \in J = [0, b], \ t \ne t_k, \ k = 1, \dots, m \tag{10.9}$$

$$\Delta y|_{t=t_k} \in I_k(y(t_k^-)), \ \ k = 1, \dots, m \tag{10.10}$$

$$y(t) = \phi(t), \ \ t \in (-\infty, 0], \tag{10.11}$$

where $F : J \times D \to 2^E$ is a compact and convex valued multi-valued map, $g : J \times D \to E$ is a given function, $\phi \in D$ where D is the phase space that will be specified later $I_k \in C(E, E), (k = 1, 2, \dots, m)$ are bounded valued multi-valued maps, $\mathcal{P}(E)$ is the collection of all E-subsets, $A : D(A) \subset E \to E$ is a non-densely defined closed linear operator on E, and E a real separable Banach space with norm $|.|$.

10.3.1   Mild Solutions

We shall consider the space

$$D_b = \{y : (-\infty, b] \to E | \; y \in PC \cap D\},$$

and let $\| \cdot \|_b$ be the seminorm in D_b defined by

$$\|y\|_b := \|y_0\|_D + \sup\{|y(t)| : 0 \le s \le b\}, \; y \in D_b.$$

Assume that F is compact and convex valued multi-valued map.

Let us start by defining what we mean by a solution of problem (10.9)–(10.11).

Definition 10.6. A function $y \in D_b$ is said to be a mild solution of system (10.9)–(10.11) if $y(t) = \phi(t)$ for all $t \in (-\infty, 0]$, the restriction of $y(\cdot)$ to the interval $[0, b]$ is continuous, and there exist $v(\cdot) \in L^1(J_k, E)$ and $\mathcal{I}_k \in I_k(y(t_k^-))$, such that $v(t) \in F(t, y_t)$ a.e $t \in [0, b]$, and y satisfies the integral equation,

$$y(t) = T_0(t)\,(\phi(0) - g(0, \phi(0))) + g(t, y_t) + \int_0^t T_1(t - s)v(s)ds$$

$$+ \sum_{0 < t_k < t} T_1(t - t_k)\mathcal{I}_k, \;\; t \in J. \tag{10.12}$$

Before beginning our result, we shall introduce the following hypotheses:

(10.6.1) There exists a constant M, such that:

$$\|T_1(t)\|_{B(E)} \le M; \;\; t \in J$$

(10.6.2) There exist constants $c_k \ge 0, k = 1, \ldots, m$ such that

$$H_d(I_k(y), I_k(x))| \le c_k|y - x| \;\; \text{for each} \;\; x, y \in E.$$

(10.6.3) F is L^1-Carathéodory with compact convex values.
(10.6.4) There exist a function $p \in L^1(J, \mathbb{R}_+)$ and a continuous nondecreasing function $\psi : [0, \infty) \to (0, \infty)$ such that

$$\|F(t, x)\| = \sup\{|v|/v \in F(t, x)\} \le p(t)\psi(\|x\|_D) \;\; \text{for a.e.} \; t \in J \text{ and each } x \in D,$$

with

$$\limsup_{u \to \infty} \frac{\left(1 - \alpha_1 K_b - M \sum_{k=1}^m c_k\right)u}{C_0 + M\|p\|_{L^1}\psi(K_b u + (MK_b + M_b)\|\phi\|_D)} > 1, \tag{10.13}$$

where

$$C_0 = \alpha_1 (MK_b + M_b)\|\phi\|_D + \alpha_2 + M(\alpha_1\|\phi\|_D + \alpha_2)$$

$$+M^2 \sum_{k=0}^{m} c_k\|\phi\|_D + M \sum_{k=0}^{m} |I_k(0)|.$$

(10.6.5) The function $g(t, .)$ is continuous on J and there exists a constant $l_g > 0$ such that

$$|g(t, u) - g(t, v)| \le l_g\|u - v\| \quad \text{for each} \quad u, v \in D.$$

(10.6.6) There exist constants α_1 and α_2 such that

$$\|g(t, u)\| \le \alpha_1\|u\|_D + \alpha_2 \text{ for each } (t, u) \in [0, b] \times D.$$

Theorem 10.7. *Assume that (10.6.1)–(10.6.6), $\phi \in D$ and $\phi(0) \in X_0, g(0, \phi(0)) \in X_0$ hold. If $l_g + M \sum_{1}^{m} c_k < 1$, then the IVP (10.9)–(10.11) has at least one mild solution on $(-\infty, b]$.*

Proof. Transform the problem (10.9)–(10.11) into a fixed point problem. Consider the multi-valued operator: $N : D \to \mathcal{P}(D)$ defined by

$$N(y) = \left\{ h \in D : h(t) = \begin{cases} \phi(t), & \text{if } t \le 0, \\ T_0(t)(\phi(0) - g(0, \phi(0))) + g(t, y_t) + \displaystyle\int_0^t T_1(t - s)v(s)ds \\ + \displaystyle\sum_{0 < t_k < t} T_1(t - t_k)\mathcal{I}_k, v \in S_{F,y}, \mathcal{I}_k \in I_k(y(t_k^-)) & \text{if } t \in J. \end{cases} \right\}$$

Now we shall show that the operator N has a fixed point . This fixed point is then the mild solution of the IVP (10.9)–(10.11).

For $\phi \in D$ define the function $x(\cdot) : (-\infty, b] \to E$ such that:

$$x(t) = \begin{cases} \phi(t), & \text{if } t \le 0 \\ T_0(t)\phi(0), & \text{if } t \in J. \end{cases}$$

Then x is an element of D_b, and $x_0 = \phi(0)$.

Set

$$y(t) = z(t) + x(t).$$

Obviously if y satisfies the integral equation

$$y(t) = T_0(t) \left(\phi(0) - g(0, \phi(0)) \right) + g(t, y_t) + \int_0^t T_1(t - s) v(s) ds$$
$$+ \sum_{0 < t_k < t} T_1(t - t_k) \mathcal{I}_k, \ t \in J.$$

then z satisfies $z_0 = 0$ and

$$z(t) = g(t, z_t + x_t) - T_0(t) g(0, \phi(0))$$
$$+ \int_0^t T_1(t - s) v(s) ds + \sum_{0 < t_k < t} T_1(t - t_k) \mathcal{I}_k, \ t \in J.$$

where $v(t) \in F(t, z_t + x_t)$ a.e $t \in [0, b]$. and $\mathcal{I}_k \in I_k(z(t_k^-) + x(t_k^-)$
Let

$$D_b^0 = \{ z \in D_b : z_0 = 0 \}.$$

For any $z \in D_b^0$, we have

$$\|z\|_b = \|z_0\|_D + \sup \{ |z(s)| : 0 \le s \le b \} = \sup \{ |z(s)| : 0 \le s \le b \}.$$

Thus $(D_b^0, \| \cdot \|_b)$ is a Banach space.

Let the operator $P : D_b^0 \to \mathcal{P}(D_b^0)$ defined by

$$P(z) = \left\{ h \in D_b^0 | h(t) = \begin{cases} 0 & \text{if } t \in (-\infty, 0]; \\ g(t, z_t + x_t) - T_0(t) g(0, \phi(0)) \\ + \int_0^t T_1(t - s) v(s) ds + \sum_{0 < t_k < t} T_1(t - t_k) \mathcal{I}_k, & \text{if } t \in J. \end{cases} \right\}$$

The operator N has a fixed point is equivalent to P has one, so it turns to prove that P has a fixed point. Consider these multi-valued operators:

$$\mathcal{A}, \mathcal{B} : D_b^0 \to \mathcal{P}(D_b^0)$$

defined by

$$\mathcal{A}(z) := \left\{ h \in D_b^0 : \ h(t) = \begin{cases} 0, & \text{if } t \le 0; \\ g(t, z_t + x_t) - T_0(t) g(0, \phi(0)) \\ + \sum_{0 < t_k < t} T_1(t - t_k) \mathcal{I}_k, \ \mathcal{I}_k \in I_k(z(t_k^-) + x(t_k^-)) & \text{if } t \in J, \end{cases} \right\}$$

and

$$
\mathcal{B}(z) := \left\{ h \in D_b^0 : \ h(t) = \begin{cases} 0, & \text{if } t \leq 0; \\[2mm] \displaystyle\int_0^t T_1(t-s)v(s)ds & \\[2mm] & \text{if } t \in J, \end{cases} \right\}
$$

where

$$
v \in S_{F,z} = \{ v \in L^1([0,b], E) : v(t) \in F(t, z_t + x_t) \text{ for a.e. } t \in [0,b] \}.
$$

It's clear that

$$
P = \mathcal{A} + \mathcal{B}
$$

Then the problem of finding mild solutions of (10.9)–(10.11) is then reduced to finding mild solutions of the operator inclusion $z \in \mathcal{A}(z) + \mathcal{B}(z)$. We shall show that the operators $\mathcal{A}$ and $\mathcal{B}$ satisfy all conditions of the Theorem 1.32. The proof will be given in several steps.

Step 1: $\mathcal{A}$ is a contraction
Let $z_1, z_2 \in D_b^0$, then from (10.6.2) and 10.6.5

$$
H_d\left(\mathcal{A}(z_1), \mathcal{A}(z_2)\right) \leq \|g(t, z_{1_t} + x_t) - g(t, z_{2_t} + x_t)\|
$$

$$
+ H_d\left(\sum_{0 < t_k < t} T_1(t - t_k) I_k(z_1(t_k^-) + x(t_k^-)), \right.
$$

$$
\left. \sum_{0 < t_k < t} T_1(t - t_k) I_k(z_2(t_k^-) + x(t_k^-)) \right)
$$

$$
\leq l_g \|z_1 - z_2\| + M \sum_1^m |z_1(t_k^-) - z_2(t_k^-)|
$$

$$
\leq \left(l_g + M \sum_1^m c_k \right) \|z_1 - z_2\|,
$$

which is a contraction since

$$
l_g + M \sum_1^m c_k < 1.
$$

Step 2 $\mathcal{B}$ has compact, convex values, and it is completely continuous. This will be given in several claims.

Claim 1: $\mathcal{B}$ has compact values. The operator $\mathcal{B}$ is equivalent to the composition $\mathcal{L} \circ S_F$ of two operators on $L^1(J, E)$ $\mathcal{L} : L^1(J, E) \rightarrow D_b^0$ is the continuous operator defined by

$$\mathcal{L}(v(t)) = \int_0^t T_1(t - s)v(s)ds, \qquad t \in J$$

Then, it suffices to show that $\mathcal{L} \circ S_F$ has compact values on D_b^0.
Let $z \in D_b^0$ arbitrary, v_n a sequence in $S_{F,z}$, then by definition of S_F, $v_n(t)$ belongs to $F(t, z_t), a.e.t \in J$. Since $F(t, z_t)$ is compact, we may pass to a subsequence.
Suppose that $v_n \rightarrow v$ in $L^1(J, E)$, where $v(t) \in F(t, z_t), a.e.t \in J$.
From the continuity of $\mathcal{L}$, it follows that $\mathcal{L}v_n(t) \rightarrow \mathcal{L}v(t)$ point wise on J as $n \rightarrow \infty$.
In order to show that the convergence is uniform, we first show that $\{\mathcal{L}v_n\}$ is an equi-continuous sequence.
Let $\tau_1, \tau_2 \in J$, then we have:

$$\left| \mathcal{L}(v_n(\tau_1)) - \mathcal{L}(v_n(\tau_2)) \right| = \left| \int_0^{\tau_1} T_1(\tau_1 - s)v_n(s)ds \right.$$

$$- \left. \int_0^{\tau_2} T_1(\tau_2 - s)v_n(s)ds \right|$$

$$\leq \int_0^{\tau_1} \left| (T_1(\tau_1 - s) - T_1(\tau_2 - s)) \right| |v_n(s)| ds$$

$$+ \int_{\tau_1}^{\tau_2} |T_1(\tau_2 - s)| |v_n(s)| ds.$$

As $\tau_1 \rightarrow \tau_2$, the right-hand side of the above inequality tends to zero. Since $T_1(t)$ is a strongly continuous operator and the compactness of $T_1(t), t > 0$, implies the continuity in uniform topology. Hence $\{\mathcal{L}v_n\}$ is equi-continuous, and an application of Arzéla-Ascoli theorem implies that there exist a subsequence which is uniformly convergent. Then we have $\mathcal{L}v_{n_j} \rightarrow \mathcal{L}v \in (\mathcal{L} \circ S_F)(z)$ as $j \mapsto \infty$, and so $(\mathcal{L} \circ S_F)(z)$ is compact. Therefore $\mathcal{B}$ is a compact valued multivalued operator on D_b^0.

Claim 2: $\mathcal{B}(z)$ is convex for each $z \in D_b^0$. Let $h_1, h_2 \in \mathcal{B}(z)$, then there exists $v_1, v_2 \in S_{F,z}$ such that, for each $t \in J$ we have

$$h_i(t) = \left\{ \begin{array}{ll} 0, & \text{if } t \in (-\infty, 0], \\[2mm] \displaystyle\int_0^t T_1(t - s)v_i(s)ds & \\[2mm] & \text{if } t \in J, i = 1, 2. \end{array} \right\}$$

Let $0 \leq \delta \leq 1$. Then, for each $t \in J$, we have

$$(\delta h_1 + (1-\delta)h_2)(t) = \begin{cases} 0, & \text{if } t \in (-\infty, 0], \\ \int_0^t T_1(t-s)[\delta v_1(s) + (1-\delta)v_2(s)]ds & \\ & \text{if } t \in J, \end{cases}$$

Since $F(t, z_t)$ has convex values, one has

$$\delta h_1 + (1-\delta)h_2 \in \mathcal{B}(z).$$

Claim 3: $\mathcal{B}$ maps bounded sets into bounded sets in D_b^0 Let $B = \{z \in D_b^0; \|z\|_\infty \leq q\}, q \in \mathbb{R}^+$ a bounded set in D_b^0. We know that for each $h \in \mathcal{B}(z)$, for some $z \in B$, there exists $v \in S_{F,z}$ such that

$$h(t) = \int_0^t T_1(t-s)v(s)ds.$$

$v \in S_{F,z} = \{v \in L^1([0, b], E) : v(t) \in F(t, z_t + x_t)$
From (10.6.4) we have

$$\|z_s + x_s\|_D \leq \|z_s\|_D + \|x_s\|_D$$
$$\leq K_b q + K_b M |\phi(0)| + M_b \|\phi\|_D$$
$$= q_*.$$

Then

$$|h(t)| \leq M\psi(q_*) \int_0^t p(s)ds$$
$$\leq M\psi(q_*)\|p\|_{L^1} = l,$$

This further implies that:

$$\|h\|_\infty \leq l$$

Then, for all $h \in \mathcal{B}(z) \subset \mathcal{B}(B) = \bigcup_{z \in B} \mathcal{B}(z)$. Hence $\mathcal{B}(B)$ is bounded.

Claim 4: $\mathcal{B}$ maps bounded sets into equi-continuous sets.

Let B be, as above, a bounded set and $h \in \mathcal{B}(z)$ for some $z \in B$. Then, there exists $v \in S_{F,z}$ such that

$$h(t) = \int_0^t T(t-s)v(s)ds, \quad t \in J$$

Let $\tau_1, \tau_2 \in J\backslash\{t_1, t_2, \ldots t_m\}$, $\tau_1 < \tau_2$. Thus if $\epsilon > 0$, we have

$$|h(\tau_2) - h(\tau_1)| \leq \int_0^{\tau_1 - \epsilon} \|T_1(\tau_2 - s) - T_1(\tau_1 - s)\||v(s)|ds$$

$$+ \int_{\tau_1 - \epsilon}^{\tau_1} \|T_1(\tau_2 - s) - T_1(\tau_1 - s)\||v(s)|ds$$

$$+ \int_{\tau_1}^{\tau_2} \|T_1(\tau_2 - s)\||v(s)|ds$$

$$\leq \psi(q_*) \int_0^{\tau_1 - \epsilon} \|T_1(\tau_2 - s) - T_1(\tau_1 - s)\|_{B(E)}p(s)ds$$

$$+ \psi(q_*) \int_{\tau_1 - \epsilon}^{\tau_1} \|T_1(\tau_2 - s) - T_1(\tau_1 - s)\|_{B(E)}p(s)ds$$

$$+ \psi(q_*) \int_{\tau_1}^{\tau_2} \|T_1(\tau_2 - s)\|_{B(E)}p(s)ds.$$

As $\tau_1 \to \tau_2$ and ϵ becomes sufficiently small, the right-hand side of the above inequality tends to zero, since $T_1(t)$ is a strongly continuous operator and the compactness of $T_1(t)$ for $t > 0$ implies the continuity in the uniform operator topology.

This proves the equi-continuity for the case where $t \neq t_i, i = 1, \ldots, m + 1$. It remains to examine the equi-continuity at $t = t_i$.

First we prove the equi-continuity at $t = t_i^-$, we have for some $z \in B$, there exists $v \in S_{F,z}$ such that

$$h(t) = \int_0^t T_1(t - s)v(s)ds, \quad t \in J$$

Fix $\delta_1 > 0$ such that $\{t_k, k \neq i\} \cap [t_i - \delta_1, t_i + \delta_1] = \emptyset$. For $0 < \rho < \delta_1$, we have

$$|h(t_i - \rho) - h(t_i)| \leq \int_0^{t_i - \rho} \|T_1(t_i - \rho - s) - T_1(t_i - s)\||v(s)|ds$$

$$+ \psi(q_*)M \int_{t_i - \rho}^{t_i} p(s)ds$$

which tends to zero as $\rho \to 0$.

Define

$$\hat{h}_0(t) = h(t), \quad t \in [0, t_1]$$

and

$$\hat{h}_i(t) = \begin{cases} h(t), & \text{if } t \in (t_i, t_{i+1}] \\ h(t_i^+), & \text{if } t = t_i \end{cases}$$

Next, we prove equi-continuity at $t = t_i^+$. Fix $\delta_2 > 0$ such that $\{t_k, k \neq i\} \cap [t_i - \delta_2, t_i + \delta_2] = \emptyset$. Then

$$\hat{h}(t_i) = \int_0^{t_i} T_1(t_i - s)v(s)ds,$$

For $0 < \rho < \delta_2$, we have

$$|\hat{h}(t_i + \rho) - \hat{h}(t_i)| \leq \int_0^{t_i} \|T_1(t_i + \rho - s) - T_1(t_i - s)\| |v(s)| ds$$
$$+ \psi(q_*)M \int_{t_i}^{t_i+\rho} p(s)ds$$

The right-hand side tends to zero as $\rho \to 0$.

The equi-continuity for the cases $\tau_1 < \tau_2 \leq 0$ and $\tau_1 \leq 0 \leq \tau_2$ follows from the uniform continuity of ϕ on the interval $(-\infty, 0]$ As a consequence of Claims 1–3 together with Arzelá–Ascoli theorem it suffices to show that $\mathcal{B}$ maps B into a precompact set in E.

Let $0 < t < b$ be fixed and let ϵ be a real number satisfying $0 < \epsilon < t$. For $z \in B$, we define

$$h_\epsilon(t) = T_1(\epsilon) \int_0^{t-\epsilon} T_1(t - s - \epsilon)v(s)ds,$$

where $v \in S_{F,z}$. Since $T_1(t)$ is a compact operator, the set

$$H_\epsilon(t) = \{h_\epsilon(t) : h_\epsilon \in \mathcal{B}(z)\}$$

is precompact in E for every ϵ, $0 < \epsilon < t$. Moreover, for every $h \in \mathcal{B}(z)$ we have

$$|h(t) - h_\epsilon(t)| = \left| \int_0^t T_1(t - s)v(s)ds - T_1(\epsilon) \int_0^{t-\epsilon} T_1(t - s - \epsilon)v(s)ds \right|$$
$$= \left| \int_{t-\epsilon}^t T_1(t - s)v(s)ds \right|$$
$$\leq M\psi(q_*) \int_{t-\epsilon}^t p(s)ds.$$

Therefore, there are precompact sets arbitrarily close to the set $H(t) = \{h(t) : h \in \mathcal{B}(z)\}$. Hence the set $H(t) = \{h(t) : h \in \mathcal{B}(B)\}$ is precompact in E. Hence the operator $\mathcal{B}$ is totally bounded.

Claim 5: $\mathcal{B}$ has closed graph.

Let $z_n \to z_*$, $h_n \in \mathcal{B}(z_n)$, and $h_n \to h_*$. We shall show that $h_* \in \mathcal{B}(z_*)$. $h_n \in \mathcal{B}(z_n)$ means that there exists $v_n \in S_{F,z_n}$ such that

$$h_n(t) = \int_0^t T_1(t - s)v_n(s)ds, \quad t \in J.$$

We must prove that there exists $v_* \in S_{F,z_*}$ such that

$$h_*(t) = \int_0^t T_1(t-s)v_*(s)ds.$$

Consider the linear and continuous operator $\mathcal{K} : L^1(J, E) \to D_b^0$ defined by

$$(\mathcal{K}v)(t) = \int_0^t T_1(t-s)v(s)ds.$$

We have

$$|(h_n(t) - (h_*(t)) \leq \|h_n - h_*\|_\infty \to 0, \quad \text{as } n \mapsto \infty.$$

From Lemma 1.11 it follows that $\mathcal{K} \circ S_F$ is a closed graph operator and from the definition of $\mathcal{K}$ one has

$$h_n(t) \in \mathcal{K} \circ S_{F,z_n}.$$

As $z_n \to z_*$ and $h_n \to h_*$, there is a $v_* \in S_{F,z_*}$ such that

$$h_*(t) = \int_0^t T_1(t-s)v_*(s)ds.$$

Hence the multi-valued operator $\mathcal{B}$ is upper semi-continuous.

Step 3: A priori bounds on solutions. Now, it remains to show that the set

$$\mathcal{E} = \{z \in D_b^0|\ z \in \lambda \mathcal{A}z + \lambda \mathcal{B}z,\ 0 \leq \lambda \leq 1\}$$

is unbounded.

Let $z \in \mathcal{E}$ be any element. Then there exist $v \in S_{F,z}$ and $\mathcal{I}_k \in I_k(z(t_k^-))$ such that

$$z(t) = \lambda g(t, z_t + x_t) - \lambda T_0(t)g(0, \phi(0))$$

$$+\lambda \int_0^t T_1(t-s)v(s)ds + \lambda \sum_{0<t_k<t} T_1(t-t_k)\mathcal{I}_k.$$

Then

$$|z(t)| \leq \alpha_1\|z_t + x_t\| + \alpha_2 + M(\alpha_1\|\phi\|_D + \alpha_2)$$

$$+M\int_0^t p(s)\psi(\|z_s + x_s\|)ds + M\sum_{k=0}^m c_k|z(t_k^-) + x(t_k^-)|$$

$$+M\sum_{k=0}^{k=m} \|I_k(0)\|$$

$$\leq \alpha_1\big(K_b|z(t)| + (MK_b + M_b)\|\phi\|_D\big) + \alpha_2 + M\big(\alpha_1\|\phi\|_D + \alpha_2\big)$$

$$+M \int_0^t p(s)\psi\big(K_b|z(s)| + (MK_b + M_b)\|\phi\|_D\big))ds$$

$$+M \sum_{k=0}^m c_k|z(t)| + M \sum_{k=0}^m c_k|x(t)| + M \sum_{k=0}^{k=m} \|I_k(0)\|$$

$$\leq \alpha_1\big(K_b|z(t)| + (MK_b + M_b)\|\phi\|_D\big) + \alpha_2 + M\big(\alpha_1\|\phi\|_D + \alpha_2\big)$$

$$+M \int_0^t p(s)\psi\big(K_b|z(s)| + (MK_b + M_b)\|\phi\|_D\big)ds + M \sum_{k=0}^m c_k|z(t|)$$

$$+M \sum_{k=0}^m c_k|x(t)| + M \sum_{k=0}^{k=m} \|I_k(0)\|.$$

Then, we have:

$$|z(t)| \leq \alpha_1\big((K_b|z(t)| + (MK_b + M_b)\|\phi\|_D)\big)$$

$$+M \int_0^t p(s)\psi\big(K_b|z(s)| + (MK_b + M_b)\|\phi\|_D\big)ds + M \sum_{k=0}^m c_k|z(t|)$$

$$+M^2 \sum_{k=0}^m c_k\|\phi\|_D + M \sum_{k=0}^{k=m} \|I_k(0)\|.$$

Thus

$$\Big(1 - \alpha_1 K_b - M \sum_{k=0}^m c_k\Big)\|z\|_{D_b^0} \leq \alpha_1 + (MK_b + M_b)\|\phi\|_D + \alpha_2 + M(\alpha_1\|\phi\|_D + \alpha_2)$$

$$+M\|p\|_{L^1}\psi\big(K_b\|forz\|_{D_b^0} + (MK_b + M_b)\|\phi\|_D\big)$$

$$+M^2 \sum_{k=0}^m c_k\|\phi\|_D + M \sum_{k=0}^{k=m} \|I_k(0)\|$$

$$= C_0 + M\|p\|_{L^1}\psi\big(K_b\|z\|_{D_b^0} + (MK_b + M_b)\|\phi\|_D\big)$$

By the previous inequality we have,

$$\frac{\left(1 - \alpha_1 K_b - M \sum_{k=0}^{m} c_k\right) \|z\|_{D_b^0}}{C_0 + M\|p\|_{L^1} \psi \left(K_b \|z\|_{D_b^0} + (MK_b + M_b)\|\phi\|_D\right)} \leq 1 \tag{10.14}$$

From (10.13) it follows that there exists a constant $R > 0$ such that for each $z \in \mathcal{E}$ with $\|z\|_{D_b^0} > R$ the propriety is not satisfied. Hence $\|z\|_{D_b^0} \leq R$ for each $z \in \mathcal{E}$ which means that $\mathcal{E}$ is bounded. As a consequence of Theorem 1.32, $\mathcal{A} + \mathcal{B}$ has a fixed point z^* on the interval $(-\infty, b]$, so $y^* = z^* + x$ is a fixed point of the operator N, which is the mild solution of problem (10.9)–(10.11). $\qquad\square$

10.3.2 Extremal Mild Solutions

In this section we shall prove the existence of maximal and minimal solutions of problem (10.9)–(10.11) under suitable monotonicity conditions on the multi-valued functions involved in it. We need the following definitions in the sequel.

Definition 10.8. We say that a continuous function $\tilde{v} \in D_b$ is a lower mild solution of problem (10.9)–(10.11) if $\tilde{v}(t) = \phi(t)$, $t \in (-\infty, 0]$, and there exist $v(\cdot) \in L^1(J_k, E)$ and $\mathcal{I}_k \in I_k(\tilde{v}(t_k^-))$, such that $v(t) \in F(t, \tilde{v}_t)$ a.e $t \in [0, b]$, and $\tilde{v}$ satisfies,

$$\tilde{v}(t) \leq T_0(t)\left(\phi(0) - g(0, \phi(0))\right) + g(t, \tilde{v}_t) + \int_0^t T_1(t-s)v(s)ds$$
$$+ \sum_{0 < t_k < t} T_1(t - t_k)\mathcal{I}_k, \quad t \in J, \ t \neq t_k,$$

and $\tilde{v}(t_k^+) - \tilde{v}(t_k^-) \leq \mathcal{I}_k$ where $\mathcal{I}_k \in I_k(\tilde{v}(t_k))$, $t = t_k, k = 1, \ldots, m$ Similarly an upper mild solution $\tilde{w}$ of IVP (10.9)–(10.11) is defined by reversing the order.

Definition 10.9. A solution x_M of IVP (10.9)–(10.11) is said to be maximal if for any other solution x of IVP (10.9)–(10.11) on J, we have that $x(t) \leq x_M(t)$ for each $t \in J$.

Similarly a minimal solution of IVP (10.9)–(10.11) is defined by reversing the order of the inequalities.

We consider the following assumptions in the sequel.

(10.9.1) The multi-valued function $F(t, y)$ is strictly monotone increasing in y for almost each $t \in J$.

(10.9.2) The IVP (10.9)–(10.11) has a lower mild solution $\tilde{v}$ and an upper mild solution $\tilde{w}$ with $\tilde{v} \leq \tilde{w}$.

(10.9.3) $T_1(t)$ is preserving the order, that is $T_1(t)v \geq 0$ whenever $v \geq 0$.

(10.9.4) The functions $I_k, k = 1, \ldots, m$ are continuous and nondecreasing.

Theorem 10.10. *Assume that assumptions (10.6.1)–(10.6.6) and (10.9.1)–(10.9.4) hold. If $\phi \in D$, $\phi(0) \in X_0$ and $g(0, \phi(0)) \in X_0$, then IVP (10.9)–(10.11) has minimal and maximal solutions on D_b.*

Proof. We can write $\tilde{v}$ and $\tilde{w}$ as

$$\tilde{v}(t) = v^*(t) + x(t)$$
$$\tilde{w}(t) = w^*(t) + x(t),$$

where $v^* \in D_b^0$ and $w^* \in D_b^0$ and $x(t)$ is defined in the above section.

Then $\tilde{v}$ is a lower solution to IVP (10.9)–(10.11) if v^* satisfies

$$v^*(t) \leq -T_0(t)g(0, \phi(0)) + g(t, v_t^* + x(t)) + \int_0^t T_1(t-s)v(s)ds$$
$$+ \sum_{0 < t_k < t} T_1(t - t_k)\mathcal{I}_k, \quad t \in J, \ t \neq t_k,$$

and $v^*(t_k^+) - v^*(t_k^-) \leq \mathcal{I}_k$ such that $\mathcal{I}_k \in I_k(v^*(t_k))$, $t = t_k, k = 1, \ldots, m$. Respectively $\tilde{w}$ is upper solution to IVP (10.9)–(10.11) if w^* satisfies the reversed inequality.

It can be shown, as in the proof of Theorem 10.2, that $\mathcal{A}$ is completely continuous and $\mathcal{B}$ is a contraction on $[v^*, w^*]$. We shall show that $\mathcal{A}$ and $\mathcal{B}$ are isotone increasing on $[v^*, w^*]$.

Let $z, \bar{z} \in [v^*, w^*]$ be such that $z \leq \bar{z}$, $z \neq \bar{z}$. Then by (10.9.4), we have for each $t \in J$

$$\mathcal{A}(z) = \left\{ h \in D_b^0 : h(t) = -T_0(t)g(0, \phi(0)) + g(t, z_t + x_t) \right.$$

$$\left. + \sum_{0 < t_k < t} T_1(t - t_k)\mathcal{I}_k, \in \mathcal{I}_k z(t_k^-) \right\}$$

$$\leq \left\{ h \in D_b^0 : h(t) = -T_0(t)g(0, \phi(0)) + g(t, \bar{z}_t + x_t) \right.$$

$$\left. + \sum_{0 < t_k < t} T_1(t - t_k)\mathcal{I}_k, \mathcal{I}_k \in I_k(\bar{z}(t_k^-)) \right\}$$

$$= \mathcal{A}(\bar{z}).$$

Similarly, by (10.9.1), (10.9.3)

$$\mathcal{B}(z) = \left\{ h \in D_b^0 : h(t) = \int_0^t T_1(t-s)v(s)ds, \ v \in S_{F,z} \right\}$$

$$\leq \left\{ h \in D_b^0 : h(t) = \int_0^t T_1(t-s)v(s)ds, \ f \in S_{F,\bar{z}} \right\}$$

$$= \mathcal{B}(\bar{z}).$$

Therefore $\mathcal{A}$ and $\mathcal{B}$ are isotone increasing on $[v^*, w^*]$.

Finally, let $y \in [v^*, w^*]$ be any element. By (10.9.2), (10.9.3) we deduce that

$$v^* \leq \mathcal{A}(v^*) + \mathcal{B}(v^*) \leq \mathcal{A}(y) + \mathcal{B}(y) \leq \mathcal{A}(w^*) + \mathcal{B}(w^*) \leq w^*,$$

which shows that $\mathcal{A}(y) + \mathcal{B}(y) \in [v^*, w^*]$ for all $y \in [v^*, w^*]$. Thus, $\mathcal{A}$ and $\mathcal{B}$ satisfy all the conditions of Theorem 10.7, hence IVP (10.9)–(10.11) has maximal and minimal solutions on J. $\qquad\square$

10.3.3 Example

To apply our previous results, we consider the following impulsive partial neutral functional differential equation

$$\frac{\partial}{\partial t}\left[v(t,\xi) - \int_{-\infty}^0 K_1(\theta, v(t+\theta), \xi)d\theta \right]$$

$$= \frac{\partial^2}{\partial t^2}\left[v(t,\xi) - \int_{-\infty}^0 K_1(\theta, v(t+\theta), \xi)d\theta \right]$$

$$+ \int_{-\infty}^0 K_2(\theta)[Q_1(t, \phi(\theta,\xi)), Q_2(t, \phi(\theta,\xi))]d\theta;$$

$$t \in J = [0,b], \ t \neq t_k, \ k = 1,\ldots,m, \ 0 \leq \xi \leq 1, \tag{10.15}$$

$$v(t_k^+,\xi) - v(t_k^-,\xi) \in b_k|v(t_k^-,\xi)|\bar{B}(0,1), \ \xi \in [0,1], \ k = 1,\ldots,m \tag{10.16}$$

$$v(t,0) - \int_{-\infty}^0 K_1(\theta, v(t+\theta), 0)d\theta = 0, \ t \in J, \tag{10.17}$$

$$v(t,1) - \int_{-\infty}^0 K_1(\theta, v(t+\theta), 1)d\theta = 0, \ t \in J, \tag{10.18}$$

$$v(\theta,\xi) = v_0(\theta,\xi) - \infty < \theta \leq 0, \ 0 \leq \xi \leq 1, \tag{10.19}$$

where $b_k > 0$, $k = 1,\ldots,m$, $K_1 : (-\infty,0] \times \mathbb{R} \to \mathbb{R}$, $K_2 : (-\infty,0] \to \mathbb{R}$, Q_1, $Q_2 : J \times \mathbb{R} \to \mathbb{R}$ and $v_0 : (-\infty,0] \times [0,1] \to \mathbb{R}$ are continuous functions, $v(t_k^-) = \lim\limits_{(h,\xi)\to(0^-,\xi)} v(t_k + h, \xi)$, $v(t_k^+) = \lim\limits_{(h,\xi)\to(0^+,\xi)} v(t_k + h, \xi)$ and $\bar{B}(0,1)$ the closed unit ball. We assume that for each $t \in J$, $Q_1(t,\cdot)$ is lower semi-continuous (i.e., the set $\{y \in \mathbb{R} : Q_1(t,y) > \mu\}$ is open for each $\mu \in \mathbb{R}$), and assume that for each $t \in J$, $Q_2(t,\cdot)$ is upper semi-continuous (i.e., the set $\{y \in \mathbb{R} : Q_2(t,y) < \mu\}$ is open for each $\mu \in \mathbb{R}$).

We choose $E = C([0,1],\mathbb{R})$ endowed with the uniform topology and consider the operator $A : D(A) \subset E \to E$ defined by:

$$D(A) = \{y \in C^2([0,1],\mathbb{R}) : y(0) = y(1) = 0\} \quad Ay = y''.$$

It is well known (see [100]) that the operator A satisfies the Hille–Yosida condition with $(0,+\infty) \subset \rho(A)$, $\|(\lambda I - A)^{-1}\| \le \frac{1}{\lambda}$ for $\lambda > 0$, and

$$X_0 = \overline{D(A)} = \{y \in E : y(0) = y(1) = 0\} \ne E.$$

So the extrapolation method can be applied. We define:

$$I_k(y(t_k^-))(\xi) = b_k|v(t_k^-,\xi)|\bar{B}(0,1), \ \xi \in [0,1], \ k = 1,\ldots,m$$

$$F(t,\phi)(\xi) = \int_{-\infty}^0 K_2(\theta)[Q_1(t,\phi(\theta,\xi)), Q_2(t,\phi(\theta,\xi))]d\theta, \ t \in J, \ \xi \in [0,1],$$

$$g(t,\phi)(\xi) = \int_{-\infty}^0 K_1(\theta,\phi(\theta)(\xi))d\theta, \ t \in J, \ \xi \in [0,1],$$

$$y(t)(\xi) = v(t,\xi), \ t \in J, \ \xi \in [0,1],$$

$$\phi(\theta)(\xi) = v_0(\theta,\xi), \ \theta \le 0, \ \xi \in [0,1].$$

Then problem (10.1)–(10.3) is an abstract formulation of the problem (10.15)–(10.19) with F compact and convex values, and it is upper semi continuous (see [101]). Assume that there are $p \in C(J,\mathbb{R}^+)$ and $\psi : [0,\infty) \to (0,\infty)$ continuous and nondecreasing such that

$$\max(|Q_1(t,y)|, |Q_2(t,y)|) \le p(t)\psi(|y|), \quad t \in J, \text{ and } y \in \mathbb{R}.$$

Under suitable conditions, the problem (10.15)–(10.19) has by Theorem 10.7 a solution on $(-\infty,b] \times [0,1]$.

10.4 Controllability of Impulsive Semi-linear Differential Inclusions in Fréchet Spaces

In this section, we use the extrapolation method combined with a recent nonlinear alternative of Leray-Schauder type for multi-valued admissible contractions in Fréchet spaces to study the existence of the mild solution for a class of non-densely defined first order semi-linear impulsive functional differential inclusions with finite delay in the semi-infinite interval $J := [0, \infty)$, and with single valued jump. More precisely we consider the first order semi-linear impulsive functional differential inclusions of the form:

$$y'(t) - Ay(t) \in F(t, y_t) + Bu(t), \quad a.e. \ t \in J \setminus \{t_1, t_2, \ldots\} \tag{10.20}$$

$$\Delta y|_{t=t_k} = I_k(y(t_k^-)), \quad k = 1, \ldots, \tag{10.21}$$

$$y(t) = \phi(t), \quad t \in H, \tag{10.22}$$

where $J := [0, \infty)$, $F : J \times D \to \mathcal{P}(E)$ is a multi-valued map with compact values, $(E, |\cdot|)$, D, $\mathcal{P}(E)$, B, $u(\cdot)$, ϕ, y_t are as in the above section and $A : D(A) \subset E \to E$ is a non-densely defined closed linear operator on E

Let A_0 the dense part of A, and let $(T_0(t))_{t \geq 0}$ the strongly continuous semigroup generated by A_0 defined on $X_0 = \overline{D(A)}$ and let $(T_1(t))_{t \geq 0}$ the extrapolated semigroup of $(T_0(t))_{t \geq 0}$ whose generator is $(A_1, D(A_1))$.

10.4.1 Main Result

We shall consider the space

$$PC = \Big\{ y : [-r, \infty) \to E : \quad y(t) \text{ is continuous everywhere except for some}$$

$$t_k \quad \text{at wich } y(t_k^-), y(t_k^+) \text{ exist with } y(t_k) = y(t_k^-), k = 1, \ldots \Big\}$$

Set

$$\Omega = \{y : [-r, \infty) \to E : y \in PC \cap D\}$$

Definition 10.11. We say that the function $y \in \Omega$ is a mild solution of system (10.20)–(10.22) if $y(t) = \phi(t)$ for all $t \in [-r, 0]$, the restriction of $y(\cdot)$ to the interval $[0, \infty)$ is continuous and there exists $v(\cdot) \in L^1_{loc}([0, \infty), E)$, such that $v(t) \in F(t, y_t)$ a.e $[0, \infty)$, and such that y satisfies the integral equation,

$$y(t) = T_0(t)\phi(0) + \int_0^t T_1(t-s)v(s)ds + \int_0^t T_1(t-s)Bu_y(s)ds$$

$$+ \sum_{0 < t_k < t} T_1(t-t_k)I_k(y(t_k^-)), \quad 0 \leq t < \infty. \tag{10.23}$$

Definition 10.12. The system (10.20)–(10.22) is said to be infinite controllable on the interval $[-r, \infty) \setminus \{t_k\}, k = 1, \ldots$ if for every initial function $\phi \in D$ and every $y_1 \in E$, and for each $n \in \mathbb{N}$, there exists a control $u \in L^2([0, t_n], U)$, such that the mild solution $y(t)$ of (10.20)–(10.22) satisfies $y(t_n) = y_1$.

Let us introduce the following hypotheses:

(10.12.1) The function $F : J \times \Omega \to \mathcal{P}_{cp}(E)$ is an L^1-Carathéodory map.

(10.12.2) There exist a function $p \in L^1(J, \mathbb{R}_+)$ and a continuous nondecreasing function $\psi : [0, \infty) \to [0, \infty)$ such that

$$\|F(t, x)\| \le p(t)\psi(\|x\|_D) \quad \text{for a.e. } t \in J \text{ and each } x \in D,$$

with

$$\int_1^\infty \frac{ds}{s + \psi(s)} = \infty.$$

(10.12.3) There exists $M > 0$ such that

$$\|T_1(t)\|_{B(E)} \le M \quad \text{for each } t > 0.$$

(10.12.4) For all $R > 0$ there exists $l_R \in L^1_{loc}([0, \infty), \mathbb{R}_+)$ such that

$$H_d(F(t, x), F(t, \bar{x})) \le l_R(t)\|x - \bar{x}\|_D \quad \text{for all } x, \bar{x} \in D \text{ with } \|x\|, \|\bar{x}\| \le R,$$

and

$$d(0, F(t, 0)) \le l_R(t) \quad \text{for a.e. } t \in J.$$

(10.12.5) There exist constants $c_k \ge 0, k = 1, \ldots$, such that

$$|I_k(y) - I_k(x)| \le c_k|x - \bar{x}| \quad \text{for each } x, \bar{x} \in E.$$

(10.12.6) For every $n > 0$, the linear operator $W : L^2(J_n, U) \to E$ $(J_n = [0, t_n])$, defined by

$$Wu = \int_0^{t_n} T(t_n - s)Bu(s)ds,$$

has a bounded inverse operator W^{-1} which takes values in $L^2(J_n, U) \setminus KerW$, and there exist positive constants $\overline{M}, \overline{M}_1$ such that $\|B\| \le \overline{M}$ and $\|W^{-1}\| \le \overline{M}_1$.

Theorem 10.13. *Assume that hypotheses (10.12.1)–(10.12.6) hold. If $M \sum_{k=1}^\infty c_k < 1$, then the IVP (10.20)–(10.22) is infinite controllable on $[-r, \infty)$.*

Proof. Using hypothesis (10.12.6) for each $y(\cdot)$ define the control

$$u_y(t) = W^{-1}\left[y_1 - T(t_n)\phi(0) - \int_0^{t_n} T(t_n - s)v(s)ds \right.$$

$$\left. - \sum_{0<t_k<s} T(s - t_k)I_k(y(t_k^-)) \right](t),$$

where

$$v \in S_{F,y} = \{v \in L^1(J, E) : v(t) \in F(t, y_t) \text{ a.e } t \in J\},$$

We shall now show that when using this control, the operator $N : \Omega \to \mathcal{P}(\Omega)$ defined by

$$N(y) = \left\{ h \in \Omega : h(t) = \begin{cases} \phi(t), & \text{if } t \in H, \\[2mm] T_0(t)\phi(0) + \displaystyle\int_0^t T_1(t - s)v(s)ds \\[2mm] \quad + \displaystyle\int_0^t T_1(t - s)(Bu_y)(s)ds \\[2mm] \quad + \displaystyle\sum_{0<t_k<t} T_1(t - t_k)I_k(y(t_k^-)), & \text{if } t \in J, \end{cases} \right\}$$

has a fixed point. This fixed point is then the mild solution of the IVP (10.20)–(10.22)

We define on Ω a family of semi-norms, thus rendering Ω into Fréchet space. Let τ be sufficiently large, then $\forall n \in \mathbb{N}$ we define in Ω the semi-norm:

$$\|y\|_n = \sup\{e^{-\tau L_n(t)}|y(t)| : -r \le t \le t_n\},$$

where

$$L_n = \int_{-r}^t \hat{l}_n(s)ds,$$

with

$$\hat{l}_n(t) = \begin{cases} 0, & \text{if } t \in H, \\[2mm] l_n(t)[M^2\overline{M}\,\overline{M_1}t_n + M] + M^2\overline{M}\,\overline{M_1}\displaystyle\sum_{k=0}^t c_k, & \text{if } t \in [0, t_n], \end{cases}$$

Thus $\Omega = \bigcup_{n\geq 1} \Omega_n$ where

$$\Omega_n = \left\{y : [-r, t_n] \to E : \quad y \in \mathcal{D} \bigcap PC_n(J, E)\right\}$$

and

$$PC_n = \left\{y : [0, t_n] \to E : \qquad y(t) \text{ is continuous everywhere except for some}\right.$$
$$\left. t_k \quad \text{at wich } y(t_k^-), y(t_k^+) \text{ exist with } y(t_k) = y(t_k^-), k = 1, \ldots n-1\right\}$$

Then Ω is a Fréchet space with the family of the semi-norms $\{\|.\|_n\}_{n\in I\!N}$.

Now, using the Frigon alternative, we are able to prove that the operator N has a fixed point.

Let $y \in \lambda N(y)$ for some $\lambda \in [0.1]$, and for some $v \in S_{F,y}$. For each $n \in I\!N$ and $t \in [0, t_n]$ we have:

$$y(t) = \lambda \left[T_0(t)\phi(0) + \int_0^t T_1(t - s)v(s)ds + \int_0^t T_1(t - s)Bu_y(s)ds \right.$$
$$\left. + \sum_{0<t_k<t} T_1(t - t_k)I_k(y(t_k^-)) \right]$$

then, we have

$$|y(t)| \leq \|T_0\|\|\phi\|_D + \int_0^t |T_1(t - s)|\|v(s)\|ds + \int_0^t |T_1(t - s)||Bu_y(s)|ds$$
$$+ \sum_{0<t_k<t} |T_1(t - t_k)||I_k(y(t_k^-))|$$
$$\leq M\|\phi\|_D + M\int_0^t p(s)\psi(\|y_s\|)ds + M\int_0^t \|B\|\|u_y(s)\|ds$$
$$+M\sum_{k=1}^n |I_k(y(t_k^-))|$$
$$\leq M\|\phi\|_D + M\int_0^t p(s)\psi(\|y_s\|)ds + M\overline{M}\int_0^t \left|W^{-1}\left[y_1 - T_0(t_n)\phi(0)\right.\right.$$
$$\left.\left. - \int_0^{t_n} T_1(t_n - \tau)v(\tau)d\tau - \sum_{0<t_k<\tau} T_1(\tau - t_k)I_k(y(t_k^-)) \right](s)\right|ds$$
$$+M\sum_{k=1}^n |I_k(y(t_k^-))|$$

$$\leq M\|\phi\|_D + M \int_0^t p(s)\psi(\|y_s\|)ds + M\overline{M}\,\overline{M}_1 \int_0^t \Bigg[|y_1| + \|T_0(t_n)\|\|\phi(0)\|$$

$$+M \int_0^{t_n} p(\tau)\psi(\|y_\tau\|)d\tau + M \sum_{0<t_k<s} |I_k(y(t_k^-))| \Bigg] ds$$

$$+M \sum_{k=1}^n |I_k(y(t_k^-))|$$

$$\leq M\|\phi\|_D + M \int_0^t p(s)\psi(\|y_s\|)ds + M\overline{M}\,\overline{M}_1 t_n |y_1| + M\overline{M}\,\overline{M}_1 t_n M\|\phi\|_D$$

$$+M^2\overline{M}\,\overline{M}_1 t_n \int_0^t p(\tau)\psi(\|y_\tau\|)d\tau + M^2\overline{M}\,\overline{M}_1 \int_0^t \sum_{0<t_k<s} |I_k(y(t_k^-))|ds$$

$$+M \sum_{k=1}^n |I_k(y(t_k^-))|.$$

It follows that

$$|y(t)| \leq M\overline{M}\,\overline{M}_1 t_n |y_1| + \Big[M + M\overline{M}\,\overline{M}_1 t_n \Big]\|\phi\|_D$$

$$+\Big[M + M^2\overline{M}\,\overline{M}_1 t_n \Big] \int_0^t p(s)\psi(\|y_s\|)ds$$

$$+M^2\overline{M}\,\overline{M}_1 \int_0^t \sum_{0<t_k<s} |I_k(y(t_k^-))|ds + M \sum_{k=1}^n |I_k(y(t_k^-))|$$

$$\leq M\overline{M}\,\overline{M}_1 t_n |y_1| + \Big[M + M\overline{M}\,\overline{M}_1 t_n \Big]\|\phi\|_D$$

$$+\Big[M + M^2\overline{M}\,\overline{M}_1 t_n \Big] \int_0^t p(s)\psi(\|y_s\|)ds$$

$$+M^2\overline{M}\,\overline{M}_1 \int_0^t \sum_{0<t_k<s} \big(|I_k(y(t_k^-)) - I_k(0)| + |I_k(0)| \big)ds$$

$$+M \sum_{k=1}^n \big(|I_k(y(t_k^-)) - I_k(0)| + |I_k(0)| \big)$$

$$\leq M\overline{M}\,\overline{M}_1 t_n |y_1| + \Big[M + M\overline{M}\,\overline{M}_1 t_n \Big]\|\phi\|_D$$

$$+\Big[M + M^2\overline{M}\,\overline{M}_1 t_n \Big] \int_0^t p(s)\psi(\|y_s\|)ds$$

$$+M^2\overline{M}\,\overline{M}_1 \int_0^t \sum_{0<t_k<s} |I_k(y(t_k^-)) - I_k(0)|ds$$

$$+M \sum_{k=1}^{n} |I_k(y(t_k^-)) - I_k(0)| + M^2\overline{M}\,\overline{M}_1 \int_0^t \sum_{0<t_k<s} |I_k(0)| ds$$

$$+M \sum_{k=1}^{n} \|I_k(0)\|$$

$$\leq M\overline{M}\,\overline{M}_1 t_n |y_1| + \left[M + M\overline{M}t_n\overline{M}_1\right] \|\phi\|_D$$

$$+\left[M + M^2\overline{M}\,\overline{M}_1 t_n\right] \sum_{k=1^n} \|I_k(0)\|$$

$$+\left[M + M^2\overline{M}t_n\overline{M}_1\right] \int_0^t p(s)\psi(\|y_s\|)ds + M^2\overline{M}\,\overline{M}_1 \int_0^t \sum_{k=1}^{n} c_k |y(t_k^-)| ds$$

$$+M \sum_{k=1}^{n} c_k |y(t_k^-)|.$$

Set

$$C = M\overline{M}\,\overline{M}_1 t_n |y_1| + \left[M + M\overline{M}t_n\overline{M}_1\right] \|\phi\|_D + \left[M + M^2\overline{M}t_n\overline{M}_1\right] \sum_{k=1}^{n} \|I_k(0)\|.$$

Now, we consider the function μ defined by:

$$\mu(t) = \sup\Big\{|y(s)| : -r \leq s \leq t\Big\}, t \leq t_n$$

Let $t^* \in [-r, t_n]$ such that $\mu(t) = |y(t^*|$

It is clear that:

if $t^* \in H$, then $\mu(t) = \|\phi\|_D$

if $t^* \in [0, t_n]$, we have for each $t \in [0, t_n]$

$$\mu(t) \leq C + \left[M + M^2\overline{M}t_n\overline{M}_1\right] \int_0^t p(s)\psi((\mu_s))ds$$

$$+M^2\overline{M}\,\overline{M}_1 \int_0^t \sum_{0<t_k<s} c_k\mu(s)ds + M \sum_{k=1}^{n} c_k\mu(t).$$

Then

$$\left[1 - M \sum_{k=1}^{n} c_k\right]\mu(t) \leq C + \left[M + M^2\overline{M}t_n\overline{M}_1\right] \int_0^t p(s)\psi(\mu(s))ds$$

$$+M^2\overline{M}\,\overline{M}_1 \int_0^t \sum_{0<t_k<s} c_k\mu(s)ds.$$

Thus we have

$$\mu(t) \le C_1 + \int_0^t \hat{M}(s)\big(\mu(s) + \psi(\mu(s))\big),$$

where

$$C_1 = \frac{C}{1 - M \sum_{k=1}^n c_k},$$

and

$$\hat{M}(s) = \frac{1}{1 - M \sum_{k=1}^n c_k} \left[\left[M + M^2 \overline{M}\, \overline{M}_1 t_n \right] p(s) + M^2 \overline{M}\, \overline{M}_1 \sum_{0<t_k<s} c_k \right].$$

Let us take the right-hand side of the above inequality as $v(t)$, then we have:

$$v(0) = C_1, \quad \mu(t) \le v(t) \qquad \forall t \in [0, t_n]$$

and

$$v'(t) = \hat{M}(t)(\mu(t) + \psi(\mu(t))).$$

Using the nondecreasing character of ψ, we get:

$$v'(t) \le \hat{M}(t)(v(t) + \psi(v(t)))a.et \in [0, t_n].$$

This implies that for each $t \in [0, t_n]$

$$\int_{v(0)}^{v(t)} \frac{ds}{s + \psi(s)} \le \int_0^{t_n} \hat{M}(s)ds \le \int_{v(0)}^\infty \frac{ds}{s + \psi(s)}.$$

Thus from (10.12.2) there exists a constant M_n such that

$$v(t) \le M_n, \qquad \forall t \in [0, t_n]$$

From the definition of μ, we conclude that

$$\sup\{|y(t)|, \quad t \in [0, t_n]\} \le M_n$$

Set

$$U_0 = \left\{ y \in \Omega_n, \quad \|y\|_n \le M_n + 1 \right\}$$

Clearly U_0 is a closed subset of Ω_n. We shall show that $N : U_0 \to \mathcal{P}(U_0)$ is a contraction and an admissible operator.

First, we prove that N is a contraction; that is, there exists $\gamma < 1$, such that

$$H_d(N(y), N(y^*)) \le \gamma \|y - y^*\|_n, \quad \text{for each } y, y^* \in U_0.$$

Let $y, y^* \in U_0$ and $h \in N(y)$. Then there exists $v(t) \in F(t, y_t)$ such that for each $t \in [0, t_n]$

$$h(t) = T_0(t)\phi(0) + \int_0^t T_1(t - s)v(s)ds + \int_0^t T_1(t - s)Bu_y(s)ds$$
$$+ \sum_{0 < t_k < t} T_1(t - t_k)I_k(y(t_k^-)).$$

From (10.12.4) it follows that

$$H_d(F(t, y_t), F(t, y_t^*)) \le l_n(t)\|y_t - y_t^*\|_D.$$

Hence there exists $v^* \in F(t, y_t^*)$ such that

$$|v(t) - v^*(t)| \le l_n(t)\|y_t - y_t^*\|_D, \quad \forall t \in [0, t_n].$$

Let us define $\forall t \in [0, t_n]$

$$h^*(t) = T_0(t)\phi(0) + \int_0^t T_1(t - s)v^*(s)ds + \int_0^t T_1(t - s)Bu_{y^*}(s)ds$$
$$+ \sum_{0 < t_k < t} T_1(t - t_k)I_k(y^*(t_k^-)).$$

Then we have

$$|h(t) - h^*(t)| = \left| \int_0^t T_1(t - s)[v(s) - v^*(s)]ds + \int_0^t T_1(t - s)[B(u_y - u_{y^*})(s)]ds \right.$$
$$\left. + \sum_{0 < t_k < t} T_1(t - t_k)[I_k(y(t_k^-)) - I_k(y^*(t_k^-))] \right|$$

$$\leq M \int_0^t l_n(s) \|y_s - y_s^*\|_D ds + M\overline{M} \int_0^t |u_y(s) - u_{y^*}(s)| ds$$

$$+ M \sum_{0 < t_k < t} c_k |y(t_k^-) - y^*(t_k^-)|$$

$$\leq M \int_0^t l_n(s) \|y_s - y_s^*\|_D ds$$

$$+ M\overline{M} \int_0^t \left| W^{-1} \left[\int_0^{t_n} T_1(t - \tau)[v(\tau) - v^*(\tau)] d\tau \right.\right.$$

$$\left.\left. + \sum_{0 < t_k < \tau} T_1(t - \tau) I_k(y(t_k^-) - y^*(t_k^-)) \right] \right| ds$$

$$+ M \sum_{0 < t_k < t} c_k |y(t_k^-) - y^*(t_k^-)|$$

$$\leq M \int_0^t l_n(s) \|y_s - y_s^*\|_D ds$$

$$+ M^2 \overline{M}\, \overline{M}_1 t_n \int_0^t l_n(s) \|y_s - y_s^*\|_D ds$$

$$+ M^2 \overline{M}\, \overline{M}_1 \int_0^t \sum_{0 < t_k < s} c_k |y(t_k^-) - y^*(t_k^-)| ds$$

$$+ \sum_{0 < t_k < t} c_k |y(t_k^-) - y^*(t_k^-)| ds$$

$$\leq \left[M + M^2 \overline{M}\, \overline{M}_1 t_n \right] \int_0^t l_n(s) \|y_s - y_s^*\|_D ds$$

$$+ M^2 \overline{M}\, \overline{M}_1 \int_0^t \sum_{0 < t_k < s} c_k |y(t_k^-) - y^*(t_k^-)| ds$$

$$+ \sum_{0 < t_k < t} c_k |y(t_k^-) - y^*(t_k^-)| ds$$

$$\leq \left[M + M^2 \overline{M}\, \overline{M}_1 t_n \right] \int_0^t l_n(s) e^{\tau L_n(s)} \|y - y^*\|_n ds$$

$$+ M^2 \overline{M}\, \overline{M}_1 \int_0^t e^{\tau L_n(s)} \sum_{0 < t_k < s} c_k \|y - y^*\|_n ds$$

$$+ M e^{\tau L_n(t)} \sum_{0 < t_k < t} c_k \|y - y^*\|_n ds,$$

which gives:

$$|h(t) - h^*(t)| \leq \left[3 \int_0^t \hat{l}_n(s) e^{\tau L_n(s)} ds + M e^{\tau L_n(t)} \sum_{0 < t_k < t} c_k \right] \|y - y^*\|_n$$

$$\leq \left[\frac{3}{\tau} e^{\tau L_n(s)} |_0^t + M e^{\tau L_n(t)} \sum_{k=1}^n c_k \right] \|y - y^*\|_n$$

$$\leq \left[\frac{3}{\tau} e^{\tau L_n(t)} - \frac{3}{\tau} + M e^{\tau L_n(t)} \sum_{k=1}^n c_k \right] \|y - y^*\|_n.$$

As τ is sufficiently large, thus

$$|h(t) - h^*(t)| \leq \left[\frac{3}{\tau} + M \sum_{k=1}^n c_k \right] e^{\tau L_n(t)} \|y - y^*\|_n.$$

Then, it follows

$$|h(t) - h^*(t)| e^{-\tau L_n(t)} \leq \left[\frac{3}{\tau} + M \sum_{k=1}^n c_k \right] \|y - y^*\|_n.$$

Therefore,

$$\|h - h^*\|_n \leq \left[\frac{3}{\tau} + M \sum_{k=1}^n c_k \right] \|y - y^*\|_n.$$

By an analogous relation, obtained by interchanging the roles of y and y^*, it follows that

$$H_d(N(y), N(y^*)) \leq \left(\frac{3}{\tau} + M \sum_{k=1}^n c_k \right) \|y - y^*\|_n.$$

So, N is a contraction.

Now, $N : \Omega_n \to \mathcal{P}_{cp}(\Omega_n)$ is given by

$$N(y) = \left\{ h \in \Omega_n : h(t) = \begin{cases} 0, & \text{if } t \in H, \\ \displaystyle \int_0^t T_1(t-s)v(s)ds \\ + \displaystyle \int_0^t T_1(t-s)(Bu_y^n)(s)ds \\ + \displaystyle \sum_{0 < t_k < t} T_1(t-t_k)I_k(y(t_k^-)), & \text{if } t \in [0, t_n], \end{cases} \right\}$$

where $v \in S^n_{F,y} = \{u \in L^1([0, t_n], E) : u \in F(t, y_t) \text{ a.e. } t \in [0, t_n]\}$. From (10.12.4)–(10.12.6) and since F is compact valued, we can prove that for every $y \in \Omega_n$, $N(y) \in \mathcal{P}_{cp}(\Omega_n)$, and there exists $y^* \in \Omega_n$ such that $y^* \in N(y^*)$. Let $h \in \Omega_n$, $y^* \in U_0$ and $\varepsilon > 0$. Now, if $\bar{y} \in N(y^*)$, then we have

$$\|y^* - \bar{y}\|_n \leq \|y^* - h\|_n + \|\bar{y} - h\|_n.$$

Since h is arbitrary we may suppose that $h \in B(\bar{y}, \varepsilon) = \{k \in \Omega_n : \|k - \bar{y}\|_n \leq \varepsilon\}$. Therefore,

$$\|y^* - \bar{y}\|_n \leq \|y^* - N(y^*)\|_n + \varepsilon.$$

On the other hand, if $\bar{y} \notin N(y^*)$, then $\|\bar{y} - N(y^*)\| \neq 0$. Since $N(y^*)$ is compact, there exists $x \in N(y^*)$ such that $\|\bar{y} - N(y^*)\|_n = \|\bar{y} - x\|_n$. Then we have

$$\|y^* - x\|_n \leq \|y^* - h\|_n + \|x - h\|_n.$$

Therefore,

$$\|y^* - x\|_n \leq \|y^* - N(y^*)\|_n + \varepsilon.$$

So, N is an admissible operator contraction. Finally, by Lemma 1.27, N has at least one fixed point, y, which is a mild solution to (10.20)–(10.22).   □

10.4.2   Example

As an application of our above result, we consider the following impulsive partial functional inclusion,

$$\frac{\partial z(t, x)}{\partial t} - d\Delta z(t, x) \in F(t - r, x) + Bu(t), \text{ a.e. } t \in J\backslash\{t_1, t_2, \ldots\}, x \in \Omega \quad (10.24)$$

$$b_k z(t_k^-, x) = z(t_k^+) - z(t_k^-), \quad k = 1, \ldots, x \in \partial\Omega \quad (10.25)$$

$$z(t, x) = 0, \; t \in [0, \infty)\backslash\{t_1, t_2, \ldots\}, x \in \overline{\Omega} \quad (10.26)$$

$$z(t, x) = \phi(t, x), \quad t \in H, x \in \overline{\Omega} \quad (10.27)$$

where d, r, b_k are positive constants, Ω is a bounded open in $I\!R^n$ with regular boundary $\partial\Omega$, $\Delta = \sum_{i=1}^{n} \frac{\partial^2}{\partial x_i^2}$, $\phi \in D = \{\psi : H \times \overline{\Omega} \to I\!R; \psi$ is continuous everywhere except for a countable number of points at which $\psi(s^-), \psi(s^+)$ exist with $\psi(s^-) = \psi(s)$, and $|\psi(\theta, x)| < \infty\}$, $0 = t_0 < t_1 < t_2 < \cdots < t_m < \cdots$, $z(t_k^+) = \lim_{(h,x) \to (0^+, x)} z(t_k + h, x)$, $z(t_k^-) = \lim_{(h,x) \to (0^-, x)} z(t_k - h, x)$, $F : [0, \infty) \times I\!R^n \to \mathcal{P}(I\!R^n)$ is a multi-valued map with compact values.

Consider $E = C(\overline{\Omega}, IR^n)$ the Banach space of continuous functions on $\overline{\Omega}$ with values in IR^n, $y(t) = z(t, .)$. Let A the operator defined in E by $Ay = d\Delta y$, $I_k :$ $E \to \overline{D(A)}$ such that $I_k(y(t_k^-)) = b_k y(t_k^-)$, then the problem (10.24)–(10.27) can be written as

$$y'(t) - Ay(t) \in F(t, y_t) + Bu(t) \qquad a.e.\ t \in J\backslash\{t_1, t_2, \ldots\}, \tag{10.28}$$

$$\Delta y|_{t=t_k} = I_k(y(t_k^-)), k \in \{1, 2, \ldots\} \tag{10.29}$$

$$y(t) = \phi(t)\ t \in H \tag{10.30}$$

We have,

$$D(A) = \{y : \quad y \in E, \quad \Delta y \in E \quad and \quad y|_{\partial\Omega} = 0\},$$

and

$$X_0 = \overline{D(A)} = \{y : \quad y \in E, \quad y|_{\partial\Omega} = 0\} \neq E$$

So, we can apply the extrapolation method.

It is well known from [100] that Δ satisfies the properties:

i) $(0, \infty) \subset \rho(\Delta)$
ii) $\|R(\lambda, \Delta)\| \leq \frac{1}{\lambda}$ for some $\lambda > 0$

It follows that Δ satisfies (Hy).

Also from [106], the family

$$T_0(t)f(s) = (4\pi)^{\frac{-n}{2}} \int_{IR^n} e^{\frac{-|s-\tau|^2}{4t}} f(\tau) d\tau$$

for $t > 0$, $s \in \mathbb{R}^n$, and $f \in X_0$ with $T(0) = I$, define a strongly continuous semigroup on E, its generator A_0 coincides with the closure of the Laplacian operator with domain X_0, and there exist constants $N_0 > 0, \omega > 0$ such that $\|T_0\| \leq N_0 e^{\omega t}$ for $t > 0$.

Thus under appropriate conditions on the function F and the operator B as those mentioned in hypotheses (10.12.1)–(10.12.6) the problem (10.24)–(10.27) has at least one mild solution.

10.5   Notes and Remarks

The results of Chap. 10 are taken from Abada et al. [1, 3]. Other results may be found in [54, 74, 107].

Chapter 11
Functional Differential Inclusions with Multi-valued Jumps

11.1 Introduction

In this chapter, we are concerned by the existence of mild solutions of functional differential inclusions with delay and multi-valued jumps in a Banach space.

11.2 Semi-linear Functional Differential Inclusions with State-Dependent Delay and Multi-valued Jump

11.2.1 Introduction

In this section, we shall be concerned with the existence of integral solutions defined on a compact real interval for first order impulsive semi-linear functional inclusions with state-dependent delay in a separable Banach space of the form:

$$y'(t) \in Ay(t) + F(t, y_{\rho(t,y_t)}), \quad t \in I = [0, b], \tag{11.1}$$

$$\Delta y(t_i) \in I_k(y_{t_k}), \quad k = 1, 2, \ldots, m, \tag{11.2}$$

$$y(t) = \phi, \quad t \in (-\infty, 0], \tag{11.3}$$

where $F : J \times \mathcal{D} \to E$ is a given multi-valued function, $\mathcal{D} = \{\psi : (-\infty, 0] \to E, \psi$ is continuous everywhere except for a finite number of points s at which $\psi(s^-), \psi(s^+)$ exist and $\psi(s^-) = \psi(s)\}, \phi \in \mathcal{D}$,where $\mathcal{D}$ is the phase space that will be specified later $(0 < r < \infty), 0 = t_0 < t_1 < \cdots < t_m < t_{m+1} = b$,

© Springer International Publishing Switzerland 2015
S. Abbas, M. Benchohra, *Advanced Functional Evolution Equations and Inclusions*, Developments in Mathematics 39,
DOI 10.1007/978-3-319-17768-7_11

$I_k : \mathcal{D} \to E$ $(k = 1, 2, \ldots, m)$, $\rho : I \times \mathcal{D} \to (-\infty, b]$, $A : D(A) \subset E \to E$ is a non-densely defined closed linear operator on E, and E a real separable Banach space with norm $|.|$.

11.2.2 Existence of Integral Solutions

In this section, we will employ an axiomatic definition for the phase space $\mathcal{D}$ which is similar to those introduced in [142]. Specifically, $\mathcal{D}$ will be a linear space of functions mapping $(-\infty, 0]$ into E endowed with a semi norm $\|.\|_{\mathcal{D}}$, and satisfies the following axioms introduced at first by Hale and Kato in [132]:

(A1) There exist a positive constant H and functions $K(\cdot)$, $M(\cdot) : \mathbb{R}^+ \to \mathbb{R}^+$ with K continuous and M locally bounded, such that for any $b > 0$, if $y : (-\infty, b] \to E$, $y \in \mathcal{D}$, and $y(\cdot)$ is continuous on $[0, b]$, then for every $t \in [0, b]$ the following conditions hold:

  (i) y_t is in $\mathcal{D}$;
  (ii) $|y(t)| \leq H\|y_t\|_{\mathcal{D}}$;
  (iii) $\|y_t\|_{\mathcal{D}} \leq K(t) \sup\{|y(s)| : 0 \leq s \leq t\} + M(t)\|y_0\|_{\mathcal{D}}$, and H, K and M are independent of $y(\cdot)$.

  Denote

$$K_b = \sup\{K(t) : t \in J\} \text{ and } M_b = \sup\{M(t) : t \in J\}.$$

(A) The space $\mathcal{D}$ is complete.

11.2.3 Main Results

Before starting and proving our main theorem for the initial value problem (11.1)–(11.3), we give the definition of the integral solution.

Definition 11.1. We say that $y : (-\infty, b] \to E$ is an integral solution of (11.1)–(11.3) if $y(t) = \phi(t)$ for all $t \in (-\infty, 0]$, the restriction of $y(\cdot)$ to the interval $[0, b]$ is continuous, and there exist $v(\cdot) \in L^1(J_k, E)$ and $\mathcal{I}_k \in I_k(y(t_k))$, such that $v(t) \in F(t, y_{\rho(t, y_t)})$ a.e $t \in [0, b]$, and y satisfies the integral equation,

(i) $y(t) = \phi(0) + A \displaystyle\int_0^t y(s)ds + \int_0^t v(s)ds + \sum_{0 < t_k < t} S'(t - t_k)\mathcal{I}_k, \quad t \in J.$

(ii) $\displaystyle\int_0^t y(s)ds \in D(A)$ for $t \in J$,

From the definition it follows that $y(t) \in \overline{D(A)}$, for each $t \geq 0$, in particular $\phi(0) \in \overline{D(A)}$. Moreover, y satisfies the following variation of constants formula:

$$y(t) = S'(t)\phi(0) + \frac{d}{dt}\int_0^t S(t-s)v(s)ds + \sum_{0<t_k<t} S'(t-t_k)\mathcal{I}_k \quad t \geq 0. \qquad (11.4)$$

We notice also that if y satisfies (11.4), then

$$y(t) = S'(t)\phi(0) + \lim_{\lambda\to\infty}\int_0^t S'(t-s)B_\lambda v(s)ds + \sum_{0<t_k<t} S'(t-t_k)\mathcal{I}_k, \quad t \geq 0.$$

we always assume that $\rho : I \times \mathcal{D} \to (-\infty, b]$ is continuous. Additionally, we introduce following hypotheses:

(Hφ)   The function $t \to \varphi_t$ is continuous from $\mathcal{R}(\rho^-) = \{\rho(s,\varphi) : (s,\varphi) \in J \times \mathcal{D}, \rho(s,\varphi) \leq 0\}$ into $\mathcal{D}$ and there exists a continuous and bounded function $L^\phi : \mathcal{R}(\rho^-) \to (0,\infty)$ such that $\|\phi_t\|_\mathcal{D} \leq L^\phi(t)\|\phi\|_\mathcal{D}$ for every $t \in \mathcal{R}(\rho^-)$.

(11.1.1) A satisfies Hille–Yosida condition;

(11.1.2)   There exist constants $c_k \geq 0$, $k = 1,\ldots,m$ such that

$$H_d(I_k(y), I_k(x))| \leq c_k|y-x| \quad \text{for each} \quad x,y \in \mathcal{D}.$$

(11.1.3)   The valued multi-valued map $F : J \times \mathcal{D} \to E$ is convex and Carathéodory;

(11.1.4)   the operator $S'(t)$ is compact in $\overline{D(A)}$ wherever $t > 0$;

(11.1.5)   There exist a function $p \in L^1(J, \mathbb{R}_+)$ and a continuous nondecreasing function $\psi : [0,\infty) \to (0,\infty)$ such that

$$\|F(t,x)\|_\mathcal{P} = \sup\{|v| : v \in F(t,x)\} \leq p(t)\psi(\|x\|_\mathcal{D}) \quad \text{for a.e. } t \in J \text{ and each } x \in \mathcal{D},$$

with $\displaystyle\int_0^b e^{-\omega s}p(s)ds < \infty$,

$$\limsup_{u\to+\infty} \frac{[(M_b + L^\phi + MK_b)\|\phi\|_\mathcal{D} + K_b]u}{c_1 + c_2\int_0^t e^{-\omega s}p(s)\psi(K_b u + (M_b + L^\phi + MK_b)\|\phi\|_\mathcal{D})\,ds} > 1, \qquad (11.5)$$

where

$$c_1 = \frac{ce^{\omega b}K_b}{1 - Me^{\omega b}K_b\sum_{k=1}^m c_k} + (M_b + L^\phi + MK_b)\|\phi\|_\mathcal{D}, \qquad (11.6)$$

and

$$c = \sum_{k=1}^{m} \left[|I_k(0)| + c_k \left(M_b + L^\phi + MK_b \right) \|\phi\|_{\mathcal{D}} \right]. \tag{11.7}$$

$$c_2 = \frac{MK_b e^{\omega b}}{1 - Me^{\omega b} K_b \sum\limits_{k=1}^{m} c_k}. \tag{11.8}$$

The next result is a consequence of the phase space axioms.

Lemma 11.2 ([139], Lemma 2.1).   *If $y : (-\infty, b] \to E$ is a function such that $y_0 = \phi$ and $y|_J \in PC(J : D(A))$, then*

$$\|y_s\|_{\mathcal{D}} \le (M_a + L^\phi)\|\phi\|_{\mathcal{D}} + K_a \sup\{\|y(\theta)\|; \ \theta \in [0, \max\{0, s\}]\}, \quad s \in \mathcal{R}(\rho^-) \cup J,$$

where $L^\phi = \sup_{t \in \mathcal{R}(\rho^-)} L^\phi(t)$, $M_a = \sup_{t \in J} M(t)$ and $K_a = \sup_{t \in J} K(t)$.

Theorem 11.3. *Assume that $(H\varphi)$ and $(11.1.1)$–$(11.1.5)$ hold. If*

$$Me^{\omega b} K_b \sum_{k=1}^{m} c_k < 1, \tag{11.9}$$

then the problem (11.1)–(11.3) has at least one integral solution on $(-\infty, b]$.

Proof. Consider the multi-valued operator:
$$N : PC\left((-\infty, b], \overline{D(A)}\right) \to \mathcal{P}(PC\left((-\infty, b], \overline{D(A)}\right)) \text{ defined by}$$

$$N(y) = \left\{ h \in PC\left((-\infty, b], \overline{D(A)}\right) : h(t) = \begin{cases} \phi(t), & \text{if } t \le 0, \\[2mm] S'(t)\phi(0) + \dfrac{d}{dt}\displaystyle\int_0^t S(t-s)v(s)ds \\[2mm] \quad + \displaystyle\sum_{0 < t_k < t} S'(t - t_k)\mathcal{I}_k, v \in S_{F, y_{\rho(s, y_s)}}, \mathcal{I}_k \in I_k(y(t_k^-)) \\[2mm] \qquad\qquad\qquad \text{if } t \in J, \end{cases} \right\}$$

For $\phi \in \mathcal{D}$ define the function $\tilde{\phi} : (-\infty, b] \to E$ such that:

$$\tilde{\phi}t = \begin{cases} \phi(t), & \text{if } t \le 0 \\[2mm] S'(t)\phi(0), & \text{if } t \in J. \end{cases}$$

Then $\tilde{\phi}_0 = \phi$. For each $x \in \mathcal{B}_b$ with $x(0) = 0$, we denote by $\bar{x}$ the function defined by

$$\bar{x}(t) = \begin{cases} 0, & t \in (-\infty, 0], \\ x(t), \, t \in J, \end{cases}$$

We can decompose it as $y(t) = \tilde{\phi}(t) + x(t), \;\; 0 \le t \le b$, which implies $y_t = \tilde{\phi}_t + x_t$, for every $0 \le t \le b$ and the function $x(.)$ satisfies

$$x(t) = \frac{d}{dt} \int_0^t S(t-s)v(s)ds$$
$$+ \sum_{0 < t_k < t} S'(t-t_k)\mathcal{I}_k \;\; t \in J,$$

where:
$$v(s) \in S_{F,x_{\rho(s,x_s + \tilde{\phi}_s)} + \tilde{\phi}_{\rho(s,x_s + \tilde{\phi}_s)}} \text{ and } \mathcal{I}_k \in I_k\left(x_{t_k} + \tilde{\phi}_{t_k}\right) \text{ Let}$$

$$\mathcal{B}_b^0 = \{x \in \mathcal{B}_b : x_0 = 0 \in \mathcal{D}\}.$$

For any $x \in \mathcal{B}_b^0$ we have

$$\|x\|_b = \|x_0\|_{\mathcal{D}} + \sup\{|x(s)| : 0 \le s \le b\} = \sup\{|x(s)| : 0 \le s \le b\}.$$

Thus $(\mathcal{B}_b^0, \|\cdot\|_b)$ is a Banach space. We define the two multi-valued operators $\mathcal{A}, \mathcal{B} : \mathcal{B}_b^0 \to \mathcal{P}(\mathcal{B}_b^0)$ by:

$$\mathcal{A}(x) := \left\{ h \in \mathcal{B}_b^0 : \; h(t) = \begin{cases} 0, & \text{if } t \in (-\infty, 0]; \\ \sum_{0 < t_k < t} S'(t-t_k)\mathcal{I}_k, & \mathcal{I}_k \in I_k\left(x_{t_k} + \tilde{\phi}_{t_k}\right), & \text{if } t \in J, \end{cases} \right\}$$

and

$$\mathcal{B}(x) := \left\{ h \in \mathcal{B}_b^0 : h(t) = \begin{cases} 0, & \text{if } t \in (-\infty, 0]; \\ \frac{d}{dt} \int_0^t S(t-s)v(s)ds, v(s) \in S_{F,x_{\rho(s,x_s + \tilde{\phi}_s)} + \tilde{\phi}_{\rho(s,x_s + \tilde{\phi}_s)}} & \text{if } t \in J. \end{cases} \right\}$$

Obviously to prove that the multi-valued operator N has a fixed point is reduced that the operator inclusion $x \in \mathcal{A}(x) + \mathcal{B}(x)$ has one, so it turns to show that the multi-valued operators $\mathcal{A}$ and $\mathcal{B}$ satisfy all conditions of Theorem 1.32. For better readability, we break the proof into a sequence of steps.

Step 1: *$\mathcal{A}$ is a contraction.* Let $x_1, x_2 \in \mathcal{B}_b^0$. Then for $t \in J$

$$H_d\left(\mathcal{A}(x_1), \mathcal{A}(x_2)\right) = H_d\left(\sum_{0<t_k<t} S'(t-t_k)I_k(x_{t_k}^1 + \tilde{\phi}_{t_k}),\right.$$

$$\left.\sum_{0<t_k<t} S'(t-t_k)I_k(x_{t_k}^2 + \tilde{\phi}_{t_k})\right)$$

$$\leq Me^{\omega b} \sum_{0\leq t_k \leq t} \left| I_k(x_{t_k}^1) - I_k(x_{t_k}^2) \right|$$

$$\leq Me^{\omega b} \sum_{k=1}^{m} c_k \| x_{t_k}^1 - x_{t_k}^2 \|_{\mathcal{D}}$$

$$\leq Me^{\omega b} K_b \sum_{k=1}^{m} c_k \| x_1 - x_2 \|_{\mathcal{D}}.$$

Hence by (11.9) $\mathcal{A}$ is a contraction.

Step 2: *$\mathcal{B}$ has compact, convex values, and it is completely continuous.* This will be given in several claims.

Claim 1: *$\mathcal{B}$ has compact values.* The operator $\mathcal{B}$ is equivalent to the composition $\mathcal{L} \circ S_F$ on $L^1(J, E)$, where $\mathcal{L} : L^1(J, E) \to \mathcal{B}_b^0$ is the continuous operator defined by

$$\mathcal{L}(v)(t) = \frac{d}{dt} \int_0^t S(t-s)v(s)ds, \ t \in J.$$

Then, it suffices to show that $\mathcal{L} \circ S_F$ has compact values on $\mathcal{B}_b^0$.

Let $x \in \mathcal{B}_b^0$ arbitrary and v_n a sequence such that $v_n(t) \in S_{F, x_{\rho(t,x_t+\tilde{\phi}_t)} + \tilde{\phi}_{\rho(t,x_t+\tilde{\phi}_t)}}$, a.e. $t \in J$. Since $F(t, x_{\rho(t,x_t+\tilde{\phi}_t)} + \tilde{\phi}_{\rho(t,x_t+\tilde{\phi}_t)})$ is compact, we may pass to a subsequence. Suppose that $v_n \to v$ in $L_w^1(J, E)$ (the space endowed with the weak topology), where $v(t) \in F(t, x_{\rho(t,x_t+\tilde{\phi}_t)} + \tilde{\phi}_{\rho(t,x_t+\tilde{\phi}_t)})$, a.e. $t \in J$. An application of Mazur's theorem [185] implies that the sequence v_n converges strongly to v and hence $v(t) \in S_{F, x_{\rho(t,x_t+\tilde{\phi}_t)} + \tilde{\phi}_{\rho(t,x_t+\tilde{\phi}_t)}}$. From the continuity of $\mathcal{L}$, it follows that $\mathcal{L}v_n(t) \to \mathcal{L}v(t)$ pointwise on J as $n \to \infty$. In order to show that the convergence is uniform, we first show that $\{\mathcal{L}v_n\}$ is an equi-continuous sequence. Let $\tau_1, \tau_2 \in J$, then we have:

$$|\mathcal{L}(v_n(\tau_1)) - \mathcal{L}(v_n(\tau_2))| = \left| \frac{d}{dt} \int_0^{\tau_1} S(\tau_1 - s)v_n(s)ds \right.$$

$$\left. - \frac{d}{dt} \int_0^{\tau_2} S(\tau_2 - s)v_n(s)ds \right|$$

$$\leq \left| \lim_{\lambda \to \infty} \int_0^{\tau_1} [S'(\tau_1 - s) - S'(\tau_2 - s)] B_\lambda v_n(s)) ds \right|$$

$$+ \left| \lim_{\lambda \to \infty} \int_{\tau_1}^{\tau_2} S'(\tau_2 - s) B_\lambda v_n(s) ds \right|.$$

As $\tau_1 \to \tau_2$, the right-hand side of the above inequality tends to zero. Since $S'(t)$ is a strongly continuous operator and the compactness of $S'(t), t > 0$, implies the uniform continuity (see [16, 168]). Hence $\{\mathcal{L}v_n\}$ is equi-continuous, and an application of Arzéla-Ascoli theorem implies that there exists a subsequence which is uniformly convergent. Then we have $\mathcal{L}v_{n_j} \to \mathcal{L}v \in (\mathcal{L} \circ S_F)(x)$ as $j \mapsto \infty$, and so $(\mathcal{L} \circ S_F)(x)$ is compact. Therefore $\mathcal{B}$ is a compact valued multi-valued operator on $\mathcal{B}_b^0$.

Claim 2: $\mathcal{B}(x)$ *is convex for each* $z \in D_b^0$. Let $h_1, h_2 \in \mathcal{B}(x)$, then there exist $v_1, v_2 \in S_{F, x_{\rho(t, x_t + \tilde{\phi}_t)} + \tilde{\phi}_{\rho(t, x_t + \tilde{\phi}_t)}}$, such that, for each $t \in J$ we have

$$h_i(t) = \begin{cases} 0, & \text{if } t \in (-\infty, 0], \\ \dfrac{d}{dt} \displaystyle\int_0^t S(t - s) v_i(s) ds & \text{if } t \in J, \end{cases} \right\}, \; i = 1, 2.$$

Let $0 \leq \delta \leq 1$. Then, for each $t \in J$, we have

$$(\delta h_1 + (1 - \delta)h_2)(t) = \begin{cases} 0, & \text{if } t \in (-\infty, 0], \\ \dfrac{d}{dt} \displaystyle\int_0^t S(t - s)[\delta v_1(s) + (1 - \delta)v_2(s)] ds & \text{if } t \in J. \end{cases} \right\}$$

Since F has convex values, one has

$$\delta h_1 + (1 - \delta)h_2 \in \mathcal{B}(x).$$

Claim 3: $\mathcal{B}$ *maps bounded sets into bounded sets in* $\mathcal{B}_b^0$
Let $B_q = \{x \in \mathcal{B}_b^0 : \|x\|_b \leq q\} \; q \in \mathbb{R}^+$ a bounded set in $\mathcal{B}_b^0$.

It is equivalent to show that there exists a positive constant l such that for each $x \in B_q$ we have $\|\mathcal{B}(x)\|_b \leq l$. So choose $x \in B_q$, then for each $h \in \mathcal{B}(x)$, and each $x \in B_q$, there exists $v \in S_{F, x_{\rho(t, x_t + \tilde{\phi}_t)} + \tilde{\phi}_{\rho(t, x_t + \tilde{\phi}_t)}}$. such that

$$h(t) = \frac{d}{dt} \int_0^t S(t - s) v(s) ds.$$

From (A) we have

$$\|x_{\rho(t, x_t + \tilde{\phi}_t)} + \tilde{\phi}_{\rho(t, x_t + \tilde{\phi}_t)}\|_{\mathcal{D}} \leq K_b q + (M_b + L^\phi)\|\phi\|_{\mathcal{D}} + K_b M |\phi(0)| = q_*$$

Then by (11.1.6) we have

$$|h(t)| \le M e^{\omega b} \psi(q_*) \int_0^t e^{-\omega s} p(s) ds := l.$$

This further implies that

$$\|h\|_{\mathcal{B}_b^0} \le l.$$

Hence $\mathcal{B}(B)$ is bounded.

Claim 4: $\mathcal{B}$ *maps bounded sets into equi-continuous sets.*

Let B_q be, as above, a bounded set and $h \in \mathcal{B}(x)$ for some $x \in B$. Then, there exists $v \in S_{F,x_{\rho(t,x_t + \tilde{\phi}_t)} + \tilde{\phi}_{\rho(t,x_t + \tilde{\phi}_t)}}$. such that

$$h(t) = \frac{d}{dt} \int_0^t S(t-s) v(s) ds, \quad t \in J.$$

Let $\tau_1, \tau_2 \in J \backslash \{t_1, t_2, \dots, t_m\}$, $\tau_1 < \tau_2$. Thus if $\epsilon > 0$, we have

$$
\begin{aligned}
|h(\tau_2) - h(\tau_1)| \le\ & \left| \lim_{\lambda \to \infty} \int_0^{\tau_1 - \epsilon} [S'(\tau_2 - s) - S'(\tau_1 - s)] B_\lambda v(s) ds \right| \\
& + \left| \lim_{\lambda \to \infty} \int_{\tau_1 - \epsilon}^{\tau_1} [S'(\tau_2 - s) - S'(\tau_1 - s)] B_\lambda v(s) ds \right| \\
& + \left| \lim_{\lambda \to \infty} \int_{\tau_1}^{\tau_2} S'(\tau_2 - s) B_\lambda v(s) ds \right| \\
\le\ & \psi(q_*) \int_0^{\tau_1 - \epsilon} \|S'(\tau_2 - s) - S'(\tau_1 - s)\|_{B(E)} p(s) ds \\
& + \psi(q_*) \int_{\tau_1 - \epsilon}^{\tau_1} \|S'(\tau_2 - s) - S'(\tau_1 - s)\|_{B(E)} p(s) ds \\
& + M e^{\omega b} \psi(q_*) \int_{\tau_1}^{\tau_2} e^{-\omega s} p(s) ds.
\end{aligned}
$$

As $\tau_1 \to \tau_2$ and ϵ becomes sufficiently small, the right-hand side of the above inequality tends to zero, since $S'(t)$ is a strongly continuous operator and the compactness of $S't)$ for $t > 0$ implies the uniform continuity. This proves the equi-continuity for the case where $t \ne t_i, i = 1, \dots, m + 1$. It remains to examine the equi-continuity at $t = t_i$. First we prove the equi-continuity at $t = t_i^-$, we have for some $x \in B_q$, there exists $v \in S_{F,x_{\rho(t,x_t + \tilde{\phi}_t)} + \tilde{\phi}_{\rho(t,x_t + \tilde{\phi}_t)}}$, such that

$$h(t) = \frac{d}{dt} \int_0^t S(t-s) v(s) ds, \quad t \in J.$$

Fix $\delta_1 > 0$ such that $\{t_k, k \neq i\} \cap [t_i - \delta_1, t_i + \delta_1] = \emptyset$. For $0 < \mu < \delta_1$, we have

$$|h(t_i - \mu) - h)(t_i)| \leq \lim_{\lambda \to \infty} \int_0^{t_i - \mu} \| \left(S'(t_i - \mu - s) - S'(t_i - s) \right) B_\lambda v(s) \| ds$$

$$+ M e^{\omega b} \psi(q_*) \int_{t_i - \mu}^{t_i} e^{-\omega s} p(s) \, ds;$$

which tends to zero as $\rho \to 0$. Define

$$\hat{h}_0(t) = h(t), \quad t \in [0, t_1]$$

and

$$\hat{h}_i(t) = \left\{ \begin{array}{ll} h(t), & \text{if } t \in (t_i, t_{i+1}] \\ h(t_i^+), & \text{if } t = t_i. \end{array} \right\}$$

Next, we prove equi-continuity at $t = t_i^+$. Fix $\delta_2 > 0$ such that $\{t_k, k \neq i\} \cap [t_i - \delta_2, t_i + \delta_2] = \emptyset$. Then

$$\hat{h}(t_i) = \int_0^{t_i} T(t_i - s) v(s) ds.$$

For $0 < \mu < \delta_2$, we have

$$|\hat{h}(t_i + \mu) - \hat{h}(t_i)| \leq \lim_{\lambda \to \infty} \int_0^{t_i} \| \left(S'(t_i + \mu - s) - S'(t_i - s) \right) B_\lambda v(s) ds$$

$$+ M e^{\omega b} \psi(q_*) \int_{t_i}^{t_i + \mu} e^{-\omega s} p(s) \, ds;$$

The right-hand side tends to zero as $\mu \to 0$. The equi-continuity for the cases $\tau_1 < \tau_2 \leq 0$ and $\tau_1 \leq 0 \leq \tau_2$ follows from the uniform continuity of ϕ on the interval $(-\infty, 0]$ As a consequence of Claims 1–3 together with Arzelá–Ascoli theorem it suffices to show that $\mathcal{B}$ maps B into a precompact set in E.

Let $0 < t < b$ be fixed and let ϵ be a real number satisfying $0 < \epsilon < t$. For $x \in B_q$, we define

$$h_\epsilon(t) = S'(\epsilon) \lim_{\lambda \to \infty} \int_0^{t - \epsilon} S'(t - s - \epsilon) B_\lambda v(s) ds,$$

where $v \in S_{F, x_{\rho(t, x_t + \tilde{\phi}_t)} + \tilde{\phi}_{\rho(t, x_t + \tilde{\phi}_t)}}$. Since

$$\left| \lim_{\lambda \to \infty} \int_0^{t - \epsilon} S'(t - s - \epsilon) B_\lambda v(s) ds \right| \leq M e^{\omega b} \psi(q_*) \int_0^{t - \epsilon} e^{-\omega s} p(s) ds.$$

the set

$$\left\{ \lim_{\lambda \to \infty} \int_0^{t-\epsilon} S'(t-s-\epsilon) B_\lambda v(s) ds : v \in S_{F, x_{\rho(t,x_t+\tilde{\phi}_t)}+\tilde{\phi}_{\rho(t,x_t+\tilde{\phi}_t)}}, x \in B_q \right\}$$

is bounded

Since $S'(t)$ is a compact operator for $t > 0$, the set

$$H_\epsilon(t) = \{ h_\epsilon(t) : \; h_\epsilon \in \mathcal{B}(x) \}$$

is precompact in E for every ϵ, $0 < \epsilon < t$. Moreover, for every $h \in \mathcal{B}(x)$ we have

$$|h(t) - h_\epsilon(t)| \le Me^{\omega b} \psi(q_*) \int_t^{t-\epsilon} e^{-\omega s} p(s) ds.$$

Therefore, there are precompact sets arbitrarily close to the set $H(t) = \{ h(t) : h \in \mathcal{B}(x) \}$. Hence the set $H(t) = \{ h(t) : h \in \mathcal{B}(B_q) \}$ is precompact in E. Hence the operator $\mathcal{B}$ is totally bounded.

Step 3: A priori bounds.

Now it remains to show that the set

$$\mathcal{E} = \{ x \in \mathcal{B}_b^0 : x \in \lambda \mathcal{A}(x) + \lambda \mathcal{B}(x) \quad \text{for some } 0 < \lambda < 1 \}$$

is bounded.

Let $x \in \mathcal{E}$. Then there exist $v \in S_{F, x_{\rho(t,x_t+\tilde{\phi}_t)}+\tilde{\phi}_{\rho(t,x_t+\tilde{\phi}_t)}}$ and $\mathcal{I}_k \in I_k \left(x_{t_k} + \tilde{\phi}_{t_k} \right)$ such that for each $t \in J$,

$$x(t) = \lambda \frac{d}{dt} \int_0^t S(t-s) v(s) + \lambda \sum_{0 < t_k < t} S'(t - t_k) \mathcal{I}_k.$$

This implies by (11.1.2), (11.1.5) that, for each $t \in J$, we have

$$|x(t)| \le \lambda Me^{\omega t} \int_0^t e^{-\omega s} p(s) \psi (\|x_{\rho(s,x_s+\tilde{\phi}_s)} + \tilde{\phi}_{\rho(s,x_s+\tilde{\phi}_s)}\|_{\mathcal{D}}) ds$$

$$+ \lambda Me^{\omega t} \sum_{k=1}^m \left| I_k \left(x_{t_k} + \tilde{\phi}_{t_k} \right) \right|$$

$$\le \lambda Me^{\omega t} \int_0^t e^{-\omega s} p(s) \psi \left(K_b |x(s)| + (M_b + L^\phi + MK_b) \|\phi\|_{\mathcal{D}} \right) ds$$

$$+ \lambda M e^{\omega t} \sum_{k=1}^{m} \left| I_k \left(x_{t_k} + \tilde{\phi}_{t_k} \right) - I_k(0) \right|$$

$$+ \lambda M e^{\omega t} \sum_{k=1}^{m} |I_k(0)|$$

$$\leq \lambda M e^{\omega t} \int_0^t e^{-\omega s} p(s) \psi \left(K_b |x(s)| + (M_b + L^\phi + M K_b) \|\phi\|_{\mathcal{D}} \right) ds$$

$$+ \lambda M e^{\omega t} \sum_{k=1}^{m} |I_k(0)| + \lambda M e^{\omega t} \sum_{k=1}^{m} c_k \left(K_b |x(s)| + (M_b + L^\phi + M K_b) \|\phi\|_{\mathcal{D}} \right)$$

$$\leq c e^{\omega t} + M e^{\omega t} \left[\int_0^t e^{-\omega s} p(s) \psi \left(K_b |x(s)| + (M_b + L^\phi + M K_b) \|\phi\|_{\mathcal{D}} \right) ds \right.$$

$$\left. + K_b \sum_{k=1}^{m} c_k |x(t)| \right].$$

Hence from (11.6) to (11.8) we have

$$(M_b + L^\phi + M K_b) \|\phi\|_{\mathcal{D}} + K_b |x(s)| \leq c_1 + c_2 \int_0^t e^{-\omega s} p(s) \psi \left(K_b |x(t)| \right.$$

$$\left. + (M_b + L^\phi + M K_b) \|\phi\|_{\mathcal{D}} \right) ds.$$

Thus

$$\frac{(M_b + L^\phi + M K_b) \|\phi\|_{\mathcal{D}} + K_b \|x\|_{\mathcal{B}_b^0}}{c_1 + c_2 \int_0^t e^{-\omega s} p(s) \psi \left(K_b |x(t)| + (M_b + L^\phi + M K_b) \|\phi\|_{\mathcal{D}} \right) ds.} \leq 1. \qquad (11.10)$$

From (11.5) it follows that there exists a constant $R > 0$ such that for each $x \in \mathcal{E}$ with $\|x\|_{\mathcal{B}_b^0} > R$ the condition (11.10) is violated. Hence $\|x\|_{\mathcal{B}_b^0} \leq R$ for each $x \in \mathcal{E}$, which means that the set $\mathcal{E}$ is bounded. As a consequence of Theorem 1.32, $\mathcal{A} + \mathcal{B}$ has a fixed point x^* on the interval $(-\infty, b]$, so $y^* = x^* + \tilde{\phi}$ is a fixed point of the operator N which is the mild solution of problem (11.1)–(11.3). $\qquad \square$

11.2.4   Example

To illustrate our previous result we consider the partial functional differential equations with state dependent delay of the form

$$\frac{\partial}{\partial t}v(t,\xi) = -\frac{\partial}{\partial \xi}v(t,\xi) + m(t)a(v(t - \sigma(v(t,0)),\xi)),$$

$$\xi \in [0,\pi], \ t \in [0,b], \tag{11.11}$$

$$v(t,0) = v(t,\pi) = 0, \ t \in [0,b], \tag{11.12}$$

$$v(\theta,\xi) = v_0(\theta,\xi), \ \xi \in [0,\pi], \ \theta \in (-\infty,0], \tag{11.13}$$

$$\Delta v(t_i)(\xi) = \int_{-\infty}^{t_i} \gamma_i(t_i - s)v(s,\xi)ds, \tag{11.14}$$

where $v_0 : (-\infty,0] \times [0,\pi] \to \mathbb{R}$ is an appropriate function, $\gamma_i \in C[0,\infty),\mathbb{R})$, $0 < t_1 < t_2 < \cdots < t_n < b$. the function $m : [0,b] \to \mathbb{R}, a : \mathbb{R} \times J \to \mathbb{R}, \sigma : \mathbb{R} \to \mathbb{R}^+$ are continuous and we assume the existence of positive constants b_1, b_2 such that $|b(t)| \leq b_1|t| + b_2$ for every $t \in \mathbb{R}$.

Let A be the operator defined on $E = C([0,\pi],\mathbb{R})$ by

$$D(A) = \{g \in C^1([0,\pi],\mathbb{R}) : g(0) = 0\}; \ Ag = g'.$$

Then

$$\overline{D(A)} = C_0([0,\pi],\mathbb{R}) = \{g \in C([0,\pi],\mathbb{R}) : g(0) = 0\}.$$

It is well known from [100] that A is sectorial, $(0,+\infty) \subseteq \rho(A)$ and for $\lambda > 0$

$$\|R(\lambda,A)\|_{B(E)} \leq \frac{1}{\lambda}.$$

It follows that A generates an integrated semigroup $(S(t))_{t \geq 0}$ and that $\|S'(t)\|_{B(E)} \leq e^{-\mu t}$ for $t \geq 0$ for some constant $\mu > 0$ and A satisfied the Hille–Yosida condition.

Set $\gamma > 0$. For the phase space, we choose $\mathcal{D}$ to be defined by

$$\mathcal{D} = C_\gamma = \{\phi \in C((-\infty,0],E) : \lim_{\theta \to -\infty} e^{\gamma\theta}\phi(\theta) \text{ exists in } E\}$$

with norm

$$\|\phi\|_\gamma = \sup_{\theta \in (-\infty,0]} e^{\gamma\theta}|\phi(\theta)|, \ \phi \in C_\gamma.$$

By making the following change of variables

$$y(t)(\xi) = v(t, \xi), \ t \geq 0, \ \xi \in (0, \pi],$$
$$\phi(\theta)(\xi) = v_0(\theta, \xi), \ \theta \leq 0, \ \xi \in [0, 1],$$
$$F(t, \varphi)(\xi) = m(t)b(\varphi(0, \xi)), \ \xi \in [0, \pi], \ \phi \in C_\gamma$$
$$\rho(t, \varphi) = t - \sigma(\varphi(0, 0))$$
$$I_k(y_{t_k}) = \int_{-\infty}^{0} \gamma_k(s)v(s, \xi)ds,$$

the problem (11.11)–(11.14) takes the abstract form (11.1)–(11.3). Moreover, a simple estimates shows that

$$\|f(t, \varphi)\| \leq m(t)[b_1\|\varphi\|_{\mathcal{D}} + b_2\pi^{1/2}] \quad \text{for all} \ (t, \varphi) \in I \times \mathcal{D},$$

with

$$\int_{1}^{\infty} \frac{ds}{\psi(s)} = \int_{1}^{\infty} \frac{ds}{b_1 s + b_2\pi^{1/2}} = +\infty,$$

and

$$d_k = \left(\int_{-\infty}^{0} \frac{(\gamma_k(s))^2}{e^{\theta s}} ds \right)^{1/2} < \infty.$$

Theorem 11.4. *Let* $\varphi \in \mathcal{D}$ *be such that* H_φ *is valid and* $t \to \varphi_t$ *is continuous on* $\mathcal{R}(\rho^-)$, *then there exists a integral solution of (11.11)–(11.14) whenever*

$$\left(1 + \left(\int_{-\infty}^{0} e^{\theta s} ds \right)^{1/2} \right) \sum_{k=1}^{m} d_k < 1.$$

11.3 Impulsive Evolution Inclusions with Infinite Delay and Multi-valued Jumps

11.3.1 Introduction

In this section, we are concerned by the existence of mild solution of impulsive semi-linear functional differential inclusions with infinite delay and multi-valued jumps in a Banach space E. More precisely, we consider the following class of semi-linear impulsive differential inclusions:

$$x'(t) \in A(t)x(t) + F(t, x_t), \quad t \in J = [0, b], \ t \neq t_k, \tag{11.15}$$

$$\Delta x\big|_{t=t_k} \in \mathcal{I}_k(x(t_k^-)), \quad k = 1, \ldots, m \tag{11.16}$$

$$x(t) = \phi(t), \quad t \in (-\infty, 0], \tag{11.17}$$

where $\{A(t) : t \in J\}$ is a family of linear operators in Banach space E generating an evolution operator, F be a Carathéodory type multi-function from $J \times \mathcal{B}$ to the collection of all nonempty compact convex subset of E, $\mathcal{B}$ is the phase space defined axiomatically which contains the mapping from $(-\infty, 0]$ into E, $\phi \in \mathcal{B}$, $0 = t_0 < t_1 < \cdots < t_m < t_{m+1} = b$, $\mathcal{I}_k : E \to \mathcal{P}(E)$, $k = 1, \ldots, m$ are multi-valued maps with closed, bounded and convex values, $x(t_k^+) = \lim_{h \to 0+} x(t_k + h)$ and $x(t_k^-) = \lim_{h \to 0+} x(t_k - h)$ represent the right and left limits of $x(t)$ at $t = t_k$. Finally $\mathcal{P}(E)$ denotes the family of nonempty subsets of E.

11.3.2 Existence Results

Definition 11.5. A function $x \in \Omega$ is said to be a mild solution of system (11.15)–(11.17) if there exists a function $f \in L^1(J, E)$ such that $f \in F(t, x_t)$ for a.e. $t \in J$

(i) $x(t) = T(t, 0)\phi(0) + \int_0^t T(t, s)f(s)ds + \sum_{0 < t_k < t} T(t, t_k)I_k(x(t_k)); \quad$ a.e. $t \in J$, $k = 1, \ldots, m$

(ii) $x(t) = \phi(t), \quad t \in (-\infty, 0],$

with $I_k \in \mathcal{I}_k(x(t_k^+))$.

We will need to introduce the following hypothesis which are assumed hereafter.

(A) $\{A(t) : t \in J\}$ be a family of linear (not necessarily bounded) operators, $A(t) : D(A) \subset E \to E$, $D(A)$ not depending on t and dense subset of E and $T : \Delta = \{(t, s) : 0 \le s \le t \le b\} \to \mathcal{L}(E)$ be the evolution operator generated by the family $\{A(t) : t \in J\}$.

(11.5.1) The multi-function $F(., x)$ has a strongly measurable selection for every $x \in \mathcal{B}$.

(11.5.2) The multi-function $F : (t, .) \to P_{cv,k}(E)$ is upper semi-continuous for a.e. $t \in J$.

(11.5.3) There exists a function $\alpha \in L^1(J, \mathbb{R}^+)$ such that

$$\|F(t, \psi)\| \le \alpha(t)(1 + \|\psi\|_{\mathcal{B}}) \quad \text{for a.e. } t \in J;$$

(11.5.4) There exists a function $\beta \in L^1(J, \mathbb{R}^+)$ such that for all $\Omega \subset \mathcal{B}$, we have

$$\chi(F(t, D)) \le \beta(t) \sup_{-\infty \le s \le 0} \chi(\Omega(s)) \quad \text{for a.e. } t \in J,$$

where $\Omega(s) = \{x(s); x \in \Omega\}$ and χ is the Hausdorff MNC.

(11.5.5) There exist constants $a_k > 0, k = 1, \ldots, m$ such that

$$\|I_k\| \leq a_k, \quad \text{where } I_k \in \mathcal{I}_k(x(t_k^+)).$$

Remark 11.6. Under conditions (11.5.1)–(11.5.3) for every piecewise continuous function $v : J \to \mathcal{B}$ the multi-function $F(t, v(t))$ admits a Bochner integrable selection (see [144]).

Let

$$\Omega_b = \{x \in \Omega : x_0 = 0\}.$$

For any $x \in \Omega_b$ we have

$$\|x\|_b = \|x\|_{\mathcal{B}} + \sup_{0 \leq s \leq b} \|x\| = \sup_{0 \leq s \leq b} \|x\|.$$

Thus $(\Omega_b, \|.\|_b)$ is a Banach space.

We note that from assumptions (11.5.1) and (11.5.3) it follows that the superposition multi-operator $S_F^1 : \Omega_b \to \mathcal{P}(L^1(J, E))$ defined by

$$S_F^1 = \{f \in L^1(J, E) : f(t) \in F(t, x_t), \quad \text{a.e. } t \in J\}$$

is nonempty set (see [144]) and is weakly closed in the following sense.

Lemma 11.7. *If we consider the sequence* $(x^n) \in \Omega_b$ *and* $\{f_n\}_{n=1}^{+\infty} \subset L^1(J, E)$, *where* $f_n \in S_{F(.,x_n)}^1$ *such that* $x^n \to x^0$ *and* $f_n \to f^0$ *then* $f^0 \in S_F^1$.

Now we state and prove our main result.

Theorem 11.8. *Under assumptions (A) and (11.5.1)–(11.5.5), the problem (11.15)–(11.17) has at least one mild solution.*

Proof. To prove the existence of a mild solution for (11.15)–(11.17) we introduce the integral multi-operator $N : \Omega_b \longrightarrow \mathcal{P}(\Omega_b)$, defined as

$$N(x) = \begin{cases} y : y(t) = T(t, 0)\phi(0) + \int_0^t T(t, s)f(s)ds \\ \sum_{0 < t_k < t} T(t, t_k)I_k(x(t_k)), & t \in J \\ y(t) = \phi(t), & t \in (-\infty, 0], \end{cases} \quad (11.18)$$

where S_F^1 and $I_k \in \mathcal{I}_k(x)$.

It is clear that the integral multi-operator N is well defined and the set of all mild solution for the problem (11.15)–(11.17) on J is the set $\mathcal{F}ixN = \{x : x \in N(x)\}$.

We shall prove that N satisfies all the hypotheses of Lemma 1.40. The proof will be given in several steps.

Step 1. Using in fact that the maps F and $\mathcal{I}$ has a convex values it easy to check that N has convex values.

Step 2. N has closed graph.

Let $\{x^n\}_{n=1}^{+\infty}$, $\{z^n\}_{n=1}^{+\infty}$, $x^n \to x^*$, $z^n \in N((x^n), n \geq 1)$ and $z^n \to z^*$. Moreover, let $\{f_n\}_{n=1}^{+\infty} \subset L^1(J, E)$ an arbitrary sequence such that $f_n \in S_F^1$ for $n \geq 1$.

Hypothesis (11.5.3) implies that the set $\{f_n\}_{n=1}^{+\infty}$ integrably bounded and for a.e. $t \in J$ the set $\{f_n(t)\}_{n=1}^{+\infty}$ relatively compact, we can say that $\{f_n\}_{n=1}^{+\infty}$ is semi-compact sequence. Consequently $\{f_n\}_{n=1}^{+\infty}$ is weakly compact in $L^1(J, E)$, so we can assume w.l.g that $f_n \rightharpoonup f^*$.

From Lemma 1.43 we know that the generalized Cauchy operator on the interval J, $G : L^1(J, E) \to C(J, E)$, defined by

$$Gf(t) = \int_0^t T(t, s)f(s)ds, \quad t \in J. \tag{11.19}$$

satisfies properties (G1) and (G2) on J.

Note that set $\{f_n\}_{n=1}^{+\infty}$ is also semi-compact and sequence $(f_n)_{n=1}^{+\infty}$ weakly converges to f^* in $L^1(J, E)$. Therefore, by applying Lemma 1.44 for the generalized Cauchy operator G of (11.19) we have in $C(J, E)$ the convergence $Gf_n \to Gf$. By means of (11.19) and (11.18), for all $t \in J$ we can write

$$z_n(t) = T(t, 0)\phi(0) + \int_0^t T(t, s)f_n(s)ds + \sum_{0 < t_k < t} T(t, t_k)I_k(x^n(t_k))$$

$$= T(t, 0)\phi(0) + \int_0^t T(t, s)f_n ds + \sum_{0 < t_k < t} T(t, t_k)I_k(x^n(t_k))$$

$$= T(t, 0)\phi(0) + Gf_n(t) + \sum_{0 < t_k < t} T(t, t_k)I_k(x^n(t_k)))$$

where S_F^1, and $I_k \in \mathcal{I}_k(x)$. By applying Lemma 1.43, we deduce

$$z_n \to T(., 0)\phi(0) + Gf + T(., t)I_k(x^*(t_k))$$

in $C(J, E)$ and by using in fact that the operator S_F^1 is closed, we get $f^* \in S_F^1$. Consequently

$$z^*(t) \to T(t, 0)\phi(0) + Gf + T(t, t)I_k(x^*(t_k)),$$

therefore $z^* \in N(x^*)$. Hence N is closed.

With the same technique, we obtain that N has compact values.

Step 3. We consider the MNC defined in the following way. For every bounded subset $\Omega \subset \mathbf{\Omega}_b$

$$v_1(\Omega) = \max_{\Omega \in \Delta(\Omega)} (\gamma_1(\Omega), \mod {}_C(\Omega)), \tag{11.20}$$

where $\Delta(\Omega)$ is the collection of all the denumerable subsets of Ω;

$$\gamma_1(\Omega) = \sup_{t \in J} e^{-Lt} \chi(\{x(t) : x \in \Omega\}); \tag{11.21}$$

where $\mod {}_C(\Omega)$ is the modulus of equi-continuity of the set of functions Ω given by the formula

$$\mod {}_C(\Omega) = \lim_{\delta \to 0} \sup_{x \in \Omega} \max_{|t_1 - t_2| \le \delta} \|x(t_1) - x(t_2)\|; \tag{11.22}$$

and $L > 0$ is a positive real number chosen such that

$$q := M \left(2 \sup_{t \in J} \int_0^t e^{-L(t-s)} \beta(s) ds + e^{Lt} \sum_{k=1}^m c_k \right) < 1 \tag{11.23}$$

where $M = \sup_{(t,s) \in \Delta} \|T(t,s)\|$.

From the Arzelá–Ascoli theorem, the measure v_1 gives a nonsingular and regular MNC (see [144]).

Let $\{y_n\}_{n=1}^{+\infty}$ be the denumerable set which achieves that maximum $v_1(N(\Omega))$, i.e.;

$$v_1(N(\Omega)) = (\gamma_1(\{y_n\}_{n=1}^{+\infty}), \mod {}_C(\{y_n\}_{n=1}^{+\infty})).$$

Then there exists a set $\{x_n\}_{n=1}^{+\infty} \subset \Omega$ such that $y_n \in N(x_n), n \ge 1$. Then

$$y_n(t) = T(t,0)\phi(0) + \int_0^t T(t,s)f(s)ds + \sum_{0 < t_k < t} T(t,t_k)I_k(x(t_k)), \tag{11.24}$$

where $f \in S_F^1$ and $I_k \in \mathcal{I}_k(x_n)$, so that

$$\gamma_1(\{y_n\}_{n=1}^{+\infty}) = \gamma_1(\{Gf_n\}_{n=1}^{+\infty}).$$

We give an upper estimate for $\gamma_1(\{y_n\}_{n=1}^{+\infty})$.

Fixed $t \in J$ by using condition (11.5.4), for all $s \in [0, t]$ we have

$$\chi(\{f_n(s)\}_{n=1}^{+\infty}) \le \chi(F(s, \{x_n(s)\}_{n=1}^{+\infty}))$$

$$\le \chi(\{F(s, x_n(s))\}_{n=1}^{+\infty})$$

$$\le \beta(s)\chi(\{x_n(s)\}_{n=1}^{+\infty})$$

$$\le \beta(s)e^{Ls} \sup_{t \in J} e^{-Lt} \chi(\{x_n(t)\}_{n=1}^{+\infty})$$

$$= \beta(s)e^{Ls} \gamma_1(\{x_n\}_{n=1}^{+\infty}).$$

By using condition (11.5.3), the set $\{f_n\}_{n=1}^{+\infty}$ is integrably bounded. In fact, for every $t \in J$, we have

$$\|f_n(t)\| \le \|F(t, x_n(t))\|$$
$$\le \alpha(t)(1 + \|x_n(t)\|).$$

The integrably boundedness of $\{f_n\}_{n=1}^{+\infty}$ follows from the continuity of x in J_k and the boundedness of set $\{x_n\}_{n=1}^{+\infty} \subset \Omega$. By applying Lemma 1.45, it follows that

$$\chi(\{Gf_n(s)\}_{n=1}^{+\infty}) \le 2M \int_0^s \beta(t) e^{Lt} (\gamma_1(\{x_n\}_{n=1}^{+\infty})) dt$$

$$= 2M\gamma_1(\{x_n\}_{n=1}^{+\infty}) \int_0^s \beta(t) e^{Lt}.$$

Thus, we get

$$\gamma_1(\{x_n\}_{n=1}^{+\infty}) \le \gamma_1(\{y_n\}_{n=1}^{+\infty}) = \gamma_1(\{Gf_n(s)\}_{n=1}^{+\infty})$$

$$= \sup_{t \in J} e^{-Lt} 2M\gamma_1(\{x_n\}_{n=1}^{+\infty}) \int_0^s \beta(t) e^{Lt} M\gamma_1(\{x_n\}_{n=1}^{+\infty}) e^{Lt} \sum_{k=1}^m c_k$$

$$\le q\gamma_1(\{x_n\}_{n=1}^{+\infty}),$$

$$(11.25)$$

and hence $\gamma_1(\{x_n\}_{n=1}^{+\infty}) = 0$, then $\gamma_1(\{x_n(t)\}_{n=1}^{+\infty}) = 0$, for every $t \in J$. Consequently

$$\gamma_1(\{y_n\}_{n=1}^{+\infty}) = 0.$$

By using the last equality and hypotheses (11.5.3) and (11.5.4), we can prove that set $\{f_n\}_{n=1}^{+\infty}$ is semi-compact. Now, by applying Lemmas 1.43 and 1.44, we can conclude that set $\{Gf_n\}_{n=1}^{+\infty}$ is relatively compact. The representation of y_n given by (11.24) yields that set $\{y_n\}_{n=1}^{+\infty}$ is also relatively compact in Ω_b, therefore $\nu_1(\Omega) = (0, 0)$. Then Ω is a relatively compact set.

Step 4. A priori bounds.

We will demonstrate that the solutions set is a priori bounded. Indeed, let $x \in N$. Then there exists $f \in S^1_{F(.,x_t(.))}$ and $I_k \in \mathcal{I}_k(x)$ such that for every $t \in J$ we have

$$\|x(t)\| = \left\| T(t, 0)\phi(0) + \int_0^t T(t, s)f(s)ds + \sum_{0 < t_k < t} T(t, t_k)I_k(x(t_k)) \right\|$$

$$\le M(\|\phi(0)\| + \sum_{0 < t_k < t} \|a_k\|) + M \int_0^t f(s)ds$$

$$\le M(\|\phi(0)\| + \sum_{0 < t_k < t} \|a_k\|) + M \int_0^b \alpha(s)(1 + \|x[\phi]_s\|)ds.$$

Using condition (A1) we have

$$\|x(t)\| \leq M(\|\phi(0)\| + \sum_{0 < t_k < t} \|a_k\|) + M \int_0^t \alpha(s)(1 + N_b\|\phi\|_{\mathcal{B}} + K_b \sup_{0 \leq \theta \leq s} \|x(\theta)\|)ds$$

$$\leq M(\|\phi(0)\| + \sum_{k=1}^m \|a_k\|) + M(1 + N_b\|\phi\|_{\mathcal{B}})\|\alpha\|_{L^1(J)}$$

$$+ MK_b \int_0^t \alpha(s) \sup_{0 \leq \theta \leq s} \|x(\theta)\|ds.$$

Since the last expression is a nondecreasing function of t, we have that

$$\sup_{0 \leq \theta \leq t} \|x(\theta)\| \leq \|M(\|\phi(0)\| + \sum_{k=1}^m \|a_k\|) + M(1 + N_b\|\phi\|_{\mathcal{B}})\|\alpha\|_{L^1(J)}$$

$$+ MK_b \int_0^t \alpha(s) \sup_{0 \leq \theta \leq s} \|x(\theta)\|ds.$$

Invoking Gronwall's inequality, we get

$$\sup_{0 \leq \theta \leq t} \|x(\theta)\| \leq \zeta e^{MK_b\|\alpha\|_{L^1}},$$

where

$$\zeta = M(\|\phi(0)\| + \sum_{k=1}^m \|a_k\|) + M(1 + N_b\|\phi\|_{\mathcal{B}})\|\alpha\|_{L^1(J)}. \qquad \square$$

11.3.3 An Example

As an application of our results we consider the following impulsive partial functional differential equation of the form

$$\frac{\partial}{\partial t}z(t,x) \in a(t,x)\frac{\partial^2}{\partial x^2}z(t,x) + \int_{-\infty}^0 P(\theta)r(t,z(t+\theta,x))d\theta,$$

$$x \in [0,\pi], \ t \in [0,b], t \neq t_k, \qquad (11.26)$$

$$z(t_k^+,x) - z(t_k^-,x) \in [-b_k|z(t_k^-,x), b_k|z(t_k^-,x)],$$

$$x \in [0,\pi], \ k = 1,\ldots,m, \qquad (11.27)$$

$$z(t,0) = z(t,\pi) = 0, \quad t \in J := [0,b], \qquad (11.28)$$

$$z(t,x) = \phi(t,x), \quad -\infty < t \leq 0, \ x \in [0,\pi], \qquad (11.29)$$

where $a(t, x)$ is continuous function and uniformly Hölder continuous in t, $b_k > 0$, $k = 1, \ldots, m$, $\phi \in \mathcal{D}$, $\mathcal{D} = \{\overline{\psi} : (-\infty, 0] \times [0, \pi] \to \mathbb{R}; \overline{\psi}$ is continuous everywhere except for a countablenumber of points at which $\overline{\psi}(s^-), \overline{\psi}(s^+)$ exist with $\overline{\psi}(s^-) = \overline{\psi}(s)\}$, $0 = t_0 < t_1 < t_2 < \cdots < t_m < t_{m+1} = b$, $z(t_k^+) = \lim_{(h,x)\to(0^+,x)} z(t_k + h, x)$, $z(t_k^-) = \lim_{(h,x)\to(0^-,x)} z(t_k + h, x)$, $P : (-\infty, 0] \to \mathbb{R}$ a continuous function, $r : \mathbb{R} \times \mathbb{R} \to \mathcal{P}_{cv,k}(\mathbb{R})$ a Carathéodory multi-valued map.

Let

$$y(t)(x) = z(t, x), \quad x \in [0, \pi], \ t \in J = [0, b],$$

$$\mathcal{I}_k(y(t_k^-))(x) = [-b_k|z(t_k^-, x), b_k|z(t_k^-, x)], \quad x \in [0, \pi], \ k = 1, \ldots, m,$$

$$F(t, \phi)(x) = \int_{-\infty}^0 P(\theta) r(t, z(t + \theta, x)) d\theta$$

$$\phi(\theta)(x) = \phi(\theta, x), \quad -\infty < t \le 0, \ x \in [0, \pi].$$

Consider $E = L^2[0, \pi]$ and define $A(t)$ by $A(t)w = a(t, x)w''$ with domain

$$D(A) = \{w \in E : w, w' \text{ are absolutely continuous}, \ w'' \in E, \ w(0) = w(\pi) = 0\}.$$

Then $A(t)$ generates an evolution system $U(t, s)$ satisfying assumptions (11.5.1) and (11.5.3). We can show that problem (11.26)–(11.29) is an abstract formulation of problem (11.15)–(11.17). Under suitable conditions, the problem (11.15)–(11.17) has at least one mild solution.

11.4 Impulsive Semi-linear Differential Evolution Inclusions with Non-convex Right-Hand Side

11.4.1 Introduction

In this section, we shall be concerned by the existence of mild solution of impulsive semi-linear functional differential inclusions with infinite delay in a separable Banach space E. First, we consider the following class of semi-linear impulsive differential inclusions:

$$x'(t) \in A(t)x(t) + F(t, x_t), \quad t \in J = [0, b], \ t \ne t_k, \tag{11.30}$$

$$\Delta x\big|_{t=t_k} \in \mathcal{I}_k(x(t_k^-)), \quad k = 1, \ldots, m \tag{11.31}$$

$$x(t) = \phi(t), \quad t \in (-\infty, 0] \tag{11.32}$$

where $\{A(t) : t \in J\}$ is a family of linear operators in Banach space E generating an evolution operator, F be a lower semi-continuous multi-function from $J \times \mathcal{B}$ to the collection of all nonempty closed compact subset of E, $\mathcal{B}$ is the phase space defined axiomatically which contains the mapping from $(-\infty, 0]$ into E, $\phi \in \mathcal{B}$,

$0 = t_0 < t_1 < \cdots < t_m < t_{m+1} = b$, $\mathcal{I}_k : E \to \mathcal{P}(E)$, $k = 1, \ldots, m$ are multi-valued maps with closed and bounded values, $x(t_k^+) = \lim_{h\to 0+} x(t_k + h)$ and $x(t_k^-) = \lim_{h\to 0-} x(t_k + h)$ represent the right and left limits of $x(t)$ at $t = t_k$. Finally $\mathcal{P}(E)$ denotes the family of nonempty subsets of E. We mention that the model with multi-valued jump sizes may arise in a control problem where we want to control the jump sizes in order to achieve given objectives.

11.4.2 Existence Results

In this section, we give our main existence result for problem (11.30)–(11.32). Before stating and proving this result, we give the definition of the mild solution.

Definition 11.9. A function $x \in \Omega$ is said to be a mild solution of system (11.15)–(11.17) if there exist a function $f \in L^1(J, E)$ such that $f \in F(t, x_t)$ for a.e. $t \in J$ and $I_k \in \mathcal{I}_k(x(t_k^+))$

(i) $x(t) = T(t, 0)\phi(0) + \int_0^t T(t, s)f(s)ds + \sum_{0 < t_k < t} T(t, t_k)I_k$,   a.e. $t \in J$, $k = 1, \ldots, m$

(ii) $x(t) = \phi(t)$, $t \in (-\infty, 0]$,

We will assume the following hypothesis

(A) $\{A(t) : t \in J\}$ be a family of linear (not necessarily bounded) operators, $A(t) : D(A) \subset E \to E$, $D(A)$ not depending on t and dense subset of E and $T : \Delta = \{(t, s) : 0 \le s \le t \le b\} \to \mathcal{L}(E)$ be the evolution operator generated by the family $\{A(t) : t \in J\}$.

Let F be a multi-function defined from $J \times \mathcal{B}$ to the family of nonempty closed convex subsets of E such that

(11.9.1) $(t, x) \mapsto F(., x)$ is $\mathcal{L} \otimes \mathcal{B}_b$-measurable ($\mathcal{B}_b$ is Borel measurable).

(11.9.2) The multi-function $F : (t, .) \to P_k(E)$ is lower semi-continuous for a.e. $t \in J$.

(11.9.3) there exists a function $\alpha \in L^1(J, \mathbb{R}^+)$ such that

$$\|F(t, \psi)\| \le \alpha(t), \quad \text{for a.e. } t \in J, \ \forall \psi \in \mathcal{B};$$

(11.9.4) There exists a function $\beta \in L^1(J, \mathbb{R}^+)$ such that for all $D \subset \mathcal{B}$, we have

$$\chi(F(t, D)) \le \beta(t) \sup_{-\infty \le s \le 0} \chi(D(s)) \quad \text{for a.e. } t \in J,$$

where, $D(s) = \{x(s); x \in D\}$ and χ is the Hausdorff MNC.

(11.9.5) There exist constants $a_k, c_k \ge 0$, $k = 1, \ldots, m$, such that

$$\|I_k\| \le a_k\|x\| + b_k, \quad \text{where } I_k \in \mathcal{I}_k(x(t_k^+)).$$
$$\chi(I_k(D)) \le c_k\chi(I_k(D)).$$
with $1 - M \sum_{0 < t_k < t} \|a_k\| > 0$.

Remark 11.10. Under conditions (11.9.1)–(11.9.3) for every for every piecewise continuous function $v : [0, b] \to \mathcal{B}$ the multi-function $F(t, v(t))$ admits a Bochner integrable selection (see [144]).

Now we state and prove our main result.

Theorem 11.11. *Under assumptions (A) and (11.9.1)–(11.9.5), the problem (11.30)–(11.32) has at least one mild solution.*

We note that from assumptions (11.9.1) and (11.9.3) it follows that the superposition multi-operator

$$S_F^1 : \Omega \to \mathcal{P}(L^1(J, E)),$$

defined by

$$S_{F(.,x)}^1 = S_F^1 = \{f \in L^1(J, E) : f(t) \in F(t, x_t), \quad \text{a.e. } t \in J\}$$

is nonempty set (see [144]).

Proof. We break the proof into a sequence of steps.

Step 1. The Mönch's condition holds. Suppose that $\Omega \subseteq B_r$ is countable and $\Omega \subseteq \overline{co}(\{0\} \cup N(\Omega))$ We will prove that Ω is relatively compact.

We consider the MNC defined in the following way. For every bounded subset $\Omega \subset \mathbf{\Omega}$

$$\nu_1(\Omega) = \max_{D \in \Delta(\Omega)} (\gamma_1(D), \mod {}_C(D)), \tag{11.33}$$

where $\Delta(\Omega)$ is the collection of all the denumerable subsets of Ω;

$$\gamma_1(D) = \sup_{t \in J} e^{-Lt} \chi(\{x(t) : x \in D\}); \tag{11.34}$$

where $\mod {}_C(D)$ is the modulus of equi-continuity of the set of functions D given by the formula

$$\mod {}_C(D) = \lim_{\delta \to 0} \sup_{x \in D} \max_{|t_1 - t_2| \leq \delta} \|x(t_1) - x(t_2)\|; \tag{11.35}$$

and $L > 0$ is a positive real number chosen so that

$$q := M \left(2 \sup_{t \in J} \int_0^t e^{-L(t-s)} \beta(s) ds + e^{Lt} \sum_{k=0}^m c_k \right) < 1 \tag{11.36}$$

where $M = \sup_{(t,s)\in\Delta} \|T(t,s)\|$.

From the Arzelá–Ascoli theorem, the measure ν_1 give a nonsingular and regular MNC (see [144]).

Let $\{y_n\}_{n=1}^{+\infty}$ be the denumerable set which achieves that maximum $\nu_1(N(\Omega))$, i.e.;

$$\nu_1(N(\Omega)) = (\gamma_1(\{y_n\}_{n=1}^{+\infty}), \mod c(\{y_n\}_{n=1}^{+\infty})).$$

Then there exists a set $\{x_n\}_{n=1}^{+\infty} \subset \Omega$ such that $y_n \in N(x_n)$, $n \geq 1$. Then

$$y_n(t) = T(t,0)\phi(0) + \int_0^t T(t,s)f(s)ds + \sum_{0<t_k<t} T(t,t_k)I_k, \qquad (11.37)$$

where $f \in S_F^1$ and $I_k \in \mathcal{I}_k(x)$, so that

$$\gamma_1(\{y_n\}_{n=1}^{+\infty}) = \gamma_1(\{Gf_n\}_{n=1}^{+\infty}).$$

We give an upper estimate for $\gamma_1(\{y_n\}_{n=1}^{+\infty})$.

Fixed $t \in J$ by using condition (11.9.4), for all $s \in [0,t]$ we have

$$\chi(\{f_n(s)\}_{n=1}^{+\infty}) \leq \chi(F(s, \{x_n(s)\}_{n=1}^{+\infty}))$$

$$\leq \beta(s)\chi(\{x_n(s)\}_{n=1}^{+\infty})$$

$$\leq \beta(s)e^{Ls} \sup_{t\in J} e^{-Lt}\chi(\{x_n(t)\}_{n=1}^{+\infty})$$

$$= \beta(s)e^{Ls}\gamma_1(\{x_n\}_{n=1}^{+\infty}).$$

By using condition (11.9.3), the set $\{f_n\}_{n=1}^{+\infty}$ is integrably bounded. In fact, for every $t \in J$, we have

$$\|f_n(t)\| \leq \|F(t, x_n(t))\|$$

$$\leq \alpha(t).$$

By applying Lemma 1.45, it follows that

$$\chi(\{Gf_n(s)\}_{n=1}^{+\infty}) \leq 2M \int_0^s \beta(t)e^{Lt}(\gamma_1(\{x_n\}_{n=1}^{+\infty}))dt$$

$$= 2M\gamma_1(\{x_n\}_{n=1}^{+\infty}) \int_0^s \beta(t)e^{Lt}dt.$$

Thus, we get

$$\gamma_1(\{x_n\}_{n=1}^{+\infty}) \leq \gamma_1(\{y_n\}_{n=1}^{+\infty})$$

$$= \sup_{t \in J} e^{-Lt} 2M\gamma_1(\{x_n\}_{n=1}^{+\infty}) \int_0^t \beta(s)e^{Ls}ds + M\gamma_1(\{x_n\}_{n=1}^{+\infty}e^{Lt}\sum_{k=1}^{m} c_k$$

$$\leq q\gamma_1(\{x_n\}_{n=1}^{+\infty}).$$

(11.38)

Therefore, we have that

$$\gamma_1(\{x_n\}_{n=1}^{+\infty}) \leq \gamma_1(\Omega) \leq \gamma_1(\{0\} \cup N(\Omega))\gamma_1(\{y_n\}_{n=1}^{+\infty}) \leq q\gamma_1(\{x_n\}_{n=1}^{+\infty}).$$

From (11.36), we obtain that

$$\gamma_1(\{x_n\}_{n=1}^{+\infty}) = \gamma_1(\Omega) = \gamma_1(\{y_n\}_{n=1}^{+\infty})$$

Coming back to the definition of γ_1, we can see

$$\chi(\{x_n\}_{n=1}^{+\infty}) = \chi(\{y_n\}_{n=1}^{+\infty}) = 0.$$

By using the last equality and hypotheses (11.9.3) and (11.9.4) we can prove that set $\{f_n\}_{n=1}^{+\infty}$ is semi-compact. Now, by applying Lemmas 1.43 and 1.44, we can conclude that set $\{Gf_n\}_{n=1}^{+\infty}$ is relatively compact in $C(J, E)$.

The representation of y_n given by (11.37) yields that set $\{y_n\}_{n=1}^{+\infty}$ is also relatively compact in $C(J, E)$, since ν_1 is a monotone, nonsingular, regular MNC, we have that

$$\nu_1(\Omega) \leq \nu_1(\overline{co}(\{0\} \cup N(\Omega))) \leq \nu_1(N(\Omega)) = \nu_1(\{y_n\}_{n=1}^{+\infty}) = (0,0).$$

Therefore, Ω is relatively compact.

Step 2. It is clear that the superposition multioperator S_F^1 has closed and decomposable values. Following the lines of [144], we may verify that S_F^1 is l.s.c.

Applying Lemma 1.46 to the restriction of S_F^1 on Ω we obtain that there exists a continuous selection

$$w : \Omega \to L^1(J, E)$$

We consider a map $N : \Omega \to \Omega$ defined as

$$x(t) = T(t,0)\phi(0) + \int_0^t T(t,s)w(x)(s)ds.$$

Since the Cauchy operator is continuous, the map N is also continuous; therefore, it is a continuous selection of the integral multi-operator.

Step 3. A priori bounds.

We will demonstrate that the solutions set is a priori bounded. Indeed, let $x \in \lambda N_1$ and $\lambda \in (0,1)$. There exists $f \in S_F^1$ and $I_k \in \mathcal{I}_k(x)$ such that for every $t \in J$ we have

$$\|x(t)\| = \left\| \lambda T(t,0)\phi(0) + \lambda \int_0^t T(t,s)f(s)ds + \lambda \sum_{0<t_k<t} T(t,t_k)I_k \right\|$$

$$\leq M \left(\|\phi(0)\| + \|x\| \sum_{k=1}^m \|a_k\| + \sum_{k=1}^m \|b_k\| \right) + M \int_0^t \alpha(s)ds,$$

hence,

$$\left(1 - M \sum_{k=1}^m \|a_k\|\right) \|x\| \leq M \left(\|\phi(0)\| + \|\alpha\| + \sum_{k=1}^m \|b_k\| \right).$$

Consequently

$$\|x\| \leq \frac{M(\|\phi(0)\| + \|\alpha\| + \sum_{k=1}^m \|b_k\|)}{1 - M \sum_{k=1}^m \|a_k\|} = C.$$

So, there exists N^* such that $\|x\| \neq N^*$, set

$$U = \{x \in \Omega : \quad \|x\| < N^*\}.$$

From the choice of U there is no $x \in \partial U$ such that $x = \lambda N_1 x$ for some $\lambda \in (0,1)$. Thus, we get a fixed point of N_1 in $\bar{U}$ due to the Mönch's Theorem. $\qquad \square$

11.4.3 An Example

As an application of our results we consider the following impulsive partial functional differential equation of the form

$$\frac{\partial}{\partial t}z(t,x) \in a(t,x)\frac{\partial^2}{\partial x^2}z(t,x) + \int_{-\infty}^0 P(\theta)r(t,z(t+\theta,x))d\theta,$$

$$x \in [0,\pi], \ t \in [0,b], t \neq t_k, \tag{11.39}$$

$$z(t_k^+,x) - z(t_k^-,x) \in [-b_k|z(t_k^-,x), b_k|z(t_k^-,x)],$$

$$x \in [0,\pi], \ k = 1,\ldots,m, \tag{11.40}$$

$$z(t,0) = z(t,\pi) = 0, \quad t \in J := [0,b], \tag{11.41}$$

$$z(t,x) = \phi(t,x), \quad -\infty < t \leq 0, \ x \in [0,\pi], \tag{11.42}$$

where $a(t,x)$ is continuous function and uniformly Hölder continuous in t, $b_k > 0$, $k = 1,\ldots,m$, $\phi \in \mathcal{D}$.

$\mathcal{D} = \{\overline{\psi} : (-\infty,0] \times [0,\pi] \to \mathbb{R}; \overline{\psi}$ is continuous everywhere except for a countable number of points at which $\overline{\psi}(s^-), \overline{\psi}(s^+)$ exist with $\overline{\psi}(s^-) = \overline{\psi}(s)\}$, $0 = t_0 < t_1 < t_2 < \cdots < t_m < t_{m+1} = b$, $z(t_k^+) = \lim_{(h,x)\to(0^+,x)} z(t_k + h,x)$, $z(t_k^-) = \lim_{(h,x)\to(0^-,x)} z(t_k + h,x)$, $P : (-\infty,0] \to \mathbb{R}$ a continuous function, $r : \mathbb{R} \times \mathbb{R} \to \mathcal{P}_{cv,k}(\mathbb{R})$ a multi-valued map.

Let

$$y(t)(x) = z(t,x), \quad x \in [0,\pi], \ t \in J = [0,b],$$

$$\mathcal{I}_k(y(t_k^-))(x) = [-b_k|z(t_k^-,x), b_k|z(t_k^-,x)], \quad x \in [0,\pi], \ k = 1,\ldots,m,$$

$$F(t,\phi)(x) = \int_{-\infty}^0 P(\theta)r(t, z(t+\theta,x))d\theta$$

$$\phi(\theta)(x) = \phi(\theta,x), \quad -\infty < t \le 0, \ x \in [0,\pi].$$

Consider $E = L^2[0,\pi]$ and define $A(t)$ by $A(t)w = a(t,x)w''$ with domain

$$D(A) = \{w \in E : w, w' \text{ are absolutely continuous, } w'' \in E, \ w(0) = w(\pi) = 0\}.$$

Then $A(t)$ generates an evolution system $U(t,s)$ satisfying assumption (11.9.1) and (11.9.3). We can show that problem (11.39)–(11.42) is an abstract formulation of problem (11.30)–(11.32). Under suitable conditions, the problem (11.30)–(11.32) has at least one mild solution.

11.5 Impulsive Evolution Inclusions with State-Dependent Delay and Multi-valued Jumps

11.5.1 Introduction

In this section, we are concerned by the existence of mild solution of impulsive semi-linear functional differential inclusions with state-dependent delay and multi-valued jumps in a Banach space E. More precisely, we consider the following class of semi-linear impulsive differential inclusions:

$$x'(t) \in A(t)x(t) + F(t, x_{\rho(t,x_t)}), \quad t \in J = [0,b], \ t \ne t_k, \tag{11.43}$$

$$\Delta x\big|_{t=t_k} \in \mathcal{I}_k(x(t_k^-)), \quad k = 1,\ldots,m \tag{11.44}$$

$$x(t) = \phi(t), \quad t \in (-\infty,0], \tag{11.45}$$

where $\{A(t) : t \in J\}$ is a family of linear operators in Banach space E generating an evolution operator, F be a Carathéodory type multi-function from $J \times \mathcal{B}$ to the collection of all nonempty compact convex subset of E, $\mathcal{B}$ is the phase space, which contains the mapping from $(-\infty, 0]$ into E, $\phi \in \mathcal{B}$, $0 = t_0 < t_1 < \cdots < t_m < t_{m+1} = b$, $\mathcal{I}_k : E \to \mathcal{P}(E)$, $k = 1, \ldots, m$ are multi-valued maps with closed, bounded and convex values, $x(t_k^+) = \lim_{h\to 0^+} x(t_k + h)$ and $x(t_k^-) = \lim_{h\to 0^+} x(t_k - h)$ represent the right and left limits of $x(t)$ at $t = t_k$. $\rho : J \times \mathcal{B} \to (-\infty, b]$.

Our goal here is to give existence results for the problem (11.43)–(11.45) without any compactness assumption. We prove existence and compactness of solutions set for problem (11.43)–(11.45), and we provide a conditions which guarantee the existence of a mild solution by using a fixed point theorem du to Mönch [162].

11.5.2 Existence Results for the Convex Case

In this section we shall prove the existence of mild solutions of problem (11.43)–(11.45). We assume that the multi-valued nonlinearity of upper Carathéodory semi-continuous type satisfies a regularity condition expressed in terms of the measures of non-compactness. We apply the theory of condensing multi-valued maps to obtain global and compactness of the solutions set.

We need the following definition in the sequel.

Definition 11.12. A function $x \in \Omega$ is said to be a mild solution of system (11.15)–(11.17) if there exist a function $f \in L^1(J, E)$ such that $f \in F(t, x_{\rho(t,x_t)})$ for a.e. $t \in J$

(i) $x(t) = T(t, 0)\phi(0) + \int_0^t T(t, s)f(s)ds + \sum_{0 < t_k < t} T(t, t_k)I_k(x(t_k))$,   a.e. $t \in J$, $k = 1, \ldots, m$

(ii) $x(t) = \phi(t)$, $t \in (-\infty, 0]$,

with $I_k \in \mathcal{I}_k(x(t_k^+))$.

we introduce the following hypotheses.

  (A) $\{A(t) : t \in J\}$ be a family of linear (not necessarily bounded) operators, $A(t) : D(A) \subset E \to E$, $D(A)$ not depending on t and dense subset of E and $T : \Delta = \{(t, s) : 0 \leq s \leq t \leq b\} \to \mathcal{L}(E)$ be the evolution operator generated by the family $\{A(t) : t \in J\}$.

  (Hϕ) The function $t \to \phi_t$ is continuous from $\mathcal{R}(\rho^-) = \{\rho(s, \varphi) : (s, \varphi) \in J \times \mathcal{B}, \rho(s, \varphi) \leq 0\}$ into $\mathcal{B}$ and there exists a continuous and bounded function $L^\phi : \mathcal{R}(\rho^-) \to (0, \infty)$ such that $\|\phi_t\|_\mathcal{B} \leq L^\phi(t)\|\phi\|_\mathcal{B}$ for every $t \in \mathcal{R}(\rho^-)$.

(11.12.1) The multi-function $F(., x)$ has a strongly measurable selection for every $x \in \mathcal{B}$.

(11.12.2) The multi-function $F : (t, .) \to P_{cv,k}(E)$ is upper semi-continuous for a.e. $t \in J$.

(11.12.3) there exists a function $\alpha \in L^1(J, \mathbb{R}^+)$ such that

$$\|F(t, \psi)\| \leq \alpha(t)(1 + \|\psi\|_\mathcal{B}) \quad \text{for a.e. } t \in J;$$

(11.12.4) There exists a function $\beta \in L^1(J, \mathbb{R}^+)$ such that for all $\Omega \subset \mathcal{B}$, we have

$$\chi(F(t, \Omega)) \le \beta(t) \sup_{-\infty \le s \le 0} \chi(\Omega(s)) \quad \text{for a.e. } t \in J,$$

where, $\Omega(s) = \{x(s); x \in \Omega\}$ and χ is the Hausdorff MNC.

(11.12.5) There exist constants $a_k, c_k > 0, k = 1, \ldots, m$ such that
  1) $\|I_k\| \le a_k$,   where $I_k \in \mathcal{I}_k(x(t_k^+))$.
  2) $\chi(I_k(D)) \le c_k \chi(D)$ for each bounded subset D of E.

The next result is a consequence of the phase space axioms.

Lemma 11.13 ([139], Lemma 2.1). *If $y : (-\infty, b] \to \mathbb{R}$ is a function such that $y_0 = \phi$ and $y|_J \in PC(J, \mathbb{R})$, then*

$$\|y_s\|_{\mathcal{B}} \le (M_b + L^\phi)\|\phi\|_{\mathcal{B}} + K_b \sup\{\|y(\theta)\|; \theta \in [0, \max\{0, s\}]\}, \quad s \in \mathcal{R}(\rho^-) \cup J,$$

where

$$L^\phi = \sup_{t \in \mathcal{R}(\rho^-)} L^\phi(t).$$

Remark 11.14. We remark that condition (H_ϕ) is satisfied by functions which are continuous and bounded. In fact, if the space $\mathcal{B}$ satisfies axiom C_2 in [142] then there exists a constant $L > 0$ such that $\|\phi\|_{\mathcal{B}} \le L \sup\{\|\phi(\theta)\| : \theta \in (-\infty, 0]\}$ for every $\phi \in \mathcal{B}$ that is continuous and bounded (see [142], Proposition 7.1.1) for details. Consequently,

$$\|\phi_t\|_{\mathcal{B}} \le L \frac{\sup_{\theta \le 0} \|\phi(\theta)\|}{\|\phi\|_{\mathcal{B}}} \|\phi\|_{\mathcal{B}}, \quad \text{for every } \phi \in \mathcal{B} \setminus \{0\}.$$

Remark 11.15. Under conditions $(H\phi)$ and (11.12.1)–(11.12.3) for every piecewise continuous function $v : J \to \mathcal{B}$ the multi-function $F(t, v(t))$ admits a Bochner integrable selection (see [144]).

Let

$$\Omega_b = \{x \in \Omega : x_0 = 0\}.$$

For any $x \in \Omega_b$ we have

$$\|x\|_b = \|x\|_{\mathcal{B}} + \sup_{0 \le s \le b} \|x\| = \sup_{0 \le s \le b} \|x\|.$$

Thus $(_b, \|.\|_b)$ is a Banach space.

We note that from assumptions (11.12.1) and (11.12.3) it follows that the superposition multi-operator $S_F^1 : \Omega_b \to \mathcal{P}(L^1(J, E))$ defined by

$$S_F^1 = \{f \in L^1(J, E) : f(t) \in F(t, x_{\rho(t, x_t)}), \quad \text{a.e. } t \in J\}$$

is nonempty set (see [144]) and is weakly closed in the following sense.

Lemma 11.16. *If we consider the sequence* $(x^n) \in \Omega_b$ *and* $\{f_n\}_{n=1}^{+\infty} \subset L^1(J, E)$, *where* $f_n \in S_{F(\cdot, x_{\rho(\cdot, x^n)}^n)}^1$ *such that* $x^n \to x^0$ *and* $f_n \to f^0$ *then* $f^0 \in S_F^1$.

Now we state and prove our main result.

Theorem 11.17. *Under assumptions (A)–(Hϕ) and (11.12.1)–(11.12.5), the problem (11.43)–(11.45) has at least one mild solution.*

Proof. To prove the existence of a mild solution for (11.43)–(11.45) we introduce the integral multi-operator $N : \Omega_b \longrightarrow \mathcal{P}(\Omega_b)$, defined as

$$N(x) = \begin{cases} y : y(t) = T(t, 0)\phi(0) + \int_0^t T(t, s)f(s)ds \\ \quad \sum_{0 < t_k < t} T(t, t_k)I_k(x(t_k)), & t \in J \\ y(t) = \phi(t), & t \in (-\infty, 0], \end{cases} \quad (11.46)$$

where S_F^1 and $I_k \in \mathcal{I}_k(x)$.

It is clear that the integral multioperator N is well defined and the set of all mild solution for the problem (11.43)–(11.45) on J is the set $\mathcal{F}ixN = \{x : x \in N(x)\}$.

We shall prove that the integral multioperator N satisfies all the hypotheses of Lemma 1.40. The proof will be given in several steps.

Step 1. Using in fact that the maps F and $\mathcal{I}$ has a convex values it easy to check that N has convex values.

Step 2. N has closed graph.

Let $\{x^n\}_{n=1}^{+\infty} \subset \Omega_b$, $\{z^n\}_{n=1}^{+\infty}$, $x^n \to x^*$, $z^n \in N((x^n), n \geq 1)$ and $z^n \to z^*$. Moreover, let $\{f_n\}_{n=1}^{+\infty} \subset L^1(J, E)$ an arbitrary sequence such that $f_n \in S_F^1$ for $n \geq 1$.

Hypothesis (11.12.3) implies that the set $\{f_n\}_{n=1}^{+\infty}$ integrably bounded and for a.e. $t \in J$ the set $\{f_n(t)\}_{n=1}^{+\infty}$ relatively compact, we can say that $\{f_n\}_{n=1}^{+\infty}$ is semi-compact sequence. Consequently $\{f_n\}_{n=1}^{+\infty}$ is weakly compact in $L^1(J, E)$, so we can assume that $f_n \to f^*$.

From Lemma 1.43 we know that the generalized Cauchy operator on the interval J, $G : L^1(J, E) \to \Omega_b$, defined by

$$Gf(t) = \int_0^t T(t, s)f(s)ds, \quad t \in J \quad (11.47)$$

satisfies properties (G1) and (G2) on J.

Note that set $\{f_n\}_{n=1}^{+\infty}$ is also semi-compact and sequence $(f_n)_{n=1}^{+\infty}$ weakly converges to f^* in $L^1(J, E)$. Therefore, by applying Lemma 1.44 for the generalized Cauchy operator G of (11.19) we have the convergence $Gf_n \to Gf$. By means of (11.19) and (11.18), for all $t \in J$ we can write

$$z_n(t) = T(t, 0)\phi(0) + \int_0^t T(t, s)f_n(s)ds + \sum_{0 < t_k < t} T(t, t_k)I_k(x^n(t_k))$$

$$= T(t,0)\phi(0) + \int_0^t T(t,s)f_n ds + \sum_{0<t_k<t} T(t,t_k)I_k(x^n(t_k))$$

$$= T(t,0)\phi(0) + Gf_n(t) + \sum_{0<t_k<t} T(t,t_k)I_k(x^n(t_k))),$$

where S_F^1, and $I_k \in \mathcal{I}_k(x)$.

By applying Lemma 1.43, we deduce

$$z_n \to T(.,0)\phi(0) + Gf + T(.,t)I_k(x^*(t_k))$$

in Ω_b and by using in fact that the operator S_F^1 is closed, we get $f^* \in S_F^1$. Consequently

$$z^*(t) \to T(t,0)\phi(0) + Gf + T(t,t)I_k(x^*(t_k)),$$

therefore $z^* \in N(x^*)$. Hence N is closed.

With the same technique, we obtain that N has compact values.

Step 3. We consider the MNC defined in the following way. For every bounded subset $\Omega \subset \Omega_b$

$$\nu_1(\Omega) = \max_{\Omega \in \Delta(\Omega)} (\gamma_1(\Omega), \mod_C(\Omega)), \qquad (11.48)$$

where $\Delta(\Omega)$ is the collection of all the denumerable subsets of Ω;

$$\gamma_1(\Omega) = \sup_{t \in J} e^{-Lt} \chi(\{x(t) : x \in \Omega\}); \qquad (11.49)$$

where $\mod_C(\Omega)$ is the modulus of equi-continuity of the set of functions Ω given by the formula

$$\mod_C(\Omega) = \lim_{\delta \to 0} \sup_{x \in \Omega} \max_{|t_1-t_2|\le\delta} \|x(t_1) - x(t_2)\|, \qquad (11.50)$$

and $L > 0$ is a positive real number chosen such that

$$q := M \left(2\sup_{t \in J} \int_0^t e^{-L(t-s)} \beta(s)ds + e^{Lt} \sum_{k=1}^m c_k \right) < 1 \qquad (11.51)$$

where $M = \sup_{(t,s)\in\Delta} \|T(t,s)\|$.

From the Arzelá–Ascoli theorem, the measure ν_1 gives a nonsingular and regular MNC (see [144]).

Let $\{y_n\}_{n=1}^{+\infty}$ be the denumerable set which achieves that maximum $\nu_1(N(\Omega))$, i.e.;

$$\nu_1(N(\Omega)) = (\gamma_1(\{y_n\}_{n=1}^{+\infty}), \mod c(\{y_n\}_{n=1}^{+\infty})).$$

Then there exists a set $\{x_n\}_{n=1}^{+\infty} \subset \Omega$ such that $y_n \in N(x_n), n \geq 1$. Then

$$y_n(t) = T(t,0)\phi(0) + \int_0^t T(t,s)f(s)ds + \sum_{0<t_k<t} T(t,t_k)I_k(x(t_k)), \qquad (11.52)$$

where $f \in S_F^1$ and $I_k \in \mathcal{I}_k(x_n)$, so that

$$\gamma_1(\{y_n\}_{n=1}^{+\infty}) = \gamma_1(\{Gf_n\}_{n=1}^{+\infty}).$$

We give an upper estimate for $\gamma_1(\{y_n\}_{n=1}^{+\infty})$.

Fixed $t \in J$ by using condition (11.12.4), for all $s \in [0,t]$ we have

$$\chi(\{f_n(s)\}_{n=1}^{+\infty}) \leq \chi(F(s,\{x_n(s)\}_{n=1}^{+\infty}))$$
$$\leq \chi(\{F(s,x_n(s))\}_{n=1}^{+\infty})$$
$$\leq \beta(s)\chi(\{x_n(s)\}_{n=1}^{+\infty})$$
$$\leq \beta(s)e^{Ls}\sup_{t \in J} e^{-Lt}\chi(\{x_n(t)\}_{n=1}^{+\infty})$$
$$= \beta(s)e^{Ls}\gamma_1(\{x_n\}_{n=1}^{+\infty}).$$

By using condition (11.12.3), the set $\{f_n\}_{n=1}^{+\infty}$ is integrably bounded. In fact, for every $t \in J$, we have

$$\|f_n(t)\| \leq \|F(t,x_n(t))\|$$
$$\leq \alpha(t)(1 + \|x_n(t)\|).$$

The integrably boundedness of $\{f_n\}_{n=1}^{+\infty}$ follows from the continuity of x in J_k and the boundedness of set $\{x_n\}_{n=1}^{+\infty} \subset \Omega$. By applying Lemma 1.45, it follows that

$$\chi(\{Gf_n(s)\}_{n=1}^{+\infty}) \leq 2M \int_0^s \beta(t)e^{Lt}(\gamma_1(\{x_n\}_{n=1}^{+\infty}))dt$$
$$= 2M\gamma_1(\{x_n\}_{n=1}^{+\infty}) \int_0^s \beta(t)e^{Lt}.$$

Thus, we get

$$\gamma_1(\{x_n\}_{n=1}^{+\infty}) \leq \gamma_1(\{y_n\}_{n=1}^{+\infty}) = \gamma_1(\{Gf_n(s)\}_{n=1}^{+\infty})$$

$$= \sup_{t\in J} e^{-Lt} 2M\gamma_1(\{x_n\}_{n=1}^{+\infty}) \int_0^s \beta(t) e^{Lt} M\gamma_1(\{x_n\}_{n=1}^{+\infty}) e^{Lt} \sum_{k=1}^m c_k$$

$$\leq q\gamma_1(\{x_n\}_{n=1}^{+\infty}),$$

$$(11.53)$$

and hence $\gamma_1(\{x_n\}_{n=1}^{+\infty}) = 0$, then $\gamma_1(\{x_n(t)\}_{n=1}^{+\infty}) = 0$, for every $t \in J$. Consequently

$$\gamma_1(\{y_n\}_{n=1}^{+\infty}) = 0.$$

By using the last equality and hypotheses (11.12.3) and (11.12.4) we can prove that set $\{f_n\}_{n=1}^{+\infty}$ is semi-compact. Now, by applying Lemmas 1.43 and 1.44, we can conclude that set $\{Gf_n\}_{n=1}^{+\infty}$ is relatively compact. The representation of y_n given by (11.24) yields that set $\{y_n\}_{n=1}^{+\infty}$ is also relatively compact in $\boldsymbol{\Omega}_b$, therefore $\nu_1(\Omega) = (0,0)$. Then Ω is a relatively compact set.

Step 4. A priori bounds.

We will demonstrate that the solutions set is a priori bounded. Indeed, let $x \in N$. Then there exists $f \in S_F^1$ and $I_k \in \mathcal{I}_k(x)$ such that for every $t \in J$ we have

$$\|x(t)\| = \left\| T(t,0)\phi(0) + \int_0^t T(t,s)f(s)ds + \sum_{0<t_k<t} T(t,t_k)I_k(x(t_k)) \right\|$$

$$\leq M\left(\|\phi(0)\| + \sum_{k=1}^m a_k \right) + M \int_0^t f(s)ds$$

$$\leq M\left(\|\phi(0)\| + \sum_{k=1}^m a_k \right) + M \int_0^b \alpha(s)(1 + \|x[\phi]_{\rho(t,x_t)}\|)ds.$$

Using Lemma 11.13, we have

$$\|x(t)\| \leq M\left(\|\phi(0)\| + \sum_{k=1}^m a_k \right) + M \int_0^t \alpha(s)\left(1 + (M_b + L^\phi)\|\phi\|_\mathcal{B} \right.$$

$$\left. + K_b \sup_{0\leq\theta\leq s} \|x(\theta)\| \right)ds$$

$$\leq M\left(\|\phi(0)\| + \sum_{k=1}^m a_k \right) + M(1 + (M_b + L^\phi)\|\phi\|_\mathcal{B})\|\alpha\|_{L^1(J)}$$

$$+ MK_b \int_0^t \alpha(s) \sup_{0\leq\theta\leq s} \|x(\theta)\|ds.$$

Since the last expression is a nondecreasing function of t, we have that

$$\sup_{0\le\theta\le t} \|x(\theta)\| \le M\left(\|\phi(0)\| + \sum_{k=1}^{m} a_k\right) + M(1 + (M_b + L^\phi)\|\phi\|_\mathcal{B})\|\alpha\|_{L^1(J)}$$
$$+ MK_b \int_0^t \alpha(s) \sup_{0\le\theta\le s} \|x(\theta)\| ds.$$

Invoking Gronwall's inequality, we get

$$\sup_{0\le\theta\le b} \|x(\theta)\| \le \zeta e^{MK_b\|\alpha\|_{L^1[0,b]}},$$

where

$$\zeta = M\left(\|\phi(0)\| + \sum_{k=1}^{m} a_k\right) + M(1 + (M_b + L^\phi)\|\phi\|_\mathcal{B})\|\alpha\|_{L^1(J)}.$$

$\square$

11.5.3 Existence Results for the Non-convex Case

This section is devoted to proving the existence of solutions for (11.43)–(11.45) with a non-convex valued right-hand side. Our result is based on Mönch's fixed point theorem combined with a selection theorem due to Bressan and Colombo (see [86]). We will assume the following hypotheses: Let F be a multi-function defined from $J \times \mathcal{B}$ to the family of nonempty closed convex subsets of E such that

(11.18.1) $(t, x) \mapsto F(., x)$ is $\mathcal{L} \otimes \mathcal{B}_b$-measurable ($\mathcal{B}_b$ is Borel measurable).

(11.18.2) The multi-function $F : (t, .) \to P_k(E)$ is lower semi-continuous for a.e. $t \in J$,

(11.18.3) there exists a function $\alpha \in L^1(J, \mathbb{R}^+)$ such that

$$\|F(t, \psi)\| \le \alpha(t), \quad \text{for a.e. } t \in J, \ \forall \psi \in \mathcal{B};$$

(11.18.4) There exists a function $\beta \in L^1(J, \mathbb{R}^+)$ such that for all $\Omega \subset \mathcal{B}$, we have

$$\chi(F(t, \Omega)) \le \beta(t) \sup_{-\infty\le s\le 0} \chi(\Omega(s)) \quad \text{for a.e. } t \in J,$$

where, $\Omega(s) = \{x(s); x \in \Omega\}$ and χ is the Hausdorff MNC,

(11.18.5) There exist constants $a_k, b_k, c_k \ge 0$, $k = 1, \dots, m$, such that

1) $\|I_k\| \le a_k\|x\| + b_k$, where $I_k \in \mathcal{I}_k(x(t_k^+))$.

2) $\chi(I_k(D)) \le c_k\chi(D)$ for each bounded subset D of E.

Now we state and prove our main result.

Theorem 11.18. *Assume that (A)–(Hϕ) and (11.18.1)–(11.18.5) hold. If*

$$M \sum_{k=1}^{m} a_k < 1.$$

Then the problem (11.43)–(11.45) has at least one mild solution.

Proof. We note that from assumptions (11.18.1) and (11.18.3) it follows that the superposition multi-operator

$$S_F^1 : \Omega_b \to \mathcal{P}(L^1(J, E)),$$

defined by

$$S_F^1 = \{f \in L^1(J, E) : f(t) \in F(t, x_{\rho(t, x_t)}), \quad \text{a.e. } t \in J\}$$

is nonempty set (see [144]).

Clearly, fixed points of the operator N are mild solutions of the problem (11.43)–(11.45).

The proof will be given in several steps.

Step 1. The Mönch's condition holds.

Suppose that $\Omega \subseteq B_r$ is countable and $\Omega \subseteq \overline{co}(\{0\} \cup N(\Omega))$ We will prove that Ω is relatively compact. We consider the MNC defined in (11.48) and $L > 0$ is a positive real number chosen such that

$$q := M \left(2 \sup_{t \in J} \int_0^t e^{-L(t-s)} \beta(s) ds + e^{Lt} \sum_{k=1}^{m} c_k \right) < 1 \tag{11.54}$$

where $M = \sup_{(t,s) \in \Delta} \|T(t, s)\|$.

From the Arzelá–Ascoli theorem, the measure ν_1 give a nonsingular and regular MNC (see [144]).

Let $\{y_n\}_{n=1}^{+\infty}$ be the denumerable set which achieves that maximum $\nu_1(N(\Omega))$, i.e.;

$$\nu_1(N(\Omega)) = (\gamma_1(\{y_n\}_{n=1}^{+\infty}), \mod c(\{y_n\}_{n=1}^{+\infty})).$$

Then there exists a set $\{x_n\}_{n=1}^{+\infty} \subset \Omega$ such that $y_n \in N(x_n), n \geq 1$. Then

$$y_n(t) = T(t, 0)\phi(0) + \int_0^t T(t, s)f(s)ds + \sum_{0 < t_k < t} T(t, t_k)I_k, \tag{11.55}$$

where $f \in S_F^1$ and $I_k \in \mathcal{I}_k(x_n)$, so that

$$\gamma_1(\{y_n\}_{n=1}^{+\infty}) = \gamma_1(\{Gf_n\}_{n=1}^{+\infty}).$$

We give an upper estimate for $\gamma_1(\{y_n\}_{n=1}^{+\infty})$.
  Fixed $t \in J$ by using condition (11.18.4), for all $s \in [0, t]$ we have

$$\begin{aligned}
\chi(\{f_n(s)\}_{n=1}^{+\infty}) &\leq \chi(F(s, \{x_n(s)\}_{n=1}^{+\infty})) \\
&\leq \beta(s)\chi(\{x_n(s)\}_{n=1}^{+\infty}) \\
&\leq \beta(s)e^{Ls} \sup_{t \in J} e^{-Lt}\chi(\{x_n(t)\}_{n=1}^{+\infty}) \\
&= \beta(s)e^{Ls}\gamma_1(\{x_n\}_{n=1}^{+\infty}).
\end{aligned}$$

By using condition (11.18.3), the set $\{f_n\}_{n=1}^{+\infty}$ is integrably bounded. In fact, for every $t \in J$, we have

$$\begin{aligned}
\|f_n(t)\| &\leq \|F(t, x_n(t))\| \\
&\leq \alpha(t).
\end{aligned}$$

By applying Lemma 1.45, it follows that

$$\begin{aligned}
\chi(\{Gf_n(s)\}_{n=1}^{+\infty}) &\leq 2M \int_0^s \beta(t)e^{Lt}(\gamma_1(\{x_n\}_{n=1}^{+\infty}))dt \\
&= 2M\gamma_1(\{x_n\}_{n=1}^{+\infty}) \int_0^s \beta(t)e^{Lt}dt.
\end{aligned}$$

Thus, we get

$$\begin{aligned}
\gamma_1(\{x_n\}_{n=1}^{+\infty}) &\leq \gamma_1(\{y_n\}_{n=1}^{+\infty}) \\
&= \sup_{t \in J} e^{-Lt}2M\gamma_1(\{x_n\}_{n=1}^{+\infty}) \int_0^t \beta(s)e^{Ls}ds + M\gamma_1(\{x_n\}_{n=1}^{+\infty})e^{Lt}\sum_{k=1}^m c_k \\
&\leq q\gamma_1(\{x_n\}_{n=1}^{+\infty}).
\end{aligned}$$

(11.56)

Therefore, we have that

$$\gamma_1(\{x_n\}_{n=1}^{+\infty}) \leq \gamma_1(\Omega) \leq \gamma_1(\{0\} \cup N(\Omega))\gamma_1(\{y_n\}_{n=1}^{+\infty}) \leq q\gamma_1(\{x_n\}_{n=1}^{+\infty}).$$

From (11.51), we obtain that

$$\gamma_1(\{x_n\}_{n=1}^{+\infty}) = \gamma_1(\Omega) = \gamma_1(\{y_n\}_{n=1}^{+\infty})$$

Coming back to the definition of γ_1, we can see

$$\chi(\{x_n\}_{n=1}^{+\infty}) = \chi(\{y_n\}_{n=1}^{+\infty}) = 0$$

By using the last equality and hypotheses (11.18.3) and (11.18.4) we can prove that set $\{f_n\}_{n=1}^{+\infty}$ is semi-compact. Now, by applying Lemmas 1.43 and 1.44, we can conclude that set $\{Gf_n\}_{n=1}^{+\infty}$ is relatively compact.

The representation of y_n yields that set $\{y_n\}_{n=1}^{+\infty}$ is also relatively compact in $\mathbf{\Omega}_b$, since ν_1 is a monotone, nonsingular, regular MNC, we have that

$$\nu_1(\Omega) \leq \nu_1(\overline{co}(\{0\} \cup N(\Omega))) \leq \nu_1(N(\Omega)) = \nu_1(\{y_n\}_{n=1}^{+\infty}) = (0,0).$$

Therefore, Ω is relatively compact.

Step 2. It is clear that the superposition multioperator S_F^1 has closed and decomposable values. Following the lines of [144], we may verify that S_F^1 is l.s.c..

Applying Lemma 1.46 to the restriction of S_F^1 on $\mathbf{\Omega}_b$ we obtain that there exists a continuous selection

$$w : \mathbf{\Omega}_b \to L^1(J, E)$$

We consider a map $N : \mathbf{\Omega}_b \to \mathbf{\Omega}_b$ defined as

$$x(t) = T(t,0)\phi(0) + \int_0^t T(t,s)w(x)(s)ds$$

Since the Cauchy operator is continuous, the map N is also continuous, therefore, it is a continuous selection of the integral multi-operator.

Step 3. A priori bounds.

We will demonstrate that the solutions set is a priori bounded. Indeed, let $x \in \lambda N_1$ and $\lambda \in (0,1)$. There exists $f \in S_F^1$ and $I_k \in \mathcal{I}_k(x)$ such that for every $t \in J$ we have

$$\|x(t)\| = \left\| \lambda T(t,0)\phi(0) + \lambda \int_0^t T(t,s)f(s)ds + \lambda \sum_{0<t_k<t} T(t,t_k)I_k \right\|,$$

$$\leq M\left(\|\phi(0)\| + \|x\| \sum_{k=1}^m a_k + \sum_{k=1}^m b_k \right) + M \int_0^t \alpha(s)ds,$$

hence,

$$\left(1 - M \sum_{k=1}^m a_k\right) \|x\| \leq M\left(\|\phi(0)\| + \|\alpha\|_{L^1} + \sum_{k=1}^m b_k \right).$$

Consequently

$$\|x\| \leq \frac{M(\|\phi(0)\| + \|\alpha\|_{L^1} + \sum_{k=1}^m b_k)}{1 - M \sum_{k=1}^m a_k} = C.$$

So, there exists N^* such that $\|x\| \neq N^*$, set

$$U = \{x \in \mathbf{\Omega}_b : \quad \|x\| < N^*\}.$$

From the choice of U there is no $x \in \partial U$ such that $x = \lambda Nx$ for some $\lambda \in (0, 1)$. Thus, we get a fixed point of N_1 in $\bar{U}$ due to the Mönch's Theorem.   □

11.5.4 An Example

As an application of our results we consider the following impulsive partial functional differential equation of the form

$$\frac{\partial}{\partial t} z(t, x) \in a(t, x) \frac{\partial^2}{\partial x^2} z(t, x) + m(t) b(t, z(t - \sigma(z(t, 0))), x),$$

$$x \in [0, \pi], \ t \in [0, b], t \neq t_k, \tag{11.57}$$

$$z(t_k^+, x) - z(t_k^-, x) \in [-b_k|z(t_k^-, x), b_k|z(t_k^-, x)],$$

$$x \in [0, \pi], \ k = 1, \ldots, m, \tag{11.58}$$

$$z(t, 0) = z(t, \pi) = 0, \quad t \in J := [0, b], \tag{11.59}$$

$$z(t, x) = \phi(t, x), \quad -\infty < t \leq 0, \ x \in [0, \pi], \tag{11.60}$$

where $a(t, x)$ is continuous function and uniformly Hölder continuous in t, $b_k > 0$, $k = 1, \ldots, m$, $\phi \in \mathcal{D}$,

$\mathcal{D} = \{\overline{\psi} : (-\infty, 0] \times [0, \pi] \to \mathbb{R}; \ \overline{\psi}$ is continuous everywhere except for a countable number of points at which $\overline{\psi}(s^-), \overline{\psi}(s^+)$ exist with $\overline{\psi}(s^-) = \overline{\psi}(s)\}$,

$0 = t_0 < t_1 < t_2 < \cdots < t_m < t_{m+1} = b$, $z(t_k^+) = \lim_{(h,x) \to (0+, x)} z(t_k + h, x)$, $z(t_k^-) = \lim_{(h,x) \to (0-, x)} z(t_k + h, x)$, $b : \mathbb{R} \times \mathbb{R} \to \mathcal{P}_{cv,k}(\mathbb{R})$ a Carathéodory multivalued map, $\sigma : \mathbb{R} \to \mathbb{R}_+$.

Let

$$y(t)(x) = z(t, x), \quad x \in [0, \pi], \ t \in J = [0, b],$$

$$\mathcal{I}_k(y(t_k^-))(x) = [-b_k|z(t_k^-, x), b_k|z(t_k^-, x)], \quad x \in [0, \pi], \ k = 1, \ldots, m,$$

$$F(t, \phi)(x) = b(t) a(t, z(t - \sigma(z(t, 0))), x)$$

$$\phi(\theta)(x) = \phi(\theta, x), \quad -\infty < t \leq 0, \ x \in [0, \pi],$$

$$\rho(t, \phi) = t - \sigma(\phi(0, 0)).$$

Consider $E = L^2[0, \pi]$ and define $A(t)$ by $A(t)w = a(t, x)w''$ with domain

$$D(A) = \{w \in E : w, w' \text{ are absolutely continuous, } w'' \in E, \ w(0) = w(\pi) = 0\}.$$

Then $A(t)$ generates an evolution system $U(t, s)$ satisfying assumption (11.12.1) and (11.12.3). For the phase space, we choose $\mathcal{B} = \mathcal{B}_\gamma$ defined by

$$\mathcal{B}_\gamma = \left\{ \phi \in \mathcal{D} : \ \lim_{\theta \to -\infty} e^{\gamma\theta}\phi(\theta) \text{ exists} \right\}$$

with the norm

$$\|\phi\|_\gamma = \sup_{\theta \in (-\infty, 0]} e^{\gamma\theta}\|\phi(\theta)\|.$$

Notice that the phase space $\mathcal{B}_\gamma$ satisfies axioms (A1) and (A3) (see [142] for more details).

We can show that problem (11.57)–(11.60) is an abstract formulation of problem (11.43)–(11.45). Under suitable conditions, the problem (11.43)–(11.45) has at least one mild solution.

11.6 Controllability of Impulsive Differential Evolution Inclusions with Infinite Delay

11.6.1 Introduction

In this section, we are concerned by a controllability problem for a system governed by a semi-linear functional differential inclusion in a separable Banach space E in the presence of impulse effects and infinite delay.

$$x'(t) \in A(t)x(t) + F(t, x_t) + (Bu)(t), \quad t \in J = [0, b], \ t \neq t_k, \quad (11.61)$$

$$\Delta x\big|_{t=t_k} \in \mathcal{I}_k(x(t_k^-)), \quad k = 1, \ldots, m \quad (11.62)$$

$$x(t) = \phi(t), \quad t \in (-\infty, 0]. \quad (11.63)$$

Assuming that the compactness of the evolution operator generated by the linear part (11.61)–(11.61) is not required. Our aim here is to give global existence and controllability results for the above problem.

11.6.2   Existence and Controllability Results

In this section, we shall establish sufficient conditions for the controllability of the first order functional semi-linear differential inclusions (11.61)–(11.63).

We define what we mean by an mild solution of problem (11.61)–(11.63).

Definition 11.19. A function $x \in \Omega$ is said to be a mild solution of the system (11.15)–(11.17) if there exist a function $f \in L^1(J, E)$ such that $f \in F(t, x_t)$ for a.e. $t \in J$

(i) $x(t) = T(t, 0)\phi(0) + \int_0^t T(t, s) \left[(B_x u)(s) + f(s) \right] ds + \sum_{0 < t_k < t} T(t, t_k) I_k(x(t_k))$,
  a.e. $t \in J$, $k = 1, \ldots, m$

(ii) $x(t) = \phi(t)$, $\quad t \in (-\infty, 0]$,

with $I_k \in \mathcal{I}_k(x(t_k^+))$.

Let us introduce the following hypotheses:

(A) $\{A(t) : t \in J\}$ be a family of linear (not necessarily bounded) operators, $A(t) : D(A) \subset E \to E$, $D(A)$ not depending on t and dense subset of E and $T : \Delta = \{(t, s) : 0 \le s \le t \le b\} \to \mathcal{L}(E)$ be the evolution operator generated by the family $\{A(t) : t \in J\}$.

(11.19.1) The multi-function $F(., x)$ has a strongly measurable selection for every $x \in \mathcal{B}$.

(11.19.2) The multi-function $F : (t, .) \to P_{cv,k}(E)$ is upper semi-continuous for a.e. $t \in J$.

(11.19.3) there exists a function $\alpha \in L^1(J, \mathbb{R}^+)$ such that

$$\|F(t, \psi)\| \le \alpha(t)(1 + \|\psi\|_{\mathcal{B}}) \quad \text{for a.e. } t \in J;$$

(11.19.4) There exists a function $\beta \in L^1(J, \mathbb{R}^+)$ such that for all bounded $\Omega \subset \mathcal{B}$, we have

$$\chi(F(t, D)) \le \beta(t) \sup_{-\infty \le s \le 0} \chi(\Omega(s)) \quad \text{for a.e. } t \in J,$$

where, $\Omega(s) = \{x(s); x \in \Omega\}$ and χ is the Hausdorff MNC.

(11.19.5) There exist constants $a_k, k = 1, \ldots, m$ such that
  1) $\|I_k\| \le a_k$, where $I_k \in \mathcal{I}_k(x(t_k^+))$.
  2) I_k are completely continuous.

Remark 11.20. Under conditions (11.19.1)–(11.19.3) and $(\bar{A}1)$ for every piecewise continuous function $v : J \to \mathcal{B}$ the multi-function $F(t, v(t))$ admits a Bochner integrable selection (see [144]).

Before stating and proving our main result in this section, we define controllability on the interval J.

Definition 11.21. The system (11.61)–(11.63) is said to be controllable on the interval J, if for every $x_0, x_1 \in E$ there exists a control $u \in L^2(J, U)$, such that there exists a mild solution $x(t)$ of (11.61)–(11.63) satisfying $x(b) = x_1$.

Let

$$\Omega_b = \{x \in \Omega : x_0 = 0\}.$$

For any $x \in \Omega_b$ we have

$$\|x\|_b = \|x\|_{\mathcal{B}} + \sup_{0 \leq s \leq b} \|x\| = \sup_{0 \leq s \leq b} \|x\|.$$

Thus $(\Omega_b, \|.\|_b)$ is a Banach space .

We note that from assumptions (11.19.1) and (11.19.3) it follows that the superposition multi-operator $S_F^1 : \Omega_b \to \mathcal{P}(L^1(J, E))$ defined by

$$S_F^1 = \{f \in L^1(J, E) : f(t) \in F(t, x_t), \quad \text{a.e. } t \in J\}$$

is nonempty set (see [144]) and is weakly closed in the following sense.

Lemma 11.22. *If we consider the sequence* $(x^n) \in \Omega_b$ *and* $\{f_n\}_{n=1}^{+\infty} \subset L^1(J, E)$, *where* $f_n \in S_{F(.,x^n_.)}^1$ *such that* $x^n \to x^0$ *and* $f_n \to f^0$ *then* $f^0 \in S_F^1$.

Now, we are able to state and prove our main theorem:

Theorem 11.23. *Assume that hypotheses (11.19.1)–(11.19.5) hold. Moreover we suppose that*

(C1) *B is a continuous operator from U to X and the linear operator* $W :$ $L^2(J, U) \to X$, *defined by*

$$Wu = \int_0^b T(t, s)Bu(s)\, ds,$$

has a bounded inverse operator $W^{-1} : X \to L^2(J, U)/KerW$ *such that* $\|B\| \leq M_1$ *and* $\|W^{-1}\| \leq M_2$, *for some positive constants* M_1, M_2.

(C2) *There exists a function* $\zeta \in L^1(J, \mathbb{R}^+)$ *such that for all* $\Omega \in \mathcal{P}_b(E)$, *we have*

$$\chi_U(W^{-1}(\Omega)(t)) \leq \zeta(t)\chi_E(\Omega) \quad \text{for a.e. } t \in J,$$

Then the problem (11.61)–(11.63) has at least one mild solution.

To prove the controllability of the problem. Using hypothesis (C1) for an arbitrary function $x(\cdot)$ define the control

$$u_x(t) = W^{-1}\Big[x_1 - T(b, 0)\phi(0) - \sum_{k=1}^N T(b, t_k)I_k(x(t_k)) - \int_0^b T(b, \tau)f(\tau)\, ds\Big](t),$$

where $u_x(\cdot) \in L^2(J, E)$. We introduce the integral multi-operator $N : \Omega_b \longrightarrow \mathcal{P}(\Omega_b)$, defined as

$$
N(x) = \begin{cases}
y : y(t) = T(t, 0)\phi(0) + \int_0^t T(t, s)[f(s) + Bu_x(s)]\, ds \\
\quad + \sum_{0 < t_k < t} T(t, t_k) I_k(x(t_k)), & t \in J \\
y(t) = \phi(t), & t \in (-\infty, 0],
\end{cases}
\tag{11.64}
$$

where S_F^1 and $I_k \in \mathcal{I}_k(x)$.

It is clear that the integral multi-operator N is well defined and the set of all mild solution for the problem (11.61)–(11.63) on J is the set $FixN = \{x : x \in N(x)\}$.

Consider now the operator $G^* : L^1(J, E) \to C(J, E)$ defined by

$$
(G^*f)(t) = \int_0^t T(t, s) B W^{-1}\left[x_1 - T(b, 0)\phi(0) - \int_0^b T(b, \tau) f(\tau)\, ds \right](t) \tag{11.65}
$$

Lemma 11.24 ([84]). *The operator G^* satisfies the proprieties (G1')–(G2).*

We shall prove that the integral multi-operator N satisfies all the hypotheses of Lemma 1.40.

Proof. We break the proof into a sequence of steps.

Step 1. Using the fact that the maps F and $\mathcal{I}$ has a convex values it easy to check that N has convex values.

Step 2. N has closed graph.

Let $\{x^n\}_{n=1}^{+\infty} \subset \Omega_b$, $\{z^n\}_{n=1}^{+\infty}$, $x^n \to x^*$, $z^n \in N((x^n), n \geq 1)$ and $z^n \to z^*$. Moreover, let $\{f_n\}_{n=1}^{+\infty} \subset L^1(J, E)$ an arbitrary sequence such that $f_n \in S_F^1$ for $n \geq 1$.

Hypothesis (11.19.3) implies that the set $\{f_n\}_{n=1}^{+\infty}$ is integrably bounded and for a.e. $t \in J$ the set $\{f_n(t)\}_{n=1}^{+\infty}$ is relatively compact, we can say that $\{f_n\}_{n=1}^{+\infty}$ is semi-compact sequence. Consequently $\{f_n\}_{n=1}^{+\infty}$ is weakly compact in $L^1(J, E)$, so we can assume that $f_n \rightharpoonup f^*$.

From Lemma 1.43 we know that the generalized Cauchy operators, $G, G^* : L^1(J, E) \to \Omega_b$, defined by

$$
Gf(t) = \int_0^t T(t, s) f(s)\, ds, \quad t \in J \tag{11.66}
$$

$$
(G^*f)(t) = \int_0^t T(t, s) B W^{-1}\left[x_1 - T(b, 0)\phi(0) - \int_0^b T(b, \tau) f(\tau)\, ds \right](t) \tag{11.67}
$$

satisfies properties (G1) and (G2) on J.

Note that set $\{f_n\}_{n=1}^{+\infty}$ is also semi-compact and sequence $(f_n)_{n=1}^{+\infty}$ weakly converges to f^* in $L^1(J, E)$. Therefore, by applying Lemmas 1.44 and 11.67 we have the convergence $Gf_n \to Gf$ and $G^*f_n \to G^*f$. By means (11.18), for all $t \in J$ we can write

$$z_n(t) = T(t, 0)\phi(0) + Gf_n(t) + G^*f_n(t)$$

$$- \int_0^t T(t, s)BW^{-1}\left(\sum_{k=1}^N T(b, t_k)I_k(x(t_k))\right) + \sum_{0 < t_k < t} T(t, t_k)I_k(x^n(t_k))),$$

$$(11.68)$$

where $S_F^1, I_k \in \mathcal{I}_k(x)$ and

By applying Lemma 1.43, we deduce

$$z_n(t) = T(., 0)\phi(0) + Gf_n(.) + G^*f_n(.) - GBW^{-1}\left(\sum_{k=1}^N T(b, t_k)I_k(x^n(t_k))\right)$$

$$+ \sum_{0 < t_k < t} T(t, t_k)I_k(x^n(t_k)))$$

$$z_n \to T(., 0)\phi(0) + Gf + G^*f - GBW^{-1}\left(\sum_{k=1}^N T(b, t_k)I_k(x^*(t_k))\right)$$

$$+ \sum_{0 < t_k < t} T(., t)I_k(x^*(t_k))$$

in $\mathbf{\Omega}_b$ and by using in fact that the operator S_F^1 is closed, we get $f^* \in S_F^1$. Consequently

$$z^*(t) = T(t, 0)\phi(0) + Gf + G^*f - GBW^{-1}\left(\sum_{k=1}^N T(b, t_k)I_k(x^*(t_k))\right)$$

$$+ \sum_{0 < t_k < t} T(t, s)I_k(x^*(t_k)),$$

therefore $z^* \in N(x^*)$. Hence N is closed.

With the same technique, we obtain that N has compact values.

Step 3. We consider the MNC defined in the following way. For every bounded subset $\Omega \subset \mathbf{\Omega}_b$

$$\nu(\Omega) = (\gamma(\Omega), \text{mod}_C(\Omega)),$$

$$(11.69)$$

γ is the modulus of fiber non-compactness

$$\gamma(\Omega) = \sup_{t \in J} \chi_E(\{x(t) : x \in \Omega\}), \tag{11.70}$$

where $\Omega(t) = \{y(t) : y \in \Omega\}$.

$\mathrm{mod}_C(\Omega)$ is the modulus of equi-continuity of the set of functions Ω given by the formula

$$\mathrm{mod}_C(\Omega) = \lim_{\delta \to 0} \sup_{x \in \Omega} \max_{|t_1 - t_2| \le \delta} \|x(t_1) - x(t_2)\|; \tag{11.71}$$

and $\bar{q} > 0$ is a positive real number chosen such that

$$\bar{q} := \left(\bar{M} + \bar{M}^2 \bar{M}_1 \int_0^b \zeta(s)ds\right)\left(\int_0^b \beta(s)ds\right) < 1. \tag{11.72}$$

From the Arzelá–Ascoli theorem, the measure ν give a nonsingular and regular MNC , see [144].

Let $\Omega \subset \boldsymbol{\Omega}_b$ be a bounded subset such that

$$\nu(N(\Omega)) \ge \nu(\Omega). \tag{11.73}$$

For any $t \in [0, b]$ we have

$$N(\Omega)(t) \subset T(.,0)\phi(0) + (G + G^*) \circ S^1_{F(\Omega)}.$$

From the boundedness of the operators $\{T(t,s)\}_{0 \le s,t \le b}$ and B. Obviously there exist constants $\bar{M}, \bar{M}_1$ such that

$$\|T(t,s)\|^{(\chi)} \le \bar{M} \le M, \quad \text{for every } 0 \le s, t \le b \tag{11.74}$$

$$\|B\|^{(\chi)} \le \bar{M}_1 \le M_1. \tag{11.75}$$

We give an upper estimate for $\nu(N(\Omega))$. By using (11.6.4) and (11.74)–(11.75)

$$\chi(\{T(t,s)f(s)\}) \le \bar{M}\beta(s) \sup_{-\infty \le s \le 0} \chi(\Omega[\phi]_s)$$

$$\le \bar{M}\beta(s)\gamma(\Omega).$$

Where $f \in S^1_{F(\Omega)}$ and $\Omega[\phi]_s = \{x[\phi]_s : x \in \Omega\}$.

Applying the Proposition 1.12, we have

$$\chi(G \circ S^1_{F(\Omega)}(t)) \le \bar{M}\gamma(\Omega) \int_0^t \beta(s)ds$$

$$\le \bar{M}\gamma(\Omega) \int_0^b \beta(s)ds. \tag{11.76}$$

By using the condition (C2) and the estimates (11.74)–(11.76) we obtain

$$\chi\left(\left\{T(t,s)BW^{-1}[x_1 - T(b,0)\phi(0) - \int_0^b T(t,\tau)f(\tau)d\tau]\right\}\right)$$

$$\leq \bar{M}\bar{M}_1\zeta(s)\chi\left(\left\{\int_0^b T(t,\tau)f(\tau)d\tau\right\}\right)$$

$$\leq \bar{M}^2\bar{M}_1\gamma(\Omega)\left(\int_0^b \beta(s)ds\right)\zeta(s)$$

$f \in S^1_{F(\Omega)}$. By Proposition 1.12, we have

$$\chi\left(G^* \circ S^1_{F(\Omega)}(t)\right) \leq \bar{M}^2\bar{M}_1\gamma(\Omega)\left(\int_0^b \beta(s)ds\right)\left(\int_0^b \zeta(s)ds\right).$$

Thus, we get

$$\chi_E(N(\Omega)(t)) \leq \left(\bar{M} + \bar{M}^2\bar{M}_1\int_0^b \zeta(s)ds\right)\left(\int_0^b \beta(s)ds\right)\gamma(\Omega).$$

Then

$$\gamma(N(\Omega)) \leq \bar{q}\gamma(\Omega).$$

Where $\bar{q}$ is the constant in (11.72), consequently $\gamma(\Omega) = 0$.

By using the last equality and hypotheses (11.19.3) and (11.19.4) we can prove that set $\{f_n\}_{n=1}^{+\infty}$ is semi-compact. Now, by applying Lemmas 1.43 and 1.44, we can conclude that set $\{(G + G^*)f_n\}_{n=1}^{+\infty}$ is relatively compact. Therefore $\nu(\Omega) = (0,0)$ and $\mathrm{mod}_C(\Omega) = 0$. Then Ω is a relatively compact set.

Step 4. A priori bounds.

Let $x \in \lambda N(x); 0 < \lambda \leq 1$, then we have

$$\|x(t)\| \leq \left\|T(t,0)\phi(0) + \int_0^t T(t,s)[f(s) + Bu_x(s)]ds\right.$$

$$\left. + \sum_{0<t_k<t} T(t,t_k)I_k(x(t_k))\right\|$$

$$\leq M\|\phi(0)\| + M\sum_{0<t_k<t}\|I_k\| + M\int_0^t \|f(s)\|ds + MM_1\int_0^t \|u_x(s)\|ds$$

$$= M\|\phi(0)\| + M\sum_{0<t_k<t}\|I_k\| + M\int_0^t \|f(s)\|ds$$

$$+ MM_1 \int_0^t \left\| W^{-1} \left[x_1 - T(b,0)\phi(0) - \sum_{k=1}^N T(b,t_k)I_k(x(t_k)) \right.\right.$$

$$\left.\left. - \int_0^b T(b,\tau)f(\tau)\,ds \right](s) \right\| ds$$

$$\leq M\|\phi(0)\| + M \sum_{0<t_k<t} \|I_k\| + M \int_0^t \|f(s)\| ds$$

$$+ MM_1 \left\| W^{-1} \left[x_1 - T(b,0)\phi(0) - \sum_{k=1}^N T(b,t_k)I_k(x(t_k)) \right.\right.$$

$$\left.\left. - \int_0^b T(b,\tau)f(\tau)\,ds \right](s) \right\|_{L^1(J,U)}$$

$$\leq M\|\phi(0)\| + M \sum_{0<t_k<t} \|I_k\| + M \int_0^t \|f(s)\| ds$$

$$+ MM_1\sqrt{b} \left\| W^{-1} \left[x_1 - T(b,0)\phi(0) - \sum_{k=1}^N T(b,t_k)I_k(x(t_k)) \right.\right.$$

$$\left.\left. - \int_0^b T(b,\tau)f(\tau)\,ds \right](s) \right\|_{L^2(J,U)}$$

$$\leq M\|\phi(0)\| + M \sum_{k=1}^m a_k + M \int_0^t \|f(s)\| ds$$

$$+ MM_1 M_2 \sqrt{b} \left[\|x_1\| + M\|\phi(0)\| + M \sum_{k=1}^m a_k + M \int_0^b \|f(\tau)\| d\tau \right]$$

Using the hypothesis (11.19.3) and the condition (A1) we have

$$\|x(t)\| \leq M\|\phi(0)\| + M \sum_{k=1}^m a_k + M \int_0^t \alpha(s)(1 + \|x[\phi]_s\|)ds$$

$$+ MM_1 M_2 \sqrt{b} \left[\|x_1\| + M\|\phi(0)\| + M \sum_{k=1}^m a_k + M \int_0^b \alpha(s)(1 + \|x[\phi]_s\|)ds \right]$$

$$\leq M\|\phi(0)\| + M \sum_{k=1}^m a_k + M \int_0^t \alpha(s)(1 + N_b\|\phi\|_{\mathcal{B}} + K_b \sup_{0\leq\theta\leq s} \|x(\theta)\|)ds$$

$$+ MM_1 M_2 \sqrt{b} \left[\|x_1\| + M\|\phi(0)\| + M \sum_{k=1}^m a_k + M \int_0^b \alpha(s)(1 + N_b\|\phi\|_{\mathcal{B}} \right.$$

$$+K_b \sup_{0 \le \theta \le s} \|x(\theta)\|)ds \Bigg]$$

$$\le C^* + M \int_0^t \alpha(s) \sup_{0 \le \theta \le s} \|x(\theta)\|)ds + M^2 M_1 M_2 \sqrt{b} \int_0^b \alpha(s) \sup_{0 \le \theta \le s} \|x(\theta)\|)ds,$$

where

$$C_1 = M(1 + MM_1 M_2 \sqrt{b}) \sum_{k=1}^m a_k,$$

$$C_2 = M \left(\|\phi(0)\| + M_1 M_2 \sqrt{b})[\|x_1\| + M\|\phi(0)\|] \right),$$

$$C_3 = M \left(1 + N_b \|\phi\|_\mathcal{B} \right) \left(1 + MM_1 M_2 \sqrt{b} \right) \|\alpha\|_{L^1(J)},$$

and $C^* = C_1 + C_2 + C_3$. Consider the function $\mu(t) = \sup_{0 \le \theta \le t} \|x(\theta)\|$, so the function

$$v(t) = \int_0^t \alpha(s)\mu(s)ds \tag{11.77}$$

is nondecreasing and we have

$$v'(t) = \alpha(t)\mu(t), \quad \text{for a.e. } t \in J. \tag{11.78}$$

Applying the last inequality

$$v'(t) \le \alpha(t)(C^* + M^2 M_1 M_2 \sqrt{b}v(b) + Mv(t)), \tag{11.79}$$

multiplying both sides of (11.79) by the function $L(t) = \exp\left(-M \int_0^t \alpha(s)ds\right)$, we obtain

$$v'(t)L(t) \le \alpha(t)L(t)(C^* + M^2 M_1 M_2 \sqrt{b}v(b) + Mv(t)).$$

It follows that

$$(v(t)L(t))' \le \alpha(t)L(t)(C^* + M^2 M_1 M_2 \sqrt{b}v(b)). \tag{11.80}$$

Integrating from 0 to b both sides of (11.80) we get

$$v(b) \exp\left(-M \int_0^b \alpha(s)ds\right) \le (C^* + M^2 M_1 M_2 \sqrt{b}v(b)) \int_0^b \alpha(t)L(t)dt.$$

Thus

$$\ell v(b) \le C^* \int_0^b \alpha(t)L(t)dt,$$

where ℓ be a constant such that

$$\ell = \exp\left(-M \int_0^b \alpha(s)ds\right) - M^2 M_1 M_2 \sqrt{b} \int_0^b \alpha(t)L(t)dt > 0.$$

Consequently

$$v(b) \le \frac{C^* \int_0^b \alpha(t)L(t)dt}{\ell} = \overline{C}. \tag{11.81}$$

Using the nondecreasing character of v, we obtain

$$v(t) \le \overline{C}, \quad \text{for all } t \in J.$$

Then

$$\|x(t)\| \le C^* + M\left(1 + MM_1 M_2 \sqrt{b}\right)\overline{C}.$$

$\square$

11.6.3 An Example

As an application of our results we consider the following impulsive partial functional differential equation of the form

$$\frac{\partial}{\partial t}z(t,x) \in a(t,x)\frac{\partial^2}{\partial x^2}z(t,x) + m(t)\int_{-\infty}^t \mu_*(t,x,s-t)ds,$$
$$x \in [0,\pi], \ t \in [0,b], t \ne t_k, \tag{11.82}$$
$$z(t_k^+,x) - z(t_k^-,x) \in [-b_k|z(t_k^-,x), b_k|z(t_k^-,x)],$$
$$x \in [0,\pi], \ k = 1,\ldots,m, \tag{11.83}$$
$$z(t,0) = z(t,\pi) = 0, \quad t \in J := [0,b], \tag{11.84}$$
$$z(t,x) = \phi(t,x), \quad -\infty < t \le 0, \ x \in [0,\pi], \tag{11.85}$$

where $a(t,x)$ is continuous function and uniformly Hölder continuous in t, $b_k > 0$, $k = 1,\ldots,m$, $\phi \in \mathcal{D}$,

$\mathcal{D} = \{\overline{\psi} : (-\infty, 0] \times [0, \pi] \to \mathbb{R}; \ \overline{\psi}$ is continuous everywhere except for a countable number of points at which $\overline{\psi}(s^-), \overline{\psi}(s^+)$ exist with $\overline{\psi}(s^-) = \overline{\psi}(s)\}$,

$0 = t_0 < t_1 < t_2 < \cdots < t_m < t_{m+1} = b, \ z(t_k^+) = \lim_{(h,x) \to (0^+, x)} z(t_k + h, x),$
$z(t_k^-) = \lim_{(h,x) \to (0^-, x)} z(t_k + h, x), \ b : \mathbb{R} \times \mathbb{R} \to \mathcal{P}_{cv,k}(\mathbb{R})$ a Carathéodory multi-valued map, $\sigma : \mathbb{R} \to \mathbb{R}_+$.

Let

$$y(t)(x) = z(t, x), \quad x \in [0, \pi], \ t \in J = [0, b],$$

$$\mathcal{I}_k(y(t_k^-))(x) = [-b_k|z(t_k^-, x), b_k|z(t_k^-, x)], \quad x \in [0, \pi], \ k = 1, \dots, m,$$

$$F(t, \phi)(x) = m(t) \int_{-\infty}^t \mu_*(t, x, s - t) ds$$

$$\phi(\theta)(x) = \phi(\theta, x), \quad -\infty < t \leq 0, \ x \in [0, \pi].$$

Consider $E = L^2[0, \pi]$ and define $A(t)$ by $A(t)w = a(t, x)w''$ with domain $D(A) = \{w \in E : w, w'$ are absolutely continuous, $w'' \in E, \ w(0) = w(\pi) = 0\}$. Then $A(t)$ generates an evolution system $U(t, s)$ satisfying assumption (11.19.1) and (11.19.3).

Assume that $B : U \to Y, \ U \subseteq J$ is a bounded linear operator and the operator

$$Wu = \int_0^b T(t, s) Bu(s) \, ds,$$

has a bounded inverse operator $W^{-1} : E \to L^2(J, U)/\ker W$.

For the phase space, we choose $\mathcal{B} = \mathcal{B}_\mathfrak{h}$ defined by $\mathcal{B}_\mathfrak{h} = \{\phi : (-\infty, 0] \to E :$ For $a > 0 \ \phi(\Theta)$ is bounded and measurable function on $[-a, 0]$, and $\int_{-\infty}^0 h(s) \sup_{s \leq \theta \leq 0} |\phi(\Theta)| d\Theta\}$ where $h : (-\infty, 0] \to (0, +\infty)$ is a continuous function with

$$l = \int_{-\infty}^0 h(s) ds < +\infty,$$

endowed with the norm

$$\|\phi\|_\mathfrak{h} = \int_{-\infty}^0 h(s) \sup_{s \leq \theta \leq 0} |\phi(\Theta)| d\Theta.$$

Notice that the phase space $\mathcal{B}_\mathfrak{h}$ is Banach space (see [142] for more details).

We can show that problem (11.26)–(11.29) is an abstract formulation of problem (11.15)–(11.17). Under suitable conditions, the problem (11.61)–(11.63) has at least one mild solution.

11.7   Notes and Remarks

The results of Chap. 11 are taken from Benchohra et al. [68, 70]. Other results may be found in [7].

Chapter 12
Functional Differential Equations and Inclusions with Delay

12.1 Introduction

In this chapter, we shall prove the existence of solutions of some classes of functional differential equations and inclusions. Our investigations will be situated in the Banach space of real functions which are defined, continuous, and bounded on the real axis $\mathbb{R}$. We will use some fixed point theorems combined with the semigroup theory.

12.2 Global Existence for Functional Differential Equations with State-Dependent Delay

12.2.1 Introduction

In this section we will use Schauder's fixed point theorem combined with the semigroup theory to have the existence of solutions of the following functional differential equation with state-dependent delay:

$$y'(t) = Ay(t) + f(t, y_{\rho(t,y_t)}), \quad \text{a.e. } t \in J := \mathbb{R}_+ \tag{12.1}$$

$$y(t) = \phi(t), \quad t \in (-\infty, 0], \tag{12.2}$$

where $f : J \times \mathcal{B} \to E$ is given function, $A : D(A) \subset E \to E$ is the infinitesimal generator of a strongly continuous semigroup $T(t), t \in J$, $\mathcal{B}$ is the phase space to be specified later, $\phi \in \mathcal{B}$, $\rho : J \times \mathcal{B} \to \mathbb{R}$, and $(E, |.|)$ is a real Banach space.

© Springer International Publishing Switzerland 2015
S. Abbas, M. Benchohra, *Advanced Functional Evolution Equations and Inclusions*, Developments in Mathematics 39,
DOI 10.1007/978-3-319-17768-7_12

12.2.2   Existence of Mild Solutions

Now we give our main existence result for problem (12.1)–(12.2). Before starting and proving this result, we give the definition of the mild solution.

Definition 12.1. We say that a continuous function $y : (-\infty, +\infty) \to E$ is a mild solution of problem (12.1)–(12.2) if $y(t) = \phi(t)$, $t \in (-\infty, 0]$ and the restriction of $y(.)$ to the interval $\mathbb{R}_+$ is continuous and satisfies the following integral equation:

$$y(t) = T(t)\phi(0) + \int_0^t T(t-s)f(s, y_{\rho(s,y_s)})ds, \ t \in J.$$

Set

$$\mathcal{R}(\rho^-) = \{\rho(s, \phi) : (s, \phi) \in J \times \mathcal{B}, \ \rho(s, \phi) \le 0\}.$$

We always assume that $\rho : J \times \mathcal{B} \to \mathbb{R}$ is continuous. Additionally, we introduce the following hypothesis:

(H_ϕ) The function $t \to \phi_t$ is continuous from $\mathcal{R}(\rho^-)$ into $\mathcal{B}$ and there exists a continuous and bounded function $\mathcal{L}^\phi : \mathcal{R}(\rho^-) \to (0, \infty)$ such that

$$\|\phi_t\| \le \mathcal{L}^\phi(t)\|\phi\| \quad \text{for every } t \in \mathcal{R}(\rho^-).$$

Remark 12.2. The condition (H_ϕ) is frequently verified by functions continuous and bounded. For more details, see for instance [142].

Lemma 12.3 ([140], Lemma 2.4). *If* $y : \mathbb{R} \to E$ *is a function such that* $y_0 = \phi$, *then*

$$\|y_s\|_{\mathcal{B}} \le (M + \mathcal{L}^\phi)\|\phi\|_{\mathcal{B}} + l \sup\{|y(\theta)| : \theta \in [0, max\{0, s\}]\}, \ s \in \mathcal{R}(\rho^-) \cup J,$$

where $\mathcal{L}^\phi = \sup_{t \in \mathcal{R}(\rho^-)} \mathcal{L}^\phi(t)$.

Let us introduce the following hypotheses:

(12.3.1) $A : D(A) \subset E \to E$ is the infinitesimal generator of a strongly continuous semigroup $T(t), t \in J$ which is compact for $t > 0$ in the Banach space E. Let $M' = \sup\{\|T\|_{B(E)} : t \ge 0\}$.

(12.3.2) The function $f : J \times \mathcal{B} \to E$ is Carathéodory.

(12.3.3) There exists a continuous function $k : J \to \mathbb{R}_+$ such that:

$$|f(t, u) - f(t, v)| \le k(t)\|u - v\|_{\mathcal{B}}, \ t \in J, \ u, v \in \mathcal{B},$$

and

$$k^* := \sup_{t \in J} \int_0^t k(s)\,ds < \infty. \tag{12.3}$$

(12.3.4) The function $t \to f(t, 0) = f_0 \in L^1(J, \mathbb{R}_+)$ with $F^* = \|f_0\|_{L^1}$.

Theorem 12.4. *Assume that (12.3.1)–(12.3.4), (H_ϕ) hold. If $k^* M' l < 1$, then the problem (12.1)–(12.2) has at least one mild solution on BC.*

Proof. Consider the operator $N : BC \to BC$ defined by:

$$(Ny)(t) = \begin{cases} \phi(t), & \text{if } t \in (-\infty, 0], \\ T(t)\,\phi(0) + \displaystyle\int_0^t T(t-s)\,f(s, y_{\rho(s,y_s)})\,ds, & \text{if } t \in J. \end{cases}$$

Let $x(.) : \mathbb{R} \to E$ be the function defined by:

$$x(t) = \begin{cases} \phi(t), & \text{if } t \in (-\infty, 0]; \\ T(t)\,\phi(0), & \text{if } t \in J, \end{cases}$$

then $x_0 = \phi$. For each $z \in BC$ with $z(0) = 0$, we denote by $\bar{z}$ the function

$$\bar{z}(t) = \begin{cases} 0, & \text{if } t \in (-\infty, 0]; \\ z(t), & \text{if } t \in J. \end{cases}$$

If y satisfies $y(t) = (Ny)(t)$, we can decompose it as $y(t) = z(t) + x(t)$, $t \in J$, which implies $y_t = z_t + x_t$ for every $t \in J$ and the function $z(.)$ satisfies

$$z(t) = \int_0^t T(t-s) f(s, z_{\rho(s,z_s+x_s)} + x_{\rho(s,z_s+x_s)})\,ds, \ t \in J.$$

Set

$$BC_0' = \{z \in BC' : z(0) = 0\}$$

and let

$$\|z\|_{BC_0'} = \sup\{|z(t)| : t \in J\}, \ z \in BC_0'.$$

BC_0' is a Banach space with the norm $\|.\|_{BC_0'}$. We define the operator $\mathcal{A} : BC_0' \to BC_0'$ by:

$$\mathcal{A}(z)(t) = \int_0^t T(t-s) f(s, z_{\rho(s,z_s+x_s)} + x_{\rho(s,z_s+x_s)})\,ds, \ t \in J.$$

We shall show that the operator $\mathcal{A}$ satisfies all conditions of Schauder's fixed point theorem. The operator A maps BC_0' into BC_0', indeed the map $\mathcal{A}(z)$ is continuous on $\mathbb{R}_+$ for any $z \in BC_0'$, and for each $t \in J$ we have

$$|\mathcal{A}(z)(t)| \le M' \int_0^t |f(s, z_{\rho(s,z_s+x_s)} + x_{\rho(s,z_s+x_s)}) - f(s,0) + f(s,0)| ds$$

$$\le M' \int_0^t |f(s,0)| ds + M' \int_0^t k(s) \|z_{\rho(s,z_s+x_s)} + x_{\rho(s,z_s+x_s)}\|_{\mathcal{B}} ds$$

$$\le M' F^* + M' \int_0^t k(s)(l|z(s)| + (m + \mathcal{L}^\phi + lM'H)\|\phi\|_{\mathcal{B}}) ds.$$

Set

$$C := (m + \mathcal{L}^\phi + lM'H)\|\phi\|_{\mathcal{B}}.$$

Then, we have

$$|\mathcal{A}(z)(t)| \le M' F^* + M' C \int_0^t k(s) ds + M' \int_0^t l|z(s)|k(s) ds$$

$$\le M' F^* + M' C k^* + M' l \|z\|_{BC_0'} k^*.$$

Hence, $\mathcal{A}(z) \in BC_0'$.

Moreover, let $r > 0$ be such that

$$r \ge \frac{M' F^* + M' C k^*}{1 - M' k^* l},$$

and B_r be the closed ball in BC_0' centered at the origin and of radius r. Let $z \in B_r$ and $t \in \mathbb{R}_+$. Then

$$|\mathcal{A}(z)(t)| \le M' F^* + M' C k^* + M' k^* l r.$$

Thus

$$\|\mathcal{A}(z)\|_{BC_0'} \le r,$$

which means that the operator $\mathcal{A}$ transforms the ball B_r into itself.

Now we prove that $\mathcal{A} : B_r \to B_r$ satisfies the assumptions of Schauder's fixed theorem. The proof will be given in several steps.

Step 1: $\mathcal{A}$ is continuous in B_r.

Let $\{z_n\}$ be a sequence such that $z_n \to z$ in B_r. At the first, we study the convergence of the sequences $(z^n_{\rho(s,z_s^n)})_{n\in\mathbb{N}}, s \in J$.

If $s \in J$ is such that $\rho(s, z_s) > 0$, then we have,

$$\|z^n_{\rho(s,z^n_s)} - z_{\rho(s,z_s)}\|_\mathcal{B} \leq \|z^n_{\rho(s,z^n_s)} - z_{\rho(s,z^n_s)}\|_\mathcal{B} + \|z_{\rho(s,z^n_s)} - z_{\rho(s,z_s)}\|_\mathcal{B}$$

$$\leq L\|z_n - z\|_\mathcal{B} + \|z_{\rho(s,z^n_s)} - z_{\rho(s,z_s)}\|_\mathcal{B},$$

which proves that $z^n_{\rho(s,z^n_s)} \to z_{\rho(s,z_s)}$ in $\mathcal{B}$ as $n \to \infty$ for every $s \in J$ such that $\rho(s, z_s) > 0$. Similarly, is $\rho(s, z_s) < 0$, we get

$$\|z^n_{\rho(s,z^n_s)} - z_{\rho(s,z_s)}\|_\mathcal{B} = \|\phi^n_{\rho(s,z^n_s)} - \phi_{\rho(s,z_s)}\|_\mathcal{B} = 0$$

which also shows that $z^n_{\rho(s,z^n_s)} \to z_{\rho(s,z_s)}$ in $\mathcal{B}$ as $n \to \infty$ for every $s \in J$ such that $\rho(s, z_s) < 0$. Combining the pervious arguments, we can prove that $z^n_{\rho(s,z_s)} \to \phi$ for every $s \in J$ such that $\rho(s, z_s) = 0$. Finally,

$$|\mathcal{A}(z_n)(t) - \mathcal{A}(z)(t)|$$

$$\leq M' \int_0^t |f(s, z^n_{\rho(s,z^n_s+x_s)} + x_{\rho(s,z^n_s+x_s)}) - f(s, z_{\rho(s,z_s+x_s)} + x_{\rho(s,z_s+x_s)})| ds$$

$$\leq M' \int_0^t |f(s, z_{\rho(s,z^n_s+x_s)} + x_{\rho(s,z^n_s+x_s)}) - f(s, z_{\rho(s,z_s+x_s)} + x_{\rho(s,z_s+x_s)})| ds.$$

Then by (12.3.2) we have

$$f(s, z^n_{\rho(s,z^n_s+x_s)} + x_{\rho(s,z^n_s+x_s)}) \to f(s, z_{\rho(s,z_s+x_s)} + x_{\rho(s,z_s+x_s)}), \quad \text{as } n \to \infty,$$

and by the Lebesgue dominated convergence theorem we get,

$$\|\mathcal{A}(z_n) - \mathcal{A}(z)\|_{BC'_0} \to 0, \quad \text{as } n \to \infty.$$

Thus $\mathcal{A}$ is continuous.

Step 2: $\mathcal{A}(B_r) \subset B_r$ this is clear.

Step 3: $\mathcal{A}(B_r)$ is equi-continuous on every compact interval $[0, b]$ of $\mathbb{R}_+$ for $b > 0$. Let $\tau_1, \tau_2 \in [0, b]$ with $\tau_2 > \tau_1$, we have:

$$|\mathcal{A}(z)(\tau_2) - \mathcal{A}(z)(\tau_1)|$$

$$\leq \int_0^{\tau_1} \|T(\tau_2 - s) - T(\tau_1 - s)\|_{B(E)} |f(s, z_{\rho(s,z_s+x_s)} + x_{\rho(s,z_s+x_s)})| ds$$

$$+ \int_{\tau_1}^{\tau_2} \|T(\tau_2 - s)\|_{B(E)} |f(s, z_{\rho(s,z_s+x_s)} + x_{\rho(s,z_s+x_s)})| ds$$

$$\leq \int_0^{\tau_1} \|T(\tau_2 - s) - T(\tau_1 - s)\|_{B(E)} |f(s, z_{\rho(s,z_s+x_s)} + x_{\rho(s,z_s+x_s)}) - f(s, 0)| ds$$

$$+ \int_0^{\tau_1} \|T(\tau_2 - s) - T(\tau_1 - s)\|_{B(E)} |f(s, 0)| ds$$

$$+ \int_{\tau_1}^{\tau_2} \|T(\tau_2 - s)\|_{B(E)} |f(s, z_{\rho(s, z_s + x_s)} + x_{\rho(s, z_s + x_s)}) - f(s, 0)| ds$$

$$+ \int_{\tau_1}^{\tau_2} \|T(\tau_2 - s)\|_{B(E)} |f(s, 0)| ds$$

$$\leq C \int_0^{\tau_1} \|T(\tau_2 - s) - T(\tau_1 - s)\|_{B(E)} k(s) ds$$

$$+ rL \int_0^{\tau_1} \|T(\tau_2 - s) - T(\tau_1 - s)\|_{B(E)} k(s) ds$$

$$+ \int_0^{\tau_1} \|T(\tau_2 - s) - T(\tau_1 - s)\|_{B(E)} |f(s, 0)| ds$$

$$+ C \int_{\tau_1}^{\tau_2} \|T(\tau_2 - s)\|_{B(E)} k(s) ds$$

$$+ rL \int_{\tau_1}^{\tau_2} \|T(\tau_2 - s)\|_{B(E)} k(s) ds$$

$$+ \int_{\tau_1}^{\tau_2} \|T(\tau_2 - s)\|_{B(E)} |f(s, 0)| ds.$$

When $\tau_2 \to \tau_1$, the right-hand side of the above inequality tends to zero, since $T(t)$ is a strongly continuous operator and the compactness of $T(t)$ for $t > 0$ implies the continuity in the uniform operator topology (see [168]), this proves the equi-continuity.

Step 4: $\mathcal{A}(B_r)(t)$ is relatively compact on every compact interval of $t \in [0, \infty)$. Let $t \in [0, b]$ for $b > 0$ and let ε be a real number satisfying $0 < \varepsilon < t$. For $z \in B_r$ we define

$$\mathcal{A}_\varepsilon(z)(t) = T(\varepsilon) \int_0^{t-\varepsilon} T(t - s - \varepsilon) f(s, z_{\rho(s, z_s + x_s)} + x_{\rho(s, z_s + x_s)}) ds.$$

Note that the set

$$\left\{ \int_0^{t-\varepsilon} T(t - s - \varepsilon) f(s, z_{\rho(s, z_s + x_s)} + x_{\rho(s, z_s + x_s)}) ds : z \in B_r \right\}$$

is bounded.
Since $T(t)$ is a compact operator for $t > 0$, the set,

$$\{\mathcal{A}_\varepsilon(z)(t) : z \in B_r\}$$

is precompact in E for every ε, $0 < \varepsilon < t$. Moreover, for every $z \in B_r$ we have

$$|\mathcal{A}(z)(t) - \mathcal{A}_\varepsilon(z)(t)|$$

$$\leq \int_{t-\varepsilon}^t T(t-s)f(s, z_{\rho(s,z_s+x_s)} + x_{\rho(s,z_s+x_s)})ds$$

$$\leq M'F^*\varepsilon + M'C\int_{t-\varepsilon}^t k(s)ds + rM'\int_{t-\varepsilon}^t lk(s)ds,$$

$$\to 0 \quad \text{as} \quad \varepsilon \to 0.$$

Therefore, the set $\{\mathcal{A}(z)(t) : z \in B_r\}$ is precompact, i.e., relatively compact.

Step 5: $\mathcal{A}(B_r)$ is equi-convergent.

Let $t \in \mathbb{R}_+$ and $z \in B_r$, we have,

$$|\mathcal{A}(z)(t)| \leq M'\int_0^t |f(s, z_{\rho(s,z_s+x_s)} + x_{\rho(s,z_s+x_s)})|ds$$

$$\leq M'F^* + M'C\int_0^t k(s)ds + M'r\int_0^t Lk(s)ds$$

$$\leq M'F^* + M'C\int_0^t k(s)ds + M'rl\int_0^t k(s)ds.$$

Then by (12.3.4), we have

$$|\mathcal{A}(z)(t)| \to M^* \leq M'F^* + M'Ck^* + M'rlk^*, \quad \text{as } t \to +\infty.$$

Hence,

$$|\mathcal{A}(z)(t) - \mathcal{A}(z)(+\infty)| \to 0, \quad \text{as } t \to +\infty.$$

As a consequence of Steps 1–4, with Lemma 1.26, we can conclude that $\mathcal{A} : B_r \to B_r$ is continuous and compact. From Schauder's theorem, we deduce that $\mathcal{A}$ has a fixed point z^*. Then $y^* = z^* + x$ is a fixed point of the operators N, which is a mild solution of the problem (12.1)–(12.2). □

12.2.3 An Example

Consider the following functional partial differential equation

$$\frac{\partial}{\partial t}z(t,x) - \frac{\partial^2}{\partial x^2}z(t,x) = e^{-t}\int_{-\infty}^0 z\left(s - \sigma_1(t)\sigma_2\left(\int_0^\pi a(\theta)|z(t,\theta)|^2 d\theta\right),x\right)ds,$$

$$x \in [0,\pi], \ t \in \mathbb{R}_+ \tag{12.4}$$

$$z(t,0) = z(t,\pi) = 0, \ t \in \mathbb{R}_+, \tag{12.5}$$

$$z(\theta,x)) = z_0(\theta,x), \ t \in (-\infty,0], \ x \in [0,\pi], \tag{12.6}$$

where $z_0 \not\equiv 0$. Set

$$f(t, \psi)(x) = \int_{-\infty}^{0} e^{-t} \psi(s, x) ds,$$

and

$$\rho(t, \psi) = t - \sigma_1(t)\sigma_2 \left(\int_0^\pi a_2(\theta) |\psi(t, \theta)|^2 d\theta \right),$$

$\sigma_i : \mathbb{R}^+ \to \mathbb{R}^+, i = 1, 2$ and $a : \mathbb{R} \to \mathbb{R}$ are continuous functions.

Take $E = L^2[0, \pi]$ and define $A : E \to E$ by $A\omega = \omega''$ with domain

$$D(A) = \{\omega \in E, \omega, \omega' \text{ are absolutely continuous, } \omega'' \in E, \omega(0) = \omega(\pi) = 0\}.$$

Then

$$A\omega = \sum_{n=1}^{\infty} n^2 (\omega, \omega_n)\omega_n, \omega \in D(A)$$

where $\omega_n(s) = \sqrt{\frac{2}{\pi}} \sin ns, n = 1, 2, \ldots$ is the orthogonal set of eigenvectors in A. It is well known (see [168]) that A is the infinitesimal generator of an analytic semigroup $T(t), t \geq 0$ in E and is given by

$$T(t)\omega = \sum_{n=1}^{\infty} \exp(-n^2 t)(\omega, \omega_n)\omega_n, \omega \in E.$$

Since the analytic semigroup $T(t)$ is compact, there exists a positive constant M such that

$$\|T(t)\|_{B(E)} \leq M.$$

Let $\mathcal{B} = BCU(\mathbb{R}^-, E)$ and $\phi \in \mathcal{B}$, then (H_ϕ).

The function $f(t, \psi)(x)$ is Carathéodory, and

$$|f(t, \psi_1)(x) - f(t, \psi_2)(x)| \leq e^{-t}|\psi_1(t, x) - \psi_2(t, x)|,$$

thus $k(t) = e^{-t}$, moreover we have

$$k^* = \sup \left\{ \int_0^t e^{-s} ds, t \in \mathbb{R}_+ \right\} = 1, f_0 \equiv 0.$$

Then the problem (12.1)–(12.2) is an abstract formulation of the problem (12.4)–(12.6), and conditions (12.3.1)–(12.3.4), (H_ϕ) are satisfied. Theorem 12.4 implies that the problem (12.4)–(12.6) has at least one mild solutions on BC.

12.3   Global Existence Results for Neutral Functional Differential Equations with State-Dependent Delay

12.3.1   Introduction

In this section we prove the existence solutions of a functional differential equation. Our investigations will be situated in the Banach space of real functions which are defined, continuous, and bounded on the real axis $\mathbb{R}$. We will use Schauder's fixed point theorem combined with the semigroup theory to have the existence of solutions of the following functional differential equation with state-dependent delay:

$$\frac{d}{dt}[y(t) - g(t, y_{\rho(t, y_t)})] = A[y(t) - g(t, y_{\rho(t, y_t)})] + f(t, y_{\rho(t, y_t)}), \quad \text{a.e. } t \in J := \mathbb{R}_+$$

$$(12.7)$$

$$y(t) = \phi(t), \quad t \in (-\infty, 0], \tag{12.8}$$

where $f, g : J \times B \to E$ are given functions, $A : D(A) \subset E \to E$ is the infinitesimal generator of a strongly continuous semigroup $T(t), t \in J$, B is the phase space to be specified later, $\phi \in B$, $\rho : J \times B \to \mathbb{R}$, and $(E, |.|)$ is a real Banach space.

12.3.2   Existence of Mild Solutions

Definition 12.5. We say that a continuous function $y : (-\infty, +\infty) \to E$ is a mild solution of problem (11.15)–(11.16) if $y(t) = \phi(t)$, $t \in (-\infty, 0]$ and the restriction of $y(.)$ to the interval $\mathbb{R}_+$ is continuous and satisfies the following integral equation:

$$y(t) = T(t)[\phi(0) - g(0, \phi(0))] + g(t, y_{\rho(t, y_t)}) + \int_0^t T(t - s)f(s, y_{\rho(s, y_s)})ds, \quad t \in J.$$

Set

$$\mathcal{R}(\rho^-) = \{\rho(s, \phi) : (s, \phi) \in J \times B, \ \rho(s, \phi) \le 0\}.$$

We always assume that $\rho : J \times \mathcal{B} \to \mathbb{R}$ is continuous. Additionally, we introduce the following hypothesis:

(H_ϕ) The function $t \to \phi_t$ is continuous from $\mathcal{R}(\rho^-)$ into $\mathcal{B}$ and there exists a continuous and bounded function $\mathcal{L}^\phi : \mathcal{R}(\rho^-) \to (0, \infty)$ such that

$$\|\phi_t\| \leq \mathcal{L}^\phi(t)\|\phi\| \quad \text{for every } t \in \mathcal{R}(\rho^-).$$

Remark 12.6. The condition (H_ϕ) is frequently verified by functions continuous and bounded.

Lemma 12.7 ([139]). *If $y : \mathbb{R} \to E$ is a function such that $y_0 = \phi$, then*

$$\|y_s\|_{\mathcal{B}} \leq (M + \mathcal{L}^\phi)\|\phi\|_{\mathcal{B}} + l \sup\{|y(\theta)| : \theta \in [0, max\{0, s\}]\}, \ s \in \mathcal{R}(\rho^-) \cup J,$$

where $\mathcal{L}^\phi = \sup\limits_{t \in \mathcal{R}(\rho^-)} \mathcal{L}^\phi(t)$.

Let us introduce the following hypotheses:

(12.7.1) $A : D(A) \subset E \to E$ is the infinitesimal generator of a strongly continuous semigroup $T(t), t \in J$ which is compact for $t > 0$ in the Banach space E. Let $M' = \sup\{\|T\|_{B(E)} : t \geq 0\}$.
(12.7.2) The function $f : J \times \mathcal{B} \to E$ is Carathéodory.
(12.7.3) There exists a continuous function $k : J \to \mathbb{R}_+$ such that:

$$|f(t, u) - f(t, v)| \leq k(t)\|u - v\|_{\mathcal{B}}, \ t \in J, \ u, v \in \mathcal{B}$$

and

$$k^* := \sup_{t \in J} \int_0^t k(s)ds < \infty.$$

(12.7.4) The function $t \to f(t, 0) = f_0 \in L^1(J, \mathbb{R}_+)$ with $F^* = \|f_0\|_{L^1}$.
(12.7.5) The function $g(t, \cdot)$ is continuous on J and there exists a constant $k_g > 0$ such that

$$|g(t, u) - g(t, v)| \leq k_g\|u - v\|_{\mathcal{B}}, \text{ for each, } u, v \in \mathcal{B}$$

and

$$g^* := \sup_{t \in J} |g(t, 0)| < \infty.$$

(12.7.6) For each $t \in J$ and any bounded set $B \subset \mathcal{B}$, the set $\{g(t, u) : u \in B\}$ is relatively compact in E
(12.7.7) For any bounded set $B \subset \mathcal{B}$, the function $\{t \to g(t, y_t) : y \in B\}$ is equi-continuous on each compact interval of $\mathbb{R}_+$.

Remark 12.8. By the condition (12.7.3), (12.7.4) we deduce that

$$|f(t,y)| \le k(t)\|u\|_{\mathcal{B}} + F^*, \ t \in J, \ u \in \mathcal{B},$$

and by (12.7.5) we deduce that:

$$|g(t,u)| \le k_g \|u\|_{\mathcal{B}} + g^* \ t \in J, \ u \in \mathcal{B}.$$

Theorem 12.9. *Assume that (12.7.1)–(12.7.7) and* (H_ϕ) *hold. If* $l(M'k^* + \alpha_1) < 1$, *then the problem (12.1)–(12.2) has at least one mild solution on BC.*

Proof. Transform the problem (12.1)–(12.2) into a fixed point problem. Consider the operator $N : BC \to BC$ defined by:

$$(Ny)(t) = \begin{cases} \phi(t); & \text{if } t \in (-\infty, 0], \\ T(t)\,[\phi(0) - g(0,\phi(0))] \\ \quad + g(t, y_{\rho(t,y_t)}) + \displaystyle\int_0^t T(t-s)\,f(s, y_{\rho(s,y_s)})\,ds; & \text{if } t \in J. \end{cases}$$

Let $x(.) : \mathbb{R} \to E$ be the function defined by:

$$x(t) = \begin{cases} \phi(t); & \text{if } t \in (-\infty, 0]; \\ T(t)\,\phi(0); & \text{if } t \in J, \end{cases}$$

then $x_0 = \phi$. For each $z \in BC$ with $z(0) = 0$, we denote by $\bar{z}$ the function

$$\bar{z}(t) = \begin{cases} 0; & \text{if } t \in (-\infty, 0]; \\ z(t); & \text{if } t \in J. \end{cases}$$

If y satisfies $y(t) = (Ny)(t)$, we can decompose it as $y(t) = z(t) + x(t)$, $t \in J$, which implies $y_t = z_t + x_t$ for every $t \in J$ and the function $z(.)$ satisfies

$$z(t) = g(t, z_{\rho(t,z_t+x_t)} + x_{\rho(t,z_t+x_t)}) - T(t)g(0,\phi(0))$$
$$+ \int_0^t T(t-s)f(s, z_{\rho(s,z_s+x_s)} + x_{\rho(s,z_s+x_s)})ds, t \in J.$$

Set

$$BC_0' = \{z \in BC' : z(0) = 0\}$$

and let

$$\|z\|_{BC_0'} = \sup\{|z(t)| : t \in J\}, \ z \in BC_0'.$$

BC_0' is a Banach space with the norm $\|.\|_{BC_0'}$. We define the operator $\mathcal{A} : BC_0' \to BC_0'$ by:

$$\mathcal{A}(z)(t) = g(t, z_{\rho(t,z_t+x_t)} + x_{\rho(t,z_t+x_t)}) - T(t)g(0, \phi(0))$$

$$+ \int_0^t T(t-s)f(s, z_{\rho(s,z_s+x_s)} + x_{\rho(s,z_s+x_s)})ds, \ t \in J.$$

We shall show that the operator $\mathcal{A}$ satisfies all conditions of Schauder's fixed point theorem. The operator A maps BC_0' into BC_0', indeed the map $\mathcal{A}(z)$ is continuous on $\mathbb{R}_+$ for any $z \in BC_0'$, and for each $t \in J$ we have

$$|\mathcal{A}(z)(t)| \le |g(t, z_{\rho(t,z_t+x_t)} + x_{\rho(t,z_t+x_t)})| + M'|g(0, \phi(0))|$$

$$+ M' \int_0^t |f(s, z_{\rho(s,z_s+x_s)} + x_{\rho(s,z_s+x_s)}) - f(s, 0) + f(s, 0)|ds$$

$$\le M'(k_g\|\phi\|_{\mathcal{B}} + g^*) + k_g\|z_{\rho(t,z_t+x_t)} + x_{\rho(t,z_t+x_t)}\|_{\mathcal{B}} + g^*$$

$$+ M' \int_0^t |f(s, 0)|ds + M' \int_0^t k(s)\|z_{\rho(s,z_s+x_s)} + x_{\rho(s,z_s+x_s)}\|_{\mathcal{B}}ds$$

$$\le M'(k_g\|\phi\|_{\mathcal{B}} + g^*) + k_g(l|z(t)| + (m + \mathcal{L}^\phi + lM'H)\|\phi\|_{\mathcal{B}}) + g^*$$

$$+ M'F^* + M' \int_0^t k(s)(l|z(s)| + (m + \mathcal{L}^\phi + lM'H)\|\phi\|_{\mathcal{B}})ds.$$

Set

$$C_1 := (m + \mathcal{L}^\phi + lM'H)\|\phi\|_{\mathcal{B}}.$$

$$C_2 := M'(k_g\|\phi\|_{\mathcal{B}} + g^*) + k_g(m + \mathcal{L}^\phi + lM'H)\|\phi\|_{\mathcal{B}} + g^* + M'F^*.$$

Then, we have

$$|\mathcal{A}(z)(t)| \le C_2 + k_g l|z(t)| + M'C_1 \int_0^t k(s)ds + M' \int_0^t l|z(s)|k(s)ds$$

$$\le C_2 + k_g l\|z\|_{BC_0'} + M'Ck^* + M'l\|z\|_{BC_0'}k^*.$$

Hence, $\mathcal{A}(z) \in BC_0'$.

Moreover, let $r > 0$ be such that

$$r \ge \frac{C_2 + M'Ck^*}{1 - l(M'k^* + \alpha_1)},$$

and B_r be the closed ball in BC_0' centered at the origin and of radius r. Let $z \in B_r$ and $t \in \mathbb{R}_+$. Then

$$|\mathcal{A}(z)(t)| \leq C_2 + k_g l r + M'Ck^* + M'k^* l r.$$

Thus

$$\|\mathcal{A}(z)\|_{BC_0'} \leq r,$$

which means that the operator $\mathcal{A}$ transforms the ball B_r into itself.

Now we prove that $\mathcal{A} : B_r \to B_r$ satisfies the assumptions of Schauder's fixed theorem. The proof will be given in several steps.

Step 1: $\mathcal{A}$ is continuous in B_r.

Let $\{z_n\}$ be a sequence such that $z_n \to z$ in B_r. At the first, we study the convergence of the sequences $(z^n_{\rho(s,z^n_s)})_{n \in \mathbb{N}}, s \in J$.

If $s \in J$ is such that $\rho(s, z_s) > 0$, then we have,

$$\|z^n_{\rho(s,z^n_s)} - z_{\rho(s,z_s)}\|_{\mathcal{B}} \leq \|z^n_{\rho(s,z^n_s)} - z_{\rho(s,z^n_s)}\|_{\mathcal{B}} + \|z_{\rho(s,z^n_s)} - z_{\rho(s,z_s)}\|_{\mathcal{B}}$$

$$\leq l\|z_n - z\|_{B_r} + \|z_{\rho(s,z^n_s)} - z_{\rho(s,z_s)}\|_{\mathcal{B}},$$

which proves that $z^n_{\rho(s,z^n_s)} \to z_{\rho(s,z_s)}$ in $\mathcal{B}$ as $n \to \infty$ for every $s \in J$ such that $\rho(s, z_s) > 0$. Similarly, is $\rho(s, z_s) < 0$, we get

$$\|z^n_{\rho(s,z^n_s)} - z_{\rho(s,z_s)}\|_{\mathcal{B}} = \|\phi^n_{\rho(s,z^n_s)} - \phi_{\rho(s,z_s)}\|_{\mathcal{B}} = 0$$

which also shows that $z^n_{\rho(s,z^n_s)} \to z_{\rho(s,z_s)}$ in $\mathcal{B}$ as $n \to \infty$ for every $s \in J$ such that $\rho(s, z_s) < 0$. Combining the pervious arguments, we can prove that $z^n_{\rho(s,z_s)} \to \phi$ for every $s \in J$ such that $\rho(s, z_s) = 0$. Finally,

$$|\mathcal{A}(z_n)(t) - \mathcal{A}(z)(t)|$$

$$\leq |g(t, z^n_{\rho(t,z^n_t)+x_t} + x_{\rho(t,z^n_t+x_t)}) - g(t, z_{\rho(t,z_t+x_t)} + x_{\rho(t,z_t+x_t)})|$$

$$+ M' \int_0^t |f(s, z^n_{\rho(s,z^n_s)+x_s} + x_{\rho(s,z^n_s+x_s)}) - f(s, z_{\rho(s,z_s+x_s)} + x_{\rho(s,z_s+x_s)})|ds$$

$$\leq |g(t, z^n_{\rho(s,z^n_s)+x_s} + x_{\rho(s,z^n_s+x_s)}) - g(t, z_{\rho(s,z_s+x_s)} + x_{\rho(s,z_s+x_s)})|$$

$$+ M' \int_0^t |f(s, z_{\rho(s,z^n_s+x_s)} + x_{\rho(s,z^n_s+x_s)}) - f(s, z_{\rho(s,z_s+x_s)} + x_{\rho(s,z_s+x_s)})|ds.$$

Then by (12.7.2), (12.7.5) we have

$$f(s, z^n_{\rho(s,z^n_s+x_s)} + x_{\rho(s,z^n_s+x_s)}) \to f(s, z_{\rho(s,z_s+x_s)} + x_{\rho(s,z_s+x_s)}), \text{ as } n \to \infty,$$

$$g(t, z^n_{\rho(t,z^n_t+x_t)} + x_{\rho(t,z^n_t+x_t)}) \to g(t, z_{\rho(t,z_t+x_t)} + x_{\rho(t,z_t+x_t)}), \text{ as } n \to \infty,$$

and by the Lebesgue dominated convergence theorem we get,

$$\|\mathcal{A}(z_n) - \mathcal{A}(z)\|_{BC_0'} \to 0, \text{ as } n \to \infty.$$

Thus $\mathcal{A}$ is continuous.

Step 2: $A(B_r) \subset B_r$. This is clear.
Step 3: $\mathcal{A}(B_r)$ is equi-continuous on every compact interval $[0, b]$ of $\mathbb{R}_+$ for $b > 0$. Let $\tau_1, \tau_2 \in [0, b]$ with $\tau_2 > \tau_1$, we have:

$$|A(z)(\tau_2) - A(z)(\tau_1)|$$

$$\leq |g(\tau_2, z_{\rho(\tau_2, z_{\tau_2} + x_{\tau_2})} + x_{\rho(\tau_2, z_{\tau_2} + x_{\tau_2})}) - g(\tau_1, z_{\rho(\tau_1, z_{\tau_1} + x_{\tau_1})} + x_{\rho(\tau_1, z_{\tau_1} + x_{\tau_1})})|$$

$$+ \|T(\tau_2) - T(\tau_1)\|_{B(E)} |g(0, \phi(0))|$$

$$+ \int_0^{\tau_1} \|T(\tau_2 - s) - T(\tau_1 - s)\|_{B(E)} |f(s, z_{\rho(s, z_s + x_s)} + x_{\rho(s, z_s + x_s)})| ds$$

$$+ \int_{\tau_1}^{\tau_2} \|T(\tau_2 - s)\|_{B(E)} |f(s, z_{\rho(s, z_s + x_s)} + x_{\rho(s, z_s + x_s)})| ds$$

$$\leq |g(\tau_2, z_{\rho(\tau_2, z_{\tau_2} + x_{\tau_2})} + x_{\rho(\tau_2, z_{\tau_2} + x_{\tau_2})}) - g(\tau_1, z_{\rho(\tau_1, z_{\tau_1} + x_{\tau_1})} + x_{\rho(\tau_1, z_{\tau_1} + x_{\tau_1})})|$$

$$+ \|T(\tau_2) - T(\tau_1)\|_{B(E)} (k_g \|\phi\|_{\mathcal{B}} + g^*)$$

$$+ \int_0^{\tau_1} \|T(\tau_2 - s) - T(\tau_1 - s)\|_{B(E)} |f(s, z_{\rho(s, z_s + x_s)} + x_{\rho(s, z_s + x_s)}) - f(s, 0)| ds$$

$$+ \int_0^{\tau_1} \|T(\tau_2 - s) - T(\tau_1 - s)\|_{B(E)} |f(s, 0)| ds$$

$$+ \int_{\tau_1}^{\tau_2} \|T(\tau_2 - s)\|_{B(E)} |f(s, z_{\rho(s, z_s + x_s)} + x_{\rho(s, z_s + x_s)}) - f(s, 0)| ds$$

$$+ \int_{\tau_1}^{\tau_2} \|T(\tau_2 - s)\|_{B(E)} |f(s, 0)| ds$$

$$\leq k_g |g(\tau_2, z_{\rho(\tau_2, z_{\tau_2} + x_{\tau_2})} + x_{\rho(\tau_2, z_{\tau_2} + x_{\tau_2})}) - g(\tau_1, z_{\rho(\tau_1, z_{\tau_1} + x_{\tau_1})} + x_{\rho(\tau_1, z_{\tau_1} + x_{\tau_1})})|$$

$$+ \|T(\tau_2) - T(\tau_1)\|_{B(E)} (k_g \|\phi\|_{\mathcal{B}} + g^*)$$

$$+ C_1 \int_0^{\tau_1} \|T(\tau_2 - s) - T(\tau_1 - s)\|_{B(E)} k(s) ds$$

$$+ rL \int_0^{\tau_1} \|T(\tau_2 - s) - T(\tau_1 - s)\|_{B(E)} k(s) ds$$

$$+ \int_0^{\tau_1} \|T(\tau_2 - s) - T(\tau_1 - s)\|_{B(E)} |f(s, 0)| ds$$

$$+C_1 \int_{\tau_1}^{\tau_2} \|T(\tau_2 - s)\|_{B(E)} k(s) ds$$

$$+rL \int_{\tau_1}^{\tau_2} \|T(\tau_2 - s)\|_{B(E)} k(s) ds$$

$$+ \int_{\tau_1}^{\tau_2} \|T(\tau_2 - s)\|_{B(E)} |f(s, 0)| ds.$$

When $\tau_2 \to \tau_1$, the right-hand side of the above inequality tends to zero. Since (12.7.7) and $T(t)$ is a strongly continuous operator and the compactness of $T(t)$ for $t > 0$, implies the continuity in the uniform operator topology (see [168]), this proves the equi-continuity.

Step 4: The set $\mathcal{A}(B_r)(t)$ is relatively compact on every compact interval of $[0, \infty)$. Let $t \in [0, b]$ for $b > 0$ and let ε be a real number satisfying $0 < \varepsilon < t$. For $z \in B_r$ we define

$$\mathcal{A}_\varepsilon(z)(t) = g(t, z_{\rho(t,z_t+x_t)} + x_{\rho(t,z_t+x_t)}) - T(\varepsilon)(T(t-\varepsilon)g(0, \phi(0)))$$

$$+ T(\varepsilon) \int_0^{t-\varepsilon} T(t - s - \varepsilon) f(s, z_{\rho(s,z_s+x_s)} + x_{\rho(s,z_s+x_s)}) ds.$$

Note that the set

$$\{g(t, z_{\rho(t,z_t+x_t)} + x_{\rho(t,z_t+x_t)}) - T(t-\varepsilon)g(0, \phi(0))$$

$$+ \int_0^{t-\varepsilon} T(t - s - \varepsilon) f(s, z_{\rho(s,z_s+x_s)} + x_{\rho(s,z_s+x_s)}) ds : z \in B_r \}$$

is bounded.

$$|g(t, z_{\rho(t,z_t+x_t)} + x_{\rho(t,z_t+x_t)}) - T(t-\varepsilon)g(0, \phi(0)).$$

$$+ \int_0^{t-\varepsilon} T(t - s - \varepsilon) f(s, z_{\rho(s,z_s+x_s)} + x_{\rho(s,z_s+x_s)}) ds| \le r$$

Since $T(t)$ is a compact operator for $t > 0$, and (12.7.6) we have that the set,

$$\{\mathcal{A}_\varepsilon(z)(t) : z \in B_r\}$$

is precompact in E for every ε, $0 < \varepsilon < t$. Moreover, for every $z \in B_r$ we have

$$|\mathcal{A}(z)(t) - \mathcal{A}_\varepsilon(z)(t)|$$

$$\le \int_{t-\varepsilon}^t T(t - s) f(s, z_{\rho(s,z_s+x_s)} + x_{\rho(s,z_s+x_s)}) ds$$

$$\leq M'F^*\varepsilon + M'C \int_{t-\varepsilon}^{t} k(s)ds + rM' \int_{t-\varepsilon}^{t} lk(s)ds,$$

$$\to 0 \quad \text{as} \quad \varepsilon \to 0.$$

Therefore, the set $\{\mathcal{A}(z)(t) : z \in B_r\}$ is precompact, i.e., relatively compact.

Step 5: $\mathcal{A}(B_r)$ is equi-convergent.

Let $t \in \mathbb{R}_+$ and $z \in B_r$, we have,

$$|\mathcal{A}(z)(t)| \leq |g(t, z_{\rho(t,z_t+x_t)} + x_{\rho(t,z_t+x_t)})| + M'|g(0, \phi(0))|$$

$$+ M' \int_0^t |f(s, z_{\rho(s,z_s+x_s)} + x_{\rho(s,z_s+x_s)})|ds$$

$$\leq C_2 + k_g lr + M'C \int_0^t k(s)ds + M'rl \int_0^t k(s)ds.$$

Then we have

$$|\mathcal{A}(z)(t)| \to C_3 \leq C_2 + k_g lr + M'Ck^* + M'lrk^*, \quad \text{as } t \to +\infty.$$

Hence,

$$|\mathcal{A}(z)(t) - \mathcal{A}(z)(+\infty)| \to 0, \quad \text{as } t \to +\infty.$$

As a consequence of Steps 1–5, we can conclude that $\mathcal{A} : B_r \to B_r$ is continuous and compact. From Schauder's theorem, we deduce that $\mathcal{A}$ has a fixed point z^*. Then $y^* = z^* + x$ is a fixed point of the operators N, which is a mild solution of the problem (12.7)–(12.8). □

12.3.3   An Example

Consider the following neutral functional partial differential equation:

$$\frac{\partial}{\partial t}[z(t,x) - g(t, z(t - \sigma(t, z(t,0)), x))] = \frac{\partial^2}{\partial x^2}[z(t,x) - g(t, z(t - \sigma(t, z(t,0)), x))]$$

$$f(t, z(t - \sigma(t, z(t,0)), x)), \ x \in [0, \pi], \ t \in \mathbb{R}_+ \tag{12.9}$$

$$z(t, 0) = z(t, \pi) = 0, \ t \in \mathbb{R}_+, \tag{12.10}$$

$$z(\theta, x) = z_0(\theta, x), \ t \in (-\infty, 0], \ x \in [0, \pi], \tag{12.11}$$

where f, g is a given functions, and $\sigma : \mathbb{R} \to \mathbb{R}^+$. Take $E = L^2[0, \pi]$ and define $A : E \to E$ by $A\omega = \omega''$ with domain

$$D(A) = \{\omega \in E, \omega, \omega' \text{are absolutely continuous}, \ \omega'' \in E, \ \omega(0) = \omega(\pi) = 0\}.$$

Then

$$Aw = \sum_{n=1}^{\infty} n^2 (\omega, \omega_n) \omega_n, \omega \in D(A),$$

where $\omega_n(s) = \sqrt{\frac{2}{\pi}} \sin ns, n = 1, 2, \ldots$ is the orthogonal set of eigenvectors in A. It is well known that A is the infinitesimal generator of an analytic semigroup $T(t), t \geq 0$ in E and is given by

$$T(t)\omega = \sum_{n=1}^{\infty} \exp(-n^2 t)(\omega, \omega_n)\omega_n, \omega \in E.$$

Since the analytic semigroup $T(t)$ is compact for $t > 0$, there exists a positive constant M such that

$$\|T(t)\|_{B(E)} \leq M.$$

Let $\mathcal{B} = BCU(\mathbb{R}^-, E)$ and $\phi \in \mathcal{B}$, then (H_ϕ), where $\rho(t, \varphi) = t - \sigma(\varphi)$.

Hence, the problem (12.1)–(12.2) is an abstract formulation of the problem (12.9)–(12.11), and if the conditions (12.3.1)–(12.3.6), (H_ϕ) are satisfied. Theorem 12.9 implies that the problem (12.9)–(12.11) has at least one mild solutions on BC.

12.4 Global Existence Results for Functional Differential Inclusions with Delay

12.4.1 Introduction

In this section we are going to prove the existence of solutions of a class of semi-linear functional evolution inclusion with delay. Our investigations will be situated in the Banach space of real continuous and bounded functions on the real half axis $\mathbb{R}_+$. We will use Bohnenblust–Karlin's fixed theorem, combined with the Corduneanu's compactness criteria. More precisely, we will consider the following problem

$$y'(t) - Ay(t) \in F(t, y_t), \quad \text{a.e. } t \in J := \mathbb{R}_+ \tag{12.12}$$

$$y(t) = \phi(t), \quad t \in H, \tag{12.13}$$

where $F : J \times C(H, E) \to \mathcal{P}(E)$ is a multi-valued map with nonempty compact values, $\mathcal{P}(E)$ is the family of all nonempty subsets of E, $A : D(A) \subset E \to E$ is the infinitesimal generator of a strongly continuous semigroup $T(t), t \in J, \phi : H \to E$ is given continuous function, and $(E, |.|)$ is a real Banach space.

12.4.2 Existence of Mild Solutions

Let us introduce the following hypotheses:

(12.5.1) $A : D(A) \subset E \to E$ is the infinitesimal generator of a strongly continuous semigroup $T(t), t \in J$ which is compact for $t > 0$ in the Banach space E. Let $M = \sup\{\|T(t)\|_{B(E)} : t \geq 0\}$.

(12.5.2) The multi-function $F : J \times C(H, E) \longrightarrow \mathcal{P}(E)$ is Carathéodory with compact and convex values.

(12.5.3) There exists a continuous function $k : J \to \mathbb{R}_+$ such that:

$$H_d(F(t, u), F(t, v)) \leq k(t)\|u - v\|,$$

for each $t \in J$ and for all $u, v \in C(H, E)$ and

$$d(0, F(t, 0)) \leq k(t),$$

with

$$k^* := \sup_{t \in J} \int_0^t k(s)ds < \infty. \tag{12.14}$$

Theorem 12.10. *Assume that (12.5.1)–(12.5.3) hold. If $k^*M < 1$, then the problem (12.12)–(12.13) has at least one mild solution on BC.*

Proof. Consider the multi-valued operator $N : BC \to \mathcal{P}(BC)$ defined by:

$$N(y) := \left\{ h \in BC : h(t) = \begin{cases} \phi(t), & \text{if } t \in H; \\ T(t)\phi(0) \\ + \int_0^t T(t - s)f(s)\,ds, & f \in S_{F,y} \quad \text{if } t \in J. \end{cases} \right\} \tag{12.15}$$

The operator N maps BC into BC; for any $y \in BC$, and $h \in N(y)$ and for each $t \in J$, we have

$$|h(t)| \leq M\|\phi\| + M\int_0^t |f(s)|ds$$

$$\leq M\|\phi\| + M\int_0^t (k(s)\|y_s\| + \|F(s, 0)\|)ds$$

$$\leq M\|\phi\| + M\int_0^t k(s)(\|y_s\| + 1)ds$$

$$\leq M\|\phi\| + M(\|y\|_{BC} + 1)k^* := c.$$

Hence, $h(t) \in BC$.

Moreover, let $r > 0$ be such that $r \geq \frac{M\|\phi\| + Mk^*}{1 - Mk^*}$, and B_r be the closed ball in BC centered at the origin and of radius r. Let $y \in B_r$ and $t \in \mathbb{R}_+$. Then,

$$|h(t)| \leq M\|\phi\| + Mk^* + Mk^* r.$$

Thus,

$$\|h\|_{BC} \leq r,$$

which means that the operator N transforms the ball B_r into itself.

Now we prove that $N : B_r \to B_r$ satisfies the assumptions of Bohnenblust–Karlin's fixed theorem. The proof will be given in several steps.

Step 1: We shall show that the operator N is closed and convex. This will be given in two claims.

Claim 1: $N(y)$ is closed for each $y \in B_r$. Let $(h_n)_{n \geq 0} \in N(y)$ such that $h_n \to \tilde{h}$ in B_r. Then for $h_n \in B_r$ there exists $f_n \in S_{F,y}$ such that:

$$h_n(t) = T(t)\phi(0) + \int_0^t T(t-s)f_n(s)ds.$$

Since F has compact and convex values and from hypotheses (12.5.2), (12.5.3), an application of Mazur's theorem [185] implies that we may pass to a subsequence if necessary to get that f_n converges to $f \in L^1(J, E)$ and hence $f \in S_{F,y}$. Then for each $t \in J$,

$$h_n(t) \to \tilde{h}(t) = T(t)\phi(0) + \int_0^t T(t-s)f(s)ds.$$

So, $\tilde{h} \in N(y)$.

Claim 2: $N(y)$ is convex for each $y \in B_r$.

Let $h_1, h_2 \in N(y)$, the there exists $f_1, f_2 \in S_{F,y}$ such that, for each $t \in J$ we have:

$$h_i(t) = T(t)\phi(0) + \int_0^t T(t-s)f_i(s)ds, i = 1, 2.$$

Let $0 \leq \delta \leq 1$. Then, we have for each $t \in J$:

$$(\delta h_1 + (1 - \delta)h_2)(t) = T(t)\phi(0) + \int_0^t T(t-s)[\delta f_1(s) + (1 - \delta)f_2(s)]ds.$$

Since $F(t, y)$ is convex, one has

$$\delta h_1 + (1 - \delta)h_2 \in N(y).$$

Step 2: $N(B_r) \subset B_r$ this is clear.

Step 3: $N(B_r)$ is equi-continuous on every compact interval $[0, b]$ of $\mathbb{R}_+$ for $b > 0$. Let $\tau_1, \tau_2 \in [0, b]$ with $\tau_2 > \tau_1$, we have

$$
\begin{aligned}
|h(\tau_2) - h(\tau_1)| \leq{}& \|T(\tau_2 - s) - T(\tau_1 - s)\|_{B(E)} \|\phi\| \\
& + \int_0^{\tau_1} \|T(\tau_2 - s) - T(\tau_1 - s)\|_{B(E)} |f(s)| ds \\
& + \int_{\tau_1}^{\tau_2} \|T(\tau_2 - s)\|_{B(E)} |f(s)| ds \\
\leq{}& \|T(\tau_2 - s) - T(\tau_1 - s)\|_{B(E)} \|\phi\| \\
& + \int_0^{\tau_1} \|T(\tau_2 - s) - T(\tau_1 - s)\|_{B(E)} (k(s)\|y_s\| + |F(s, 0)|) ds \\
& + \int_{\tau_1}^{\tau_2} \|T(\tau_2 - s)\|_{B(E)} (k(s)\|y_s\| + |F(s, 0)|) ds \\
\leq{}& \|T(\tau_2 - s) - T(\tau_1 - s)\|_{B(E)} \|\phi\| \\
& + (r + 1) \int_0^{\tau_1} \|T(\tau_2 - s) - T(\tau_1 - s)\|_{B(E)} k(s) ds \\
& + (r + 1) \int_{\tau_1}^{\tau_2} \|T(\tau_2 - s)\|_{B(E)} k(s) ds.
\end{aligned}
$$

When $\tau_2 \to \tau_1$, the right-hand side of the above inequality tends to zero, since $T(t)$ is a strongly continuous operator and the compactness of $T(t)$ for $t > 0$, implies the continuity in the uniform operator topology (see [168]). This proves the equi-continuity.

Step 4: $N(B_r)$ is relatively compact on every compact interval of $\mathbb{R}_+$. Let $t \in [0, b]$ for $b > 0$ and let ε be a real number satisfying $0 < \varepsilon < t$. For $y \in B_r$, let $h \in N(y)$, $f \in S_{F,y}$ and define

$$
h_\varepsilon(t) = T(t)\phi(0) + T(\varepsilon) \int_0^{t-\varepsilon} T(t - s - \varepsilon) f(s) ds.
$$

Note that the set

$$
\left\{ T(t)\phi(0) + \int_0^{t-\varepsilon} T(t - s - \varepsilon) f(s) ds : y \in B_r \right\}
$$

is bounded.

$$
\left| T(t)\phi(0) + \int_0^{t-\varepsilon} T(t - s - \varepsilon) f(s) ds \right| \leq r.
$$

Since $T(t)$ is a compact operator for $t > 0$, the set,

$$H_\varepsilon(t) = \{h_\varepsilon(t) : h_\varepsilon \in N(y), y \in B_r\}$$

is precompact in E for every ε, $0 < \varepsilon < t$. Moreover, for every $y \in B_r$ we have

$$|h(t) - h_\varepsilon(t)| \le M \int_{t-\varepsilon}^t |f(s)|ds$$

$$\le M \int_{t-\varepsilon}^t (k(s)\|y_s\| + |F(s,0)|)ds$$

$$\le M(1+r) \int_{t-\varepsilon}^t k(s)ds$$

$$\to 0 \quad \text{as} \quad \varepsilon \to 0.$$

Therefore, the set $H(t) = \{h(t) : h \in N(y), y \in B_r\}$ is precompact, i.e., relatively compact. Hence the set $H(t) = \{h(t) : h \in N(B_r)\}$ is relatively compact.

Step 5: N has closed graph.

Let $\{y_n\}$ be a sequence such that $y_n \to y_*$, $h_n \in N(y_n)$ and $h_n \to h_*$. We shall show that $h_* \in N(y_*)$. $h_n \in N(y_n)$ means that there exists $f_n \in S_{F,y_n}$ such that

$$h_n(t) = T(t)\,\phi(0) + \int_0^t T(t-s) f_n(s)\,ds, \ t \in J.$$

We must prove that there exists f_*

$$h_*(t) = T(t)\,\phi(0) + \int_0^t T(t-s) f_*(s)\,ds, \ t \in J.$$

Consider the linear and continuous operator $K : L^1(J,E) \to BC$ defined by

$$K(v)(t) = \int_0^t T(t-s)v(s)ds.$$

We have

$$|K(f_n)(t) - K(f_*)(t)| =$$
$$|(h_n(t) - T(t)\,\phi(0)) - (h_*(t) - T(t)\,\phi(0))| = |h_n(t) - h_*(t)|$$
$$\le \|h_n - h_*\|_\infty \to 0, \ as \ n \to \infty.$$

From Lemma 1.11 it follows that $K \circ S_F$ is a closed graph operator and from the definition of K has

$$h_n(t) - T(t)\phi(0) \in K \circ S_{F,y_n}.$$

As $y_n \to y_*$ and $h_n \to h_*$, there exist $f_* \in S_{F,y_*}$ such that:

$$h_*(t) - T(t)\phi(0) = \int_0^t T(t-s) f_*(s).$$

Hence the multi-valued operator N has closed graph, which implies that it is upper semi-continuous.

Step 6: $N(B_r)$ is equi-convergent. Let $h \in N(y)$, there exists $f \in S_{F,y}$ such that for each $t \in \mathbb{R}_+$ and $y \in B_r$ we have

$$|h(t)| \leq M\|\phi\| + M \int_0^t |f(s)|ds$$

$$\leq M\|\phi\| + Mk^* + Mr \int_0^t k(s)ds$$

$$\leq M\|\phi\| + Mk^* + Mrk^*.$$

Then,

$$|h(t)| \to l \leq M\|\phi\| + Mk^*(1 + r), \quad \text{as } t \to +\infty.$$

Hence,

$$|h(t) - h(+\infty)| \to 0, \quad \text{as } t \to +\infty.$$

As a consequence of Steps $1 - 6$, and Lemma 1.26, we conclude from Bohnenblust–Karlin's theorem that N has a fixed point y which is a mild solution of the problem (12.12)–(12.13).

$\square$

12.4.3 An Example

Consider the functional partial differential inclusion

$$\frac{\partial}{\partial t}z(t, x) - \frac{\partial^2}{\partial x^2}z(t, x) \in F(t, z(t - r, x)), \ x \in [0, \pi], \ t \in J := \mathbb{R}_+, \qquad (12.16)$$

$$z(t, 0) = z(t, \pi) = 0, \ t \in J, \qquad (12.17)$$

$$z(t, x) = \phi(t), \ t \in H, \ x \in [0, \pi], \qquad (12.18)$$

where F is a given multi-valued map. Take $E = L^2[0, \pi]$ and define $A : E \to E$ by $A\omega = \omega''$ with domain

$$D(A) = \{\omega \in E; \omega, \omega' \text{are absolutely continuous}, \ \omega'' \in E, \ \omega(0) = \omega(\pi) = 0\}.$$

Then,

$$Aw = \sum_{n=1}^{\infty} n^2(\omega, \omega_n)\omega_n, \quad \omega \in D(A)$$

where $\omega_n(s) = \sqrt{\frac{2}{\pi}} \sin ns, n = 1, 2, \ldots$, is the orthogonal set of eigenvectors in A. It is well known (see [168]) that A is the infinitesimal generator of an analytic semigroup $T(t), t \geq 0$ in E and is given by

$$T(t)\omega = \sum_{n=1}^{\infty} \exp(-n^2 t)(\omega, \omega_n)\omega_n, \quad \omega \in E.$$

Since the analytic semigroup $T(t)$ is compact for $t > 0$, there exists a positive constant M such that

$$\|T(t)\|_{B(E)} \leq M.$$

Then the problem (12.12)–(12.13) is the abstract formulation of the problem (12.16)–(12.18). If conditions (12.5.1)–(12.5.3) are satisfied, Theorem 12.10 implies that the problem (12.16)–(12.18) has at least one global mild solution on BC.

12.5   Global Existence Results for Functional Differential Inclusions with State-Dependent Delay

12.5.1   Introduction

In this section we are going to prove the existence of solutions of a functional differential inclusion. Our investigations will be situated in the Banach space of real functions which are defined, continuous, and bounded on the real axis $\mathbb{R}$. We will use Bohnenblust–Karlin's fixed theorem, combined with the Corduneanu's compactness criteria. More precisely we will consider the following problem:

$$y'(t) - Ay(t) \in F(t, y_{\rho(t,y_t)}), \quad \text{a.e. } t \in J := \mathbb{R}_+ \tag{12.19}$$

$$y(t) = \phi(t), \quad t \in (-\infty, 0], \tag{12.20}$$

where $F : J \times \mathcal{B} \to \mathcal{P}(E)$ is a multi-valued map with nonempty compact values, $\mathcal{P}(E)$ is the family of all nonempty subsets of E, $A : D(A) \subset E \to E$ is the infinitesimal generator of a strongly continuous semigroup $T(t), t \in J$, and $(E, |.|)$ is a real Banach space. $\mathcal{B}$ is the phase space, $\phi \in \mathcal{B}, \rho : J \times \mathcal{B} \to \mathbb{R}$.

12.5.2 Existence of Mild Solutions

Now we give our main existence result for problem (12.19)–(12.20). Before starting and proving this result, we give the definition of the mild solution.

Definition 12.11. We say that a continuous function $y : (-\infty, +\infty) \to E$ is a mild solution of problem (12.19)–(12.20) if $y(t) = \phi(t)$ for all $t \in (-\infty, 0]$, and the restriction of $y(\cdot)$ to the interval J is continuous and there exists $f(\cdot) \in L^1(J, E)$: $f(t) \in F(t, y_{\rho(t, y_t)})$ a.e. in J such that y satisfies the following integral equation

$$y(t) = T(t)\phi(t) - \int_0^t T(t-s) f(s) \, ds \quad \text{for each } t \in J. \tag{12.21}$$

Set

$$\mathcal{R}(\rho^-) = \{\rho(s, \phi) : (s, \phi) \in J \times \mathcal{B}, \ \rho(s, \phi) \le 0\}.$$

We always assume that $\rho : J \times \mathcal{B} \to \mathbb{R}$ is continuous. Additionally, we introduce the following hypothesis:

(H_ϕ) The function $t \to \phi_t$ is continuous from $\mathcal{R}(\rho^-)$ into $\mathcal{B}$ and there exists a continuous and bounded function $\mathcal{L}^\phi : \mathcal{R}(\rho^-) \to (0, \infty)$ such that

$$\|\phi_t\| \le \mathcal{L}^\phi(t)\|\phi\| \quad \text{for every } t \in \mathcal{R}(\rho^-).$$

Remark 12.12. The condition (H_ϕ), is frequently verified by functions continuous and bounded.

Let us introduce the following hypotheses:

(12.11.1) $A : D(A) \subset E \to E$ is the infinitesimal generator of a strongly continuous semigroup $T(t), t \in J$ which is compact for $t > 0$ in the Banach space E. Let $M' = \sup\{\|T\|_{B(E)} : t \ge 0\}$.

(12.11.2) The multi-function $F : J \times \mathcal{B} \longrightarrow \mathcal{P}(E)$ is Carathéodory with compact and convex values.

(12.11.3) There exists a continuous function $k : J \to \mathbb{R}_+$ such that:

$$H_d(F(t, u), F(t, v)) \le k(t) \, \|u - v\|_\mathcal{B}$$

for each $t \in J$ and for all $u, v \in \mathcal{B}$ and

$$d(0, F(t, 0)) \le k(t)$$

with

$$k^* := \sup_{t \in J} \int_0^t k(s) ds < \infty. \tag{12.22}$$

Theorem 12.13. *Assume that (12.11.1)–(12.11.3),(H_ϕ) hold. If $k^* M' L < 1$, then the problem (12.19)–(12.20) has at least one mild solution on BC.*

Proof. Consider the multi-valued operator $N : BC \to \mathcal{P}(BC)$ defined by:

$$N(y) := \left\{ h \in BC : h(t) = \begin{cases} \phi(t), & \text{if } t \in (-\infty, 0]; \\ T(t)\,\phi(0) + \displaystyle\int_0^t T(t-s) f(s)\, ds, & \text{if } t \in J, \end{cases} \right\}$$

where $f \in S_{F, y_{\rho(s, y_s)}}$.

Let $x(\cdot) : \mathbb{R} \to E$ be the function defined by:

$$x(t) = \begin{cases} \phi(t), & \text{if } t \in (-\infty, 0]; \\ T(t)\,\phi(0), & \text{if } t \in J. \end{cases}$$

Then $x_0 = \phi$. For each $z \in BC$ with $z(0) = 0$, we denote by $\bar{z}$ the function

$$\bar{z}(t) = \begin{cases} 0, & \text{if } t \in (-\infty, 0]; \\ z(t), & \text{if } t \in J, \end{cases}$$

if $y(\cdot)$ satisfies (12.21), we can decompose it as $y(t) = z(t) + x(t)$, $t \in J$, which implies $y_t = z_t + x_t$ for every $t \in J$ and the function $z(\cdot)$ satisfies

$$z(t) = \int_0^t T(t-s) f(s)\, ds, \quad t \in J,$$

where $f \in S_{F, z_{\rho(s, z_s + x_s)} + x_{\rho(s, z_s + x_s)}}$.

Set

$$BC'_0 = \{ z \in BC' : z(0) = 0 \}$$

and let

$$\| z \|_{BC'_0} = \sup\{ |z(t)| : t \in J \}, \quad z \in BC'_0.$$

BC'_0 is a Banach space with the norm $\| \cdot \|_{BC'_0}$.

We define the operator $\mathcal{A} : BC'_0 \to \mathcal{P}(BC'_0)$ by:

$$\mathcal{A}(z) := \left\{ h \in BC'_0 : h(t) = \begin{cases} 0, & \text{if } t \le 0; \\ \displaystyle\int_0^t T(t-s) f(s)\, ds, & \text{if } t \in J, \end{cases} \right\}$$

where $f \in S_{F, z_{\rho(s, z_s + x_s)} + x_{\rho(s, z_s + x_s)}}$.

The operator A maps BC_0' into BC_0', indeed the map $\mathcal{A}(z)$ is continuous on $\mathbb{R}_+$ for any $z \in BC_0'$, $h \in \mathcal{A}(z)$ and for each $t \in J$ we have

$$|h(t)| \leq M' \int_0^t |f(s)| ds$$

$$\leq M' \int_0^t (k(s) \| z_{\rho(s, z_s + x_s)} + x_{\rho(s, z_s + x_s)} \|_\mathcal{B} + |F(s, 0)|) ds$$

$$\leq M' \int_0^t k(s) ds + M' \int_0^t k(s)(L|z(s)| + (M + \mathcal{L}^\phi + LM'H) \|\phi\|_\mathcal{B}) ds$$

$$\leq M' k^* + M' \int_0^t k(s)(L|z(s)| + (M + \mathcal{L}^\phi + LM'H) \|\phi\|_\mathcal{B}) ds.$$

Set

$$C := (M + \mathcal{L}^\phi + LM'H) \|\phi\|_\mathcal{B}.$$

Then, we have

$$|h(t)| \leq M' k^* + M' C \int_0^t k(s) ds + M' \int_0^t L|z(s)| k(s) ds$$

$$\leq M' k^* + M' C k^* + M' L \|z\|_{BC_0'} k^*.$$

Hence, $\mathcal{A}(z) \in BC_0'$.

Moreover, let $r > 0$ be such that

$$r \geq \frac{M' k^* + M' C k^*}{1 - M' k^* L},$$

and B_r be the closed ball in BC_0' centered at the origin and of radius r. Let $z \in B_r$ and $t \in \mathbb{R}_+$. Then

$$|h(t)| \leq M' k^* + M' C k^* + M' k^* L r.$$

Thus

$$\|h\|_{BC_0'} \leq r,$$

which means that the operator $\mathcal{A}$ transforms the ball B_r into itself.

Now we prove that $\mathcal{A} : B_r \to \mathcal{P}(B_r)$ satisfies the assumptions of Bohnenblust–Karlin's fixed theorem. The proof will be given in several steps.

Step 1: We shall show that the operator $\mathcal{A}$ is closed and convex. This will be given in several claims.

Claim 1: $\mathcal{A}(z)$ is closed for each $z \in B_r$.

Let $(h_n)_{n \geq 0} \in \mathcal{A}(z)$ such that $h_n \to \tilde{h}$ in B_r. Then for $h_n \in B_r$ there exists $f_n \in S_{F, z\rho(s, z_s + x_s) + x\rho(s, z_s + x_s)}$ such that for each $t \in J$,

$$h_n(t) = \int_0^t T(t - s) f_n(s) ds.$$

Using the fact that F has compact values and from hypotheses (12.11.2), (12.11.3) we may pass a subsequence if necessary to get that f_n converges to $f \in L^1(J, E)$ and hence $f \in S_{F, z\rho(s, z_s + x_s) + x\rho(s, z_s + x_s)}$. Then for each $t \in J$,

$$h_n(t) \to \tilde{h}(t) = \int_0^t T(t - s) f(s) ds.$$

So, $\tilde{h} \in \mathcal{A}(z)$.

Claim 2: $\mathcal{A}(z)$ is convex for each $z \in B_r$.

Let $h_1, h_2 \in \mathcal{A}(z)$, the there exists $f_1, f_2 \in S_{F, z\rho(s, z_s + x_s) + x\rho(s, z_s + x_s)}$ such that, for each $t \in J$ we have:

$$h_i(t) = \int_0^t T(t - s) f_i(s) ds, i = 1, 2.$$

Let $0 \leq \delta \leq 1$. Then, we have for each $t \in J$:

$$(\delta h_1 + (1 - \delta) h_2)(t) = \int_0^t T(t - s)[\delta f_1(s) + (1 - \delta) f_2(s)] ds.$$

Since F has convex values, one has

$$\delta h_1 + (1 - \delta) h_2 \in \mathcal{A}(z)$$

Step 2: $\mathcal{A}(B_r) \subset B_r$ this is clear.

Step 3: $\mathcal{A}(B_r)$ is equi-continuous on every compact interval $[0, b]$ of $\mathbb{R}_+$ for $b > 0$. Let $\tau_1, \tau_2 \in [0, b], h \in \mathcal{A}(z)$ with $\tau_2 > \tau_1$, we have:

$$|h(\tau_2) - h(\tau_1)|$$

$$\leq \int_0^{\tau_1} \|T(\tau_2 - s) - T(\tau_1 - s)\|_{B(E)} |f(s)| ds$$

$$+ \int_{\tau_1}^{\tau_2} \|T(\tau_2 - s)\|_{B(E)} |f(s)| ds$$

$$\leq \int_0^{\tau_1} \|T(\tau_2 - s) - T(\tau_1 - s)\|_{B(E)} (k(s)\|z_{\rho(s,z_s+x_s)} + x_{\rho(s,z_s+x_s)}\|_B + |F(s,0)|)ds$$

$$+ \int_{\tau_1}^{\tau_2} \|T(\tau_2 - s)\|_{B(E)} (k(s)\|z_{\rho(s,z_s+x_s)} + x_{\rho(s,z_s+x_s)}\|_B + |F(s,0)|)ds$$

$$\leq C \int_0^{\tau_1} \|T(\tau_2 - s) - T(\tau_1 - s)\|_{B(E)} k(s)ds$$

$$+ rL \int_0^{\tau_1} \|T(\tau_2 - s) - T(\tau_1 - s)\|_{B(E)} k(s)ds$$

$$+ \int_0^{\tau_1} \|T(\tau_2 - s) - T(\tau_1 - s)\|_{B(E)} k(s)ds$$

$$+ C \int_{\tau_1}^{\tau_2} \|T(\tau_2 - s)\|_{B(E)} k(s)ds$$

$$+ rL \int_{\tau_1}^{\tau_2} \|T(\tau_2 - s)\|_{B(E)} k(s)ds$$

$$+ \int_{\tau_1}^{\tau_2} \|T(\tau_2 - s)\|_{B(E)} k(s)ds.$$

When $\tau_2 \to \tau_2$, the right-hand side of the above inequality tends to zero, since $T(t)$ is a strongly continuous operator and the compactness of $T(t)$ for $t > 0$ implies the continuity in the uniform operator topology (see [168]), this proves the equi-continuity.

Step 4: $\mathcal{A}(B_r)$ is relatively compact on every compact interval of $[0, \infty)$.

Let $t \in [0, b]$ for $b > 0$ and let ε be a real number satisfying $0 < \varepsilon < t$. For $z \in B_r$ we define

$$h_\varepsilon(t) = T(\varepsilon) \int_0^{t-\varepsilon} T(t - s - \varepsilon)f(s)ds.$$

Note that the set

$$\left\{ \int_0^{t-\varepsilon} T(t - s - \varepsilon)f(s)ds : z \in B_r \right\}$$

is bounded.

$$\left| \int_0^{t-\varepsilon} T(t - s - \varepsilon)f(s)ds \right| \leq r.$$

Since $T(t)$ is a compact operator for $t > 0$, the set,

$$\{h_\varepsilon(t) : z \in B_r\}$$

is precompact in E for every ε, $0 < \varepsilon < t$. Moreover, for every $z \in B_r$ we have

$$|h(t) - h_\varepsilon(t)|$$

$$\leq M' \int_{t-\varepsilon}^{t} |f(s)| ds$$

$$\leq M' \int_{t-\varepsilon}^{t} k(s) ds + M' C \int_{t-\varepsilon}^{t} k(s) ds + r M' \int_{t-\varepsilon}^{t} L k(s) ds,$$

$$\to 0 \quad \text{as} \quad \varepsilon \to 0.$$

Therefore, the set $\{h(t) : z \in B_r\}$ is precompact, i.e., relatively compact.

Step 5: $\mathcal{A}$ has closed graph.

Let $\{z_n\}$ be a sequence such that $z_n \to z_*, h_n \in \mathcal{A}(z_n)$ and $h_n \to h_*$. We shall show that $h_* \in \mathcal{A}(z_*)$. $h_n \in \mathcal{A}(z_n)$ means that there exists $f_n \in S_{F,z_{\rho(s,z_s^n + x_s)}^n + x_{\rho(s,z_s^n + x_s)}}$ such that

$$h_n(t) = \int_0^t T(t-s) f_n(s)\, ds,$$

we must prove that there exists f_*

$$h_*(t) = \int_0^t T(t-s) f_*(s)\, ds.$$

Consider the linear and continuous operator $K : L^1(J, E) \to B_r$ defined by

$$K(v)(t) = \int_0^t T(t-s) v(s) ds.$$

we have

$$|K(f_n)(t) - K(f_*)(t)| = |h_n(t) - h_*(t)| \leq \|h_n - h_*\|_\infty \to 0, \text{ as } n \to \infty$$

From Lemma 2.2 it follows that $K \circ S_F$ is a closed graph operator and from the definition of K has

$$h_n(t) \in K \circ S_{F,z_{\rho(s,z_s^n + x_s)}^n + x_{\rho(s,z_s^n + x_s)}}.$$

As $z_n \to z_*$ and $h_n \to h_*$, there exist $f_* \in S_{F,z_{\rho(s,z_s^* + x_s)}^* + x_{\rho(s,z^* + x_s)}}$ such that:

$$h_*(t) = \int_0^t T(t-s) f_*(s) ds.$$

Hence the multi-valued operator $\mathcal{A}$ is upper semi-continuous.

Step 6: $\mathcal{A}(B_r)$ is equi-convergent.

Let $z \in B_r$, we have, for $h \in \mathcal{A}(z)$:

$$|h(t)| \leq M' \int_0^t |f(s)|ds$$

$$\leq M'k^* + M'C \int_0^t k(s)ds + M'r \int_0^t Lk(s)ds$$

$$\leq M'k^* + M'C \int_0^t k(s)ds + M'rL \int_0^t k(s)ds.$$

Then by (12.22), we have

$$|h(t)| \to l \leq M'k^*(1 + C + rL), \quad \text{as } t \to +\infty.$$

Hence,

$$|h(t) - h(+\infty)| \to 0, \quad \text{as } t \to +\infty.$$

As a consequence of Steps 1–4, with Lemma 1.26, we can conclude that $\mathcal{A} : B_r \to \mathcal{P}(B_r)$ is continuous and compact. From Bohnenblust–Karlin's fixed theorem, we deduce that $\mathcal{A}$ has a fixed point z^*. Then $y^* = z^* + x$ is a fixed point of the operators N, which is a mild solution of the problem (12.19)–(12.20). □

12.5.3 An Example

Consider the following functional partial differential equation

$$\frac{\partial}{\partial t}z(t, x) - \frac{\partial^2}{\partial x^2}z(t, x) \in F(t, z(t - \sigma(t, z(t, 0)), x))$$

$$x \in [0, \pi], \ t \in \mathbb{R}_+ \tag{12.23}$$

$$z(t, 0) = z(t, \pi) = 0, \ t \in \mathbb{R}_+, \tag{12.24}$$

$$z(\theta, x) = z_0(\theta, x), \ t \in (-\infty, 0], \ x \in [0, \pi], \tag{12.25}$$

where F is a given multi-valued map, and $\sigma : \mathbb{R} \to \mathbb{R}^+$ is continuous.

Take $E = L^2[0, \pi]$ and define $A : E \to E$ by $A\omega = \omega''$ with domain

$$D(A) = \{\omega \in E, \omega, \omega' \text{are absolutely continuous}, \omega'' \in E, \omega(0) = \omega(\pi) = 0\}.$$

Then

$$A\omega = \sum_{n=1}^{\infty} n^2(\omega, \omega_n)\omega_n, \omega \in D(A)$$

where $\omega_n(s) = \sqrt{\frac{2}{\pi}} \sin ns, n = 1, 2, \ldots$ is the orthogonal set of eigenvectors in A. It is well known (see [168]) that A is the infinitesimal generator of an analytic semigroup $T(t), t \geq 0$ in E and is given by

$$T(t)\omega = \sum_{n=1}^{\infty} \exp(-n^2 t)(\omega, \omega_n)\omega_n, \omega \in E.$$

Since the analytic semigroup $T(t)$ is compact, there exists a positive constant M such that

$$\|T(t)\|_{B(E)} \leq M.$$

12.6 Notes and Remarks

The results of Chap. 12 are taken from [2, 5, 46–49, 70]. Other results may be found in [139, 171].

Chapter 13
Second Order Functional Differential Equations with Delay

13.1 Introduction

In this chapter, we present some existence of global mild solutions for some classes of second order semi-linear functional equations with delay.

13.2 Global Existence Results of Second Order Functional Differential Equations with Delay

13.2.1 Introduction

In this section we provide sufficient conditions for the existence of global mild solutions for two classes of second order semi-linear functional equations with delay. Our investigations will be situated in the Banach space of real continuous and bounded functions on the real half axis $\mathbb{R}_+$. First, we will consider the following problem

$$y''(t) = Ay(t) + f(t, y_t); \quad \text{a.e. } t \in J := \mathbb{R}_+ \tag{13.1}$$

$$y(t) = \phi(t); \quad t \in H, \quad y'(0) = \varphi, \tag{13.2}$$

where $f : J \times C(H, E) \to E$ is given function, $A : D(A) \subset E \to E$ is the infinitesimal generator of a strongly continuous cosine family of bounded linear

© Springer International Publishing Switzerland 2015
S. Abbas, M. Benchohra, *Advanced Functional Evolution Equations and Inclusions*, Developments in Mathematics 39,
DOI 10.1007/978-3-319-17768-7_13

operators $(C(t))_{t \in \mathbb{R}}$, on E, $\phi : H \to E$ is given continuous function, and $(E, |.|)$ is a real Banach space. Later, we consider the following problem

$$y''(t) = Ay(t) + f(t, y_{\rho(t, y_t)}); \quad \text{a.e. } t \in J := \mathbb{R}_+ \qquad (13.3)$$

$$y(t) = \phi(t) \in \mathcal{B}; \quad y'(0) = \varphi, \qquad (13.4)$$

where $f : J \times \mathcal{B} \to E$ is given function, $A : D(A) \subset E \to E$ as in problem (11.25)–(7.1), $\phi \in \mathcal{B}$, $\rho : J \times \mathcal{B} \to \mathbb{R}$, and $(E, |.|)$ is a real Banach space. The main results are based upon Schauder's fixed theorem combined with the family of cosine operators.

Our purpose in this section is to consider a simultaneous generalization of the classical second order abstract Cauchy problem studied by Travis and Weeb in [179, 180]. Additionally, we observe that the ideas and techniques in this section permit the reformulation of the problems studied in [38, 67] to the context of partial second order differential equations.

13.2.2 Existing Result for the Finite Delay Case

In this section by $BC := BC([-r, +\infty))$ we denote the Banach space of all bounded and continuous functions from $[-r, +\infty)$ into $\mathbb{R}$ equipped with the standard norm

$$\|y\|_{BC} = \sup_{t \in [-r, +\infty)} |y(t)|.$$

Now we give our main existence result for problem (13.1)–(13.2). Before starting and proving this result, we give the definition of a mild solution.

Definition 13.1. We say that a continuous function $y : [-r, +\infty) \to E$ is a mild solution of problem (11.25)–(7.1) if $y(t) = \phi(t)$, $t \in H$, $y(.)$ and $y'(0) = \varphi$, and

$$y(t) = C(t)\phi(0) + S(t)\varphi + \int_0^t C(t-s)f(s, y_s)ds, \ t \in J.$$

Let us introduce the following hypotheses:

(13.1.1) $C(t)$ is compact for $t > 0$ in the Banach space E. Let

$$M = \sup\{\|C\|_{B(E)} : t \geq 0\}, \text{ and } M' = \sup\{\|S\|_{B(E)} : t \geq 0\}.$$

(13.1.2) The function $f : J \times C(H, E) \to E$ is Carathéodory.
(13.1.3) There exists a continuous function $k : J \to \mathbb{R}_+$ such that:

$$|f(t, u) - f(t, v)| \leq k(t)\|u - v\|, \ t \in J, \ u, v \in C(H, E)$$

and

$$k^* := \sup_{t \in J} \int_0^t k(s)ds < \infty.$$

(13.1.4) The function $t \to f(t, 0) = f_0 \in L^1(J, \mathbb{R}_+)$ with $F^* = \|f_0\|_{L^1}$.
(13.1.5) For each bounded $B \subset BC$ and $t \in J$ the set:

$$\left\{ C(t)\phi(0) + S(t)\varphi + \int_0^t C(t-s)f(s, y_t)ds : y \in B \right\}$$

is relatively compact in E.

Theorem 13.2. *Assume that (13.1.1)–(13.1.5) hold. If $K^*M < 1$, then the problem (13.1)–(13.2) has at least one mild solution on BC.*

Proof. Let the operator: $N : BC \to BC$ be defined by:

$$(Ny)(t) = \begin{cases} \phi(t), & \text{if } t \in H, \\ C(t)\,\phi(0) + S(t)\varphi + \displaystyle\int_0^t C(t-s)\,f(s, y_s)\,ds, & \text{if } t \in J. \end{cases}$$

The operator N maps BC into BC; indeed the map $N(y)$ is continuous on $[-r, +\infty)$ for any $y \in BC$, and for each $t \in J$, we have

$$|(Ny)(t)| \leq M\|\phi\| + M'\|\varphi\| + M \int_0^t |f(s, y_s) - f(s, 0) + f(s, 0)| ds$$

$$\leq M\|\phi\| + M'\|\varphi\| + M \int_0^t |f(s, 0)| ds + M \int_0^t k(s)\|y_s\| ds$$

$$\leq M\|\phi\| + M'\|\varphi\| + MF^* + M \int_0^t k(s)\|y_s\| ds$$

$$\leq M\|\phi\| + M'\|\varphi\| + MF^* + M\|y\|_{BC} k^* := c.$$

Let

$$C = M\|\phi\| + M'\|\varphi\|.$$

Hence, $N(y) \in BC$.

Moreover, let $r > 0$ be such that $r \geq \frac{C + MF^*}{1 - Mk^*}$, and B_r be the closed ball in BC centered at the origin and of radius r. Let $y \in B_r$ and $t \in \mathbb{R}_+$. Then,

$$|(Ny)(t)| \leq C + MF^* + Mk^*r.$$

Thus,

$$\|N(y)\|_{BC} \leq r,$$

which means that the operator N transforms the ball B_r into itself.

Now we prove that $N : B_r \to B_r$ satisfies the assumptions of Schauder's fixed theorem. The proof will be given in several steps.

Step 1: N is continuous in B_r.

Let $\{y_n\}$ be a sequence such that $y_n \to y$ in B_r. We have

$$|(Ny_n)(t) - (Ny)(t)| \le M \int_0^t |f(s, y_{s_n}) - f(s, y_s)| ds.$$

Then by (13.1.2) we have $f(s, y_{s_n}) \to f(s, y_s)$, as $n \to \infty$, for a.e. $s \in J$, and by the Lebesgue dominated convergence theorem we have

$$\|(Ny_n) - (Ny)\|_{BC} \to 0, \text{ as } n \to \infty.$$

Thus, N is continuous.

Step 2: $N(B_r) \subset B_r$ this is clear.

Step 3: $N(B_r)$ is equi-continuous on every compact interval $[0, b]$ of $\mathbb{R}_+$ for $b > 0$. Let $\tau_1, \tau_2 \in [0, b]$ with $\tau_2 > \tau_1$, we have

$$|N(y)(\tau_2) - N(y)(\tau_1)|$$

$$\le \|C(\tau_2 - s) - C(\tau_1 - s)\|_{B(E)} \|\phi\| + \|S(\tau_2 - s) - S(\tau_1 - s)\|_{B(E)} \|\varphi\|$$

$$+ \int_0^{\tau_1} \|C(\tau_2 - s) - C(\tau_1 - s)\|_{B(E)} |f(s, y_s)| ds$$

$$+ \int_{\tau_1}^{\tau_2} \|C(\tau_2 - s)\|_{B(E)} |f(s, y_s)| ds$$

$$\le \|C(\tau_2 - s) - C(\tau_1 - s)\|_{B(E)} \|\phi\| + \|S(\tau_2 - s) - S(\tau_1 - s)\|_{B(E)} \|\varphi\|$$

$$+ \int_0^{\tau_1} \|C(\tau_2 - s) - C(\tau_1 - s)\|_{B(E)} |f(s, y_s) - f(s, 0) + f(s, 0)| ds$$

$$+ \int_{\tau_1}^{\tau_2} \|C(\tau_2 - s)\|_{B(E)} |f(s, y_s) - f(s, 0) + f(s, 0)| ds$$

$$\le \|C(\tau_2 - s) - C(\tau_1 - s)\|_{B(E)} \|\phi\| + \|S(\tau_2 - s) - S(\tau_1 - s)\|_{B(E)} \|\varphi\|$$

$$+ r \int_0^{\tau_1} \|C(\tau_2 - s) - C(\tau_1 - s)\|_{B(E)} k(s) ds$$

$$+ \int_0^{\tau_1} \|C(\tau_2 - s) - C(\tau_1 - s)\|_{B(E)} |f(s, 0)| ds$$

$$+ r \int_{\tau_1}^{\tau_2} \|C(\tau_2 - s)\|_{B(E)} k(s) ds$$

$$+ \int_{\tau_1}^{\tau_2} \|C(\tau_2 - s)\|_{B(E)} |f(s, 0)| ds.$$

When $\tau_2 \to \tau_2$, the right-hand side of the above inequality tends to zero, since $C(t), S(t)$ are a strongly continuous operator and the compactness of $C(t), S(t)$ for $t > 0$, implies the continuity in the uniform operator topology (see [179, 180]). This proves the equi-continuity.

Step 4:$N(B_r)$ is relatively compact on every compact interval of $[0, \infty)$ by (13.1.5).

Step 5: $N(B_r)$ is equi-convergent.

Let $y \in B_r$, we have:

$$|(Ny)(t)| \leq M\|\phi\| + M'\|\varphi\| + M \int_0^t |f(s, y_s)| ds$$

$$\leq C + MF^* + Mr \int_0^t k(s) ds$$

$$\leq C + MF^* + Mr \int_0^t k(s) ds.$$

Then

$$|(Ny)(t)| \to C_1 \leq C + MF^* + Mk^* r, \quad \text{as } t \to +\infty.$$

Hence,

$$|(Ny)(t) - (Ny)(+\infty)| \to 0, \quad \text{as } t \to +\infty.$$

As a consequence of Steps 1–5, with Lemma 1.26, we can conclude that $N : B_r \to B_r$ is continuous and compact. From Schauder's theorem, we deduce that N has a fixed point y^* which is a mild solution of the problem (13.1)–(13.2). $\square$

13.2.3 Existing Results for the State-Dependent Delay Case

In this section by $BC := BC(\mathbb{R})$ we denote the Banach space of all bounded and continuous functions from $\mathbb{R}$ into E equipped with the standard norm

$$\|y\|_{BC} = \sup_{t \in \mathbb{R}} |y(t)|.$$

Finally, by $BC' := BC'(\mathbb{R}_+)$ we denote the Banach space of all bounded and continuous functions from $\mathbb{R}_+$ into E equipped with the standard norm

$$\|y\|_{BC'} = \sup_{t \in \mathbb{R}_+} |y(t)|.$$

Now we give our main existence result for problem (13.3)–(13.4). Before starting and proving this result, we give the definition of a mild solution.

Definition 13.3. We say that a continuous function $y : (-\infty, +\infty) \to E$ is a mild solution of problem (11.15)–(11.16) if $y(t) = \phi(t)$, $t \in (-\infty, 0]$, $y(.)$ is continuously differentiable and $y'(0) = \varphi$ and

$$y(t) = C(t)\phi(0) + S(t)\varphi + \int_0^t C(t-s)f(s, y_{\rho(t, y_t)})ds, \ t \in J.$$

Set

$$\mathcal{R}(\rho^-) = \{\rho(s, \phi) : (s, \phi) \in J \times \mathcal{B}, \ \rho(s, \phi) \leq 0\}.$$

We always assume that $\rho : J \times \mathcal{B} \to \mathbb{R}$ is continuous. Additionally, we introduce the following hypothesis:

(H_ϕ) The function $t \to \phi_t$ is continuous from $\mathcal{R}(\rho^-)$ into $\mathcal{B}$ and there exists a continuous and bounded function $\mathcal{L}^\phi : \mathcal{R}(\rho^-) \to (0, \infty)$ such that

$$\|\phi_t\| \leq \mathcal{L}^\phi(t)\|\phi\| \quad \text{for every } t \in \mathcal{R}(\rho^-).$$

Remark 13.4. The condition (H_ϕ) is frequently verified by functions continuous and bounded.

Let us introduce the following hypotheses:

(13.3.1) $C(t), S(t)$ are compact for $t > 0$ in the Banach space E. Let $M = \sup\{\|C\|_{B(E)} : t \geq 0\}$, and $M' = \sup\{\|S\|_{B(E)} : t \geq 0\}$.

(13.3.2) The function $f : J \times \mathcal{B} \to E$ is Carathéodory.

(13.3.3) There exists a continuous function $k : J \to \mathbb{R}_+$ such that:

$$|f(t, u) - f(t, v)| \leq k(t)\|u - v\|, \ t \in J, \ u, v \in \mathcal{B}$$

and

$$k^* := \sup_{t \in J} \int_0^t k(s)ds < \infty.$$

(13.3.4) The function $t \to f(t, 0) = f_0 \in L^1(J, \mathbb{R}_+)$ with $F^* = \|f_0\|_{L^1}$.

(13.3.5) For each bounded $B \subset BC'$ and $t \in J$ the set:

$$\{S(t)\varphi + \int_0^t C(t-s)f(s, y_{\rho(t, y_t)})ds : y \in B\}$$

is relatively compact in E.

Theorem 13.5. *Assume that (13.3.1)–(13.3.5),(H_ϕ) hold. If $K^*Ml < 1$, then the problem (13.3)–(13.4) has at least one mild solution on BC.*

Proof. Consider the operator: $N : BC \to BC$ define by:

$$(Ny)(t) = \begin{cases} \phi(t), & \text{if } t \in (-\infty, 0], \\ C(t)\,\phi(0) + S(t)\varphi + \displaystyle\int_0^t C(t-s)\,f(s, y_{\rho(t,y_t)})\, ds, & \text{if } t \in J. \end{cases}$$

Let $x(.) : \mathbb{R} \to E$ be the function defined by:

$$x(t) = \begin{cases} \phi(t); & \text{if } t \in (-\infty, 0]; \\ C(t)\,\phi(0); & \text{if } t \in J, \end{cases}$$

then $x_0 = \phi$. For each $z \in BC$ with $z(0) = 0$, $y'(0) = \varphi = z'(0) = \varphi_1$, we denote by $\bar{z}$ the function

$$\bar{z}(t) = \begin{cases} 0; & \text{if } t \in (-\infty, 0]; \\ z(t); & \text{if } t \in J. \end{cases}$$

If y satisfies $y(t) = (Ny)(t)$, we can decompose it as $y(t) = z(t) + x(t)$, $t \in J$, which implies $y_t = z_t + x_t$ for every $t \in J$ and the function $z(.)$ satisfies

$$z(t) = S(t)\varphi_1 + \int_0^t C(t-s)\,f(s, z_{\rho(s,z_s+x_s)} + x_{\rho(s,z_s+x_s)})ds, t \in J.$$

Set

$$BC_0' = \{z \in BC' : z(0) = 0\}$$

and let

$$\|z\|_{BC_0'} = \sup\{|z(t)| : t \in J\}, \ z \in BC_0'.$$

BC_0' is a Banach space with the norm $\|.\|_{BC_0'}$. We define the operator $\mathcal{A} : BC_0' \to BC_0'$ by:

$$\mathcal{A}(z)(t) = S(t)\varphi_1 + \int_0^t C(t-s)\,f(s, z_{\rho(s,z_s+x_s)} + x_{\rho(s,z_s+x_s)})ds, \ t \in J.$$

We shall show that the operator $\mathcal{A}$ satisfies all conditions of Schauder's fixed point theorem. The operator A maps BC_0' into BC_0', indeed the map $\mathcal{A}(z)$ is continuous on $\mathbb{R}_+$ for any $z \in BC_0'$, and for each $t \in J$ we have

$$|\mathcal{A}(z)(t)| \leq M'\|\varphi_1\| + M \int_0^t |f(s, z_{\rho(s,z_s+x_s)} + x_{\rho(s,z_s+x_s)}) - f(s,0) + f(s,0)|ds$$

$$\leq M'\|\varphi_1\| + M \int_0^t |f(s,0)|ds + M \int_0^t k(s)\|z_{\rho(s,z_s+x_s)} + x_{\rho(s,z_s+x_s)}\|_\mathcal{B}ds$$

$$\leq M'\|\varphi_1\| + MF^* + M \int_0^t k(s)(l|z(s)| + (m + \mathcal{L}^\phi + lMH)\|\phi\|_\mathcal{B})ds.$$

Let

$$C = (m + \mathcal{L}^\phi + lMH)\|\phi\|_\mathcal{B}.$$

Then, we have:

$$|\mathcal{A}(z)(t)| \leq M'\|\varphi_1\| + MF^* + MC\int_0^t k(s)ds + Ml\int_0^t k(s)|z(s)|ds$$

$$\leq M'\|\varphi_1\| + MF^* + MCk^* + Ml\|z\|_{BC_0'}k^*.$$

Hence, $\mathcal{A}(z) \in BC_0'$.

Moreover, let $r > 0$ be such that $r \geq \frac{M'\|\varphi_1\|+MF^*+MCk^*}{1-Mlk^*}$, and B_r be the closed ball in BC_0' centered at the origin and of radius r. Let $y \in B_r$ and $t \in \mathbb{R}_+$. Then,

$$|\mathcal{A}(z)(t)| \leq M'\|\varphi_1\| + MF^* + MCk^* + Mlk^*r.$$

Thus,

$$\|\mathcal{A}(z)\|_{BC_0'} \leq r,$$

which means that the operator N transforms the ball B_r into itself.

Now we prove that $\mathcal{A} : B_r \to B_r$ satisfies the assumptions of Schauder's fixed theorem. The proof will be given in several steps.

Step 1: $\mathcal{A}$ is continuous in B_r.

Let $\{z_n\}$ be a sequence such that $z_n \to z$ in B_r. At the first, we study the convergence of the sequences $(z^n_{\rho(s,z_s^n)})_{n\in\mathbb{N}}, s \in J$.

If $s \in J$ is such that $\rho(s, z_s) > 0$, then we have,

$$\|z^n_{\rho(s,z_s^n)} - z_{\rho(s,z_s)}\|_\mathcal{B} \leq \|z^n_{\rho(s,z_s^n)} - z_{\rho(s,z_s^n)}\|_\mathcal{B} + \|z_{\rho(s,z_s^n)} - z_{\rho(s,z_s)}\|_\mathcal{B}$$

$$\leq l\|z_n - z\|_{B_r} + \|z_{\rho(s,z_s^n)} - z_{\rho(s,z_s)}\|_\mathcal{B},$$

which proves that $z^n_{\rho(s,z_s^n)} \to z_{\rho(s,z_s)}$ in $\mathcal{B}$ as $n \to \infty$ for every $s \in J$ such that $\rho(s, z_s) > 0$. Similarly, is $\rho(s, z_s) < 0$, we get

$$\|z^n_{\rho(s,z_s^n)} - z_{\rho(s,z_s)}\|_\mathcal{B} = \|\phi^n_{\rho(s,z_s^n)} - \phi_{\rho(s,z_s)}\|_\mathcal{B} = 0$$

which also shows that $z^n_{\rho(s,z^n_s)} \to z_{\rho(s,z_s)}$ in $\mathcal{B}$ as $n \to \infty$ for every $s \in J$ such that $\rho(s,z_s) < 0$. Combining the pervious arguments, we can prove that $z^n_{\rho(s,z_s)} \to \phi$ for every $s \in J$ such that $\rho(s,z_s) = 0$. Finally,

$$|\mathcal{A}(z_n)(t) - \mathcal{A}(z)(t)|$$

$$\leq M \int_0^t |f(s, z^n_{\rho(s,z^n_s+x_s)} + x_{\rho(s,z^n_s+x_s)}) - f(s, z_{\rho(s,z_s+x_s)} + x_{\rho(s,z_s+x_s)})| ds.$$

Then by (13.3.2) we have

$$f(s, z^n_{\rho(s,z^n_s+x_s)} + x_{\rho(s,z^n_s+x_s)}) \to f(s, z_{\rho(s,z_s+x_s)} + x_{\rho(s,z_s+x_s)}), \text{ as } n \to \infty,$$

and by the Lebesgue dominated convergence theorem we get,

$$\|\mathcal{A}(z_n) - \mathcal{A}(z)\|_{BC'_0} \to 0, \text{ as } n \to \infty.$$

Thus $\mathcal{A}$ is continuous.

Step 2: $\mathcal{A}(B_r) \subset B_r$ this is clear.

Step 3: $\mathcal{A}(B_r)$ is equi-continuous on every compact interval $[0, b]$ of $\mathbb{R}_+$ for $b > 0$. Let $\tau_1, \tau_2 \in [0, b]$ with $\tau_2 > \tau_1$, we have

$$|\mathcal{A}(z)(\tau_2) - \mathcal{A}(z)(\tau_1)|$$

$$\leq \|S(\tau_2 - s) - S(\tau_1 - s)\|_{B(E)} \|\varphi_1\|$$

$$+ \int_0^{\tau_1} \|C(\tau_2 - s) - C(\tau_1 - s)\|_{B(E)} |f(s, z^n_{\rho(s,z^n_s+x_s)} + x_{\rho(s,z^n_s+x_s)})| ds$$

$$+ \int_{\tau_1}^{\tau_2} \|C(\tau_2 - s)\|_{B(E)} |f(s, z^n_{\rho(s,z^n_s+x_s)} + x_{\rho(s,z^n_s+x_s)})| ds$$

$$\leq \|S(\tau_2 - s) - S(\tau_1 - s)\|_{B(E)} \|\varphi_1\|$$

$$+ \int_0^{\tau_1} \|C(\tau_2 - s) - C(\tau_1 - s)\|_{B(E)} |f(s, z^n_{\rho(s,z^n_s+x_s)} + x_{\rho(s,z^n_s+x_s)}) - f(s,0)| ds$$

$$+ \int_0^{\tau_1} \|C(\tau_2 - s) - C(\tau_1 - s)\|_{B(E)} f(s,0)| ds$$

$$+ \int_{\tau_1}^{\tau_2} \|C(\tau_2 - s)\|_{B(E)} |f(s, z^n_{\rho(s,z^n_s+x_s)} + x_{\rho(s,z^n_s+x_s)}) - f(s,0)| ds$$

$$+ \int_{\tau_1}^{\tau_2} \|C(\tau_2 - s)\|_{B(E)} |f(s,0)| ds$$

$$\leq \|S(\tau_2 - s) - S(\tau_1 - s)\|_{B(E)} \|\varphi_1\|$$

$$+ C \int_0^{\tau_1} \|C(\tau_2 - s) - C(\tau_1 - s)\|_{B(E)} k(s) ds$$

$$+lr \int_0^{\tau_1} \|C(\tau_2 - s) - C(\tau_1 - s)\|_{B(E)} k(s) ds$$

$$+ \int_0^{\tau_1} \|C(\tau_2 - s) - C(\tau_1 - s)\|_{B(E)} |f(s,0)| ds$$

$$+C \int_{\tau_1}^{\tau_2} \|C(\tau_2 - s)\|_{B(E)} k(s) ds$$

$$+lr \int_{\tau_1}^{\tau_2} \|C(\tau_2 - s)\|_{B(E)} k(s) ds$$

$$+ \int_{\tau_1}^{\tau_2} \|C(\tau_2 - s)\|_{B(E)} |f(s,0)| ds.$$

When $\tau_2 \to \tau_2$, the right-hand side of the above inequality tends to zero, since $C(t)$ are a strongly continuous operator and the compactness of $C(t)$ for $t > 0$ implies the continuity in the uniform operator topology (see [179, 180]). This proves the equi-continuity.

Step 4: $N(B_r)$ is relatively compact on every compact interval of $[0, \infty)$. This is satisfied from (13.1.5).

Step 5: $N(B_r)$ is equi-convergent.

Let $y \in B_r$, we have:

$$|\mathcal{A}(z)(t)| \leq M' \|\varphi_1\| + M \int_0^t |f(s, z_{\rho(s,z_s^n+x_s)}^n + x_{\rho(s,z_s^n+x_s)})| ds$$

$$\leq M' \|\varphi_1\| + MF^* + MCk^* + Mrl \int_0^t k(s) ds.$$

Then

$$|\mathcal{A}(z)(t)| \to C_1 \leq M' \|\varphi_1\| + MF^* + Mk^*(C + lr), \quad \text{as } t \to +\infty.$$

Hence,

$$|\mathcal{A}(z)(t) - \mathcal{A}(z)(+\infty)| \to 0, \quad \text{as } t \to +\infty.$$

As a consequence of Steps 1–5, with Lemma 1.26, we can conclude that $\mathcal{A} : B_r \to B_r$ is continuous and compact. we deduce that $\mathcal{A}$ has a fixed point z^*. Then $y^* = z^* + x$ is a fixed point of the operators N, which is a mild solution of the problem (13.3)–(13.4). $\square$

13.2.4 Examples

Example 1. Consider the functional partial differential equation of second order:

$$\frac{\partial^2}{\partial t^2}z(t,x) = \frac{\partial^2}{\partial x^2}z(t,x) + f(t,z(t-r,x)), \quad x \in [0,\pi], \ t \in J := \mathbb{R}_+, \quad (13.5)$$

$$z(t,0) = z(t,\pi) = 0, \ t \in \mathbb{R}_+, \quad (13.6)$$

$$z(t,x) = \phi(t), \quad \frac{\partial z(0,x)}{\partial t} = w(x), \ t \in H, \ x \in [0,\pi], \quad (13.7)$$

where f is a given map. Take $E = L^2[0,\pi]$ and define $A : E \to E$ by $A\omega = \omega''$ with domain

$$D(A) = \{\omega \in E; \omega, \omega' \text{ are absolutely continuous}, \ \omega'' \in E, \ \omega(0) = \omega(\pi) = 0\}.$$

It is well known that A is the infinitesimal generator of a strongly continuous cosine function $(C(t))_{t \in \mathbb{R}}$ on E, respectively. Moreover, A has discrete spectrum, the eigenvalues are $-n^2, n \in \mathbb{N}$ with corresponding normalized eigenvectors $z_n(\tau) := (\frac{2}{\pi})^{\frac{1}{2}} \sin n\tau$, and the following properties hold:

(a) $\{z_n : n \in \mathbb{N}\}$ is an orthonormal basis of E.
(b) If $y \in E$, then $Ay = -\sum_{n=1}^{\infty} n^2 < y, z_n > z_n$.
(c) For $y \in E$, $C(t)y = \sum_{n=1}^{\infty} \cos(nt) < y, z_n > z_n$, and the associated sine family is

$$S(t)y = \sum_{n=1}^{\infty} \frac{\sin(nt)}{n} < y, z_n > z_n$$

which implies that the operator $S(t)$ is compact for all $t > 0$ and that

$$\|C(t)\| = \|S(t)\| \leq 1, \text{ for all } t \geq 0.$$

(d) If Φ denotes the group of translations on E defined b $\Phi(t)y(\xi) = \tilde{y}(\xi + t)$ where $\tilde{y}$ is the extension of y with period 2π, then $C(t) = \frac{1}{2}(\Phi(t) + \Phi(-t)); A = B^2$, where B is the infinitesimal generator of the group Φ on

$$X = \{y \in H^1(0,\pi) : y(0) = x(\pi) = 0\}.$$

Then the problem (13.1)–(13.2) is an abstract formulation of the problem (13.5)–(13.7). If conditions (13.1.1)–(13.1.5) are satisfied. Theorem 13.2 implies that the problem (13.5)–(13.7) has at least one mild solution on BC.

Example 2. Take $E = L^2[0,\pi]; B = C_0 \times L^2(g,E)$ and define $A : E \to E$ by $A\omega = \omega''$ with domain

$$D(A) = \{\omega \in E; \omega, \omega' \text{ are absolutely continuous}, \ \omega'' \in E, \ \omega(0) = \omega(\pi) = 0\}.$$

It is well known that A is the infinitesimal generator of a strongly continuous cosine function $(C(t))_{t\in\mathbb{R}}$ on E, respectively. Moreover, A has discrete spectrum, the eigenvalues are $-n^2, n \in \mathbb{N}$ with corresponding normalized eigenvectors $z_n(\tau) := (\frac{2}{\pi})^{\frac{1}{2}} \sin n\tau$, and the following properties hold:

(a) $\{z_n : n \in \mathbb{N}\}$ is an orthonormal basis of E.
(b) If $y \in E$, then $Ay = -\sum_{n=1}^{\infty} n^2 < y, z_n > z_n$.
(c) For $y \in E$, $C(t)y = \sum_{n=1}^{\infty} \cos(nt) < y, z_n > z_n$, and the associated sine family is

$$S(t)y = \sum_{n=1}^{\infty} \frac{\sin(nt)}{n} < y, z_n > z_n$$

which implies that the operator $S(t)$ is compact for all $t > 0$ and that $\|C(t)\| = \|S(t)\| \le 1$ for all $t \in \mathbb{R}$.
(d) If Φ denotes the group of translations on E defined by

$$\Phi(t)y(\xi) = \tilde{y}(\xi + t),$$

where $\tilde{y}$ is the extension of y with period 2π, then

$$C(t) = \frac{1}{2}(\Phi(t) + \Phi(-t)), \; A = B^2,$$

where B is the infinitesimal generator of the group Φ on

$$X = \{y \in H^1(0, \pi) : y(0) = x(\pi) = 0\}.$$

Consider the functional partial differential equation of second order:

$$\frac{\partial^2}{\partial t^2} z(t, x) = \frac{\partial^2}{\partial x^2} z(t, x) + \int_{-\infty}^{0} a(s - t)z(s - \rho_1(t)\rho_2(\|z(t)\|), x)ds,$$

$$x \in [0, \pi], \; t \in J := \mathbb{R}_+, \tag{13.8}$$

$$z(t, 0) = z(t, \pi) = 0, \; t \in \mathbb{R}_+, \tag{13.9}$$

$$z(t, x) = \phi(t), \; \frac{\partial z(0, x)}{\partial t} = \omega(x), \; t \in H, \; x \in [0, \pi], \tag{13.10}$$

where $\rho_i : [0, \infty) \rightarrow [0, \infty), a; \mathbb{R} \rightarrow \mathbb{R}$ be continuous, and $L_f = \left(\int_{-\infty}^{0} \frac{a^2(s)}{g(s)}ds\right)^{\frac{1}{2}} < \infty$. Under these conditions, we define the function $f : J \times \mathcal{B} \rightarrow E, \rho : J \times \mathcal{B} \rightarrow \mathbb{R}$ by

$$f(t, \psi)(x) = \int_{-\infty}^{0} a(s)\psi(s, x)ds,$$

$$\rho(s, \psi) = s - \rho_1(s)\rho_2(\|\psi(0)\|),$$

we have $\|f(t, .)\|_{\mathcal{B}} \le L_f$.

Then the problem (13.3)–(13.4) is an abstract formulation of the problem (13.8)–(13.10). If conditions (13.1.1)–(13.1.5) are satisfied. Theorem 12.4 implies that the problem (13.8)–(13.10) has at least one mild solution on BC.

13.3 Notes and Remarks

The results of Chap. 13 are taken from Alaidarous et al. [21, 22] and Benchohra et al. [55–57, 82, 83]. Other results may be found in [111, 166].

References

1. N. Abada, M. Benchohra, H. Hammouche, A. Ouahab, Controllability of impulsive semilinear functional differential inclusions with finite delay in Fréchet spaces. Discuss. Math. Differ. Incl. Control Optim. **27**(2), 329–347 (2007)

2. N. Abada, R.P. Agarwal, M. Benchohra, H. Hammouche, Existence results for nondensely impulsive semilinear functional differential equations with state-dependente delay. Asian Eur. J. Math. **1**(4), 449–468 (2008)

3. N. Abada, M. Benchohra, H. Hammouche, Existence and controllability results for impulsive partial functional differential inclusions. Nonlinear Anal. **69**, 2892–2909 (2008)

4. N. Abada, M. Benchohra, H. Hammouche, Existence and controllability results for non-densely defined impulsive semilinear functional differential equations. Differ. Integr. Equ. **21**(5–6), 513–540 (2008)

5. N. Abada, M. Benchohra, H. Hammouche, Nonlinear impulsive partial functional differential inclusions with state-dependent delay and multivalued jumps. Nonlinear Anal. **4**, 791–803 (2010)

6. N. Abada, M. Benchohra, H. Hammouche, Existence results for semilinear differential evolution equations with impulses and delay. Cubo **12**, 1–17 (2010)

7. N. Abada, R.P. Agarwal, M. Benchohra, H. Hammouche, Impulsive semilinear neutral functional differential inclusions with multivalued jumps. Appl. Math. **56**, 227–250 (2011)

8. M. Adimy, K. Ezzinbi, A class of linear partial neutral functional differential equations with nondense domain J. Differ. Equ. **147**, 285–332 (1998)

9. M. Adimy, K. Ezzinbi, Existence and linearized stability for partial neutral functional differential equations with nondense domains. Differ. Equ. Dyn. Syst. **7**, 371–417 (1999)

10. M. Adimy, H. Bouzahir, K. Ezzinbi, Existence for a class of partial functional differential equations with infinite delay. Nonlinear Anal. **46**, 91–112 (2001)

11. M. Adimy, H. Bouzahir, K. Ezzinbi, Local existence and stability for some partial functional differential equations with infinite delay. Nonlinear Anal., **48**, 323–348 (2002)

12. M. Adimy, H. Bouzahir, K. Ezzinbi, Existence and stability for some partial neutral functional differential equations with infinite delay. J. Math. Anal. Appl. **294**, 438–461 (2004)

13. R.P. Agarwal, S.R. Grace, D. O'Regan, *Oscillation Theory for Second Order Dynamic Equations* (Taylor & Francis, London, 2003)

14. R.P. Agarwal, M. Bohner, S.H. Saker, Oscillation of second order delay dynamic equations. J. Math. Anal. Appl. **300**, 203–217 (2004)

© Springer International Publishing Switzerland 2015
S. Abbas, M. Benchohra, *Advanced Functional Evolution Equations and Inclusions*, Developments in Mathematics 39,
DOI 10.1007/978-3-319-17768-7

15. R.P. Agarwal, S. Baghli, M. Benchohra, Controllability of mild solutions on semiinfinite interval for classes of semilinear functional and neutral functional evolution equations with infinite delay. Appl. Math. Optim. **60**, 253–274 (2009)

16. N.U. Ahmed, *Semigroup Theory with Applications to Systems and Control*. Pitman Research Notes in Mathematics Series, vol. 246 (Longman Scientific & Technical, Harlow; Wiley, New York, 1991)

17. N.U. Ahmed, Measure solutions for impulsive systems in Banach spaces and their control. Dyn. Contin. Discrete Impuls. Syst. Ser. A **6**, 519–535 (1999)

18. N.U. Ahmed, Optimal control for impulsive systems in Banach spaces. Inter. J. Differ. Equ. Appl. **1**(1), 37–52 (2000)

19. N.U. Ahmed, Systems governed by impulsive differential inclusions on Hilbert spaces. Nonlinear Anal. **45**, 693–706 (2001)

20. N.U. Ahmed, *Dynamic Systems and Control with Applications* (World Scientific, Hackensack, 2006)

21. E. Alaidarous, M. Benchohra, I. Medjadj, Global existence results for neutral functional differential inclusions with state-dependent delay (submitted)

22. E. Alaidarous, M. Benchohra, I. Medjadj, Global existence results for second order functional differential equations with delay (submitted)

23. H. Amann, *Linear and Quasilinear Parabolic Problems* (Birkhäuser, Berlin, 1995)

24. B. Amir, L. Maniar, Application de la Théorie d'Extrapolation pour la Résolution des Équations Différentielles á Retard Homogènes. Extracta Math. **13**, 95–105 (1998)

25. B. Amir, L. Maniar, Composition of pseudo almost periodic functions and Cauchy problems with operator of nondense domain. Ann. Math. Blaise Pascal **6**, 1–11 (1999)

26. A. Arara, M. Benchohra, A. Ouahab, Some uniqueness results for controllability functional semilinear differential equations in Fréchet spaces. Nonlinear Oscil. **6**(3), 287–303 (2003)

27. A. Arara, M. Benchohra, S.K. Ntouyas, A. Ouahab, Existence results for boundary value problems for fourth-order differential inclusions with nonconvex valued right hand side Arch. Math. (Brno) **40**, 219–227 (2004)

28. A. Arara, M. Benchohra, L. Górniewicz, A. Ouahab, Controllability results for semilinear functional differential inclusions with unbounded delay. Math. Bull. **3**, 157–183 (2006)

29. W. Arendt, Vector valued Laplace transforms and Cauchy problems. Israel J. Math. **59**, 327–352 (1987)

30. W. Arendt, Resolvent positive operators and integrated semigroup. Proc. Lond. Math. Soc. **3**(54), 321–349 (1987)

31. O. Arino, K. Boushaba, A. Boussouar, A mathematical model of the dynamics of the phytoplankton-nutrient system. Spatial heterogeneity in ecological models (Alcalá de Henares, 1998). Nonlinear Anal. RWA **1**(1), 69–87 (2000)

32. C. Avramescu, Some remarks on a fixed point theorem of Krasnoselskii. Electron. J. Qual. Differ. Equ. **5**, 1–15 (2003)

33. S. Baghli, M. Benchohra, Uniqueness results for evolution equations with infinite delay in Fréchet spaces. Fixed Point Theory **9**, 395–406 (2008)

34. S. Baghli, M. Benchohra, Perturbed functional and neutral functional evolution equations with infinite delay in Fréchet spaces. Electron. J. Differ. Equ. **69**, 1–19 (2008)

35. S. Baghli, M. Benchohra, Multivalued evolution equations with infinite delay in Fréchet spaces. Electron. J. Qual. Theory Differ. Equ. **33**, 24 (2008)

36. S. Baghli, M. Benchohra, K. Ezzinbi, Controllability results for semilinear functional and neutral functional evolution equations with infinite delay. Surv. Math. Appl. **4**, 15–39 (2009)

37. S. Baghli, M. Benchohra, Existence results for semilinear neutral functional differential equations involving evolution operators in Fréchet spaces. Georgian Math. J. **17**, 423–436 (2010)

38. S. Baghli, M. Benchohra, Global uniqueness results for partial functional and neutral functional evolution equations with infinite delay. Differ. Integr. Equ. **23**, 31–50 (2010)

39. D.D. Bainov, P.S. Simeonov, *Systems with Impulse Effect* (Ellis Horwood Ltd., Chichister, 1989)

40. D.D. Bainov, P.S. Simeonov, *Oscillation Theory of Impulsive Differential Equations* (International Publications, Orlando, 1998)

41. K. Balachandran, E.R. Anandhi, Boundary controllability of integrodifferential systems in Banach spaces. Proc. Indian Acad. Sci. (Math. Sci.) **111**, 127–135 (2001)

42. K. Balachandran, E.R. Anandhi, Controllability of neutral integrodifferential infinite delay systems in Banach spaces. Taiwan. J. Math. **8**, 689–702 (2004)

43. K. Balachandran, J.P. Dauer, Controllability of nonlinear systems in Banach spaces: a survey. Dedicated to Professor Wolfram Stadler. J. Optim. Theory Appl. **115**, 7–28 (2002)

44. K. Balachandran, R.R. Kumar, Existence of solutions of integrodifferential evolution equations with time varying delays. Appl. Math. E **7**, 1–8 (2007)

45. K. Balachandran, A. Leelamani, Null controllability of neutral integrodifferential systems with infinite delay. Math. Problems Eng. **2006**, 1–18 (2006)

46. A. Baliki, M. Benchohra, Global existence and stability for neutral functional evolution equations with state-dependent delay. Nonautonomous Dyn. Syst. **1**(1), 112–122 (2014)

47. A. Baliki, M. Benchohra, Global existence and asymptotic behaviour for functional evolution equations. J. Appl. Anal. Comput. **4**(2), 129–138 (2014)

48. A. Baliki, M. Benchohra, Global existence and stability for neutral functional evolution equations. Rev. Roumaine Math. Pures Appl. **LX**(1), 71–82 (2015)

49. A. Baliki, M. Benchohra, Global existence and stability for second order functional evolution equations with infinite delay (submitted)

50. M. Belmekki, M. Benchohra, S.K. Ntouyas, Existence results for semilinear perturbed functional differential equations with nondensely defined operators. Fixed Point Theory Appl. **2006**, 1–13 (2006)

51. M. Belmekki, M. Benchohra, S.K. Ntouyas, Existence results for nondensely defined semilinear functional differential inclusions. J. Nonlinear Func. Anal. Differ. Equ. **1**(2), 261–279 (2007)

52. M. Belmekki, M. Benchohra, K. Ezzinbi, S.K. Ntouyas, Existence results for some partial functional differential equations with infinite delay. Nonlinear Stud. **15**(4), 373–385 (2008)

53. M. Belmekki, M. Benchohra, S.K. Ntouyas, Existence results for densely defined semilinear functional differential inclusions. J. Contemp. Math. **1**(1), 25–44 (2008)

54. M. Benchohra, L. Gorniewicz, Existence results of nondensely defined impulsive semilinear functional differential inclusions with infinite delay. J. Fixed Point Theory Appl. **2**(1), 11–51 (2007)

55. M. Benchohra, I. Medjedj, Global existence and stability results for functional differential equations with delay. Commun. Appl. Anal. **17**(2), 213–220 (2013)

56. M. Benchohra, I. Medjedj, Global existence results for functional differential inclusions with delay. Nonlinear Oscil. **17**(2), 161–169 (2014)

57. M. Benchohra, I. Medjedj, Global existence results for neutral functional differential equations with state-dependent delay. Differ. Equ. Dyn. Syst. (to appear)

58. M. Benchohra, S.K. Ntouyas, Existence of mild solutions on semiinfinite interval for first order differential equation with nonlocal condition. Comment. Math. Univ. Carolinae **41**(3), 485–491 (2000)

59. M. Benchohra, S.K. Ntouyas, Existence of mild solutions for certain delay semilinear evolution inclusions with nonlocal condition. Dyn. Syst. Appl. **9**(3), 405–412 (2000)

60. M. Benchohra, S.K. Ntouyas, Existence results on infinite intervals for neutral functional differential and integrodifferential inclusions in Banach spaces. Georgian Math. J. **7**, 609–625 (2000)

61. M. Benchohra, S.K. Ntouyas, Existence results for neutral functional differential and integrodifferential inclusions in Banach spaces. Electron. J. Differ. Equ. **2000**(20), 1–15 (2000)

62. M. Benchohra, S.K. Ntouyas, Existence results for functional differential inclusions. Electron. J. Differ. Equ. **2001**(41), 1–8 (2001)
63. M. Benchohra, S.K. Ntouyas, An existence theorem for an hyperbolic differential inclusion in Banach spaces. Discuss. Math. Differ. Incl. Control Optim. **22**, 5–16 (2002)
64. M. Benchohra, S.K. Ntouyas, Controllability results for multivalued semilinear neutral functional equations. Math. Sci. Res. J. **6**, 65–77 (2002)
65. M. Benchohra, S.K. Ntouyas, Existence of mild solutions of semilinear evolution inclusions with nonlocal conditions. Georgian Math. J. **7**(2), 221–230 (2002)
66. M. Benchohra, S.K. Ntouyas, Existence of mild solutions of second order initial value problems for delay integrodifferential inclusions with nonlocal conditions. Math. Bohem. **127**, 613–622 (2002)
67. M. Benchohra, A. Ouahab, Impulsive neutral functional differential equations with variable times. Nonlinear Anal. **55**(6), 679–693 (2003)
68. M. Benchohra, M. Ziane, Impulsive evolution inclusions with infinite delay and multivalued jumps. Surv. Math. Appl. **7**, 1–14 (2012)
69. M. Benchohra, M. Ziane, Impulsive semilinear evolution differential inclusions with nonconvex right hand side. Libertas Math. **32**, 193–206 (2012)
70. M. Benchohra, M. Ziane, Impulsive evolution inclusions with state-dependent delay and multivalued jumps. Electron. J. Qual. Theory Differ. Equ. **42**, 1–21 (2013)
71. M. Benchohra, M. Ziane, Controllability of impulsive functional evolution inclusions with infinite delay (submitted)
72. M. Benchohra, L. Górniewicz, S.K. Ntouyas, Controllability of neutral functional differential and integrodifferential inclusions in Banach spaces with nonlocal conditions. Nonlinear Anal. Forum **7**, 39–54 (2002)
73. M. Benchohra, L. Górniewicz, S.K. Ntouyas, Controllability results for multivalued semilinear differential equations with nonlocal conditions. Dyn. Syst. Appl. **11**, 403–414 (2002)
74. M. Benchohra, E. Gatsori, J. Henderson, S.K. Ntouyas, Nondensely defined evolution impulsive differential inclusions with nonlocal conditions. J. Math. Anal. Appl. **286**, 307–325 (2003)
75. M. Benchohra, E.P. Gatsori, L. Górniewicz, S.K. Ntouyas, Controllability results for evolution inclusions with non-local conditions. Z. Anal. Anwendungen **22**, 411–431 (2003)
76. M. Benchohra, L. Górniewicz, S.K. Ntouyas, *Controllability of Some Nonlinear Systems in Banach spaces: The fixed point theory approach* (Pawel Wlodkowicz University College, Plock, 2003)
77. M. Benchohra, E.P. Gastori, S.K. Ntouyas, Existence results for semi-linear integrodifferential inclusions with nonlocal conditions. Rocky Mountain J. Math. **34**, 833–848 (2004)
78. M. Benchohra, E.P. Gatsori, S.K. Ntouyas, Multivalued semilinear neutral functional differential equations with nonconvex-valued right-hand side. Abstr. Appl. Anal. **2004**(6), 525–541 (2004)
79. M. Benchohra, J. Henderson, S.K. Ntouyas, A. Ouahab, On first order impulsive dynamic equations on time scales. J. Differ. Equ. Appl. **10**, 541–548 (2004)
80. M. Benchohra, L. Górniewicz, S.K. Ntouyas, A. Ouahab, Controllability results for nondensely semilinear functional differential equations. Z. Anal. Anwendungen **25**, 311–325 (2006)
81. M. Benchohra, J. Henderson, S.K. Ntouyas, *Impulsive Differential Equations and Inclusions*, vol. 2 (Hindawi Publishing Corporation, New York, 2006)
82. M. Benchohra, I. Medjedj, J.J. Nieto, P. Prakash, Global existence for functional differential equations with state-dependent delay. J. Funct. Spaces Appl. **2013**, 863561, 7 (2013)
83. M. Benchohra, J. Henderson, I. Medjedj, Global existence results for functional differential inclusions with state-dependent delay Math. Modelling Appl. **19**(4), 524–536 (2014)
84. I. Benedetti, V. Obukhovskii, P. Zecca, Controllability for impulsive semilinear functional differential inclusions with a non-compact evolution operator. Discuss. Math. Differ. Incl. Control Optim. **31**, 39–69 (2004)

85. A. Bensoussan, G. Da Prato, M.C. Delfour, S.K. Mitter, *Representation and Control of Infinite Dimension Systems*, vol. 2, Systems and Control: Foundations and Applications (Birkhauser, Boston, 1993)

86. A. Bressan, G. Colombo, Extensions and selections of maps with decomposable values. Stud. Math. **90**, 70–85 (1988)

87. T.A. Burton, A fixed-point theorem of Krasnoselskii. Appl. Math. Lett. **11**(1), 85–88 (1998)

88. T.A. Burton, C. Kirk, A fixed point theorem of Krasnoselskii type. Math. Nachr. **189**, 23–31 (1998)

89. L. Byszewski, Theorems about the existence and uniqueness of solutions of a semilinear evolution nonlocal Cauchy problem. J. Math. Anal. Appl. **162**, 494–505 (1991)

90. L. Byszewski, Existence and uniqueness of solutions of semilinear evolution nonlocal Cauchy problem. Zeszyty Nauk. Politech. Rzeszowskiej Mat. Fiz. **18**, 109–112 (1993)

91. L. Byszewski, Existence and uniqueness of mild and classical solutions of semilinear functional-differential evolution nonlocal Cauchy problem, in *Selected Problems of Mathematics*, 25–33, 50th Anniv. Cracow Univ. Technol. Anniv. Issue, 6 (Cracow University Technology, Kraków, 1995)

92. L. Byszewski, H. Akca, On a mild solution of a semilinear functional-differential evolution nonlocal problem. J. Appl. Math. Stochastic Anal. **10**, 265–271 (1997)

93. N. Carmichael, M.D. Quinn, An approach to nonlinear control problems using the fixed point methods, degree theory and pseudo-inverses. Numer. Funct. Anal. Optim. **7**, 197–219 (1984–1985)

94. C. Castaing, M. Valadier, *Convex Analysis and Measurable Multifunctions*. Lecture Notes in Mathematics, vol. 580 (Springer, New York, 1977)

95. A. Coldbeter, Y.X. Li, Dupont, *Pulsatile Signalling in Intercellular Communication: Experimental and Theoretical Aspects*. Math. Appl. to Biology and Medicine (Werz, Winnipeg, 1993), pp. 429–439

96. C. Corduneanu, *Integral Equations and Stability of Feedback Systems* (Acedemic, New York, 1973)

97. C. Corduneanu, V. Lakshmikantham, Equations with unbounded delay. Nonlinear Anal. **4**, 831–877 (1980)

98. R. Curtain, H.J. Zwart, *An Introduction to Infinite Dimensional Linear Systems Theory* (Springer, New Yok, 1995)

99. G. Da Prato, E. Grisvard, On extrapolation spaces. Rend. Accad. Naz. Lincei. **72**, 330–332 (1982)

100. G. Da Prato, E. Sinestrari, Differential operators with non-dense domains. Ann. Scuola. Norm. Sup. Pisa Sci. **14**, 285–344 (1987)

101. K. Deimling, *Multivalued Differential Equations* (Walter de Gruyter, Berlin/New York, 1992)

102. B.C. Dhage, Fixed-point theorems for discontinuous multivalued operators on ordered spaces with applications. Comput. Math. Appl. **51**, 589–604 (2006)

103. T. Diagana, *Almost Automorphic Type and Almost Periodic Type Functions in Abstract Spaces* (Springer, New York, 2013)

104. S. Djebali, L. Gorniewicz, A. Ouahab, *Solution Sets for Differential Equations and Inclusions* (Walter de Gruyter, Berlin, 2013)

105. J. Dugundji, A. Granas, *Fixed point Theory* (Springer, New York 2003)

106. K.J. Engel, R. Nagel, *One-Parameter Semigroups for Linear Evolution Equations* (Springer, New York, 2000)

107. L.H. Erbe, W. Krawcewicz, Existence of solutions to boundary value problems for impulsive second order differential inclusions, Rocky Mountain J. Math. **22**, 519–539 (1992)

108. K. Ezzinbi, Existence and stability for some partial functional differential equations with infinite delay. Electron. J. Differ. Equ. **2003**(116), 1–13 (2003)

109. K. Ezzinbi, J. Liu, Nondensely defined evolution equations with nonlocal conditions. Math. Comput. Modelling **36**, 1027–1038 (2002)

110. K. Ezzinbi, J. Liu, Periodic solutions of non-densely defined evolution equations. J. Appl. Math. Stochastic Anal. **15**(2), 113–123 (2002)

111. H.O. Fattorini, *Second Order Linear Differential Equations in Banach Spaces*. North Holland, Mathematical Studies (North Holland, Amsterdam, 1985)

112. A. Freidman, *Partial Differential Equations* (Holt, Rinehat and Winston, New York, 1969)

113. M. Frigon, Fixed point results for generalized contractions in gauge spaces and applications. Proc. Am. Math. Soc. **128**(10), 2957–2965 (2000)

114. M. Frigon, *Fixed Point Results for Multivalued Contractions on Gauge Spaces. Set Valued Mappings with Applications in Nonlinear Analysis*. Ser. Math. Anal. Appl., vol. 4 (Taylor & Francis, London, 2002), pp. 175–181

115. M. Frigon, *Fixed Point and Continuation Results for Contractions in Metric and Gauge Spaces. Fixed Point Theory and Its Applications*, vol. 77 (Banach Center/Polish Academic Science, Warsaw, 2007), pp. 89–114

116. M. Frigon, A. Granas, Résultats de type Leray-Schauder pour des contractions sur des espaces de Fréchet. Ann. Sci. Math. Québec **22**(2), 161–168 (1998)

117. X. Fu, Controllability of neutral functional differential systems in abstract space. Appl. Math. Comput. **141**, 281–296 (2003)

118. X. Fu, Controllability of abstract neutral functional differential systems with unbounded delay. Appl. Math. Comput. **151**, 299–314 (2004)

119. X. Fu, K. Ezzinbi, Existence of solutions for neutral functional differential evolution equations with nonlocal conditions. Nonlinear Anal. **54**, 215–227 (2003)

120. E.P. Gastori, Controllability results for nondendely defined evolution differential inclusions with nonlocal conditions. J. Math. Anal. Appl., **297**, 194–211 (2004)

121. E. Gastori, S.K. Ntouyas, Y.G. Sficas, On a nonlocal cauchy problem for differential inclusions. Abstr. Appl. Anal. **2004**(5), 425–434 (2004)

122. J.A. Goldstein, *Semigroups of Linear operators and Applications* (Oxford University Press, New York, 1985)

123. L. Górniewicz, *Topological Fixed Point Theory of Multivalued Mappings*. Mathematics and its Applications, vol. 495 (Kluwer Academic, Dordrecht, 1999)

124. J.R. Graef, On the oscillation of impulsively damped halflinear oscillators, in *Proceedings of the 6th Colloquium on the Qualitative Theory of Differential Equations (Szeged, 1999)* vol. 14, 12 pp

125. J.R. Graef, J. Karsai, Oscillation and nonoscillation in nonlinear implusive system with increasing energy, in Proceeding of the Third International Conference on Dynamical systems and Differential Equations. Discrete Contin. Dyn. Syst. **7**, 161–173 (2000)

126. J.R. Graef, A. Ouahab, Some existence results and uniqueness for functional impulsive differential equations with variable times in Fréchet spaces. Dyn. Contin. Discrete Impuls. Syst. **14**, 27–45 (2007)

127. J.R. Graef, J. Henderson, A. Ouahab, *Impulsive Differential Inclusions. A Fixed Point Approach* (Walter de Gruyter, Berlin, 2013)

128. A. Granas, J. Dugundji, *Fixed Point Theory* (Springer, New York, 2003)

129. G. Guo, A class of second-order impulsive integro-differential equations on unbounded domain in a Banach space. Appl. Math. Comput. **125**, 59–77 (2002)

130. G. Guhring, F. Rabiger, W. Ruess, Linearized stability for semilinear non-autonomous evolution equations to retarded differential equations. Differ. Integr. Equ. **13**, 503–527 (2000)

131. J.K. Hale, *Theory of Functional Differential Equations* (Springer, New York, 1977)

132. J. Hale, J. Kato, Phase space for retarded equations with infinite delay. Funkcial. Ekvac. **21**, 11–41 (1978)

133. J.K. Hale, S.M. Verduyn Lunel, *Introduction to Functional Differential Equations*. Applied Mathematical Sciences, vol. 99 (Springer, New York, 1993)

134. S. Heikkila, V. Lakshmikantham, *Monotone Iterative Technique for Nonlinear Discontinuous Differential Equations* (Marcel Dekker Inc., New York, 1994)

135. J. Henderson, A. Ouahab, Existence results for nondensely defined semilinear functional differential inclusions in Fréchet spaces. Electron. J. Qual. Theory Differ. Equ. **17**, 1–17 (2005)

136. H.R. Henriquez, Existence of periodic solutions of neutral functional differential equations with unbounded delay. Proyecciones **19**(3), 305–329 (2000)

137. E. Hernandez, Regularity of solutions of partial neutral functional differential equations with unbounded delay. Proyecciones **21**(1), 65–95 (2002)

138. E. Hernandez, A Massera type criterion for a partial neutral functional differential equation. Electron. J. Differ. Equ. **2002**(40), 1–17 (2002)

139. E. Hernández, A. Prokopczyk, L. Ladeira, A note on partial functional differential equations with state-dependent delay. Nonlinear Anal. **7**, 510–519 (2006)

140. E. Hernández, R. Sakthivel, Rathinasamy, A. Tanaka, Existence results for impulsive evolution differential equations with state-dependent delay. Electron. J. Differ. Equ. **2008**(28), 1–11 (2008)

141. Y. Hino, S. Murakami, Total stability in abstract functional differential equations with infinite delay. *Electronic Journal of Qualitative Theory of Differential Equations*. Lecture Notes in Mathematics, vol. 1473 (Springer, Berlin, 1991)

142. Y. Hino, S. Murakami, T. Naito, *Functional Differential Equations with Unbounded Delay* (Springer, Berlin, 1991)

143. Sh. Hu, N. Papageorgiou, *Handbook of Multivalued Analysis*, vol. I (Kluwer, Dordrecht/Boston/London, 1997)

144. M. Kamenskii, V. Obukhovskii, P. Zecca, *Condensing Multivalued Maps and Semilinear Differential Inclusions in Banach Spaces* (Walter de Gruyter Series in Nonlinear Analysis and Applications, Berlin, 2001)

145. F. Kappel, W. Schappacher, Some considerations to the fundamental theory of infinite delay equations. J. Differ. Equ. **37**, 141–183 (1980)

146. H. Kellermann, M. Hieber, Integrated semigroup. J. Funct. Anal. **84**, 160–180 (1989)

147. M. Kisielewicz, *Differential Inclusions and Optimal Control* (Kluwer, Dordrecht, 1991)

148. V. Kolmanovskii, A. Myshkis, *Introduction to the Theory and Applications of Functional-Differential Equations*. Mathematics and its Applications, vol. 463 (Kluwer Academic, Dordrecht, 1999)

149. S.G. Krein, *Linear Differential Equations in Banach Spaces* (The American Mathematical Society, Providence, 1971)

150. V. Lakshmikantham, D.D. Bainov, P.S. Simeonov, *Theory of Impulsive Differential Equations* (World Scientific, Singapore, 1989)

151. V. Lakshmikantham, N.S. Papageorgiou, J. Vasundhara, The method of upper and lower solutions and monotone technique for impulsive differential equations with variable moments. Appl. Anal. **15**, 41–58 (1993)

152. V. Lakshmikantham, S. Leela, S.K. Kaul, Comparaison principle for impulsive differential equations with variable times and stability theory. Nonlinear Anal. **22**, 499–503 (1994)

153. L. Lasiecka, R. Triggiani, Exact controllability of semilinear abstract systems with application to waves and plates boundary control problems. Appl. Math. Optim. **23**, 109–154 (1991)

154. A. Lasota, Z. Opial, An application of the Kakutani-Ky Fan theorem in the theory of ordinary differential equations. Bull. Acad. Pol. Sci. Ser. Sci. Math. Astronom. Phys. **13**, 781–786 (1965)

155. G. Li, Sh. Song, Ch. Wu, Controllability of evolution inclusions with nonlocal conditions. J. Syst. Sci. Complexity **18**(1), 35–42 (2005)

156. G. Li, X. Xue, Controllability of evolution inclusions with nonlocal conditions. Appl. Math. Comput. **141**, 375–384 (2003)

157. X. Li, J. Yong, *Optimal Control Theory for Infinite Dimensional Systems* (Birkhauser, Berlin, 1995)

158. J. Liang, T. Xiao, The cauchy problem for nonlinear abstract fuctionnal differential with infinte delay. Comput. Math. Appl. **40** (6&7), 693–707 (2000)

159. J.H. Liu, Nonlinear impulsive evolution equations, Dyn. Contin. Discrete Impuls. Syst. Ser. A **6**, 77–85 (1999)

160. L. Maniar, A. Rhandi, Inhomogeneous retarded equation in infinite dimentional space via extrapolation spaces. Rend. Circ. Mat. Palermo **47**, 331–346 (1998)

161. S. Migorski, A. Ochal, Nonlinear impulsive evolution inclusions of second order. Dyn. Syst. Appl. **16**, 155–173 (2007)

162. H. Mönch, Boundary value problems for nonlinear ordinary differential equations of second order in Banach spaces. Nonlinear Anal. **4**, 985–999 (1980)

163. R. Nagel, E. Sinestrari, *Inhomogeneous Volterra Integrodifferential Equations for Hille-Yosida operators, In Functional Analysis*, ed. by K.D. Bierstedt, A. Pietsch, W.M. Ruess, D. Voigt (Marcel Dekker, New York, 1998) pp. 51–70

164. J. Neerven, *The Adjoint of a Semigroup of Linear Operators*. Lecture Notes in Math., vol. 1529 (Springer, New York, 1992)

165. S.K. Ntouyas, Global existence for neutral functional integrodifferential equations. Nonlinear Anal. **30**, 2133–2142 (1997)

166. S.K. Ntouyas, P.Ch. Tsamatos, Global Existence for second order functional semilinear integro-differential equations. Math. Slovaca **50**(1), 95–109 (2000)

167. V. Obukhovskii, Semilinear functional-differential inclusions in a Banach space and controlled parabolic systems. Soviet J. Autom. Inf. Sci. **24**, 71–79 (1991)

168. A. Pazy, *Semigroups of Linear Operators and Applications to Partial Differential Equations* (Springer, New York, 1983)

169. N.A. Perestyuk, V.A. Plotnikov, A.M. Samoilenko, N.V. Skripnik, *Differential Equation with Impulse Effects, Multivalued Right-hand Sides with Discontinuities* (Walter de Gruyter, Berlin/Boston, 2011)

170. M.D. Quinn, N. Carmichael, An approach to nonlinear control problem using fixed point methods, degree theory, pseudo-inverses. Numer. Funct. Anal. Optim. **7**, 197–219 (1984–1985)

171. AV. Rezounenko, J. Wu, A non-local PDE model for population dynamics with state-selective delay: local theory and global attractors. J. Comput. Appl. Math. **190**(1–2), 99–113 (2006)

172. A.M. Samoilenko, N.A. Perestyuk, *Impulsive Differential Equations* (World Scientific, Singapore, 1995)

173. K. Schumacher, Existence and continuous dependence for differential equations with unbounded delay. Arch. Ration. Mech. Anal. **64**, 315–335 (1978)

174. G. Sell, Y. You, *Dynamics of Evolution Equations* (Springer, New York, 2001)

175. E. Sinestrari, Continuous interpolation spaces and spatial regularity in nonlinear Volterra integrodifferential equations. J. Integr. Equ. **5**, 287–308 (1983)

176. R.E. Showalter, *Monotone Operators in Banach Space and Nonlinear Partial Differential Equations*. Mathematical Surveys and Monographs, vol. 49 (The American Mathematical Society, New York, 1997)

177. J. Smoller, *Shock Waves and Reaction-Diffusion Equations* (Springer, New York, 1983)

178. R. Temam, *Infinite-Dimensional Dynamical Systems in Mechanics and Physics*. Applied Mathematical Sciences, vol. 68 (Springer, New York, 1988)

179. C.C. Travis, G.F. Webb, Existence and stability for partial functional differential equations. Trans. Am. Math. Sci. **200**, 395–418 (1974)

180. C.C. Travis, G.F. Webb, Existence, stability and compactness in the α−norm for partial functionaldifferential equations. Trans. Am. Math. Sci. **240**, 129–143 (1978)

181. A.A. Tolstonogov, *Differential Inclusions in a Banach Space* (Kluwer Academic, Dordrecht, 2000)

182. A.N. Vityuk, On solutions of hyperbolic differential inclusions with a nonconvex right-hand side. (Russian) Ukran. Mat. Zh. **47**(4), 531–534 (1995); translation in Ukrainian Math. J. **47**(4), 617–621 (1995)

183. D.R. Willé, C.T.H. Baker, Stepsize control and continuity consistency for state-dependent delay-differential equations. J. Comput. Appl. Math. **53** (2), 163–170 (1994)

184. J. Wu, *Theory and Applications of Partial Functional Differential Equations*. Applied Mathematical Sciences, vol. 119 (Springer, New York, 1996)

185. K. Yosida, *Functional Analysis*, 6th edn. (Springer, Berlin, 1980)

186. J. Zabczyk, *Mathematical Control Theory* (Birkhauser, Berlin, 1992)

187. S. Zheng, *Nonlinear Evolution Equations*. Chapman & Hall/CRC Monographs and Surveys in Pure and Applied Mathematics, vol. 133 (Chapman & Hall/CRC, Boca Raton, 2004)

Index

Printed in the United States
By Bookmasters